Advances in Intelligent Systems and Computing

172

For further volumes:
http://www.springer.com/series/11156

Lorna Uden, Francisco Herrera, Javier Bajo Pérez,
and Juan Manuel Corchado Rodríguez (Eds.)

7th International Conference on Knowledge Management in Organizations: Service and Cloud Computing

Editors
Lorna Uden
FCET
Staffordshire University
The Octagon, Beaconside
UK

Javier Bajo Pérez
Faculty of Computer Science
Pontificial University of Salamanca
Salamanca
Spain

Francisco Herrera
Department Computer Science and
Artificial Intelligence
University of Granada
Granada
Spain

Juan Manuel Corchado Rodríguez
Department of Computing Science
and Control
Faculty of Science
University of Salamanca
Salamanca
Spain

ISSN 2194-5357
e-ISSN 2194-5365
ISBN 978-3-642-30866-6
e-ISBN 978-3-642-30867-3
DOI 10.1007/978-3-642-30867-3
Springer Heidelberg New York Dordrecht London

Library of Congress Control Number: 2012939635

Printed on acid-free paper

Springer is part of Springer Science+Business Media (www.springer.com)

Preface

Knowledge is increasingly recognised as the most important resource in organisations and a key differentiating factor in business today. It is increasingly being acknowledged that Knowledge Management (KM) can bring about the much needed innovation and improved business performance in organisations. The service sector now dominates the economies of the developed world. Service innovation is fast becoming the key driver of socio-economic, academic and commercial research attention. Knowledge Management plays a crucial role in the development of sustainable competitive advantage through innovation in services. There is tremendous opportunity to realise business value from service innovation by using the knowledge about services to develop and deliver new information services and business services.

Although there are several perspectives on KM, they all share the same core components, namely: People, Processes and Technology. Organisations of all sizes across nearly every industry are seeking new ways to address their knowledge management requirements. Cloud computing offers many solutions to the problems facing KM implementation. Cloud computing is an emerging technology that can provide users with all kinds of scalable services, such as channels, tools, applications, social support for users' personal knowledge amplification, personal knowledge use/reuse, and personal knowledge sharing.

The seventh International Conference on Knowledge Management in Organizations (KMO) offers researchers and developers from industry and the academic world to report on the latest scientific and technical advances on knowledge management in organisations. It provides an international forum for authors to present and discuss research focused on the role of knowledge management for innovative services in industries, to shed light on recent advances in cloud computing for KM as well as to identify future directions for researching the role of knowledge management in service innovation and how cloud computing can be used to address many of the issues currently facing KM in academia and industrial sectors. This conference provides papers that offer provocative, insightful, and novel ways of developing innovative systems through a better understanding of the role that knowledge management plays.

The KMO 2012 proceedings consist of 53 papers covering different aspects of knowledge management and service. Papers came from many different countries including Australia, Austria, Brazil, China, Chile, Colombia, Denmark, Finland, France, Gambia, India, Japan, Jordan, Netherlands, Malaysia, Malta, Mexico, Netherlands, New Zeland, Saudi Arabia, Spain, Slovakia, Slovenia, Taiwan, Turkey, United States of America and United Kingdom. We would like to thank our program committee, reviewers and authors for their contributions. Without their efforts, there would be no conference and proceedings.

Salamanca
July 2012

Lorna Uden
Francisco Herrera
Javier Bajo Pérez
Juan Manuel Corchado Rodríguez

Organization

Conference Chair

Dr. Lorna Uden	Staffordshire University, UK

Program Chair

Dr. Francisco Herrera	University of Granada, Spain

Program Committee

Dr. Senen Barro	University of Santiago de Compostela, Spain
Dr. Hilary Berger	UWIC. England. UK
Dr. Pere Botella	Polytechnic University of Catalonia – BarcelonaTech, Spain
Dr. Vicente Botti	Polytechnic University of Valencia, Spain
Dr. Senoo Dai	Tokyo Institute of Technology, Japan
Dr. Flavius Frasincar	Erasmus University Rotterdam, The Netherlands
Dr. Wu He	Old Dominion University, USA
Dr. Marjan Hericko	University of Maribo, Slovenia
Dr. George Karabatis	University of Maryland, Baltimore County, USA
Dr. Dario Liberona	University of Santiago, Chile
Dr. Remy Magnier-Watanabe	University of Tsukuba, Tokyo, Japan
Dr. Victor Hugo Medina Garcia	Universidad Distrital Francisco José de Caldas, Colombia
Dr. Marja Naaranoja	Vaasa University of Applied Sciences, Finland
Dr. Paulo Novais	University of Minho, Portugal
Dr. Takao Terano	Tokyo Institute of Technology, Japan
Dr. Derrick Ting	National University of Kaohsiung, Taiwan
Dr. Lorna Uden	Staffordshire University, UK
Dr. Michael Vallance	Future University Hakodate, Japan

Dr. William Wang	Auckland University of Technology, New Zealand
Dr. Leon S.L. Wang	National University of Kaohsiung, Taiwan
Dr. Guandong Xu	Victoria University, Australia
Dr. Ales Zivkovic	University of Maribo, Slovenia

Organization Committe

Dr. Juan M. Corchado (Chair)	University of Salamanca, Spain
Dr. Javier Bajo (Vice-chair)	Pontifical University of Salamanca, Spain
Dr. Sara Rodríguez	University of Salamanca, Spain
Dr. Francisco de Paz Santana	University of Salamanca, Spain
Dr. Emilio S. Corchado	University of Salamanca, Spain
Dr. Belén Pérez	University of Salamanca, Spain
Fernando De la Prieta	University of Salamanca, Spain
Carolina Zato	University of Salamanca, Spain
Antonio J. Sanchez	University of Salamanca, Spain

Contents

Section 1: Innovation in Knowlege Management

Section 2: Knowledge Management in Business

Section 3: Knowledge Management in Education

Section 4: Knowledge Management in the Internet Age

Evaluation of a Self-adapting Method for Resource Classification in Folksonomies

José Javier Astrain, Alberto Córdoba, Francisco Echarte, and Jesús Villadangos

Dept. Ingeniería Matemática e Informática, Universidad Pública de Navarra, Campus de Arrosadía, 31006 Pamplona, Spain
josej.astrain@unavarra.es, alberto.cordoba@unavarra.es, patxi@eslomas.com, jesusv@unavarra.es

Abstract. Nowadays, folksonomies are currently the simplest way to classify information in Web 2.0. However, such folksonomies increase continuously their amount of information without any centralized control, complicating the knowledge representation. We analyse a method to group resources of collaborative-social tagging systems in semantic categories. It is able to automatically create the classification categories to represent the current knowledge and to self-adapt to the changes of the folksonomies, classifying the resources under categories and creating/deleting them. As opposed to current proposals that require the re-evaluation of the whole folksonomy to maintain updated the categories, our method is an incremental aggregation technique which guarantees its adaptation to highly dynamic systems without requiring a full reassessment of the folksonomy.

1 Introduction

Folksonomies are nowadays a widely used system of classifying information for knowledge representation [10]. Tags made by users provide semantic information that can be used for knowledge management. As users annotate the same resources, frequently using the same tags, their semantic representations for both tags and annotated resources emerge [12, 13, 15]. Folksonomies are based on the interaction of multiple users to jointly create a "collective intelligence" that defines the semantics of the information. Although users follow an easy and simple mechanism to classify resources, knowledge representation becomes more difficult as the volume of information increases [3]. Folksonomies are difficult to be studied due to their three-dimensional structure (hypergraph) [2, 9]. Then, different two-dimensional contexts of this information are often considered [2] (tag-user, tag-resource and tag-tag).

Folksonomies are highly dynamic systems, so any proposal should be scalable and able to adapt to its evolution. In [8], a generalization of the methods used to obtain two-dimensional projections dividing them into non-incremental aggregation methods and incremental aggregation methods is proposed. The first one includes solutions similar to those proposed in [2, 9] where the incorporation of new information to the folksonomy involves the complete recalculation of the similarity matrices. The second one includes solutions in which new annotations introduced in

L. Uden et al. (Eds.): 7th International Conference on KMO, AISC 172, pp. 1–12.
springerlink.com

the folksonomy do not involve a repetition of all the calculations. Faced with these tag-centric based proposals, resource-centric approaches have been poorly studied with the aim of structuring the folksonomy resources. In this paper, we consider the improvement of the folksonomy knowledge representation by creating semantic categories, also called concepts, that group the folksonomy resources. We try to obtain the relationship between kolksonomies and ontologies obtaining the semantics from both tag and resources.

Some works in literature [1, 14] attempt to classify the resources of a folksonomy into concepts to offer improvements in user navigation. In [1] the resources of a folksonomy are classified under a set of classification concepts. These classification concepts are previously obtained through a manually search through a repository of ontologies. Once obtained the classification concepts, an algorithm classifies the tags of the folksonomy under the concepts obtained combining the co-occurrences of the tags and the results obtained after submitting certain search patterns to Google. Folksonomy resources are classified into concepts according to the tags they have been assigned. Furthermore, the authors do not propose an implementation or prototype of their proposal, so it is not possible to assess to what extent the classification aided navigation. This non-incremental method depends on the intervention of a user which defines the terms for the classification of ontologies in which resources are classified, and the use of external information sources. Authors also fail to take into account the evolution of the folksonomy and how to adapt their proposal to the incorporation of new tags and resources.

An optimization algorithm for the classification of resources under a set of concepts, using a set of predefined concepts and a set of previously classified resources, is used in [14]. In this regard, the goal of this algorithm is similar to the method described in this work, since it directly classifies the resources under concepts. However, the algorithm has some drawbacks such as: a) the categories are fixed, and their evolution are not considered; b) it requires full reassessment of the algorithm each time the folksonomy evolves since it is a non-incremental method; and c) it does not describe how resources are sorted within each concept.

This paper introduces a simple method for the automatic classification of the resources of a folksonomy into semantic concepts. The main goal is to improve the knowledge management in folksonomies, keeping the annotation method as simple as usual. Folksonomies are highly dynamic systems where new tags and resources are created continuously, so the method builds and adapts these concepts automatically to the folksonomy's evolution. Concepts can appear or disappear by grouping new resources or disaggregating existing ones and resources would be automatically assigned to those concepts. Furthermore, the method has a component-based open architecture which allows its application to folksonomies with different characteristics. It uses a reduced set of the folksonomy's tags to represent both the semantics of the resources and concepts because it requires a high classification efficiency to allow its application to real folksonomies (where the number of annotations grows very fast), and assigns automatically an appropriated name to those concepts.

The rest of the paper is organized as follows: Section 2 describes the method; Section 3 deals with the method evaluation using a real folksonomy; Acknowledgements, Conclusions and References end the paper.

2 Method Description

The method, introduced in [6], follows a component based open architecture, which allows its application to folksonomies with different characteristics. It requires a high classification efficiency to allow its application to real folksonomies, where the number of annotations grows very fast.

Given a folksonomy, the method initially creates a set of concepts where resources are grouped, and it assigns a name to each concept according to the semantic information provided by the resources grouped in each concept and their annotations. Once created the concepts, each new annotation of the folksonomy is processed updating the semantic information and adapting the concepts when necessary. The method starts with the creation of the set of representative tags (S_{rt}) and the vectorial representation of the resources of the folksonomy. The component *Representations* is in charge of these tasks. It may use different criteria to create S_{rt} like the tags more frequently used, the tags used by more users and even more. Then, each resource is assigned to subsets $R_{converged}$ or $R_{pending}$ in terms of whether they have converged or not. In order to obtain those tags that best describe the semantic of a resource, in [11], it is showed that tagging distributions of heavily tagged resources tend to stabilize into power law distributions. The component *Convergence* may use many criteria like the total amount of annotations, or the number of annotations associated to the S_{rt} set to assign the resources to $R_{converged}$.

The component *Clustering* clusters the resources of the folksonmy belonging to $R_{converged}$ in a set of concepts, generating the set of semantic concepts on which resources of the folksonomy are grouped (C) and the set of pairs (r, c) where $r \in R$ and $c \in C$, representing that resource r is grouped into the concepts c (Z). The initial clustering of the resources may follow different criteria: applying clustering techniques to the resources of the folksonomy, creating manually the classification concepts and selecting a relevant resource as seed of the classifier, or adapting some tag clustering algorithms like T-Know [1] in order to classify resources, instead of tags, under the concepts that have been previously selected. The component *MergingSplitting* analyses the concepts provided in order to evaluate the convenience of merging or splitting any of them, updating C and Z sets. It merges those concepts whose similarity values are greater than 0.75, using the cosine measure. The component *Classifier* is in charge of grouping the resources under those concepts with which they have high semantic similarity by comparing all the resources belonging to the $R_{converged}$ set with the concepts in which they are grouped in Z. The component assigns the resource to the $R_{classified}$ set according to the similarity measures between each resource and its group or keeps it on $R_{converged}$ and removes it from Z. The component *Representations* creates the vector representations for the C concepts using the Z set, and the component *Naming* assigns meaningful names to these concepts according to some criteria like the tags with higher weights.

```
1.  Representations::createSrt()
2.  Representations::createVectors()

3.  forall r ∈ R do
4.      if Convergence::hasConverged(r) then
5.          assign r to R_converged
6.          else assign r to R_pending
7.      endif
8.  endforall

9.  Clustering::create(R_converged)
10. MergingSplitting::process(C,Z)

11. forall r ∈ R_converged do
12.     if Classifier::isCorrectlyClassified(Z,r) then
13.         assign r to R_classified
14.         else drop r from Z
15.     endif
16. endforall

17. Representations::createConceptVectors(C,Z)
18. Naming::process(C,Z)

19. while true do

20.     tagging = wait(FolksonomyEvolution)
21.     if not Representations::inSrt(t) then
22.         continue
23.     endif

24.     Representations::updateVectors(Tagging)

25.     if tagging.action = create then
26.         if r ∈ R_pending
                and Convergence::hasConverged(r) then
27.             assign r to R_converged
28.         endif
29.         if r ∈ R_converged then
30.             Classifier::classify(Z,r)
31.             Representations::updateConceptVector(Z,r)
32.         endif
33.     endif

34.     if RecalculationCondition::check() then
35.         Representations::updateSrt()
36.         Clustering::update(C,Z)
37.         MergingSplitting::process(C,Z)
38.         Representations::updateVectors(C,Z)
39.         Naming::process(C,Z)
40.     endif
41. endwhile
```

Fig. 1 Method algorithm

At this point, the method has built $R_{pending}$, $R_{converged}$, $R_{classified}$, C and Z sets from the folksonomy, so it provides a set of concepts that group folksonomy resources based on their semantics. Once created these concepts, the method self-adapts to the evolution of the folksonomy taking into account the new annotations made by users. The lines 19 to 41 of the pseudocode described in Fig. 1 are responsible of processing these new annotations. The method waits for a change on the folksonomy when creating or removing an annotation. This change is represented by the method as a new *Tagging* element, which contains the annotation information (user, resource and tag), and if it has been created or deleted. If the tag used in the *Tagging* does not belong to the representative set of tags (S_{rt}), *Tagging* is ignored and it expects the reception of new annotations. If this tag belongs to the S_{rt} set, the component *Representations* updates the vectorial representation of the resource. If the resource belongs to the $R_{classified}$ set the component also updates the vectorial representation of the concept in which the resource is grouped. If the *Tagging* is of type *create*, the method checks whether the resource has converged or not, and the possibility of grouping it under any existing concept. If the resource belongs to the $R_{pending}$ set, the component *ConvergenceCriterion* checks if the resource has converged after receiving the new annotation. If so, or if the resource already previously belonged to $R_{converged}$, the component *Classifier* provides the most appropriate concept for this resource. The component compares the semantic of the resource with that of each concept in C. Based on this similarity, the component assigns the resource to the $R_{classified}$ set and creates a new entry in Z, or lets the resource continue to be assigned to $R_{converged}$. If the resource is assigned to a concept, *Representations* component updates the vectorial representation of the concept with the resource information.

The method groups the resources in concepts, so that once a resource is classified it will never return to the set $R_{converged}$. Therefore, when a *Tagging* of type *delete* is received, the method does not checks if the resource has converged or whether it must continue to exist under the current concept, the method only updates the corresponding vectorial representation. In addition to gathering new converged resources into existing concepts, the method considers the information received from the new *Taggings*, updating the S_{rt} set and the existing concepts. Thus, S_{rt} and C sets may adapt to the folksonomy's evolution, performing their adaptation for example to new users' interests. Since a unique *Tagging* does not use to significantly affect the S_{rt} set or the concepts set, and this update can be quite expensive computationally, the method uses the component *RecalculationCondition* to determine when to update both concepts and S_{rt}. The recalculation may be performed considering many criteria, for example, after a certain number of processed *Taggings*, after a given time period, or whenever a resource or a set of resources are classified. When the component determines the convenience of performing the recalculation, in a first step the S_{rt} set is updated taking into account its establishment criteria using component *Representations*. It then uses the *Clustering* component to update the existing concepts (C) and the resources grouped in them (Z). The component *MergingSplitting* reviews these concepts creating, splitting them when necessary. Once obtained the elements C and Z, *Representations* updates the concept representation vectors, and *Naming* assigns a name to each one of the concepts. Upon the completion of these tasks, the method returns to stand waiting for the arrival of new *Taggings* to the folksonomy for their processing.

3 Method Evaluation

This section is devoted to evaluate experimentally the proposed method using data retrieved from Del.icio.us. The method creates automatically a set of concepts from an existing folksonomy and groups under these concepts the resources of the folksonomy, according to their semantics. As the folksonomy receives new annotations, the method groups new resources, takes into account new relevant tags, and adapts the existing concepts.

With the aid of a page scraper we have collected a set of 15,201 resources, with 44,437,191 annotations, 1,293,351 users and 709,657 tags from Del.icio.us (15th-30th September, 2009)[1]. Those annotations, depicted in Table 1, concern a period time from January 2007 to September 2009. We use annotations prior to 2009 to simulate an initial state of the folksonomy in order to create the initial concepts. The rest of annotations correspond to January to September 2009 and they are used to simulate the folksonomy evolution by means of *Taggings* elements of type *create*. Table 2 summarizes the information concerning the initial folksonomy (t_0) and its state after including the *Tagging* elements concerning the period Jan.-Sept. (t_1 to t_9).

[1] http://www.eslomas.com/publicaciones/KMO2012/

Table 1 Annotation distribution

Year	Annotations	Year	Annotations	Year	Annotations
1998	2	2002	299	2006	3,140,591
1999	3	2003	628	2007	7,237,129
2000	7	2004	52,345	2008	13,753,922
2001	12	2005	719,216	2009	19,533,037

Table 2 Users, tags and resources in each subset of the experiment

	Increment of the number of elements				Aggregated values			
t_i	annotations	users	resources	tags	annotations	users	resources	tags
t_0	24,904,154	972,695	12,117	489,125	24,904,154	972,695	12,117	489,125
t_1	1,704,682	35,480	342	23,581	26,608,836	1,008,175	12,459	512,706
t_2	1,811,331	37,240	353	23,066	28,420,167	1,045,415	12,812	535,772
t_3	2,179,539	40,672	407	26,101	30,599,706	1,086,087	13,219	561,873
t_4	2,153,461	34,603	336	24,187	32,753,167	1,120,690	13,555	586,060
t_5	2,230,512	33,078	391	24,919	34,983,679	1,153,768	13,946	610,979
t_6	2,304,614	33,959	348	24,460	37,288,293	1,187,727	14,294	635,439
t_7	2,437,317	34,137	345	24,951	39,725,610	1,221,864	14,639	660,390
t_8	2,617,998	36,438	368	26,332	42,343,608	1,258,302	15,007	686,722
t_9	2,093,583	35,049	194	22,935	44,437,191	1,293,351	15,201	709,657

The method is configured for the experimentation using the following components. The comparison between resource and concepts vectors has been performed using the cosine measure. In the following we describe the components used to configure the method. The component *Representations* considers a S_{rt} set built using those tags with at least 1,000 annotations. The component *Convergence* fixes the convergence criterion to 100 annotations [7]. The component *Classifier* uses the method presented in [5] to classify the resources under the most similar concepts. In the evolution task, the classifier applies for a given resource each time it reaches a multiple of 50 annotations. This component has been configured to take into account the two most similar concepts (c_i, c_j) to each resource (r_i), and classify only the resource when the difference between the resource and these concepts is greater than a minimum threshold value of 0.10 ($|sim(r_i, c_i) - sim(r_i, c_j)| \geq 0.10$). As proved in [5], the method is able to classify the resources providing a high precision. The component *Recalculation* recalculates monthly both sets S_{rt} and C (after each t_i $(i : 1..9)$). The component *Clustering* uses a k-means algorithm, determining the k value by the expression $k = \sqrt{\frac{n}{2}}$, being n the number of resources in $R_{converged}$ at the concepts creation and in $R_{classified}$ when recalculating. Thus, the number of concepts can grow as the folksonomy evolves. At the initial concepts creation, the initial centroids are randomly defined since any a-priori knowledge is not considered. When recalculating, the representation vectors of C concepts are used to define the initial centroids, and if $k \geq |C|$, then some resources are randomly selected to define the $k - |C|$ new clusters. The implementation of the algorithm is performed in a distributed way where the calculus of the number of cluster changes at each iteration is performed over different PCs of a cluster using a task queue manager (Gearman). We use the k-means method instead of hierarchical techniques because we want to provide a concept cloud (see Figure 2) with a similar user experience to tag clouds. Component *MergingSplitting* merges using the cosine measure those concepts whose similarity values are greater than 0.75, once the *Clustering* component obtains C and Z sets.

Fig. 2 Concept cloud

And finally, component *Naming* assigns a name to each concept according to the most relevant tags of its vector. When the weight of several tags is greater than the 50% of the weight of the most relevant tag, the name of the concept is obtained through the concatenation of those tags after verifying, by means of the Levenshtein distance, that tags are not syntactic variations of other previous tags. So, a concept in which the two most relevant tags are *php* (weight 127,427) and *programming* (weight 39,743), is named “Php”.

Besides Gearman, Memcached has been used as a cache system in order to reduce the number of accesses to the database to obtain resources, concepts and representation vectors. The employed hardware consists of four commodity PC with Intel Core 2 Duo processors at 2.13 GHz, and 2GB of RAM memory. In order to perform the distributed tasks, 8 processes (2 on each PC) have been executed to perform the processing of the K-means slices, and 24 processes have been executed to process the *Tagging* processing tasks.

Table 3 summarizes the information concerning the evolution of the number of tags, resources and concepts, from t_0 to t_9. Regarding the tags and the S_{rt} set, it shows the evolution of the number of tags of the folksonomy and the number of tags in S_{rt}, and the ratio between them. One can note that the number of tags in S_{rt} increases as the number of annotations in the folksonomy does. The increment in the number of annotations carries with a higher number of tags exceeding the threshold of 1,000 annotations required to be part of S_{rt}. Lets note that: i) the method represents the semantics of the resources with less than 0.40% (1,939/489,125) of the existing tags; and ii) this value decreases slowly as more annotations arrive to the folksonomy (up to 0.38% after processing t_9). This makes the cost of the method, in terms of space and process, significantly lower than the required when considering all the tags of the folksonomy to represent the semantics of the resources and their concepts associated. Table 3 shows that the number of classified resources increases as the number of resources of the folksonomy does. As folksonomy evolves with new annotations, some of the $R_{pending}$ resources converge and pass to $R_{converged}$ set, while other resources receive enough annotations to determine an adequate concept, passing to $R_{classified}$. Regarding concepts, Table 3 also shows the evolution in the k value used by the *Clustering* component, and the number of concepts created at each recalculation. When recalculating, *Clustering* component tries to create a number of k concepts considering the number of $R_{classified}$ resources, however, some of these concepts may be merged by *MergingSplitting* component when their similarity are greater than the defined threshold (0.75). Note that both k and $|C|$ values gradually increases after each recalculation. Although in this experiment the number of concepts increases, the method allows the creation, splitting and merging of concepts, so this number may also decrease. In this experiment the clustering is based on k-means, with k depending on the number of resources, causing the maintenance or increase of the number of concepts. The number of concepts can decrease only if, after clustering, the *MergingSplitting* component finds that there are two or more concepts with a similarity degree greater than 0.75.

Table 3 Adaptation of the method as the folksonomy evolves

	Tags			Resources			Concepts	
t_i	$\|T\|$	$\|S_{rt}\|$	$\frac{\|S_{rt}\|}{\|T\|}$	$R_{pending}$	$R_{converged}$	$R_{classified}$	$\|C\|$	k
t_0	489.125	1.939	0,40%	337	3.098	8.682	75	77
t_1	512.706	2.037	0,40%	289	3.184	8.986	76	78
t_2	535.772	2.124	0,40%	305	3.190	9.317	77	80
t_3	561.873	2.204	0,39%	282	3.226	9.711	78	81
t_4	586.060	2.286	0,39%	269	3.237	10.049	78	82
t_5	610.979	2.362	0,39%	250	3.277	10.419	79	83
t_6	635.439	2.437	0,38%	225	3.300	10.769	80	84
t_7	660.390	2.526	0,38%	206	3.033	11.400	81	85
t_8	686.722	2.616	0,38%	139	2.915	11.953	82	86
t_9	709.657	2.716	0,38%	43	2.917	12.241	83	87

Analysing the creation of concepts and the evolution in the number of resources grouped in each concept, one can note that the creation of certain concepts produces a decrease in the number of resources grouped in other concepts. For example, after processing the *Tagging* set t_9, the new annotations received in the transition from the eighth to ninth iteration lead the concept *Business&Finance&Money* be split into two concepts: *Business* and *Finance&Money*. The initial concept (*Business*), which previously brought together 215 resources, now groups 118 resources. The new concept created groups 119 resources, of which 110 (92 %) were previously grouped under *Business&Finance&Money*. In most cases, the number of resources classified under each concept grows as the number of annotations of the folksonomy does, but in some cases, this number may decrease. This decrease occurs either when the concept under which resources were classified is split either when the arrival of new annotations allow a better definition of the classification concept. As occurs with *Web2.0&Social&Socialnetworking*, in which the number of classified resources decreases from 244 to 230 in the transition from iteration 8 to 9. In this transition, 220 resources remain classified under this concept, 24 of them are reclassified under other existing concepts (*Video*, *Business*, *Twitter*, *Blog*, *Tools&Web2.0* and *Generator&Tools&Fun*), 6 new resources are classified under this concept and the remaining 4 resources were previously classified under other concepts.

Table 4 Number of resources classified under the eight first concepts

	Concept	Statistics				Iteration									
		Avg	Dev	Max	Min	0	1	2	3	4	5	6	7	8	9
1	Hardware&Electronics&Technology	4.2	64	72	60	60	61	62	63	64	64	66	69	71	72
2	Jobs&Career&Job	85	6.4	95	75	75	77	81	83	84	84	88	90	92	95
3	Books&Literature	171	13.0	192	152	152	156	161	166	169	172	177	182	185	192
4	Video&Web2.0	214	26.9	252	175	175	183	192	201	206	221	225	237	248	252
5	Html&Webdesign&Web	85	9.2	103	75	75	77	80	80	80	81	85	88	98	103
6	Linux&Opensource	116	6.4	125	108	108	109	110	112	115	117	118	122	125	125
7	Art&Design	198	7.0	208	189	189	190	195	193	194	200	204	205	206	208
74	Howto&Diy	81	8.6	93	70	82	70	72	72	75	83	86	88	92	93

Table 5 Evolution of the resources classified

Number of changes	Initial iteration										Total
	t_0	t_1	t_2	t_3	t_4	t_5	t_6	t_7	t_8	t_9	
1	320	25	29	17	11	14	11	26	22	0	475
2	79	2	1	3	2	3	0	1	0	0	91
3	2	0	1	0	0	0	0	0	0	0	3
4	1	0	0	0	0	0	0	0	0	0	1

Table 4 shows the number of resources classified under some concepts. In the same way, the creation of the concept *Origami*, which groups 16 resources, occurs after processing the set t_1 and its resources come from *Howto& Diy*, *Art&Design* and the rest are resources classified during the processing of t_1 and therefore were not grouped into any concept after t_0. It has been found therefore that the origin of the resources of the concepts created in the evolution presents a semantic relationship to the concepts created. This indicates that as the folksonomy receives new annotations, the method is able to group all the related resources and to evolve the concepts so as to adapt to user interests: e.g., resources that were initially grouped into *Howto&Diy*, representing pages with instructions on how to do things with the appearance in the folksonomy of more information on origami causes the method to create a specific concept for this topic, bringing together new and existing resources.

Table 5 shows the evolution of the resources classified. The left side of the table shows that 11,671 of the 12,241 resources retain their original classification along the evolution of the folksonomy, 475 of the resources change their concept of classification once, 91 of them change twice, 3 of them make it thrice, and only 1 makes it 4 times. The right side shows the distribution of the number of changes among classification concepts experienced by resources according to the iteration in which they are introduced. Of the set of resources introduced in the initial iteration which change their classification concept along the evolution of the folksonomy, 320 of them make it once, 79 of them change their classification concept twice, and so on. An analysis of the 83 classification concepts involved in this process shows that 75 of them appear in the initial iteration, and the rest appear one for each iteration, except in the fifth iteration, which does not introduce any new concept.

Regarding the performance of the method, the time spent in each stage has been registered. Table 6 shows the time needed to create the initial concepts (lines 1-18 of the method described in Fig. 1).

Table 6 Creation costs for the initial concepts

Action	Time (minutes)
S_{rt} creation	1.10
Vectorial representation creation	22.83
Convergence verification	0.08
Clustering (k=76)	192.21
MergingSplitting	0.98
Assignation to $R_{classified}$	61.62
Creation of concept' vectors	0.53
Naming	0.01

Table 7 Time spent in evolution

		Time (minutes)	
t_i	Number of taggings	Tag. Processing	Recalculation
t_1	1,704,682	28.23	23.47
t_2	1,811,331	31.24	24.35
t_3	2,179,539	41.98	48.91
t_4	2,153,461	43.12	29.18
t_5	2,230,512	40.36	56.45
t_6	2,304,614	41.12	27.39
t_7	2,437,317	41.84	81.34
t_8	2,617,998	32.65	27.26
t_9	2,093,583	37.34	63.12

Table 7 shows the time spent processing the *Tagging* elements of t_1 to t_9 sets (lines 19 to 41 in Fig. 1). Regarding the *Tagging* elements processing, the method has processed them with an average throughput of 965.74 *Tagging* elements per second, which represents an average of 40.24 *Tagging* elements per second and tagging processing worker (24 workers, threads annotating concurrently). These results show the applicability of the method in real systems, processing the new annotations to adapt the classification concepts. Regarding the time spent in the recalculation, it is quite variable, depending on the number of iterations of k-means algorithm. However, the time spent by the *Clustering* component when recalculating is much lower than the time spent at the first clustering (t_0), because the representation vectors of C are used as initial centroids, while at the first clustering the centroids are randomly created. We are now working on the use of different clustering techniques, including hierarchical clustering techniques, implementing the *Clustering* component over Apache Mahout, which implements a Framework MapReduce[4] allowing the usage of different clustering techniques.

In order to validate the quality of the classification of resources obtained, we have evaluated the 12,241 resources considered with the aid of 102 computer science students (advanced and regular internet users) at the Universidad Pública de Navarra during the course 2010-2011. Each reviewer has evaluated a subset of the resource set, ensuring that each resource has been evaluated by five different reviewers. Each reviewer has evaluated its subset of resources after the initial classification, and then those new resources and those whose category of classification has changed along the different recalculations. Reviewers evaluated, for each of the resources, how well resources are classified under their category of classification, quantifying this value between 1 and 5, meaning (1) a very poor classification, (2) a poor classification, (3) a reviewer indecision, (4) a good classification and (5) a very good classification. Reviewers considered 106 resources *very poor* classified (1%), 259 *poor* classified (2%), 4,492 *good* classified and 6,313 *very good* classified. The reviewers were hesitant with 971 resources (8%).

4 Conclusions

We have proposed, implemented and analysed a simple and incremental method for the automatic and semantic creation of concepts to group the resources of a folksonomy, in order to improve the knowledge management in folksonomies, without changing the way users make their annotations. The method automatically creates these concepts and adapts them to the folksonomy evolution over time, grouping new resources and creating, merging or splitting concepts as needed. It is an incremental-aggregation technique that adapts to the folksonomy evolution, without requiring to re-evaluate the whole folksonomy.

The method uses a small subset of tags, the set of more representative tags (S_{rt}), in order to apply it to real folksonomies and their evolution without adversely affecting its performance. Furthermore, the method is based on a component based open architecture. This allows its application to folksonomies with different features and needs, and a useful way to assign names to the classification concepts.

The semantic information assigned to the resource by their annotations allows the automatic classification of the resources of a folksonomy under classification concepts without requiring the intervention of human experts. We have experimentally validated, with the aid of human experts, that the method is able to create automatically these concepts and adapt them to the folksonomy evolution over time, classifying new resources and creating new concepts to represent more accurately the semantic of the resources.

Acknowledgements. Research partially supported by the Spanish Research grants TIN2011-28347-C02-02, TIN2010-17170 and INNPACTO-370000-2010-36. The authors also wish to acknowledge their collaboration to students who have validated the proposal.

References

1. Abbasi, R., Staab, S., Cimiano, P.: Organizing resources on tagging systems using t-org. In: Bridging the Gap between Semantic Web and Web 2.0 (2007)
2. Cattuto, C., Benz, D., Hotho, A., Stumme, G.: Semantic Grounding of Tag Relatedness in Social Bookmarking Systems. In: Sheth, A.P., Staab, S., Dean, M., Paolucci, M., Maynard, D., Finin, T., Thirunarayan, K. (eds.) ISWC 2008. LNCS, vol. 5318, pp. 615–631. Springer, Heidelberg (2008)
3. Chi, E.H., Mytkowicz, T.: Understanding navigability of social tagging systems. In: Conference on Human Factors in Computing Systems (2007)
4. Dean, J., Ghemawat, S.: Mapreduce: Simplified data processing on large clusters. Communications of the ACM 51(1), 107–113 (2008)

5. Echarte, F., Astrain, J.J., Córdoba, A., Villadangos, J., Labat, A.: Acoar: a method for the automatic classification of annotated resources. In: Proc. of the 5th Int. Conference on Knowledge Capture, pp. 181–182. ACM, New York (2009)
6. Echarte, F., Astrain, J.J., Córdoba, A., Villadangos, J., Labat, A.: A self-adapting method for knowledge management in collaborative and social tagging systems. In: 6th Int. Conf. on Knowledge Capture, pp. 175–176. ACM, New York (2011)
7. Golder, S.A., Huberman, B.A.: Usage patterns of collaborative tagging systems. Journal of Information Science 32(2), 198–208 (2006)
8. Markines, B., Cattuto, C., Menczer, F., Benz, D., Hotho, A., Stumme, G.: Evaluating similarity measures for emergent semantics of social tagging. In: Proc. of the 18th International Conference on World Wide Web, pp. 641–650. ACM, New York (2009)
9. Mika, P.: Ontologies are us: A unified model of social networks and semantics. Web Semantics: Science, Services and Agents on the World Wide Web 5(1), 5–15 (2007)
10. Peters, I.: Folksonomies: indexing and retrieval in Web 2.0. Knowledge & Information: Studies in Information Science. De Gruyter/Saur (2009)
11. Robu, V., Halpin, H., Shepherd, H.: Emergence of consensus and shared vocabularies in collaborative tagging systems. ACM Transactions on the Web 3(4), 1–34 (2009)
12. Staab, S.: Emergent semantics. IEEE Intelligent Systems 17(1), 78–86 (2002)
13. Steels, L.: The origins of ontologies and communication conventions in multi-agent systems. Autonomous Agents and Multi-Agent Systems 1(2), 169–194 (1998)
14. Yin, Z., Li, R., Mei, Q., Han, J.: Exploring social tagging graph for web object classification. In: Proc. of the 15th ACM SIGKDD Int. Conf. on Knowledge Discovery and Data Mining, pp. 957–966. ACM, New York (2009)
15. Zhang, L., Wu, X., Yu, Y.: Emergent Semantics from Folksonomies: A Quantitative Study. In: Spaccapietra, S., Aberer, K., Cudré-Mauroux, P. (eds.) Journal on Data Semantics VI. LNCS, vol. 4090, pp. 168–186. Springer, Heidelberg (2006)

Emerging Concepts between Software Engineering and Knowledge Management

Sandro Javier Bolaños Castro[1], Víctor Hugo Medina García[1], and Rubén González Crespo[2]

[1] District University "Francisco José de Caldas", Bogotá, Colombia
[2] Pontificial University of Salamanca, Madrid, Spain
{sbolanos,vmedina}@udistrital.edu.co,
ruben.gonzalez@upsam.net

Abstract. Software Engineering (SE) uses different theories to empower its practices. One such theory is Knowledge Management (KM), which provides an important conceptual heritage. Our proposal establishes emerging concepts that enrich SE from KM. All these concepts are in between knowledge and software, hence we call them Softknowledge (SK), y Hardknowledge (HK); they constitute Knowledgeware (KW). In this paper we emphasize the intentionality that pertains to these concepts, which is a fundamental characteristic for the development, maintenance, and evolution of software. Additionally, we propose a nurturing environment based on the present proposal.

1 Introduction

Since the term SE was coined [1], there has been constant crisis within this discipline. This crisis can be seen when observing the discouraging figures reported by official bodies like the Standish Group [2], where high percentages of failure on the projects conducted are reported. Aiming at improving such a bleak situation, a variety of software development processes have been proposed, ranging from code and fix [3] to Waterfall [4], Spiral [5], V [6], b [7], RUP [8], among others. On the other hand, methodologies such as XP [9], Scrum [10], Crystal [11], ASD [12] have been proposed to address the development of practices, values and principles [13]. It seems that knowing the way (how), the people (who), the place (where) and the time (when), that contribute to the development, the processes and the methodology around a problem is a suitable roadmap; however, discouraging reports continue to appear.

The present paper points in a different direction, namely knowledge management and it attempts to reduce failure of the software projects. Although this approach was studied already [14], [15], [16], emerging concepts in between the theories of Software Engineering and Knowledge Management have not been proposed to strengthen the understanding and the solutions to some of the most relevant problems encountered in Software Engineering.

L. Uden et al. (Eds.): 7th International Conference on KMO, AISC 172, pp. 13–24.
springerlink.com

2 Problem Statement

Although Software Engineering has been empowered from Knowledge Management, the work done so far has adopted concepts that, from the perspective of Knowledge Management, favor Software Engineering. One of these concepts is what is known as *tacit knowledge* [17]; however, when adopting a concept, not only are the solutions brought in but also the problems associated to the framework within the providing discipline of origin. Even though the classification of both tacit knowledge and explicit knowledge [18] together with its processes framework (SECI) [19] represent a good approximation to many of the problems found in collaborative work – which is typical of software projects such as: the strengthening of knowledge generation [20], knowledge gathering [21], knowledge exchange [22] and knowledge co-creation [23] processes – it is clear that engineers always end up facing the same advantages and disadvantages that have already been identified for the same processes in the field of KM.

The tendency of researchers to narrow the gap between two theories as a means of improving knowledge frameworks causes harmful side effects, which are eventually identified in most cases. There is a special and apparent similarity between software engineering and knowledge management, but you need to check the most favorable concepts carefully and find mechanisms beyond the mere adoption and adaptation of these ideas.

We believe that, more than doing a theory transfer, it is necessary to generate emerging concepts.

3 Knowledgeware: The Missing Link between Knowledge and Software

Software is a product obtained from intellect [24], it is an extension of our thoughts [25]. When software products are made, there are as many variables as people participating in the development process – "Conway's law" [26], [27]. It could also be claimed that software is knowledge; however, such an statement is too straightforward and it would be reckless not to acknowledge knowledge as intelligence [28], an ability [29], as states of the mind [30], beliefs, commitments, intentions [18] among other features; and although software is pervaded with our reasoning, that reasoning is constrained by the conditions of the programming language and programming paradigms [31], and also by the intended purposes [32] of the documents supporting the software product. The solution to this situation is knowledgeware KW; this new species can be found between knowledge and software, and it takes its traits from both concepts.

Given that, from the perspective of knowledge management, there is a clear distinction between tacit knowledge and explicit knowledge and such a distinction points to the possibility of coding; likewise, in the case of KW, we propose to have a distinction between what we call softknowledge SK and hardknowledge HK. From the point of view of coding, tacit knowledge is orthogonal to explicit knowledge. While tacit knowledge resides in people, explicit knowledge can be stored using physical media, typically IT media. One of the most recurrent concerns is perhaps the way in which organizations gather knowledge and manage it, and so different approaches have been proposed [33], [34], [35], [36], [37], [38]. We consider that leaving tacit knowledge to the world of people is the most suitable approach, our concern should actually focus on the construction of a bridge to join tacit knowledge and software; this bridge is what we regard as SK. Likewise, there should also be a bridge between explicit knowledge and software, which is what we regard as HK.

Knowledge resides in peopleware PW [39], SK and HK reside in KW, and ultimately software resides in hardware. Knowledge management analyses PW from the point of view of the organization; among some of these frameworks we find: organizational knowledge management pillar frameworks [33], intangible asset frameworks [34], intellectual asset model [35], knowledge conversion frameworks [18], knowledge transfer model [36], knowledge management process model [37], and knowledge construction frameworks [38], among others. Regarding these models and frameworks, a lot of effort has gone into defining different types of processes that have an impact on the knowledge of the people who make up the organization. Likewise, within software engineering, the problem of conducting a project has been addressed, where projects themselves are based on organizations consisting of people who must assume the responsibility of a given process in order to obtain software. While in knowledge management processes lead to knowledge, processes in software engineering lead to software. We propose that in order to find a path from knowledge, through KW, up to software, it is necessary to conduct both a traceability process and a representation process. Fig. 1.

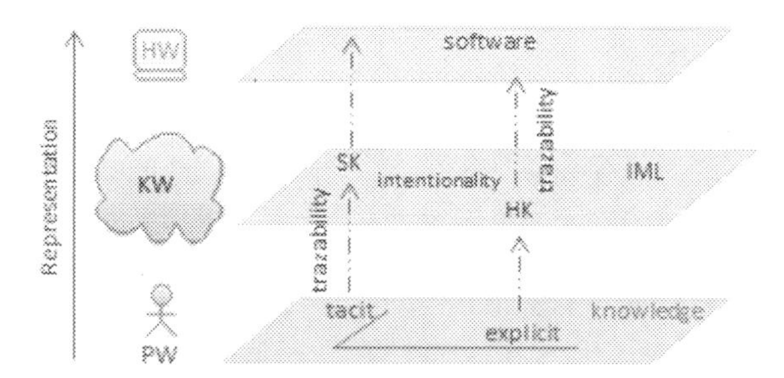

Fig. 1 Knowledge Model

3.1 Traceability

Software traceability [40] is a powerful mechanism that allows the construction of conceptual continuity. Such continuity must permit going back on every step taken in order to understand the origin of concepts [41]. The traceability we propose goes from tacit knowledge to SK, and from there on it goes all the way to end up becoming software; it also goes form explicit knowledge to HK, and form there on it ends up in software.

3.2 Representation

In order to make traceability visible, we propose carrying out representation through languages. If traceability remains over the path tacit-knowledge-to-SK-to-software, we suggest exploiting the advantages of widely-recognized modeling languages, namely archimate [42], UML [43], [44], SPEM [45], or exploiting notations such as BPMN [46] supported by international documents [47]. We also propose the creation of ontological entities supported by languages such as OWL [48]. We believe that software development, when oriented to the creation of models capable of intentional perspectives, can capture certain descriptions where software approximates the knowledge wherein it was created. If traceability also remains on the path explicit-knowledge-HK-software, we suggest exploiting the advantages of having a key between modeling and programming languages, whose perspectives and transformations [49] lead the way from the models language to the programming language.

We coined the concept of Intentional Modeling Language IML, fig. 1, whose additional characteristic over conventional modeling languages will be to provide representation mechanisms that allow expressing the models extended semantics. Some ways to achieve this might be found in the construction of an enriched profiles vocabulary that helps to tinge the models. It is possible to obtain these mechanisms as extension mechanisms in languages like UML. Fortunately, from the perspective of the programming language-transformation-model, this area has been widely developed e.g. MDA [50].

3.3 Intentionality

Most of the problems associated to software lie in the complexity introduced not only by source codes but also by the large volumes of documentation supporting such codes, not to include the typical risks of both coding and documents falling out-of-date. In an attempt to develop and maintain software, engineers often resort to keeping an artifact logbook, where specifications, architectural and design models, user manuals and other records can be found. However, it is by no means an easy task to deal with a product that has been expressed using a language that is intended for a machine as well as using other languages understood by humans. The most considerable difficulty is that the type of knowledge resulting from abstraction, which is expressed in documents and software artifacts, does not easily reflect the actual intention of such creation - of course it is very difficult to capture subjective expressions –; however, it should be possible at least to capture an approximate description. KW was defined as a subjective expression that contains knowledge and that allows itself to be captured through a graphical or textual description by using modeling and programming languages that can incorporate intentional expression mechanisms. Even in art, where so many different emotions are awaken, the piece of art itself provides information about its creation conditions and even about its creator; the piece of art is pervaded by the artist's intentions. Of course in software these intentions must be a lot more bounded; fortunately, disciplines such as software architecture and software design, which leave their mark on software, are very useful to clarify the purpose of how applications are built. In this context, architectural patterns [51], [52], families and styles [53], and also design patterns [54], [55] represent a clearly intentional guide to software.

Other intentional approaches can be used as a mechanism for profiles; in this sense, languages like UML [56] allow an extension of the models semantic meaning through such profiles. Another very valuable approach lies in onthological definitions, which allows adjusting the problem/solution domain with a clear sense of purpose, combating ambiguity. Following these ideas, KW is the traceable and re-presentable knowledge that reflects its intentions in software and therefore facilitates not only software development but software evolution.

4 Intentionality-and-Process-Centered Environment

The importance of processes for both knowledge management and Software Engineering has already been understood, there is even a wide variety of process centred software engineering environments PSEE [57] for software development, which use process modeling languages PML [57]. Nevertheless, although these proposals represent a good development guide when software evolution needs to be evidenced, the roadmaps are not loaded with the necessary intentions to clearly establish the initial purpose of using a given process. We coined the term intentionality-and-process-centered environments.

This type of environment is supported by our own PML proposal. The plus in our PML lies in the SK-and-HK visual modeling from an innovative perspective. Within innovation, we put forward the idea of enriching model semantics and so being capable of building KW. The details of the proposed PML are beyond the scope of this article, but we do illustrate the most relevant concepts from the four points of view of the proposal, namely the knowledge and communications viewpoints, which points to the "software" product; and the improvement and incidents viewpoints, which points to the development process.

4.1 Knowledge Viewpoint

This viewpoint allows modeling knowledge descriptions as well as actors and roles. In this category, a graph permits signaling the most important features from the intellectual assets involved in the development process. Fig. 2 shows the actor called "engineer" with his corresponding role, namely "tester"; the SK is represented as a round-line cloud and HK appears as a straight-line cloud. It can be seen that SK describes ability, while HK describes the possible transformation into a informatics protocol. Both the descriptions are representations of knowledge with a particular intention that allows enriching the semantics of the product.

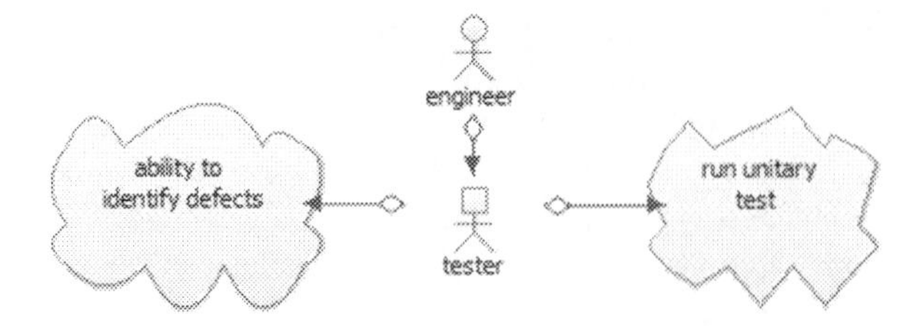

Fig. 2 Knowledge viewpoint

4.2 Communications Viewpoint

Fig. 3 shows how the architect asks the different people involved in the organization about the business nature by conducting an interview whose purpose is to know the architecture. Since the Tower of Babel, the problem of communication has been the obstacle when it comes to setting up projects [24]. Having a simple but expressive vocabulary is another way to achieve clear intentions; we propose modeling a communication interface between actors and roles that expresses the mechanism as well as the ideas arising from a given situation.

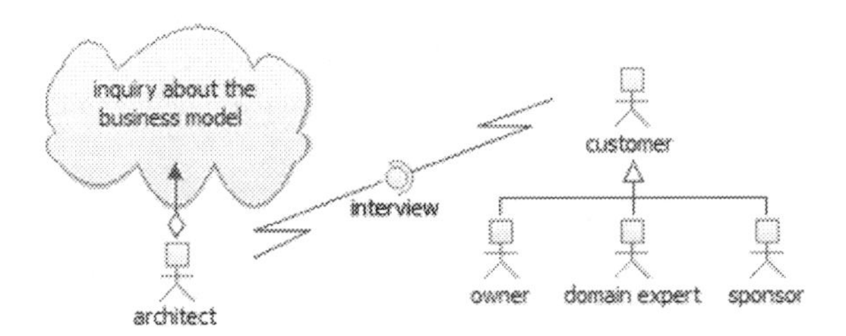

Fig. 3 Communication viewpoint

4.3 Improvement Viewpoint

One of the maxims of software engineering is "process improvement" [58], but yet it has not been directly described in the process model, therefore it is not visible, and because of this, the model and its execution end up being divergent; the person leading the process is supposed to be aware of such an improvement, but a loss of memory is suffered from one process to the other. We propose that the process be enriched with descriptions that point in the direction of improvement. These descriptions can be seen as principles, among other things. Fig. 4.

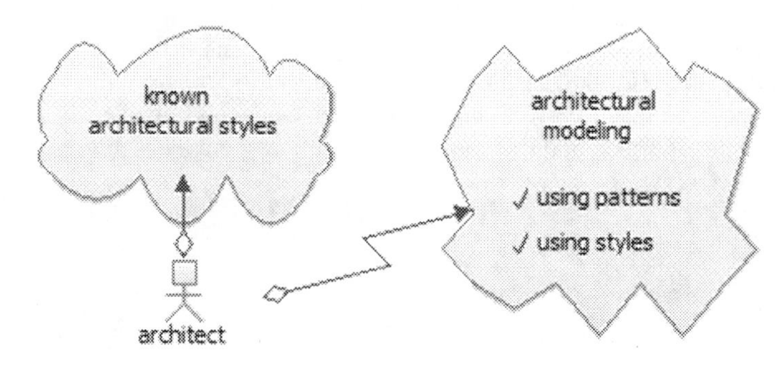

Fig. 4 Improvement viewpoint

4.4 Incidents Viewpoint

In the software development process, another fundamental criterion that must be considered as an intrinsic factor to software is risk [5]; however, like its improvement counterpart, this criterion is not directly associated to a process model either. We believe that risk together with limitations and descriptions are part of a greater set of events that we call incidents, which should be directly reflected in models, Fig. 5.

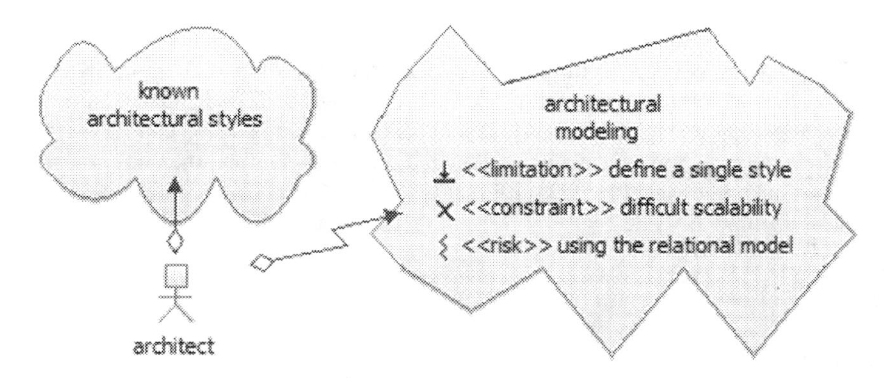

Fig. 5 Incidents viewpoint

The purpose is to provide the models with intentions by taking advantage of either the graphical or textual extensibility mechanisms.

5 Software Tools

There is a wide variety of software development environments that are process-centered, such a Oikos [69], spade [60], Dynamite [61], adele-tempo [62], PADM [63] among others. We propose to have a PSEE plus the models intentions, which we call *Coloso* and will soon be available in its open version at www.colosoft.com.co. This environment integrates Archimate (for the architectural layer) UML (for the design layer) and java (for the programming layer). The whole process is managed from our PML layer. Our proposal also integrates pattern and anti-pattern components, metrics, and process templates. Fig. 6.

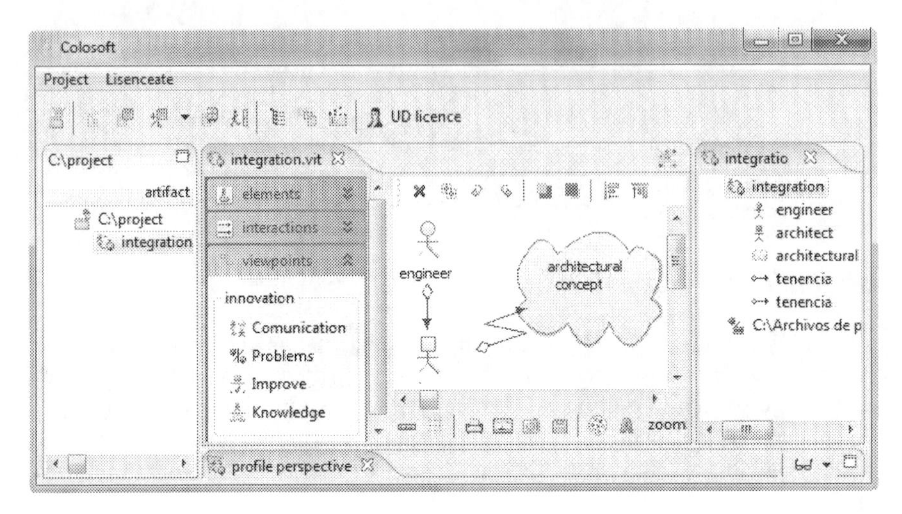

Fig. 6 Coloso

The proposed concept from the *Coloso* platform allows modeling both SK and HK together with their constituent properties. The necessary conceptual continuity between the knowledge layer and the software engineering layer can be established using traceability mechanisms, which are implemented on the platform, including the components that link different types of resources and artifacts such as models, documents, codes and so on. With this approach, it is possible to go from the idea embodied in a conceptual model to its representation expressed in a programming language.

6 Future Work

Three software projects are being developed and tested. An eight-month schedule and a five-people team are the common features of these projects, where two phases are proposed, namely a development phase and a maintenance-and-evolution phase. Our proposal will be used in two of the projects; one of these projects will experience a change in its personnel during the second phase. This event is expected to shed some light on the impact of our proposal.

Another future study addresses the creation and profile-and-ontology enrichment of the *Coloso* environment, which is expected to allow different shadings of the software problem/solution domains. This study is also intended to strengthen the emerging patterns found in between knowledge management and software engineering, which are additional to those patterns already in use [64]; this should allow handling an enriched vocabulary in software processes.

7 Conclusions

Great care is needed when applying theories with the purpose of enriching a discipline; since such theories come together with their own strengths and risks. We believe that the blend of two theories such as SE and KM will necessarily give rise to new concepts like the ones proposed in the present article. The aim of having new concepts is to integrate the strong points; we consider that knowledge is reflected in software, but its intentions will get lost due to the limitations and restrictions that pertain to programming languages unless we decide to use mechanisms like the ones proposed in this article.

Since software is not only about source codes and executable files, we must take advantage of other constructs such as models, which, by using semantic extensions, can be pervaded by the purpose, which makes it easier to perform software development tasks, software maintenance and software evolution. In this article, SK and HK are proposed as two new concepts that can tinge both knowledge and software. By using these two concepts, we propose to construct a bridge between the PW domain knowledge and the hardware domain software. SK and HK must be characterized since they are intentional, that is, they provide the purpose of their very making through different mechanisms, allowing their traceability to be evidenced through the given presentation.

References

1. Naur, P., Randell, B.: Software Engineering: Report of the Working Conference on Software Engineering. NATO Science Committee (1969)
2. Standish Group: The Extreme Chaos. Standish Group International, Inc. (2006)
3. MCconell, S.: Repid Development. Microsoft press (1996)
4. Winston, R.W.: Managing the development of large software systems. In: Proceedings, IEEE Wescon, pp. 1–9 (1970)
5. Boehm, B.: A spiral model of software development and enhancement. SIGSOFT Software Engineering Notes 11(4) (1986)
6. Forsberg, K., Mooz, H.: The relationship of system engineering to the project cycle. In: At NCOSE, Chattanooga, Tennessee (1991)
7. Ruparelia, N.B.: Software development lifecycle models. SIGSOFT Software Engineering Notes 35(3) (2010)
8. Jacobson, I., Booch, G., Rumbaugh, J.: El proceso unificado de desarrollo de software. Addison Wesley (2000)
9. Beck, K.: Embracing change with extreme programming. Computer, 70–77 (1999)
10. Schwaber, K.: Scrum Development Process. In: OOPSLA 1995 Business Object Design and Implementation Workshop. Springer (1995)
11. Cockburn, A.: Crystal Clear. A human-powered methodology for small teams. Agile software development series. Addison-Wesley (2004)
12. Highsmith, J.: Adaptive software development: A collaborative approach to managing complex systems. Dorset House, Nueva York (2000)
13. Manifesto for Agile Software Development, `http://www.agilemanifesto.org/`

14. Ward, J., Aurum, A.: Knowledge management in software engineering describing the process. In: Proceedings of 2004 Australian Software Engineering Conference, pp. 137–146 (2004)
15. Dakhli, S., Ben Chouikha, M.: The knowledge-gap reduction in software engineering. In: Third International Conference on Research Challenges in Information Science, RCIS 2009, pp. 287–294 (2009)
16. Santos, F.S., Moura, H.P.: What is wrong with the software development?: research trends and a new software engineering paradigm. In: OOPSLA 2009: Proceeding of the 24th ACM SIGPLAN Conference Companion on Object Oriented Programming Systems Languages and Applications. ACM (2009)
17. Polanyi, M.: The Tacit Dimension. Routledge, London (1967)
18. Nonaka, I.: A Dynamic Theory of Organisational Knowledge Creation. Organisation Science 5(1) (1994)
19. Nonaka, I., Takeuchi, H.: The Knowledge-Creating Company. Oxford University Press, New York (1995)
20. Les, Z., Les, M.: Shape understanding: knowledge generation and learning. IEEE Transactions on Knowledge and Data Engineering 16(3), 343–353 (2004)
21. Khankasikam, K.: Knowledge capture for Thai word segmentation by using CommonKADS. In: 2010 The 2nd International Conference on Computer and Automation Engineering (ICCAE), vol. 1, pp. 307–311 (2010)
22. Yang, L.: Knowledge, Tacit Knowledge and Tacit Knowledge Sharing: Brief Summary of Theoretical Foundation. In: International Conference on Management and Service Science, MASS 2009, pp. 1–5 (2009)
23. Yao, C.-Y.: Knowledge interaction, adaptive learning, value co-creation and business model innovation. In: Technology Management in the Energy Smart World (PICMET), Proceedings of PICMET 2011, pp. 1–8 (2011)
24. Brooks, F.P.J.: The Mythical Man-Month. Addison-Wesley (1995)
25. Weinberg, G.M.: The psychology of computer programming. Litton educatioal publishing, Inc. (1971)
26. Conway, M.: How Do Committees Invent? Datamation 14(4), 28–31 (1968)
27. Kwan, I., Cataldo, M., Damian, D.: Conway's Law Revisited: The Evidence for a Task-Based Perspective. IEEE Software 29(1), 90–93 (2012)
28. Nelson, R., Winter, S.: An Economic Theory of Evolutionary Change. Belknap Press of Harvard University Press, Cambridge (1982)
29. Wagner, R.K.: Tacit knowledge in everyday intelligence behavior. Journal of Personality and Social Psychology 52(6), 1236–1247 (1987)
30. Schubert, P., Lincke, D., Schmid, B.: A global knowledge medium asa virtual community: The net academy concept. In: The Fourth Americas Conference on Information Systems Proceedings (August 1998)
31. Tucker, A., Noonan, R.: Programming –languages, Principles and Paradigms. Ed. Mc Graw Hill (2002)
32. De Padua Albuquerque Oliveira, A., Do Prado Leite, J.C.S., Cysneiros, L.M., Cappelli, C.: Eliciting Multi-Agent Systems Intentionality: from Language Extended Lexicon to i* Models. In: XXVI International Conference of the Chilean Society of Computer Science, SCCC 2007, pp. 40–49 (2007)
33. Wiig, K.: Knowledge Management Foundations. Schema Press, Arlington (1993)
34. Sveiby, K.: The New Organizational Wealth. Berrett-Koehler, San Francisco (1997)
35. Petrash, G.: Dow's Journey to a Knowledge Value Management Culture. European Management Journal 14(4), 365–373 (1996)

36. Szulanski, G.: Exploring Internal Stickiness: Impediments to the Transfer of Best Practice Within the Firm. Strategic Management Journal 17, 27–43 (1996)
37. Alavi, M.: KPMG Peat Marwick U.S.: One Giant Brain. Harvard Business School (Case), 9-397-108, Rev. (July 11, 1997)
38. Leonard-Barton: Wellsprings of Knowledge. Harvard Business School Press, Boston (1995)
39. Demarco, T., Lister, T.: Peopleware productibe projects dn teams. Dorset House (1999)
40. Seibel, A.: From Software Traceability to Global Model Management and Back Again. In: 2011 15th European Conference on Software Maintenance and Reengineering (CSMR), pp. 381–384 (2011)
41. Bolaños, S., Medina, V., Aguilar, L.J.: Principios para la formalizaci6n de Ingeniería de Software. Revista Ingenieria 14(1) (2009)
42. Open Group: Technical Standar Archimate 1.0 Specification. Published by the open group (2009), http://www.archimate.org
43. Booch, G., Rumbaugh, J., Jacobson, I.: The Unified Modelling Language User Guide. Addison-Wesley (2005)
44. Booch, G., Rumbaugh, J., Jacobson, I.: The Unified Modelling Languages Reference Manual. Addison-Wesley (2005)
45. SPEM, Software & Systems Process Engineering Meta-Model Specification. Version 2.0. Final Adopted Specification, http://www.omg.org
46. Object Management Group (OMG): Business ProcessModeling Notation (BPMN), Version 1.2, OMG Document Number: formal/2009-01-03, OMG (January 2009)
47. Mens, K., Mens, T., Wermelinger, M.: Maintaining software through intentional source-code views. In: SEKE 2002: Proceedings of the 14th International Conference on Software Engineering and Knowledge Engineering. ACM (2002)
48. OWL: Web Ontology Language, http://www.w3.org/TR/owl-guide/
49. Zorzan, F., Riesco, D.: Transformation in QVT of Software Development Process based on SPEM to Workflows. Latin America Transactions, IEEE (Revista IEEE America Latina) 6(7), 655–660 (2008)
50. Kleppe, A., Warmer, J., Bast, W.: MDA explained the model driven architecture: practice and promise. Addison-Wesley (2003)
51. Buschmann, F., et al.: Pattern-Oriented Software Architecture a System of Pattern. Wiley (1996)
52. Dikel, D., Kane, D., Wilson, J.: Software Architecture Organizational Pinciples and Patterns. Pretince Hall (2001)
53. Shaw, M., Garlan, D.: Software Architecture. Prentice Hall (1996)
54. Gamma, E., Helm, R., Jonson, R., Vlissides, J.: Design Patterns Elements of Reusable Object-Orienyted Software. Addison-Wesley (1995)
55. Kuchana, P.: Software Architecture Design Patterns in Java. Auerbach (2004)
56. OMG, UML Profile for Schedulability, Performance, and Time Specification (September 2003)
57. Derniame, J.C., Kaba, B.A., Wastell, D.: Software process: principles, methodology, and technology. Springer, Heidelberg (1999)
58. CMMI Product Team, "Improving processes for better services–CMMI® for Services, Version 1.2", technical report, CMU/SEI-2009-TR-001, ESC-TR-2009-001 (February 2009)

59. Montangero, C., Ambriola, V.: OIKOS: Constructing Process-Centred SDEs. In: Finkelstein, A., Kramer, J., Nuseibeh, B. (eds.) Software Process Modelling and Technology, pp. 335–353. Research Studies Press, Taunton (1994)
60. Bandinelli, S., Fuggetta, A., Ghezzi, C., Lavazza, L.: SPADE: An Environment for Software Process Analysis, Design, and Enactment. In: Finkelstein, A., Kramer, J., Nuseibeh, B. (eds.) Software Process Modelling and Technology, pp. 223–247. Research Studies Press, Taunton (1994)
61. Heiman, P., Joeris, G., Krapp, C.A., Westfechtel, B.: DYNAMITE: Dynamic Task Nets for Software Process Management. In: Proceedings of the 18th International Conference on Software Engineering, pp. 331–341. IEEE Computer Press, Berlin (1996)
62. Belkhatir, N., Estublier, J., Melo, W.: ADELE-TEMPO: An Environment to Support Process Modelling and Enaction. In: Finkelstein, A., Kramer, J., Nuseibeh, B. (eds.) Software Process Modelling and Technology, pp. 187–222. Research Studies Press, Taunton (1994)
63. Bruynooghe, R.F., Greenwood, R.M., Robertson, I., Sa, J., Warboys, B.C.: PADM: Towards a Total Process Modelling System. In: Finkelstein, A., Kramer, J., Nuseibeh, B. (eds.) Software Process Modelling and Technology, pp. 293–334. Research Studies Press, Taunton (1994)
64. Bolaños, S., Medina, V.: Knowledge management patterns applied to the software development process. In: 6th International Conference, KMO (2011)

An Ecosystem Approach to Knowledge Management

Vanessa Chang[1] and Amanda Tan[2]

[1] School of Information Systems, Curtin University of Technology,
Perth, Western Australia
vanessa.chang@curtin.edu.au
[2] Deloitte Consulting, Perth, Western Australia
amtan@deloitte.com.au

Abstract. Effective use of knowledge in organisations would typically lead to improved organizational performance. Organisations that practice and leverage organizational knowledge through integration, innovation and sharing of lessons learned will continue to improve and strengthen the organisation's operation. The ecosystem approach described in this paper centred on the premise that knowledge is individualised information enriched through the process of learning, then shared and applied to practical situations. To capture better knowledge management, a collaborative knowledge management framework is used to examine the positions of knowledge management in an organisation. To highlight the complex and heterogeneous nature of knowledge management, a set of practices aimed to enhance collaboration is presented as holistic approach toward improving practical learning environments. In addition, a set of actionable knowledge management strategies to improve the relationships among the components interacting within each organisational ecosystem can be recommended as a result of using the framework.

Keywords: Knowledge Management, Organisational Learning, Learning Ecosystem, Collaborative Learning.

1 Introduction

Today's enterprises are knowledge-based entities where proactive knowledge management (KM) is vital to attaining competitiveness [15]. Nonaka and Takeuchi [22] emphasised that knowledge is a strategic corporate asset that has become "the resource, rather than [just] a resource". Consequently, KM is an integral business function as organizations [3] begin to realise that their "competitiveness hinges on [the] effective management of intellectual resources" [12].

Por [27] articulates that "knowledge exists in ecosystems, in which information, ideas and inspiration cross-fertilise and feed one another", implying that the interactions that occur among individuals, organizations and knowledge artefacts are the primary sources of learning, knowledge creation, sharing and utilisation. Drawing on this insight, this research emphasises that KM is a set of principles

L. Uden et al. (Eds.): 7th International Conference on KMO, AISC 172, pp. 25–35.
springerlink.com

and practices that aims to improve collaborative and cooperative interactions that occur within a particular organizational environment. Janz and Prasarnphanich [16] conducted a study where knowledge is intrinsically gained through cooperative learning with a view that successful organizational KM involves effectively facilitating the ongoing creation, transfer and application of knowledge [2] by creating an environment that is conducive to collaborative learning. This way, learning is direct interactions between individuals and peers in small groups. This environment stresses the importance of just-in-time socialisation or networking in face-to-face or online modes.

Organisational knowledge arises through a continuous dialogue between explicit and tacit knowledge where knowledge is developed through "unique patterns of interaction between technologies, techniques and people" [4]. Individuals continue to develop new knowledge and organizations play a critical role in articulating and amplifying that knowledge [21]. The successful codification of individuals' tacit knowledge to permanently retain as organisational knowledge relies on the effective implementation of KM.

While there has been considerable development in the areas of knowledge, KM, communities of practice and organizational learning, much of the KM research has been accomplished without substantial impact on the way organizations operate [12]. To bridge the gap between theory and practice, this study draws on Brown's [5] proposition that "an organization is a knowledge ecology" comprise of interacting components. The research advances Chang and Guetl's [6] ecosystem-based framework by drawing on the relationship between KM and collaborative organizational learning, and highlights three key categories of contributors to the dynamics of collaborative learning environments; these being (1) the learning or knowledge management utilities, (2) the learning stakeholders and (3) the internal and external influences that could impact the learning environment [6]. The Collaborative Learning Ecosystem (CLES) serves as a valuable framework for examining and improving the KM positions of organizations.

2 Knowledge Management and Collaborative Organizational Learning

McAdam and Reid [19] and Janz and Prasarnphanich [16] observed that knowledge and learning are inextricably related, each relying on and influencing the other. A learning organization is one creates and shares new knowledge, recognizing that its ability to solve problems does not rely solely on technology [12] or in the individual expertise of personnel, but rather, is a result of an interaction among components within the organizational knowledge base [9]. Alavi and Leidner [2] reinforce that bringing individuals together in a collaborative environment to enhanceboth tacit and explicit knowledge [30] is a KM imperative.

Nonaka [21] asserts that "organisational knowledge creation is a never-ending process that upgrades itself continuously"; and new concepts evolve after it is created, justified and modelled. This interactive process that occurs both intra-organizationally and inter-organizationally is facilitated by collaborative organizational learning where the state of knowledge is continually changed to enable its application to a problem or situation [11].

It is clear that effective organizational learning is dependent on the successful creation, storage, retrieval and transfer of knowledge for appropriate application. Learning, knowledge creation and sharing take place at the individual, team and organizational levels, and these processes are vital to organizational learning [11, 2]. The key focus of KM is to encourage and simplify the collaborative learning process into consideration a range of technological, cultural and social factors.

2.1 Development of a Technological Infrastructure

Information technology (IT) serve as an effective enabler [16] of various facets of KM, from the capturing of tacit or explicit knowledge to its application [29]. KM tools assist in the development of K systems involving a combination of people, technology and culture [18].

Without a stable technological infrastructure, an organization will have difficulty enabling its knowledge workers to collaborate on a large scale [13]. The selection of technology, implementation approach and delivery of content must be performed with a focus on users' needs [13]. The ongoing strategic planning and maintenance of IT applications, including ongoing IT risk management that incorporates security and authentication controls [29] and business continuity management, are integral aspects of developing a stable IT infrastructure. Users must be guided on how to use the technology to communicate and share knowledge [13]. Also, if KM is to be a strategic asset rather than a 'passing fad', it must be aligned with economic value which is attainable by "grounding KM within the context of the business strategy" [34].

2.2 A Knowledge-Friendly Culture

Organisational culture is considered the most significant input to effective KM and organizational learning in that culture shapes the values and beliefs that could encourage or impede knowledge creation, sharing and decision-making [16]. Janz and Prasarnphanich [16] suggest that a positive culture is one that promotes knowledge-related activities by giving workers the support and incentives to create an environment that favours knowledge exchange and accessibility. Alavi, Kayworth and Leidner [1] conclude that there exists a positive relationship between a 'good' knowledge culture (defined by trust, collaboration and learning) and a firm's ability to effectively manage knowledge. They express that a culture of trust and collaboration establishes a greater willingness among employees to share insights and expertise. In contrast, "value systems that emphasise individual power and competition among firm members will lead to knowledge hoarding behaviours" [1].

2.3 Communities of Practice and Social Networks

A community of practice is a "self-organised, self-directed group of people" [8], who are held together by a common purpose, shared interests and varied affiliations [5]. These communities of practice gain access to the creativity and resourcefulness of informal groups of peers, allowing its members to "initiate and contribute to projects across organizational boundaries" [27]. These platforms for interaction contribute to the amplification and development of new knowledge [21], as ideas initially formed in individuals' minds are able to be developed through the interactions based on a "passion for a joint enterprise" [8]. Therefore, communities of practice, formal and informal networks are increasingly valued as fundamental platforms for collaborative learning [9]. In today's web-based era, organizations that create an environment that supports the formation of virtual communities of practice or communities of influence can also gain significant benefits in the areas of knowledge transfer, response times and in-novation [8].

Developing a climate conducive to learning is expected to enhance organizational learning, and in turn, business performance [16]. Nunes et al. [23] state that "KM advantages have to be clear and easily attainable; otherwise organizations will continue to focus on the traditional way of working". Earl [10] and Grover and Davenport [12] assert that even if an organisation embraces the concept that well-managed knowledge could enhance performance, they often do not know how to plan and execute KM initiatives. To address the KM challenge where the application of KM is problematic and the factors requiring consideration are often ambiguous, the Collaborative Learning Ecosystem (CLES) framework has been developed to holistically examine unique organizational learning environments.

3 The Collaborative Learning Ecosystem (CLES) Framework

A vital aspect of this framework is the ability to capture the evolving nature of knowledge and hence the term 'ecosystem' was used. The term 'ecosystem' was originally defined by A.G. Tansley as "a biotic community or assemblage and its associated physical environment in a specific place" [26]. The definition implicitly highlights the existence of interactions among the biotic (living) and a-biotic (non-living) components, as well as intrinsically within various highly-complex elements. Pickett and Cadenasso's [26] insights on the applicability of the ecosystem concept to "any system of biotic and a-biotic components interacting in a particular spatial area" led Chang and Guetl [6] to apply the concept to the learning domain in developing an initial "Learning Ecosystem" (LES) framework.

Papows [25] notes that "effective KM systems enable tacit and explicit knowledge to feed off of one another in an iterative manner". It is through this collaboration among the ecosystem components that organisational knowledge is able to grow and be translated into increased value. With a view to highlight knowledge worker as both 'teacher' and 'learner' engaged in collaborative learning activities, the cur-rent research extends Chang and Guetl's [6] framework to incorporate a focus on collaboration. The model is re-named a 'Collaborative Learning Ecosystem' (CLES) to represent the framework's intended application focus.

3.1 *Overview of the CLES Framework*

The CLES framework (Fig. 1) emphasises "a holistic approach that highlights the significance of each component, their behaviour, relationship and interactions, as well as the environmental borders in order to examine an existing system or form an effective and successful system" [6].

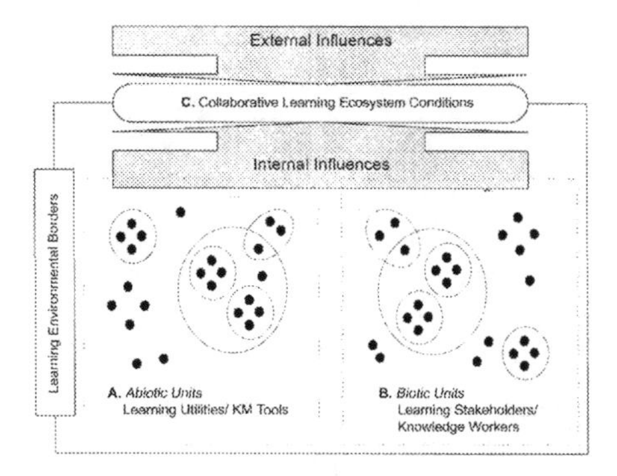

Fig. 1 Representation of the CLES (Adapted from Chang & Guetl [6])

The CLES attempts to simultaneously highlight three key components of contributors (as described below) that "consists of the stakeholders incorporating the whole chain of the [collaborative] learning processes, the learning utilities and the learning environment, within specific boundaries, called environmental borders".

A: Learning Utilities or KM Tools
These are the static and dynamic media that contain and deliver the learning content, and are represented by individual and clusters (each dot denotes a single unit and a group of units make up a cluster that function and work together to serve a purpose) of 'a-biotic units'. These typically (but not necessarily) technology-oriented utilities include all hardware, software and any other computerised platforms that carry the content to the learner.

B: Learning Stakeholders or Knowledge Workers
Learning Stakeholders comprise (i) the learning communities and (ii) other stakeholders who contribute to and/ or benefit from the ecosystem. Learning communities constitute individuals or workgroups (as denoted by clusters of 'biotic units') who can "interact and collaborate synchronously and asynchronously with one another" [6]. Other stakeholders are those who provide the learning content, or support the learning processes through the provision of expertise and services.

C: Collaborative Learning Environment (Restricted CLES Conditions)
The learning environment is dynamic due to changes in a range of internal and external influences, and the impacts of these influences are dependent on the lifecycle of the examined system. External influences include economic dynamics,

domain knowledge, competition and technology advancements [24]. Cultural and sociological influences, funding, business strategies and management support are examples of internal influences [6].

3.2 Investigating the Specifics of the Learning Utilities or KM Tools (A)

It is crucial that both IT and non-IT mediums are considered when investigating the range of learning utilities employed by a firm. The ways in which learning utilities are used by stakeholders in carrying out tasks should be examined and where appropriate, represented using Alavi and Leidner's [2] framework shown in Table 1. Once the complete set of learning utilities have been identified, the relationships, usage and implementation effectiveness of each can be investigated. The nature of the interactions among individuals and these utilities is also an issue of interest of the CLES framework.

Table 1 KM Processes and the Potential Role of ICT (Adapted from Alavi & Leidner [1])

KM Processes	Knowledge Creation	Knowledge Storage/Retrieval	Knowledge Transfer	Knowledge Application
Supporting Information Technologies	- Data Mining tools - Learning Tools	- E-Bulletin Boards - Repositories - Databases - Search and Retrieval Tools	- E-Bulletin Boards - Discussion Forums - Knowledge Directories	- Expert Systems - Workflow Systems
Platform Technologies	*Groupware and Communication Technologies* - Instant Messaging, Email, Online Forums, Web 2.0 tools			
	Intranets/ Extranets			
	Internet			

3.3 Investigating the Specifics of the Learning Stakeholders (B)

Learning communities possess learning attributes, which include unique learning styles, strategies and preferences. The learner's demographics, experience, skills, IT competence, objectives, motivations and needs are also important characteristics. Of the range of learning attributes which could influence collaborative learning, a set of characteristics considered to be central are established based on the following characteristics and existing literature.

- Learning style and preference [4]
- Background experiences and perceptions of personal contribution [4, 12, 16, 17, 20]
- Expectations of contributors and usefulness of information [16, 20]
- Motivation to learn [4, 12]
- Motivation to share knowledge [1, 14, 18, 25, 33]

3.4 *Investigating the Internal and External Environmental Influences (C)*

The application of CLES involves the investigation of the influences affecting a firm's internal and external operating environments. These influences and their impacts usually fluctuate across the business life cycle; and it is vital that the organization adapts to the conditions prevalent at a particular point in time and take action to facilitate the ongoing interaction among the stakeholders and utilities.

A range of existing frameworks can be usedto examine an organization's internal and external environments. Examples of these frameworks include the SWOT analysis, internal value chain or network analysis [28, 31)], Porter's [28] five forces model and the Political, Economic, Social and Technological model.

3.5 *Investigating the Internal Environmental Influences*

Cultural, business strategies and management support are examples of internal influences of a firm's KM implementation success. The factors and the corresponding literature are used to evaluate the specifics of each internal influence.

- Management leadership and support [7, 13, 14, 18]
- Existence of a knowledge-friendly culture [7, 13, 16, 18, 20]
- Clear knowledge strategy [7, 18, 34]
- Commitment towards an IT infrastructure [2, 7, 13, 18, 29]

3.6 *Investigating the External Environmental Influences*

Organizations operate within a broad external environment characterised by a climate that is susceptible to radical change [25]. Chang and Guetl [6] consider the industry, government policies, competition, and technology life cycles as important external influences to a firm's collaborative learning environment. The standard Political, Economic, Social and Technological (PEST) analysis could serve as an effective approach to consider the external environment of an organization.

Competition is a key external influence that has an impact on the learning environment of a firm. Porter's [28] Five Competitive Forces model provides a detailed understanding of the competitive environment and relative position of the firm in its industry. The ways in which competitive forces affect a firm's KM and collaborative learning activities can be evaluated based on this analysis.While the PEST and Porter's Five Forces models may not be necessary for every case, each model provides a comprehensive structure for thoroughly evaluating the range of external factors impacting on an organization's learning conditions.

4 Operationalizing the CLES Framework

The CLES framework is believed to provide a holistic approach to facilitate the development of collaborative learning environments. The key to maintaining a positive environment is to "improve the ecosystem as a whole" [6], which in practice, refers to incorporating user-centric collaborative learning, technological innovation, content and learning design in line with the prevalent environmental conditions. These could result in the development of a range of knowledge management practices to help learners respond to uncertain conditions. It is vital that all the components that contribute to or interact within the CLES are appropriately integrated and that a balance in the utilization of each component is achieved [6].

Fig. 2 provides an example of an organization represented as an ecosystem using the CLES framework. This small business general engineering firm operates two distinct business functions of fabrication and machining. The organization which provides specialized services has 7 staff, 2 of whom are office staff, and it is in the early stages of its business.

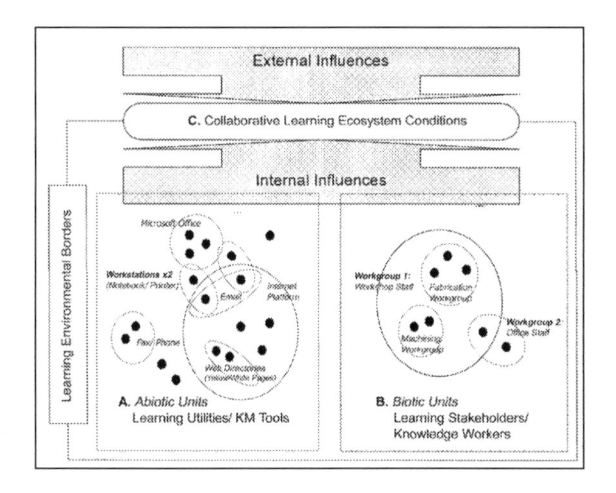

Fig. 2 Representation of a CLES

Analysis was conducted on effectiveness of each learning utility (A), the learning attributes of the learning stakeholders (B) and the levels of interaction among these to carry out work processes. The impacts of various internal and external influences were examined, and based on the analysis; key favourable and unfavourable aspects of the firm's KM practices were identified.

The favourable aspects were:

1. The users willingly adapt to and make the most of technology-oriented utilities.
2. The information required is usually easily accessible in a readily useable format and does not need to be deciphered.
3. Good use of document templates by office staff to create consistent documents.
4. Frequent use of formal face-to-face interaction platforms to facilitate knowledge transfer and collaborative learning (eg. daily workshop meetings).
5. Active leadership in building the organizational structure and culture.
6. Positive knowledge-friendly culture that is conducive to collaborative learning (staff have a genuine interest in each others' lives.

7. Minimal barriers to knowledge sharing – on both the social and work levels (atmosphere that values mutual trust, openness and collaboration).
8. Staff have a relatively high motivation to learn and consider their working experience to be a constant learning process.
9. Staff are given opportunities and the flexibility to leverage their backgrounds and experiences.
10. Capable and motivated staff members whose skills are valuable assets.

The unfavourable aspects were:

1. IT utilities are inadequately attuned to users' information needs
   a) lack of integration of IT utilities
   b) risk of human data-entry error due to minimal integrity constraints and in-efficient information sharing and transfer – due to physical separation of workstations.
2. High volume of inaccurate or incomplete documents.
3. Poor (or non-existent) backup and disaster recovery strategies.
4. Inadequate security controls –critical documents are not password protected.
5. Complacent attitudes towards internet security.
6. Significant reliance on individuals' expertise –relies on individuals' 'head knowledge' which is often not documented.
7. Substantial reliance (possibly over-reliance) on ad-hoc (verbal) communication - insufficient enforcement of the use of documents and records for information transfer.
8. Weak records-management and filing procedures.

Based on the above favourable and unfavourable analyses and in view of the apparent lack in technological integration in the firm, the following key recommendations were noted:

1. Actively discover ways to better leverage IT capabilities.
2. Proactively encourage the use of desired learning utilities to gather and accurately communicate work-related information.
3. Implement formal disaster recovery and information security controls.

Based on the premise that the key focus of KM is to encourage organisational learning by taking into account a range of technological, environmental and sociological factors, the CLES framework aims to provide a holistic perspective on the specifics of the (A) Learning Utilities/ Knowledge Management Tools; (B) Learning Stakeholders (individuals and groups of knowledge workers) and the (C) Internal and External Environmental Influences; which contribute to the dynamics of the particular organisation.

The primary aim is to examine the relationships and interactions that occur within the 'environmental borders' of the organisation, and evaluate how effectively KM practices facilitate organisational learning in the firm. Taking into account the behavioural and interaction performance gaps identified, as well as the technological, environmental and sociological factors, a viable set of KM practices to promote more effective collaborative learning is recommended.

In addition to the brief CLES application as provided in Section 5, KM studies using the CLES framework has been carried out on small-to-medium enterprises and these are reported elsewhere [32]. The examination of the firm's KM effectiveness involved a description of its key business processes, followed by an examination of the learning utilities, learning stakeholders and the impacts various environmental influences have on the organization's learning environment. Focus was placed on investigating the range of KM tools and practices employed, and effectiveness of these to support key business processes and organizational value creation; and the impacts of environmental influences on staff interaction, KM initiatives and organizational learning. The CLES allowed a range of significant aspects to be considered in order to make suitable recommendations to facilitate a more conducive collaborative learning environment.

References

1. Alavi, M., Kayworth, T.R., Leidner, D.E.: An Empirical Examination of the Influence of Organisational Culture on Knowledge Management Practices. Journal of Management Information Systems 22(3), 191–224 (2005)
2. Alavi, M., Leidner, D.E.: Review: Knowledge Management and Knowledge Management Systems: Conceptual Foundations and Research Issues. MIS Quarterly 25(1), 107–136 (2001)
3. Baskerville, R., Dulipovici, A.: The Theoretical Foundations of Knowledge Management. Knowledge Management Research & Practice 4, 83–105 (2006)
4. Bhatt, G.D.: Knowledge Management in Organizations: Examining the Interaction between Technologies, Techniques and People. Journal of Knowledge Management 5(1), 68–74 (2001)
5. Brown, J.S.: Sustaining the Ecology of Knowledge. Leader to Leader Institute (12) (Spring 1999), http://www.johnseelybrown.com/Sustaining_the_Ecology_of_Knowledge.pdf (accessed November 3, 2007)
6. Chang, V., Guetl, C.: E-Learning Ecosystem (ELES) – A Holistic Approach for the Development of more Effective Learning Environment for Small-and-Medium Sized Enterprises (SMEs). In: Inaugural IEEE International Conference on Digital Ecosystems and Technologies (DEST), Cairns, Australia (February 2007)
7. Davenport, T.H., De Long, D.W., Beers, M.C.: Successful Knowledge Management Projects. Sloan Management Review 39(8), 43–58 (1998)
8. Deloitte & Touche LLP (UK) Knowledge Management (2005), http://www.deloitte.com/dtt/article/0,1002,sid%253D5013%2526cid%253D90965,00.html (accessed September 11, 2007)
9. Dickinson, A.: Enhancing Knowledge Management in Enterprises, The Knowledge Board (2002), http://www.knowledgeboard.com/download/593ENKE_KM_Framework.pdf (accessed June 9, 2007)
10. Earl, M.: Knowledge Management Strategies: Toward a Taxonomy. Journal of Management Information Systems 18(1), 215–233 (2001)
11. Gourova, E., Ducatel, K., Gavigan, J., Di Pietrogiacomo, P., Scapolo, F.: Final Report - Enlargement Futures Project - Expert Panel on Technology. Knowledge and Learning (2001), http://forera.jrc.es/documents/eur20118en.pdf (accessed June 7, 2007)
12. Grover, V., Davenport, T.H.: General Perspectives on Knowledge Management: Fostering a Research Agenda. Journal of Management Information Systems 18(1), 5–21 (2001)
13. Hasanali, F.: Critical Success Factors of Knowledge Management (2002), http://www.providersedge.com/docs/km_articles/Critical_success_factors_of_KM.pdf (accessed October 1, 2007)

14. Holsapple, C.W., Joshi, K.D.: An Investigation of Factors that Influence the Management of Knowledge in Organizations. Journal of Strategic Information Systems, 235–261 (2000)
15. Holsapple, C.W., Joshi, K.D.: Knowledge Management: A Threefold Framework. The Information Society 18, 47–64 (2002)
16. Janz, B.D., Prasarnphanich, P.: Understanding the Antecedents of Effective Knowledge Management: The Importance of a Knowledge-Centred Culture. Decision Sciences 34(2), 351–384 (2003)
17. Lee, D.J., Ahn, J.H.: Reward Systems for Intra-Organizational Knowledge Sharing. European Journal of Operational Research 180, 938–956 (2007)
18. Liebowitz, J.: Key Ingredients to the Success of an Orgnization's Knowledge Management Strategy. Knowledge and Process Management 6(1), 37–40 (1999)
19. McAdam, R., Reid, R.: SME and Large Organisation Perceptions of Knowledge Management: Comparisons and Contrasts. Journal of Knowledge Management 5(3), 231–241 (2001)
20. McDermott, R., O'Dell, C.: Overcoming Cultural Barriers to Sharing Knowledge. Journal of Knowledge Management 5(1), 76 (2001)
21. Nonaka, I.: A Dynamic Theory of Organisational Knowledge Creation. Organisational Science 5(1), 14–37 (1994)
22. Nonaka, I., Takeuchi, H.: The Knowledge-Creating Company. Oxford University Press, New York (1995)
23. Nunes, M.B., Annansingh, F., Eaglestone, B., Wakefield, R.: Knowledge Management Issues in Knowledge-Intensive SMEs. Journal of Documentation 62(1), 101–119 (2006)
24. Nyhan, B., Cressey, P., Tomassini, M., Poell, K.: Facing up to the Learning Organisation Challenge: Key Issues from a European Perspective, vol. 1. Office for Official Publications of the European Communities, Luxembourg (2003)
25. Papows, J.: Enterprise.com. Perseus Books Group, USA (1998)
26. Pickett, S.T.A., Cadenasso, M.L.: The Ecosystem as a Multidimensional Concept: Meaning, Model, and Metaphor. Institute of Ecosystem Studies 5, 1–10 (2002)
27. Por, G.: Designing Knowledge Ecosystems for Communities of Practice. In: Advancing Organizational Capability via Knowledge Management. Community Intelligence Labs, LA (1997), http://www.co-i-l.com/coil/knowledge-garden/dkescop/ (accessed June 7, 2007)
28. Porter, M.E.: Competitive Advantage: Creating and Sustaining Superior Performance. Free Press, New York (1985)
29. Sahasrabudhe, V.: Information Technology in Support of Knowledge Management. In: Srikantaiah, T.K., Koenig, M.E.D. (eds.) Knowledge Management for the Information Professional, pp. 269–276. American Society for Information Science, USA (2001)
30. Sedita, S.R.: Back to "Tribal Fires" Explicit and Tacit Knowledge, Formal and Informal Learning, Towards a New Learning Ecosystem. In: DRUID Summer Conference 2003 on Creating, Sharing and Transferring Knowledge. The role of Geography, Institutions and Organizations, Copenhagen, Denmark (2003)
31. Stabell, C.B., Fjeldstad, O.D.: Configuring Value for Competitive Advantage: On Chains, Shops, and Networks. Strategic Management Journal 19, 413–437 (1998)
32. Tan, A., Chang, V.: An Ecosystem Approach to KM: Case Studies of SMEs. In: 19th Australasian Conference on Information Systems, Christchurch, New Zealand (December 2008)
33. Ward, J., Peppard, J.: Strategic Planning for Information Systems, 3rd edn. John Wiley & Sons Ltd., Chichester (2005)
34. Zack, M.H.: Developing a Knowledge Strategy. California Management Review 41(3) (1999)

Discourse and Knowledge Matters: Can Knowledge Management Be Saved?

Lesley Crane, David Longbottom, and Richard Self

Faculty of Business, Law and Computing, University of Derby
L.Crane1@unimail.derby.ac.uk

Abstract. The Knowledge Management (KM) literature is reviewed with a focus on theory, finding a core issue in the lack of a widely accepted and understood definition of knowledge. Theories are categorised on the bisecting continua of personal vs. organizational knowledge, and reified knowledge vs. knowledge as social action. It is argued that a fresh approach based on the Discourse Psychology framework, and its research tool of discourse analysis, would shed new light on the primary issues. Social interaction – and therefore, language – is considered by many KM theorists to be essential to knowledge sharing and creation, yet language has not been the locus of investigation. DP views language as the site of social action, and reality construction. Consequently, a study of talk in interaction is likely to reveal more about the nature of knowledge and in particular its psychological formulation, with implications for its management.

Keywords: Knowledge Management, Discourse Psychology, discourse analysis.

1 Introduction: A Sea of Theory

There can be few domains of practice and academic inquiry which attract such a broad spectrum of theories and points of debate as Knowledge Management (KM). (Despres and Chauvel, 2002) suggest there are too many theories and that this, combined with what (McFarlane, 2011) calls a lack of a unifying framework on the near horizon, presents a challenge for KM and its practitioners. As to be expected, this patchwork quilt of theories is indicative of definitional difficulties, particularly that of knowledge itself (Quintane et al., 2011). (Bouthillier and Shearer, 2002), like many before and since, construct this definitional problem as a major issue, casting KM as an ill-defined field which none-the-less makes substantial claims. Is KM at risk of drowning in its own sea of theory?

It is also not surprising that a number of studies and commentators point to the high failure rates associated with KM initiatives in organizations (Weber, 2007; Burford et al., 2011). Mainstream KM theory and practice are criticised for relying on flawed or misunderstood assumptions and structures (Virtanen, 2011), while others question what it is that KM is actually managing (Crane, 2011). Most mainstream KM theories set out to design a formula for storing, codifying, transferring

L. Uden et al. (Eds.): 7th International Conference on KMO, AISC 172, pp. 37–46.
springerlink.com

and sharing, and creating knowledge within organizations. The claim is that such a harnessing of knowledge will lead to innovation and competitive edge. However, the risk is that such an approach leads to a reification of knowledge as object (Thompson and Walsham, 2004), and that consequently "real" knowledge is simply devalued and debased. If KM is to evolve and to become an embedded organizational practice, it may be time for a different approach.

This paper argues the case for a fresh theoretical approach to KM. It is proposed that Discourse Psychology (DP) can provide both the theoretical framework, and research methodology, that will pave the way for a different understanding of knowledge and its management in organizations. From the DP perspective, knowledge is not an object to be captured, stored and passed around. Rather, knowledge is something that people do in social interaction: knowledge is constructed and shared in talk and text interaction. Notably, many existing KM theories take this general direction emphasising the concepts of knowledge as action, and the importance of language (Thompson and Walsham, 2004; Blackler, 1993; Burford et al., 2011). DP takes these concepts one step further, focusing the enquiry on how knowledge is created / shared in talk interaction, what discursive resources people call upon to achieve this, and with what consequences.

Beginning with a critical review of a sampling of KM theories, the paradigm of Discourse Psychology and its research method of discourse analysis are investigated. The discussion section considers what Discourse Psychology could bring to KM, with conclusions focusing on the question of "where next?"

2 Theories in Knowledge Management

A contemporary classification organises KM theory into the continua of organizational knowledge vs. personal knowledge, and knowledge as object vs. knowledge as social action.

This sampling of theory – which is by no means exhaustive - shows the majority of theories located along the "knowledge as social action" axis, split between a focus on personal knowledge vs. organizational knowledge. A strong advocate of personal knowledge is (Grant, 2002), who is critical of some theories) for their subscription to organizational knowledge. According to Grant, the unit of analysis must be the person. In his "knowledge-based view of the firm", Grant proposes that knowledge is the most strategically important of a firm's resources, and the purpose of firms is to coordinate teams of specialists, and to facilitate the mutual integration of knowledge. In this model, the goal is knowledge integration. Although Grant's work has been interpreted as a theoretical position, Grant questions this, describing it as a set of ideas that emphasises the role and importance of knowledge. None-the-less, according to Grant, it is the knowledge of individual persons that is the valued asset, and this can only be realised by integration through team-work.

Other personal knowledge theories follow the notion of knowledge as social action more closely. (Boisot's, 2002) I-Frame theory of knowledge creation emphasises the importance of learning as the foundation of knowledge creation, arguing that people do not share knowledge. Rather they share information that becomes knowledge once internalised to the individual. In Boisot's model, knowledge is highly personal, and relies on shared repertoires between individuals. As such it leans more towards a cognitive understanding of knowledge, and raises implications such as people's attention span, perception, short and long term memory, aswell as cognitive demand, although these do not feature in Boisot's framework. The theme that is evident here is that theories in the personal knowledge – social action axis are walking in territory that is of intense interest to the Social Psychologist.

(Blackler's, 1993) "knowledge and the theory of organizations", based on a modified version of Activity Theory, also emphases the central role of social learning. This framework straddles the personal vs. organizational knowledge continuum in it could refer to either. Blackler argues that knowledge is performative, not a possession, preferring the term "knowing" over "knowledge". He is critical of the rational-cognitive mainstream approaches to KM in their reification of knowledge, and their assumption of rationality in both organization and individual – which Blackler disputes. (Blackler, 1995) goes onto argue that there is a shift away from knowledge as situated in bodies and routines (embodied and embedded) towards knowledge as situated in brains (embrained), dialogue (encultured) and symbols (encoded). In other words, Blackler argues for and promotes a move away from knowledge as objective, tangible and routinisable, to knowledge as social action.

Moving more towards the organizational end of the personal-organizational knowledge spectrum, Brown and Duguid place communities of practice at the heart of their proposed architecture for organizational knowledge: "(T)the hard work of organizing knowledge is a critical aspect of what firms and other organizations do" (1999: p 28). They argue that knowledge is mostly collective, and that successful communities of practice are generally informal. However, their architecture is largely dominated by themes of organizational command and control. Note, for instance, Brown et al.'s emphasis on the "organizing" work of firms contrasted with (Grant's, 2002) position on the organization as co-ordinator. (Leonard and Sensiper, 2002) also pursue the notion of people sharing knowledge in group work. Leonard et al.'s theory of creative abrasion frames the different backgrounds, skills, experiences and understood social norms amongst individuals as the factors which generate the melting pot of innovation. In this implied chaotic environment, people will challenge each other leading to an abrasion of different ideas, which in turn gives rise to new ones. As a basic idea, this is not divorced entirely from the concepts of social learning advocated by (Blackler, 1993, 1995) and (Boisot, 2002).

Turning to the other end of the spectrum, the majority of theories taking the line of reifying knowledge largely cluster around a focus on organizational knowledge. The leading theory is that of the knowledge creating firm (Nonaka, 1994; Nonaka and Toyama, 2007). This introduces the SECI model, a dynamic model which explains knowledge creation as an interaction between subjectivities and objectivities. According to this model, new knowledge is created in a spiral of interaction between the processes of socialisation, externalisation (explicit knowledge), combination and internalisation (tacit knowledge).

Similarly to Brown and (Duguid, 1999), (Nonaka, 1994) describes the "informal community" as the location of emerging knowledge, but then suggests that these need to be related to the formal hierarchical structure of the organization. This implies a transformation of the informal to the formal. In his definition of the structure of knowledge, Nonaka proposes that tacit knowledge refers to future events, while explicit knowledge deals with the past, and that only tacit knowledge comprises cognitive elements. There is no evidence for this claim, and arguably, tacit knowledge – even if one takes the view that it is internalised, comprising skills, difficult to articulate, the "more than we can tell" element of knowledge – could refer to the past as well as to the future.

In this brief review of KM theory, I have shown how these are variously focused on personal or organizational knowledge, and vary in their approach to knowledge as either object or as social action. I have also shown how they largely embrace the same factors, but use different terminologies and different emphases. A further commonality across KM theory is the substantial assumptions on which they are based. These include the assumption that knowledge can be identified as a singular thing or activity; that KM outcomes can be measured in some way; that the tacit can be made explicit and vice versa; that knowledge resides in people's heads, but they must be motivated to share it. Others assume that language, communication and social interaction are important, but how is not specified; that what will work in one culture or organization will work in another; and finally, that with the right organizational structure, knowledge can be commanded and controlled.

The paper now turns to consider how Discourse Psychology represents a fresh perspective and approach to KM.

3 Discourse Psychology and Discourse Analysis: New Horizons

Discourse Analysis (DA) is becoming increasingly popular in organizational and inter-organizational research, according to (Phillips and Di Domenico, 2009). As Hardy reports, "(S)such discursive studies are playing a major role in the study of organizations and in shaping some of the key debates that frame organization and management theory," (2001: p 25). There are many different "flavours" of DA (see Phillips et al. for a comprehensive summary, and a review of their application in organization studies), but as Phillips et al. point out, they all share an interest in how social reality is constituted in talk. DA is generally framed within the postmodernist, social constructionist paradigm. This is positioned as diametrically opposed to conventional or traditional theoretical and research approaches. Whilst, as with all theoretical approaches and research methods, DA has attracted its share of criticism, for instance, for the subjective nature of its data (Zajacova, 2002), "(T)there can be no question that the legitimacy of postmodern paradigms is well established and at least equal to the legitimacy of received and conventional paradigms,"(Guba and Lincoln, 2005). The focus here is on the framework offered by Discourse Psychology.

Discourse Psychology (DP) was first introduced by Potter and Wetherell in 1987. This ground-breaking work applies the DP theoretical framework, and its distinctive research methodology of discourse analysis (DA), to the study of attitudes and behaviour as examples of studying psychological phenomena through the lens of discourse. By comparing and contrasting this to conventional theories and methods, Potter et al. highlight the weakness of mainstream approaches, and the advantages of DP in revealing the social world as constituted in language.

The core assumption of DP is that language is the site and location of the social world - human action and performance (as distinct from behaviour). People use language to create versions of the social world. Unlike other brands of DA, the data which DP focuses on includes any form of talk and text, drawn from any medium. Conventional Social Psychology, and research methods in general, approach language as a transparent medium which reflects reality as it is (Marshall, 1994). In the conventional approach, the topic of interest is what people report or say, as a means of uncovering some hidden cognitive structures (Billig, 2001) such as attitudes or beliefs, or intentions to act. Such approaches and methods have come in for considerable criticism over the last two decades (Marshall and Wetherell, 2001; Billig et al., 2003; Antaki, 2000). Rather than study what Billig describes as "...ghostly essences, lying behind and supposedly controlling what can be directly observed" (p 210), if one takes the approach that psychology is constituted in language, then it becomes possible to study the processes of thinking directly.

The assumptions that flow from this basis of language as the site of social action are what make DP singular: that talk is constructive, functional, consequential and variant (Potter and Wetherell, 1987). It is constructive in that talk involves an active selection, which may be conscious or not. This in turn implies that all language is functional in that it works to achieve some accomplishment (e.g., persuasion or argument). It is consequential in the sense that talk construction and function lead to consequences for the speaker and the co-participant(s). It is variable in that one person can describe another, or a phenomenon, action or scene in two completely different ways to two different people.

Unlike traditional approaches, DP and its method of DA are not attempting to understand what is said in talk interaction as a means of shining a torch on some underlying cognitive entity (Potter and Edwards, 2003), rather their focus is on how social reality is produced. (Marshall, 1994) offers a neat and succinct comparison of the relative perspectives on language between the post-modernist and conventional paradigms: traditional approaches view language as a mirror that unproblematically reflects reality, whereas DA approaches language as the site of study in its own right. This juxtaposition of approach immediately highlights a major problem in conventional methods. In its use of questionnaires, scales and so forth, the conventionalist searches for patterns and consistencies in their data with the aim of uncovering phenomena which can be generalised. But, how does the conventionalist accommodate for the variation in language? In this case, the conventionalist must "sort out" any variability in the data as it will impact on results. In the DA paradigm, variability in language is sought for, and studied for its consequence and function. As Potter and (Wetherell, 1987) argue, it is this feature of DP/DA which is its empirical mainstay. Phillips and Di Domenico sum this up: "(L)language and its effects are not a problem to be managed as in positivism, but become the core concern of social science," (2009; p 547).

Thus far, the paper has described the theoretical stance of DP and its research method. Next is a brief review of DA in action, with an overview of the types of insight and knowledge that the discipline has delivered (see Wood and Kroger (2000) for a more extended review). Note that while the following are all examples of DA studies, they are not all strictly in the DP convention.

Identity is a particular area of interest for the discourse analyst. In a classic work, (Abell and Stokoe, 2001)'s analysis of a television interview with Princess Diana shows how she actively constructs two separate identities, one a royal role, the other her private self, exploring the tensions between these, and how she uses her identity work to accomplish consequential actions. In another classic work, Locke and Edwards (2003) study former President Clinton's testimony before the Grand Jury, showing how he constructs his identity as caring, sincere, rational and consistent, and the effects this has on his testimony. A study of courtoom identity demonstrates how identity management can be used to potentially influence a jury (Hobbs, 2003), while another shows how discourse is used to construct the category of "older worker" and its associated meanings (Ainsworth and Hardy, 2004).

Studies of discourse in the workplace include (Alvesson's, 1998) work on how men and women construct gender in an advertising agency context. The findings indicate how identity work by men places stronger emphasis on workplace sexuality in response to the ambiguous nature of advertising work. In a fascinating study of healthworkers' actual practice vs. official policy, Marshall (1994) demonstrates the discrepancies between abstract and actual practice.

Studies of text – including online discussion forums – have also yielded fruitful and meaningful results. Crane (forthcoming), for example, investigates how knowledge practitioners construct identity as expert, creating status, trust and recognition. (Crane, 2011) also investigates how the actors in a forum construct KM, comparing this with issues and debates in the literature. In a persuasive work, (Slocum-Bradley, 2010) analyses discourse in the Danish Euro Referendum, showing how identity construction in discourse could lead to a "yes" or "no" vote.

This is just a tiny sample of the work that has been done. While many of these cases are not situated in the organizational workplace context, their implications for workplace discourse and organizational life are none-the-less valid and transferable. They show how people do the business of constructing everyday realities, and how these realities can often be shown to be at odds with assumed reality. They demonstrate how fundamentally important language is to the action of, for instance, constructing identity, accomplishing persuasion and engendering trust. These are important and influential concepts in the everyday business of the organization. The following discussion section considers some the primary implications that a DP approach might have for KM's primary issues.

4 Discussion: Discourse and Knowledge Matters

To begin, it is worth drawing attention to the fact the majority of studies in the KM domain involve conventional research methods: quantitative surveys, structured interviews and case studies. The introduction of the DP framework represents quite a departure. What will the DP-DA approach bring to KM?

First, consider that the core aim of DA is to shine a light on how people do things in their talk in interaction. Conventional methods of enquiry focus on what people do, with an inferred and subsequent why. By gaining a greater understanding of how people accomplish actions in talk, and with what consequences, the influencing and impacting circumstances can be improved (if needed), fixed (if found to be broken) and applied in ways and to activities not previously countenanced. If it is understood how things are done, these actions and the actions around them can be applied in a more powerful and effective way. To simplify this concept, take (Marshall's, 1994) study of the discourse of health workers. The findings show a discrepancy between official policy and what happens in real practice: while workers advocate and support the "good practice" required of policy, events in reality often prevent or limit such practice. As Marshall argues, such a finding would not have been possible using conventional methods: an analysis of workers' discourse reveals how they construct their work and relationships with others, as opposed to self-reports about what they do (in an ideal world). The problem is revealed, and actions can be taken to address it.

Secondly, the notion of knowledge being shared and created in social interaction with others, and the primacy of language, is a relatively widely shared notion amongst KM theorists and workers. But, language and language practices have not been the focus of enquiry. What DP brings to this agenda is a theoretical framework and a tool for focusing the enquiry on discourse. It is not suggested that, in this sense, DP is a replacement for KM theory, but rather as an extension in a direction which many have already signposted.

Finally, in framing KM's core questions of how is new knowledge created and how is knowledge shared (or how can it be shared) within the DP paradigm, one is immediately forced to consider the nature of knowledge. According to DP, knowledge can be viewed as discursive action, done in social interaction with others: it is, in this view, a social construction. That is not to suggest that an individual cannot create new knowledge on their own: but in this case, individual new knowledge can be seen as the product of their existence in a world of sensory stimulus. This notion of socially constructed knowledge is consistent with some of the KM literature (Blackler, 1993, 1995; Quintane et al., 2011; Burford et al., 2011). It is impossible to avoid asking, if knowledge is action as a social construction, how will we know it when we see it? The position argued here is that knowledge is not a single, well-bounded phenomenon, but rather a collection of psychological phenomena, including, for example, identity construction. If knowledge is approached as a psychological phenomenon or phenomena, then it is appropriate to apply the DP framework to its investigation. It is proposed that the identification of these related phenomena – and they may turn out to be completely integral to what we eventually define as "knowledge" – will lead to a much clearer and accessible understanding of knowledge, and consequently how we do knowledge.

Before turning to some conclusions, it is worth noting a major limitation of the DP-DA approach. The nature of this type of enquiry means that subjectivity – both on the part of the participant and of the researcher – must be carefully attended to (Potter and Wetherell, 1987). The role of the researcher must be included in the focus of the study. While it is important to include context as a core part of the enquiry's focus, it is equally important to manage the experiences, opinions, goals and so forth of the researcher. One method of addressing these issues is to include extracts from the data in study reports to enable the reader to make their own independent judgement of the researcher's interpretation.

5 Conclusions: Where Next?

This paper has attempted to show how KM is an important endeavour for organizations wishing to attain and maintain competitive edge and reap the benefits of innovation. But, the literature indicates there are ongoing, apparently irresolvable issues, predominantly concerned with the definition of knowledge. The paper has shown how KM theories can be categorised into how they treat knowledge: knowledge as object vs knowledge as social action, and personal knowledge vs. organizational knowledge. There is a strong sense that there are too many theories, too many opposing definitions of knowledge, and debates that surround almost all aspects of both theory and practice like a besieging army. It is suggested that a fresh approach is needed, and that this could well take the form of DP-DA. Importantly, DP is not positioned as a replacement for existing KM theories – or their research methods – but rather it constitutes a method and a means for taking these forward along a path that many are already standing at the edge of. Namely, for those who view knowledge as a social construction, with the logical inference of the importance of language, then is it not time to turn the lens of enquiry onto discourse itself as the field in which all of the actions of KM take place? The implications for organizational practice are potentially enormous, and as yet untapped. Anybody need a lifejacket?

References

Abell, J., Stokoe, E.: Broadcasting the royal role: constructing culturally situated identities in the Princess Diana Panorama interview. British Journal of Social Psychology 40, 417–435 (2001)

Ainsworth, S., Hardy, C.: Critical discourse analysis and identity: why bother? Critical Discourse Studies 1(2), 225–259 (2004)

Alvesson, M.: Gender relations and identity at work: a case of masculinitites and femininities in an advertising agency. Human Relations 51(8), 969 (1998)

Antaki, C.: Simulation versus the thing itself: commentary on Markman and Tetlock. British Journal of Social Psychology 39, 327–331 (2000)

Billig, M.: Discursive, Rhetorical and Idealogical messages. In: Wetherell, M., Taylor, S., Yates, S. (eds.) Discourse Theory and Practice: a reader. Sage, London (2001)

Blackler, F.: Knowledge and the Theory of Organizations: Organizations as activity systems and the reframing of management. Journal of Management Studies 30(6), 863–884 (1993)

Blackler, F.: Knowledge, knowledge work and organizations: an overview and interpretation. Organization Studies 16(6), 1021–1046 (1995)

Boisot, M.: The Creation and sharing of knowledge. In: Choo, C., And Bontis, N. (eds.) The Strategic Management of Intellectual Capital and Organizational Management. Oxford University Press, Oxford (2002)

Bouthillier, F., Shearer, K.: Understanding knowledge management and information management: the need for an empirical perspective. Information Research 8(1) (2002)

Brown, J., Duguid, P.: Organizing knowledge. Reflections 1(2), 28–44 (1999)

Burford, S., Kennedy, M., Ferguson, S., Blackman, D.: Discordant theories of strategic management and emergent practice in knowledge-intensive organizations. Journal of Knowledge Management Practice 12(3) (2011)

Crane, L.: What Do Knowledge Managers Manage? Practitioners' discourse in an online forum compared and contrasted with the literature. Journal of Knowledge Management Practice 12(4) (2011)

Crane, L.: Trust me, I'm an expert: identity construction and knowledge sharing. Journal of Knowledge Management (forthcoming)

Despres, C., Chauvel, D.: Knowledge, Context, and the Management of Variation. In: Choo, C., Bontis, N. (eds.) The Strategic Management of Intellectual Capital and Organizational Knowledge. Oxford University Press, Oxford (2002)

Grant, R.: The Knowledge-Based View of the Firm. In: Choo, C., Bontis, N. (eds.) The Strategic Management of Intellectual Capital and Organizational Knowledge. Oxford University Press, Oxford (2002)

Guba, E., Lincoln, Y.: Paradigmatic controversies, contraditions, and emerging confluences. In: Denzin, N., Lincoln, Y. (eds.) The Sage Handbook of Qualitative Research, 3rd edn. Sage, London (2005)

Hardy, C.: Researching organizational discourse. International Studies of Management and Organization 31(3), 25–47 (2001)

Hobbs, P.: "Is that what we're here about?": a lawyer's use of impression management in a closing argument at trial. Discourse & Society 14(3), 273–290 (2003)

Leonard, D., Sensiper, S.: The role of tacit knowledge in group innovation. In: Choo, C., Bontis, N. (eds.) The Strategic Management of Intellectual Capital and Organizational Knowledge. Oxford University Press, Oxford (2002)

Locke, A., Edwards, D.: Bill and Monica: Memory, emotion and normativity in Clinton's Grand Jury Testimony. British Journal of Social Psychology 42, 239–256 (2003)

Marshall, H.: Discourse Analysis in an Occupational Context. In: Cassell, C., Symon, G. (eds.) Qualitative Methods in Organizational Research. Sage, London (1994)

McFarlane, D.: Personal Knowledge Management (PKM): Are we really ready? Journal of Knowledge Management Practice 12(3) (September 2011)

Nonaka, I.: A Dynamic Theory of Organizational Knowledge Creation. Organization Science 5(1), 14–37 (1994)

Nonaka, I., Toyama, R.: Why do firms differ: the theory of the knowledge-creating firm. In: Ichijo, K., Nonaka, I. (eds.) Knowledge Creation and Management: New Challenges for Managers. Oxford University Press, Oxford (2007)

Phillips, N., Di Domenico, M.: Discourse Analysis in Organizational Research: Methods and Debates. In: Buchanan, D., Bryman, A. (eds.) The Sage Handbook of Organizational Research Methods. Sage, London (2009)

Potter, J., Edwards, D.: Sociolinguistics, Cognitivism and Discursive Psychology. International Journal of English Studies 3(1), 93–109 (2003)

Potter, J., Wetherall, M.: Discourse and Social Psychology: beyond attitudes and behaviour. Sage, London (1987)

Quintane, E., Casselman, R., Reiche, S., Nylund, P.: Innovation as a knowledge based outcome. Journal of Knowledge Management 15(6), 928–947 (2011)

Slocum-Bradley, N.: Identity construction in Europe: a discursive approach. International Journal of Theory and Research 10(1), 50–68 (2010)

Thompson, M., Walsham, G.: Placing knowledge management in context. Journal of Management Studies 41(5), 725–747 (2004)

Virtanen, I.: Externalization of tacit knowledge implies a simplified theory of cognition. Journal of Knowledge Management Practice 12(3) (September 2011)

Weber, R.: Addressing Failure Factors in Knowledge Management. The Electronic Journal of Knowledge Management 5(3), 333–346 (2007)

Wetherell, M.: Themes in discourse research: the case of Diana. In: Wetherell, M., Taylor, S., Yates, S. (eds.) Discourse Theory and Practice: A Reader. Sage, London (2001)

Wood, L., Kroger, R.: Doing Discourse Analysis: methods for studying action in talk and text. Sage, London (2000)

Zajacova, A.: The background of discourse analysis: a new paradigm in Social Psychology. Journal of Social Distress and the Homeless 11(1), 25–40 (2002)

An Integrated Pruning Criterion for Ensemble Learning Based on Classification Accuracy and Diversity

Bin Fu[1], Zhihai Wang[1,*], Rong Pan[2], Guandong Xu[3], and Peter Dolog[2]

[1] School of Computer and Information Technology
Beijing Jiaotong University, Beijing, 100044, China
{09112072,Zhhwang}@bjtu.edu.cn
[2] Department of Computer Science, Aalborg University, Denmark
{Rpan,Dolog}@cs.aau.dk
[3] School of Engineering & Science, Victoria University, Australia
Guandong.Xu@vu.edu.au

Abstract. Ensemble pruning is an important issue in the field of ensemble learning. Diversity is a key criterion to determine how the pruning process has been done and measure what result has been derived. However, there is few formal definitions of diversity yet. Hence, three important factors that should be further considered while designing a pruning criterion is presented, and then an effective definition of diversity is proposed. The experimental results have validated that the given pruning criterion could single out the subset of classifiers that show better performance in the process of hill-climbing search, compared with other definitions of diversity and other criteria.

Keywords: Ensemble Learning, Classification, Ensemble Pruning, Diversity of Classifiers.

1 Introduction

Ensemble learning refers to a process that learns multiple classifiers to predict the label of an unknown instance, and then combines all the predictions to produce a final prediction. Each classifier learned is also called a base classifier [1]. In the last decade, ensemble learning has gained more and more concerns, due to it could significantly enhance the generalization ability of classification models [2]. In general, the learning process mainly consists of two steps: one is building a number of more diverse and accurate classifiers in training process, the other is aggregating all of the classifiers' predictions to produce a final prediction in classification process [3].

* Corresponding author.

L. Uden et al. (Eds.): 7th International Conference on KMO, AISC 172, pp. 47–58.
springerlink.com

Various algorithms for ensemble learning have been proposed, including bagging, AdaBoost, and random subspace [4, 5, 6], which have been proven to be effective solutions in many fields. In recent years, some researchers further proposed ideas of ensemble pruning, which refers to use only a subset other than all of the base classifiers for classification [7, 8, 9]. There are two primary reasons for it, one is that the time consumption is unaffordable when using all the classifiers for prediction, the other one is that while building a large number of classifiers, not only the unnecessary, but also some bad classifiers are brought in, thus using all of them may reduce the overall performance of classification. Therefore, It is recognized as a more appropriate approach to select a subset of the classifiers with better performance and diversity.

Ensemble pruning is viewed as a classical search problem and thus consists of two key issues. Firstly, an appropriate optimization function or pruning criterion is needed to evaluate the quality of a subset of classifiers, then how to efficiently search in the space of classifiers to get the optimal subset given objective function. Various approaches based on hill-climbing or other strategy have been proposed, and they can be divided into forward search and backward cutting algorithm by the search direction [10, 11, 12]. One of the main differences among these approaches is that different objective functions are used. For example, Caruana et al. used the classification accuracy [7], while Partalas et al. used one kind of diversity as the objective function etc. [13].

Despite of the achievements, there are still some potential problems have not been solved well. Firstly, the selected subset of classifiers may fit the current instances over if accuracy is taken as the only pruning criterion, and thus the generalization ability of the classification model would be weakened. Secondly, there is no a better definition of diversity yet, in spite of having been recognized as an important measurement for a set of classifiers. Most of the existing definitions of diversity just focus on one particular aspect, but do not get a comprehensive understanding of it. Finally, It is pointed out that there might be no direct correlation between the classifiers' diversity and accuracy given the existing definitions of diversity [3], thus it's important to design a more insightful definition of diversity. To solve these problems, we firstly give three key factors that should be considered with respects to diversity, and then define a novel pruning criterion based on the concept of margin function and the proposed factors. In the experiments, the results validate that the given pruning criterion could single out the subset of classifiers that show better performance, compared with several definitions of diversity and other criteria.

The remainder of this paper is organized as follows: In section 2, we introduce the related works about existing pruning criteria. Section 3 gives the factors that should be considered in regard to pruning criterion, and describes our proposed criterion and algorithm in detail. In section 4, we compare our proposed method with other methods, and further analyze the results in detail. Conclusions and future work are given in the last section.

2 Related Work

In this section, we describe the denotations used throughout the paper. Let $X = \{(x_1,y_1),(x_2,y_2),\cdots,(x_n,y_n)\}$ be an instance set of size n, where x_i is the ith instance and y_i is its label. Let $H = \{h_1,h_2,\cdots,h_l\}$ be a set of classifiers. The relationship of predictions between pairwise classifiers h_i and h_j for unknown instances can be depicted as in Table 1, where h_i correct or h_j correct means the number of instances predicted correctly by classifier h_i or h_j, h_i wrong or h_j wrong indicates the opposite, respectively.

Table 1 Classification results between two classifiers

	h_i correct	h_j wrong
h_i correct	N_{11}	N_{10}
h_i wrong	N_{10}	N_{00}

Currently, various criteria have been proposed for ensemble pruning. These criteria fall into two main categories: accuracy-based criteria and diversity-based criteria. The accuracy-based criteria try to find out the subset of classifiers with best accuracy. These criteria include accuracy, root-mean-squared-error, mean cross-entropy, precision/recall, average accuracy, ROC curve, etc.. Caruana et al. used 10 kinds of accuracy-based criteria to prune the whole set of 1000 classifiers, and showed that the performance was improved, compared with the original set of classifiers as well as the optimal single classifier [14]. Furthermore, a ? criterion named SAR was proposed.

The diversity of ensemble learning refers to the predictions of all base classifiers for an unknown instance should be different as much as possible, especially when they predict incorrectly. Thus the wrong predictions of some classifiers would be covered and corrected by others. For example, using voting strategy, we could make correct prediction if the number of correct predictions is greater than the number of wrong predictions. Currently, various definitions have been proposed and they mainly fall into two categories: pairwise diversity and non-pairwise diversity [15].

2.1 Pairwise Diversity

For this kind of diversity, the diversities of each pairwise classifiers should be calculated firstly, and then be averaged for the whole ensemble. Skalak et al. proposed the *Dis* to calculate the ration between the instances predicted differently by pairwise classifiers and the whole instances [16], and its definition is given in formula (1). The greater the *Dis* is, the more diverse the classifiers are.

$$Dis_{i,j} = \frac{N^{10} + N^{01}}{N^{11} + N^{10} + N^{01} + N^{00}} \tag{1}$$

Giacinto et al. proposed *DF* (Double Fault) [17], as shown in formula (2). Comparing with *Dis*, *DF* only considers the situation where instances are predicted wrong by both of the classifiers. The smaller the *DF* is, the more diverse the classifiers are.

$$DF_{i,j} = \frac{N^{00}}{N^{11} + N^{10} + N^{01} + N^{00}} \tag{2}$$

2.2 *Non-pairwise Diversity*

For this kind of diversity, all the classifiers are involved simultaneously, other than calculating each pairwise classifiers. Kohavi et al. proposed the Kohavi-Wolpert variance [18]. It computed the distribution of an instance's predictions under all classifiers, as shown in formula (3). The smaller the variance is, the more diverse the classifiers are.

$$variance_x = \frac{1}{2}\left(1 - \sum_{i=1}^{c} P(y = w_i|X)^2\right) \tag{3}$$

Hashem et al. proposed $Diff$, which computed the distribution of each instance's times when it was classified correctly by all classifiers [19]. It is pointed out that if the classifiers are diverse, each instance should always be classified correctly by some classifiers, and thus the variance of distribution of each instance's times be classified correctly should be small. The definition of $Diff$ is shown in formula (4), where V_i denotes the ratio between the classifiers that classify x_i correctly and the whole classifiers, and *var* is the variance of the V_i's distribution.

$$Diff = var(V_i) \tag{4}$$

Kuncheva et al. summarized 10 kinds of definitions of diversity, and examined the relationship between diversity and accuracy of the set of classifiers [3]. Tang et al. depicted the relationship between diversity and margin, and further proved that maximizing diversity equaled to minimizing the maximum margin [20].

3 A Margin-Based Criterion for Ensemble Pruning

As mentioned above, previous definitions of diversity have been given. However, each of these definitions only focuses on a particular aspect of diversity, thus only using them might not be enough to get a better subset of classifiers. Furthermore, Kuncheva et al. have proved that there is no obvious correlation between accuracy and these existing diversities, and further research about diversity is needed [3]. In this section, we hence give the factors need consideration when designing pruning criterion and propose a novel criterion base on the concept of margin.

After analyzing previous definitions of diversity, we think the following issues should be further considered when devising the diversity or pruning criterion for ensemble pruning.

(1) How a single instance is classified by all of the classifiers: Intuitively, the more possibly each instance is classified correctly, the better a set of classifiers is. Since the aim of ensemble learning is to improve the prediction performance, thus the accuracy should also be considered when searching diverse classifiers. For example, the DF, shown in formula (2), is a criterion that considers the situation when an instance is classified incorrectly by two differently classifiers based on the criterion Dis.

(2) The distribution of the number of instances that are classified correctly by all classifiers: For a instance, the number of classifiers that could classify it should be similar with others, since a instance always could be classified correctly by some classifiers. For example, the $Diff$ shown in formula (4) computes variance of the times of each instance be classified correctly. The smaller the variance is, the more diverse the classifiers are.

(3) The distribution of the number of different wrong predictions for one instance: For a instance, the wrong predictions should also be as different as possible. Thus there would be less wrong predictions that are repeated many times, and they also could then be covered and corrected by the correct predictions more easily when using a voting method to combine the predictions. For example, the Kohavi-Wolpert variance depicted in formula (3) calculates the distribution of different predictions. However it calculates the distribution of all predictions, not only the wrong predictions.

Schapire et al. use the concept of margin to explain the good performance of AdaBoost method. The definition of the margin is shown in formula (5).

$$margin(x,y) = v_y - \max_{c=1,2,\ldots,L(c \neq y)} v_c \tag{5}$$

Where v_y is the number of classifiers that predict instance x's label as y, and $\max(v_c)$ is the maximum number of classifiers that predict other label except y. Shapire et al. have proven that the larger the averaged margin value on all instances is, the smaller the error rate of the ensemble of classifiers is [21].

In order to explain why the ensemble of classifiers could not get good performance, while they have been diverse already given existing definitions of diversity, Tang et al. relate the concepts of margin and diversity, and prove that maximizing the diversity could not always lead to maximizing the minimum margin of the ensemble of classifiers, thus could not always get the optimal subset of classifiers and good performance [20]. Therefore, the concept of margin should also be considered when devising the pruning criterion.

3.1 A New Pruning Criterion Based on Margin

Taking all the aforementioned key issues and the concept of margin into consideration, we propose a novel criterion for ensemble pruning. It is named MMMV (Maximize Margin and Minimize Variance) and its definition is shown in formula (6). This criterion is composed of 3 main components. Each one corresponds to one issue mentioned in subsection 3.1.

$$mmmv = \frac{mean(margin(X))}{var(margin(X)) * mean(var(\omega_i))} \tag{6}$$

The first component is shown as formula (7). It uses the average margin value on the test set of instances to measure how each test instance is classified by the ensemble of classifiers. The larger the average margin value is, the more possibly the instances could be classified correctly.

$$mean(margin(X)) = \frac{1}{n}\sum_{i=1}^{n} margin(x_i, y_i) \tag{7}$$

The second component is shown as formula (8). It calculates the variance of the distribution of each instance's margin value, in order to examine how the ensemble of classifiers could classify the whole test set correctly. The smaller it is, the more average of each instance's times being classified correctly is.

$$var(margin(X)) = \sqrt{\frac{\sum_{i=1}^{n} (margin(x_i, y_i) - mean(margin(X))^2}{n-1}} \tag{8}$$

The last component calculates the variance of the distribution of each classifier's times when classifies the instance incorrectly.

$$mean(var(\omega_i)) = \frac{1}{n}\sum_{i=1}^{n} \sqrt{\frac{\sum_{j=1(\omega_j \neq y_i)}^{L} \omega_{ij} mean(\omega_i))}{L-2}} \tag{9}$$

It is shown in formula (9), where ω_{ij} denotes the number of classifiers that predict label j for instance i. The smaller it is, the more average of the number of different false predictions is.

In summary, the greater the MMMV value is, the better the ensemble of classifiers should be. In the process of ensemble pruning, we should choose the subset of classifiers with the maximum MMMV value.

3.2 *Hill-Climbing Method for Ensemble Pruning*

After determining the pruning criterion, how to search the subset of classifiers is another basic issue of ensemble pruning. Among various search strategy, hill-climbing is an effective and common used method. It expects to get a global optimal result by generating the local optimal results recursively. In this paper, we simply use the forward search hill-climbing method with our proposed criterion for ensemble pruning. Firstly, the subset is set to empty, and then recursively examine each remaining classifier and add it into the subset if the subset could get the maximum MMMV value when it is added into. This recursion continues until some criterion is met. The whole process is depicted in Alg. 1.

Algorithm 1. The process of hill-climbing-based selection

Require:
  dataset D, ensemble of classifier L, size of subset n.
Ensure:
  a subset of classifier.
1: Use Bagging method to build L base classifiers;
2: **for** i=1 to L **do**
3:   create a training dataset using sampling with replacement from the dataset D
4:   build a C4.5 classifier on the new training dataset
5: **end for**
6: set the subset $C = \{\}$;
7: select a classifier randomly and add it into C;
8: **for** i=1 to n **do**
9:   find the classifier c from the remaining classifiers that can maximum the MMMV value of the subset C
10:   add c into C
11: **end for**
12: **return** C;

In the process depicted in Alg.1, the predictions of each base classifier for the instances should be stored in order decrease the time consumption. In the next section, we will compare our proposed method with other existing methods and analyze the results in detail.

4 Experimental Design and Analysis

In this section, we firstly describe the experimental setup and data sets. Then the experimental results are given and analyzed.

4.1 The Datasets and Setup

32 data sets from UCI [23] is used for the experiments. These datasets are listed in Table 2, including the name of the dataset, the number of instances, attributes and labels.

The setup of the experiments is as follows. We use C4.5 as the base classifier model; bagging method is used to build 200 base classifiers to form the original ensemble of classifiers. We then compare the accuracy of the selected subset of classifiers using MMMV DF, $Diff$ and Margin, and the whole ensemble of classifiers without pruning. The size of the selected subset of classifiers is set to 40, and a forward search method is used to find the subset. 70% of each dataset are used as the

Table 2 Description of data sets

Number	Dataset	Instances	Attributes	Labels
1	anneal.ORIG	798	38	6
2	audiology	226	69	24
3	autos	205	26	7
4	bridges-version2	108	13	6
5	car	1728	6	4
6	Cermatology	366	34	6
7	ecoli	336	7	8
8	flags	194	30	8
9	german	1000	20	2
10	glass7	214	10	7
11	hayes-roth-test	28	5	4
12	heart-h	294	13	5
13	house-votes-84	435	16	2
14	Hypothyroid	3772	19	6
15	kr-vs-kp	3196	36	2
16	led	1000	10	7
17	letter	20000	16	16
18	lung-cancer	32	56	3
19	lymph	148	19	4
20	mfeat-mor	2000	6	10
21	mfeat-pixel	2000	240	10
22	pid	768	8	2
23	primary-tumor	339	17	22
24	ptn	329	17	22
25	sbl	683	35	19
26	segment	2310	19	7
27	segment-challenge	2100	19	7
28	solarflare	1389	9	2
29	soybean	351	35	19
30	splice	3190	61	3
31	ttt	958	9	2
32	vehicle	846	18	4

training dataset, and the rest as the test dataset. All the methods are implemented on the Weka platform [22], and are executed 5 times using 10-fold cross validation on all datasets with different random seeds 1, 3, 5, 7, 11, respectively. The final results are the averaged values.

4.2 *The Result and Analysis*

We compare the accuracy of the subset of classifiers selected using different selection criteria and the original ensemble of classifiers without pruning on the 32 datasets, and the results are given in Table 3. The boldface indicates the best method.

Table 3 Description of the experimental results

Number	MMMV	*DF*	*diff*	Margn	Bagging
1	**95.5556**	90.3704	85.1852	91.1111	89.2593
2	**82.3529**	79.4118	75.0000	80.8824	80.8824
3	75.8065	67.7419	72.5806	**82.2581**	77.4194
4	84.3750	65.6250	84.3750	87.5000	**90.6250**
5	**92.8709**	90.1734	92.6782	91.7148	92.2929
6	**97.2727**	94.5455	92.7273	97.2727	97.2727
7	88.1188	83.1683	85.1485	87.1287	**89.1089**
8	64.4068	59.3220	54.2373	**69.4915**	**69.4915**
9	**83.6667**	80.6667	77.6667	81.6667	83.3333
10	**86.1538**	76.9231	83.0769	83.0769	81.5385
11	77.7778	55.5556	55.5556	77.7778	**88.8889**
12	83.1461	83.1461	86.5169	85.3933	**86.5169**
13	**96.9466**	94.6565	96.1832	96.1832	96.1832
14	**99.5583**	99.3816	99.2933	**99.5583**	**99.5583**
15	**99.1658**	97.9145	98.2273	**99.1658**	98.6444
16	**75.0000**	71.6667	73.3333	73.6667	73.6667
17	94.2167	93.1000	93.0333	94.2667	**95.0500**
18	60.0000	**70.0000**	60.0000	60.0000	**70.0000**
19	75.5556	73.3333	75.5556	**80.0000**	**80.0000**
20	77.3333	79.3333	76.6667	77.1667	**79.6667**
21	**93.1667**	84.5000	89.0000	90.3333	90.3333
22	83.5498	85.2814	83.5498	85.2814	**85.7143**
23	55.8824	51.9608	**61.7647**	58.8235	60.7843
24	52.9412	57.8431	56.8627	50.0000	**60.7843**
25	92.1951	87.3171	89.2683	92.1951	**93.6585**
26	**97.6912**	96.3925	96.5368	96.8254	96.3925
27	97.1111	96.6667	96.6667	96.8889	**97.5556**
28	**86.0911**	83.4532	84.6523	84.4125	83.9329
29	**94.1463**	87.8049	91.7073	92.6829	93.6585
30	93.7304	93.2079	93.3124	**93.8349**	93.6259
31	**89.5833**	88.5417	84.0278	86.8056	87.8472
32	**86.6142**	81.4961	81.1024	85.0394	85.4331

We firstly compare MMMV with other pruning criteria. The experimental results in Table 3 show that the MMMV we proposed in this paper performs best on 16 data sets out of the 32 data sets. For the other pruning criteria, the DF performs best on one dataset meanwhile the $diff$ on two datasets. Therefore, compared with the criteria which only focus on one aspect of diversity, considering more factors of diversity is more appropriate. In addition, we also compare MMMV method with the basic definition of margin, and it is showed that the margin is the optimal method for 7 data sets. The experimental results indicate that consideration of the variance of the marginal value and the wrong predictions' distribution could further get the subset of classifiers with better performance. Secondly, We also compare the prediction's accuracy of the subset of classifiers selected using MMMV and the original classifier set without pruning that generated using Bagging method. It can be seen from Table 3 that the original set of classifiers performs best on 16 data sets. The subset selected using MMMV has a similar performance, but it only utilizes 20% of the base classifiers after pruning.

In addition, due to the MMMV method considers the base classifier's misclassification distribution on the test dataset and the number of classifiers that could predict correctly for each instance, we also analyze the number of classes and the size of test dataset to examine their impacts on the performance of our proposed method. It is observed that on the large datasets with relatively small number of classes, such as german, mfeat-pixel, solarflare etc., MMMV performs better. However, on the small datasets with relatively large number of classes, such as primary-tumor, ptn etc., MMMV performs bad. So the MMMV criterion for pruning is much more appropriate to the relatively large datasets with small number of classes.

5 Conclusion

Ensemble pruning is a very important step in the process of ensemble learning, and some researchers have proposed a variety of pruning criteria. Although diversity is an important criterion for pruning, There is not a formal definition of it yet. Most existing definition of diversity only focus on a particular aspect, and thus the subset of classifiers selected using them may not be the most diverse actually. In addition, some researchers also have showed that there is no obvious relationship between the accuracy and diversity of the subset of classifiers. To solve above problems, we analysis several existing definitions of diversity, and point out the factors that should be taken into consideration when designing new definition of diversity or other criteria. These factors include: to what extent the subset of classifiers could classify an instance correctly; the distribution of results, when the subset of classifiers classify an instance incorrectly; the distribution of the number of instances that are classified by each base classifier in the whole data set. Moreover, we propose a novel definition of pruning criterion considering aforementioned factors. the experimental results show that our proposed criterion could find out more accurate subset of base classifiers on majority of the data sets.

One potential problem is that we just validate the superiority of our proposed method through experiment, how to theoretically prove the validity of it should be studied in our future work. In addition, hill-climbing search method could not always find out a globally optimal solution, thus design other strategy of pruning is another major focus of the future work.

Acknowledgements. This research has been partially supported by the Fundamental Research Funds for the Central Universities, China (2011YJS223), National Natural Science Fund of China (60673089, 60973011), EU FP7 ICT project M-Eco: Medical Ecosystem Personalized Event-based Surveillance (No.247829).

References

1. Dietterich, T.G.: Ensemble Methods in Machine Learning. In: Kittler, J., Roli, F. (eds.) MCS 2000. LNCS, vol. 1857, pp. 1–15. Springer, Heidelberg (2000)
2. Shipp, C., Kuncheva, L.: Relationships between combination methods and measures of diversity in combing classifiers. Information Fusion 3(2), 135–148 (2002)
3. Kuncheva, L., Whitaker, C.: Measures of diversity in classifier ensembles and their relationship with the ensemble accuracy. Machine Learning 51(2), 181–207 (2003)
4. Breiman, L.: Bagging predictors. Machine Learning 24(2), 123–140 (1996)
5. Freund, Y., Schapire, R.: Experiments with a new boosting algorithm. In: Proceedings of the Thirteenth International Conference on Machine Learning (ICML 1996), pp. 148–156 (1996)
6. Ho, T.: The random subspace method for constructing decision forests. IEEE Transactions on Pattern Analysis and Machine Intelligence 20(8), 832–844 (1998)
7. Caruana, R., Niculescu-Miil, A., Crew, G., et al.: Ensemble Selection from Libraries of Models. In: Proceedings of the Twenty-First International Conference (ICML 2004), pp. 18–25 (2004)
8. Martinez-Munoz, G., Suarez, A.: Using boosting to prune bagging ensembles. Pattern Recognition Letters 28, 156–165 (2007)
9. Tsoumakas, G., Partalas, I., Vlahavas, I.: An Ensemble Pruning Primer. In: Okun, O., Valentini, G. (eds.) Applications of Supervised and Unsupervised Ensemble Methods. SCI, vol. 245, pp. 1–13. Springer, Heidelberg (2009)
10. Martinez-Munoz, G., Suarez, A.: Aggregation ordering in bagging. In: Proceedings of the 2004 International Conference on Artificial Intelligence and Applications, pp. 258–263 (2004)
11. Yang, Y., Korb, K.B., Ting, K.M., Webb, G.I.: Ensemble Selection for SuperParent-One-Dependence Estimators. In: Zhang, S., Jarvis, R.A. (eds.) AI 2005. LNCS (LNAI), vol. 3809, pp. 102–112. Springer, Heidelberg (2005)
12. Banfield, R., Hall, L., Bowyer, K., et al.: Ensemble diversity measures and their application to thinning. Information Fusion 6(1), 49–62 (2005)
13. Partalas, I., Tsoumakas, G., Vlahavas, I.: Focused ensemble selection: a diversity-Based method for greedy ensemble selection. In: Proceedings of the 18th European Conference on Artificial Intelligence (ECAI 2008), pp. 117–121 (2008)
14. Caruana, R., Niculescu-Mizil, A.: Data mining in metric space: an empirical analysis of supervised learning performance criteria. In: Proceedings of the 10th ACM SIGKDD International Conference on Knowledge Discovery and Data Mining (KDD 2004), pp. 69–78 (2004)

15. Kapp, M., Sabourin, R., Maupin, P.: An empirical study on diversity measures and margin theory for ensembles of classifiers. Information Fusion (2007)
16. Skalak, D.: The sources of increased accuracy for two proposed boosting algorithms. In: Proceedings of the 11th AAAI Conference on Artificial Intelligence, pp. 120–125 (1996)
17. Giacinto, G., Roli, F.: Design of effective neural network ensembles for image classification processes. Image Vision and Computing Journal 19(9), 699–707 (2001)
18. Kohavi, R., Wolpert, D.: Bias plus variance decomposition for zero-one loss functions. In: Proceedings of the Thirteenth International Conference on Machine Learning, pp. 275–283 (1996)
19. Hashem, L., Salamon, P.: Neural network ensembles. IEEE Transaction on Pattern Analysis and Machine Intelligence 12(10), 993–1001 (1990)
20. Tang, E., Suganthan, P., Yao, X.: An analysis of diversity measures. Machine Learning 65(1), 241–271 (2006)
21. Schapire, R., Freund, Y., Bartlett, P., et al.: Boosting the margin: a new explanation for the effectiveness of voting methods. In: Proceedings of the Fourteenth International Conference on Machine Learning (ICML 1997), pp. 322–330 (1997)
22. Witten, I., Frank, E.: Data Mining: Practical Machine Learning Tools and Techniques with Java Implementations, pp. 365–368. Morgan Kaufmann Publishers, San Francisco (2000)
23. Asuncion A, Newman, D.: UCI Machine Learning Repository. University of California, School of Information and Computer Science, Irvine, CA (2007), `http://www.ics.uci.edu/~mlearn/MLRepository.html`

A Process-Oriented Framework for Knowledge-Centered Support in Field Experience Management

Wu He

Department of Information Technology and Decision Sciences, Old Dominion University, Norfolk, VA, 23529
whe@odu.edu

Abstract. In this paper, the author discusses an exploratory study to provide IT support for field experience management using a process-oriented field experience knowledge management framework. This framework is structured around three sequential non-lineal phases: (1) before: preparing, searching and placement; (2) during: Supervising, tracking progress and providing feedback; and (3) after: Reflecting, assessment and reporting. Based on the proposed framework, the author constructed a knowledge-centered prototype system to support the field experience tracking and administration process. The paper also presents an evaluative performance of the system and discusses the implications. The results show that the proposed framework is valuable and can be used to guide and facilitate knowledge-center support system development in the area of field experience management.

Keywords: Knowledge-Centered Support, Field Experience Management, Process-oriented Framework.

1 Introduction

Field experiences such as internships, practicum & co-ops play an important role within education by providing on-the-job experiences to students prior to graduation [3]. Students can be better prepared if their academic learning is reinforced through authentic workplace experience, where the link between theory and professional practice can be realized [14]. Many programs such as computer science, information technology, engineering, business and teacher education have made field experiences an integral part of their teaching programs for both undergraduate and graduate students [6, 11]. Field experiences are designed to provide students with the opportunity to gain experiences in the workplace, to see the field they want to enter, to apply the skills and knowledge they've gained from courses into the practice, and to become inquiring, reflective professionals [2, 12]. For example, in the field of information technology and computer science, field experiences provide students with the opportunity to develop their technical and/or

L. Uden et al. (Eds.): 7th International Conference on KMO, AISC 172, pp. 59–67.
springerlink.com

project management skills, to understand teamwork, ethical and social values in professional settings, to gain industry experience, to develop a net-work of contacts for future work, and to enhance their resumes, in order to prepare them for be-coming IT professionals [13]. The role of field experience is highly valued and considered important for students' job seeking and placement as most IT positions now require varying level of field experience.

In practice, many people are involved in creating successful field experiences. The key players in the field experience are the faculty, employers and students. Figure 1 displays the relationship among them.

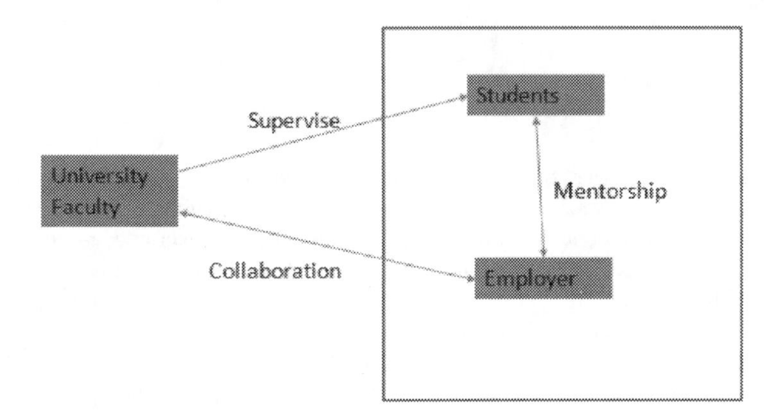

Fig. 1 Field experience relationship

Although field experience is highly valued, little attention has been given to the use of knowledge-based systems in administering the field experience process and outcomes. Administering field experiences is challenging because it presents a range of problems such as supervision of students, difficulty in developing and maintaining collaborative relationships between employers and university instructors. For example, since university instructors have few opportunities to visit or communicate with employers who provide field experience opportunities to students, this often lead to lack of clearly agreed upon goals, roles, values, requirements and responsibilities for guiding students' field experience and give rise to the tendency to blame each other when problems arise [6].

It is also noted that in many fields, a substantial percentage of the field experience is tacit knowledge, which is often context-specific and situation-dependent [1, 8]. An example of tacit knowledge is knowing how to troubleshoot problems found in an industry-specific information system if the solutions to the problems were not documented before. Some other examples of requiring tacit knowledge include maintaining or supporting obsolete applications, developing requirement specifications, and debugging complex business programs. A good way of developing tacit knowledge is to get "on-the-job" training through the field experience which is typically conducted in the work setting under the direction of one or more of the employer's supervisory personnel. However, unfortunately, the tacit knowledge students learned through their field experience or "on-the-job" training is often not documented or tracked.

With these considerations in mind and providing optimal field experience to students, we propose a process-oriented knowledge management framework outlining specific activities structured around three sequential non-linear/iterative phases which follow the typical field experience process. This framework provides a systemic understanding of the various roles, requirements and responsibilities associated with students' field experience. As many schools would like to keep students field experience information in their own database for research, regulation, privacy and documentation purposes, free open source Web 2.0 tools such as Wikis and Blogs won't be adopted. Moreover, many schools have specific requirements on the information they would like to track and store. To meet their specific needs and test the value of the framework, we conducted an exploratory study and developed a web-based prototype to support the field experience tracking and administration. This prototype was designed to help administrators more effectively manage the field experiences process and outcomes. By conducting this exploratory study, we can find what issues or challenges may exist in the proposed framework, whether the proposed approach is viable, feasible and scalable across diverse educational environment. The proposed framework is not intended to create a new knowledge management (KM) theory, but to propose a KM system to support the field experience tracking and administration.

The remainder of the paper is organized as follows. In section 2, we describe a process-oriented knowledge management framework for managing the field experience process. Section 3 describes practical considerations/challenges for the implementation of the proposed framework as well as the implementation and evaluation of a web-based field experience tracking and management system. Section 4 presents the conclusion and future research.

2 A Process-Oriented Framework for Knowledge-Centered Support in Field Experience Management

Guided by the literature in field experience and our practical experience, a process-oriented knowledge management framework outlining specific activities structured around three sequential non-linear/iterative stages is proposed as follows (See Figure 2). The framework was designed to improve capture of field experience-related knowledge, retain lessons and share/replicate successful experience to provide better field experiences to students and support more informed decisions and actions.

The following is a brief description of each stage listed in the framework.

Stage 1: Before the field experience:

In the preparation phase, To prepare for the field experience, students are encouraged to talk to their faculty advisors to clarify the goals and roles for the field experience, and figure out what value they want to gain from the field experience such as an internship, practicum or & co-ops. In the search phase, students will research information on the target fields and actively look for the field experience opportunities such as internships that are a good fit for the goals. Common search methods include talking to professors, friends, and family; research opportunities and possibilities through the web, databases, and campus career resource center; do a self-assessment of personal skills; Create or update your resume, cover letter and list of references; apply for the field experience opportunities and prepare for the interview. In the placement phase, students will be involved activities including interviewing with various employers, discussing the requirements and responsibilities, finding a match that meets the needs, accepting the offer for field experience such as an internship, determining the field experience schedule, meeting with the field experience supervisors or mentors, and keeping in touch with them if possible.

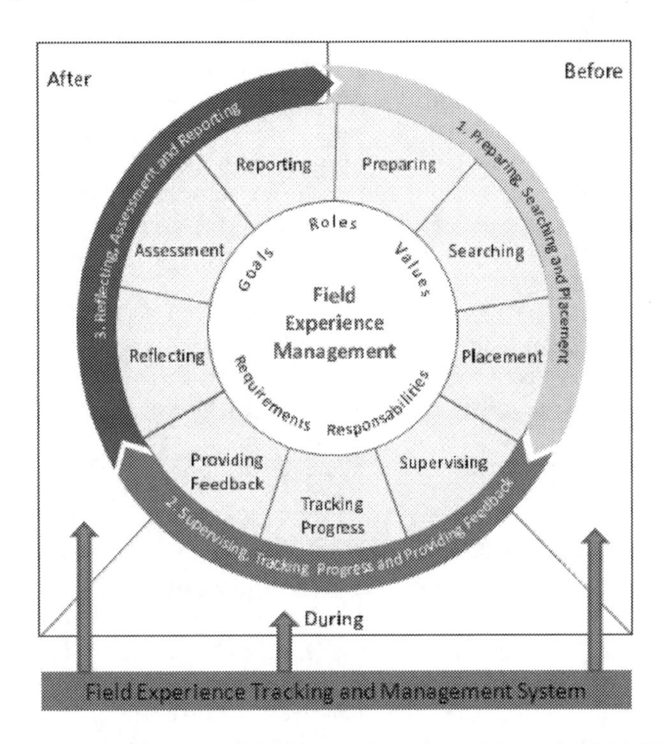

Fig. 2 A Process-oriented Framework for Knowledge-Centered Support in Field Experience Management

Stage 2: During the field experience:

During the field experience, students will carry out assigned activities and tasks in the work setting under the direction of one or more of the employer's supervisory personnel. Students will have the opportunity to interact with others, ask questions, and get to do hands-on work. Students are expected to behave professionally during their field experience.

Students are also expected to maintain ongoing communication with their faculty advisor and field experience supervisors in the workplace. Oftentimes, students will be asked to report their learning progress, outcomes and self-reflection on work performance while maintaining confidentiality of information about employers from time to time. Faculty advisor and field experience supervisors will track students' progress and collaborate to provide guidance and feedback on students' work and learning. If any conflicts or issues rise, the three parties are expected to work together to resolve the conflicts or issues.

Stage 3: After the field experience:
To ensure the quality of field experience, students are encouraged to take the time to engage in self-reflection on their overall performance and learning after the field experience for their personal professional growth and development [7]. A reflection paper is often required after the field experience. Students' performance will be evaluated by their faculty advisors and/or field experience supervisors after the field experience. A variety of assessment methods or instruments may be used to evaluate students' learning outcomes (e.g., projects, portfolios, papers and other artifacts), experience, performance and growth [9].

For the instructors and program administrators, they often compile and aggregate the data collected or generated from the field experience to develop a final field experience summary report which can be used to assist decision making on field experience strategies, share successful experience and lessons, and guide their future efforts or actions on preparing students for field experience.

To be used as a practical roadmap, the above field experience process in the proposed framework requires a supportive environment [4, 10] such as organizational culture, administrative support, and technology support. From the knowledge management (KB) perspective, it is important to ensure that people, process and technology are aligned to provide better support and achieve better outcomes [5]. Successful implementation of the framework requires a web-enabled knowledge-center support (KCS) system such as a field experience tracking and management system to track, administer and streamline the whole process, and to preserve the knowledge generated throughout the field experience process for knowledge sharing. Knowledge-Centered Support (KCS) is a practical approach to organizing and implementing KM in a support environment for organizations [5].

3 A Knowledge-Center Support System for Field Experience Tracking and Administration

To evaluate our proposed framework and help improve the field experience management in real world, we worked with a teacher education program at a large public university to implement a web-based field experience tracking and management system named iTrax. The current form of the prototype supports both stage 2 (during field experience) and stage 3(after field experience) of the field experience management process. We hope to expand the prototype to help support or facilitate some of the activities in stage 1 as well in the future.

The iTrax system was designed to allow students in a teacher education program to report their daily field experience including daily event logs for classroom activities, time spent on each activity, types of experiences (observation, one-on-one teaching, small group teaching, shadowing), assessment, issues encountered, outcomes, and reflective comments on that day. In addition, the system provides modules for instructors, mentors in the workplace, and program administrators to provide feedback, track, monitor, and manage aggregated data about students' self-reported field experience. Figure 3 is a screenshot of the instructor module of the system.

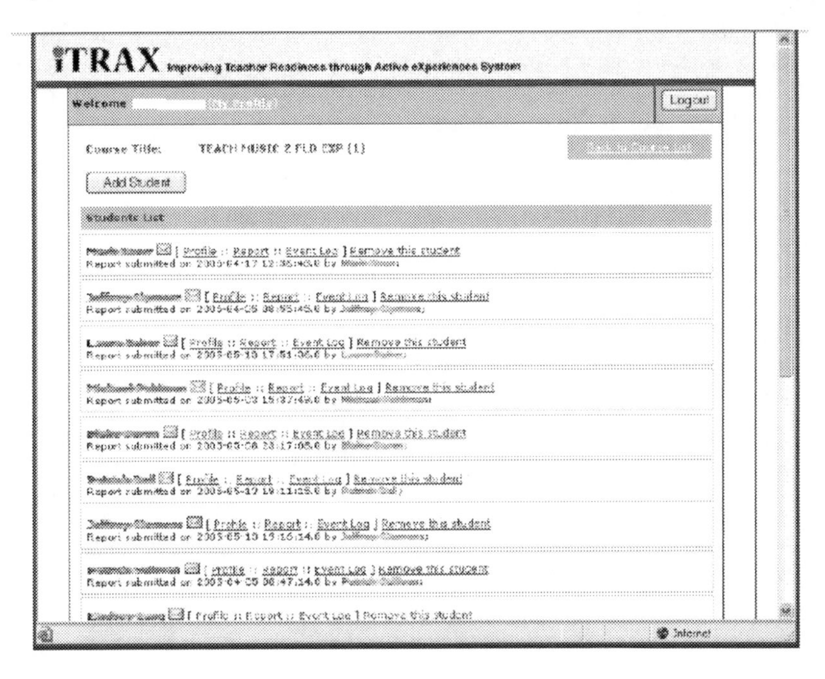

Fig. 3 A screenshot of the instructor module of the system

In essence, the system is a web-based educational knowledge management (KM) system for field experience management. The raw data collected can be used to assess students' progress throughout their field experiences and also as first-hand examples and evidence to support decision making. Instructors and program administrators can search the database and use data collected from this system to find insights, patterns, major issues, best practices and thus inform their program development efforts including teaching and policy making. For example, clustering analysis and classification can be carried out to categorize issues encountered during the field experience process.

The process of building the system was based on the classical ADDIE model which includes five major phases: analysis, design, development, implementation, and evaluation (Table1). We had multiple meetings with the teacher education program personnel in order to understand their specific needs and requirements.

We refined our system design several times based on their feedback and suggestions. An Oracle database was used as the backend database. Both Java server pages (JSP) and PHP were used as scripting languages to create different parts of the system.

Table 1 Development using ADDIE Model

ADDIE Model	Activities
Analysis	Review of current forms and processes; meeting with stakeholders
Design	Interface design; Database design
Development	System Programming
Implementation	Deployment of test system and Integration
Evaluation	Usability testing and Formative evaluation

We conducted an exploratory pilot study to evaluate the prototype by employing usability testing and formative evaluation methods. Thirteen undergraduate students, one university instructor and three host teachers (field experience supervisors in the workplace) participated in the pilot study. The students in the pilot study had all had previous field experiences and had used the paper system in those previous experiences. The pilot study included two surveys (one at mid-way through the semester, and one at the end of the semester), ongoing discussions with the university instructor, and a final discussion with the host teachers over email. Our evaluation results show that students preferred the iTRAX system to the paper-based system in terms of field experience reporting. In ongoing discussions with the university instructor, she reported that in the past, it was very difficult to manage the paperwork associated with a full class of students in various field experiences; but that with the iTRAX system, she was able to quickly see how each pre-service teacher performed, provide feedback and assessment, and complete the paperwork for the class. Host teachers (mentors in workplace) also indicated their preference for the system to track and administer the field experience. Our results also show that the prototype provides several benefits to the teacher education program including access to up-to-date information and knowledge about field experiences, ability to search and track the types of experiences over time, ability to aggregate data about the students' field experiences, and a reduction of paper printing, duplication and delivery.

Suggestions for improving the system include implementing features to allow students and host teachers to customize the interface by adding additional categories; adding other stakeholder roles (heads of departments in schools, colleagues, etc.) to the system.

4 Conclusion and Future Work

Higher education programs need to prepare their graduates for the practical challenges they can expect to face upon entering the workforce. Learning in the workplace has been institutionalized and is being seen as an integral part of the university curricula [14]. From the knowledge management perspective, it is important to develop a systematic way to streamline the management of field experience process using knowledge-centered systems. In this paper we present a process-oriented field experience knowledge management framework structured around three sequential non-lineal phases: (1) before: preparing, searching and placement; (2) during: Supervising, tracking progress and providing feedback; and (3) after: Reflecting, assessment and reporting. We have also described our experience in designing and developing a web-based knowledge-centered support system for tracking and administering field experiences. The evaluation results show that our framework is valuable and can be used to guide and facilitate knowledge-center support system development in the area of field experience management. Future work involves expanding the system to support some of the activities in stage 1 of the framework, and integrating data mining techniques with the system to classify and analyze the data collected by the system for optimal decision making.

References

1. Awad, E., Ghaziri: Knowledge management 2nd edn., International Technology Group, Ltd. (2010)
2. Brownlee, J., Carrington, S.: Opportunities for authentic experience and reflection: a teaching programme designed to change attitudes towards disability for pre-service teachers. Support for Learning 15(3), 99–105 (2000)
3. Carpenter, D.A.: Meaningful information systems internships. Journal of Information Systems Education 14(2), 201–210 (2003)
4. Gaines, B.R.: Knowledge-Based Systems, vol. 3(4), pp. 192–203 (December 1990)
5. Gilbert, P., Morse, R., Lee, M.: Enhancing IT Support with Knowledge Management. White paper (2007), `http://www.ca.com/files/whitepapers/knowledge_mgmt_wp.pdf` (retrieved on October 10, 2011)
6. He, W., Means, T., Lin, G.: Tracking and Managing Field Experiences in Teacher Development Programs. International Journal of Technology in Teaching and Learning 2(2), 137–151 (2006)
7. Kong, S., Shroff, R., Hung, H.K.: A web enabled video system for self reflection by student teachers using a guiding framework. Australasian Journal of Educational Technology 25(4), 544–558 (2009)

8. Mooradian, N.: Tacit knowledge: philosophic roots and role in KM. Journal of Knowledge Management 9(6), 104–113 (2005)
9. Puckett, J.B., Anderson, R.S.: Assessing field experience. New Directions for Teaching and Learning 91, 53–60 (2002)
10. Rai, R.K.: Knowledge management and organizational culture: a theoretical integrative framework. Journal of Knowledge Management 15(5), 779–801 (2011)
11. Rickman, R., McDonald, M., McDonald, G., Heeler, P.: Enhancing the computer networking curriculum. In: Proceedings of the 6th Annual Conference on Innovation and Technology in Computer Science Education, Canterbury, United Kingdom, pp. 157–160 (June 2001)
12. Schön, D.A.: The reflective practitioner: How professionals think in action. Basic Books, New York (1993)
13. Tan, J., Phillips, J.: Incorporating service learning into computer science courses. Journal of Computing Sciences in Colleges 20(4), 57–62 (2005)
14. Venables, A., Tan, G.: Realizing Learning in the Workplace in an Undergraduate IT Program. Journal of Information Technology Education: Innovations in Practice 8 (2009)

Towards Cross-Language Sentiment Analysis through Universal Star Ratings

Alexander Hogenboom, Malissa Bal, Flavius Frasincar, and Daniella Bal

Econometric Institute, Erasmus University Rotterdam, P.O. Box 1738,
NL-3000 DR, Rotterdam, The Netherlands
{hogenboom,frasincar}@ese.eur.nl,
{malissa.bal,daniella.bal}@xs4all.nl

Abstract. The abundance of sentiment-carrying user-generated content renders automated cross-language information monitoring tools crucial for today's businesses. In order to facilitate cross-language sentiment analysis, we propose to compare the sentiment conveyed by unstructured text across languages through universal star ratings for intended sentiment. We demonstrate that the way natural language reveals people's intended sentiment differs across languages. The results of our experiments with respect to modeling this relation for both Dutch and English by means of a monotone increasing step function mainly suggest that language-specific sentiment scores can separate universal classes of intended sentiment from one another to a limited extent.

Keywords: Sentiment analysis, cross-language analysis, star ratings.

1 Introduction

Today's Web enables people to produce an ever-growing amount of virtual utterances of opinions in any language. Anyone can write reviews and blogs, post messages on discussion forums, or publish whatever crosses one's mind on Twitter at any time. This yields a continuous flow of an overwhelming amount of multi-lingual data, containing traces of valuable information – people's sentiment with respect to products, brands, etcetera. As recent estimates indicate that one in three blog posts [15] and one in five tweets [12] discuss products or brands, the abundance of user-generated content published through such social media renders automated cross-language information monitoring tools crucial for today's businesses.

Sentiment analysis comes to answer this need. Sentiment analysis refers to a broad area of natural language processing, computational linguistics, and text mining. Typically, the goal is to determine the polarity of natural language text. An intuitive approach involves scanning a text for cues signaling its polarity. Existing approaches typically consist of language-specific parts such as sentiment lexicons, i.e., lists of words and their associated sentiment, possibly differentiated by Part-of-Speech

L. Uden et al. (Eds.): 7th International Conference on KMO, AISC 172, pp. 69–79.
springerlink.com

(POS) and/or meaning [2], or components for, e.g., identifying the POS or lemma of words. Yet, each language-specific sentiment analysis approach typically produces sentiment scores for texts in its reference language. Intuitively, these scores should be comparable across languages, irrespective of the techniques used to get these scores – provided that these techniques adhere to the same constraints in that they, e.g., produce a score on a continuous scale between -1 (negative) and 1 (positive). However, sentiment scores are not directly comparable across languages, as they tend to be affected by many different language-specific phenomena [3].

Therefore, we propose to perform cross-language sentiment analysis not by using the sentiment scores associated with natural language content per se, but by involving another way of measuring sentiment – sentiment classification by means of star ratings. In such ratings, which are commonly used in, e.g., reviews, a more positive sentiment towards the topic of a text is typically reflected by a higher number of stars associated with the text. Sentiment scores are affected by the way people express themselves in natural language, whereas star ratings are (universal) classifications of the sentiment that people actually intend to convey.

In this paper, we aim to gain insight in the relation between language-specific sentiment scores and universal star ratings in order to be able to compare sentiment scores across languages. As such, we benefit from the robust and fine-grained type of analysis that traditional, lexicon-based sentiment analysis techniques offer [24], while using universal star ratings in order to scale the obtained language-specific sentiment scores so that sentiment normalization across languages can be realized.

The remainder of this paper is structured as follows. First, we discuss related work on cross-language sentiment analysis in Sect. 2. We then propose a method for making language-specific sentiment scores comparable across languages through universal classifications of intended sentiment in Sect. 3. A discussion of insights following from an evaluation of our method is presented in Sect. 4. Last, we conclude and propose directions for future work in Sect. 5.

2 Cross-Language Sentiment Analysis

In a recent literature survey on sentiment analysis [21], the current surge of research interest in systems that deal with opinions and sentiment is attributed to the fact that, in spite of today's users' hunger for and reliance upon on-line advice and recommendations, explicit information on user opinions is often hard to find, confusing, or overwhelming. Many sentiment analysis approaches exist, yet the topic of cross-language sentiment analysis has been relatively unexplored.

Among popular bag-of-word approaches to sentiment analysis in an arbitrary language (typically English), a binary encoding of text, indicating the presence or absence of specific words, has initially proven to be an effective representation [20]. Later research has focused on different vector representations of text, including vector representations with additional features representing semantic distinctions between words [26] or vector representations with sophisticated weighting schemes for word features [19]. Such vector representations are typically used by machine learning algorithms in order to score a piece of natural language text for its associated sentiment or to classify it as either positive or negative.

The alternative lexicon-based approaches typically exhibit lower classification accuracy, but tend to be more robust across domains [24]. Also, lexicon-based approaches can be generalized relatively easily to other languages by using dictionaries [17]. Recently proposed lexicon-based sentiment analysis techniques range from rather simple [9, 10] to more sophisticated approaches that take into account structural or semantic aspects of content, for instance by means of a deeper linguistic analysis focusing on differentiating between rhetorical roles of text segments [8, 23, 24].

These existing approaches may work very well for the language they have been designed for, yet applying the state-of-the-art in sentiment analysis on entirely new languages has been shown to have its challenges, as each language may require another approach [18]. Existing research on sentiment analysis in different languages has been focused mainly on how to create new sentiment analysis methods with minimal effort, without losing too much accuracy with respect to classifying a text as either positive or negative. The focus of existing research varies from creating sentiment lexicons [11, 25] to constructing new sentiment analysis frameworks [1, 5, 6, 7, 18] for languages other than the reference language.

Moens and Boiy [18] have analyzed the creation of different sentiment analysis frameworks for different languages, while requiring minimal human effort for developing these frameworks. For any of their considered languages, Moens and Boiy [18] recommend a three-layer sentiment analysis framework in order to realize a fast way of computing sentiment scores. The first layer is very fast in its computations, yet does yield very accurate sentiment scores. When the result of a computation is not of a desired level of accuracy, the text is processed by the second, more precise, but also slower, computation layer. This process is repeated on the third layer. If still no accurate score is computed, the score of layer two is kept. The results of Moens and Boiy [18] indicate that the specifics of the configurations of such frameworks differ per language.

Rather than creating language-specific sentiment analysis frameworks, Bautin, Vijayarenu, and Skiena [4] have proposed to analyze cross-lingual sentiment by means of machine translation. They use machine translation in order to convert all considered texts into English and subsequently perform sentiment analysis on the translated results. By doing so, the authors assume that the results of the analysis on both the original text and the translated text are comparable and that the errors made by the machine translation do not significantly influence the results of the sentiment analysis.

However, the quality of machine translation may very well have an influence on the quality of the output of the sentiment analysis on the translated text, as low-quality translations do not typically form accurate representations of the original content and hence are not likely to convey the sentiment of the original text. Moreover, when focusing on developing sentiment lexicons for other languages by means of, for instance, machine translation, the need for distinct sentiment analysis approaches for different languages [18] is largely ignored.

Therefore, Bal et al. [3] have recently proposed a framework in which the sentiment in documents written in multiple languages can be assessed by means of language-specific sentiment analysis components. By means of this framework, the sentiment in documents has been compared with the sentiment conveyed by their translated counterparts. These experiments have shown that sentiment scores are not directly comparable across languages, as these scores tend to be affected by many different language-specific phenomena. In addition to this, Wierzbicka [27, 28] has argued that there is a cultural dimension to cross-language sentiment differences, as every language imposes its own classification upon human emotional experiences, thus rendering English sentiment-carrying words artifacts of the English language rather than culture-free analytical tools.

In this light, we are in need of a more universal way of capturing sentiment in order to be able to compare sentiment expressed in different languages. In this paper, we assume that star ratings can be used for this purpose. A higher number of stars associated with a piece of sentiment-carrying natural language text is typically associated with a more positive sentiment of the author towards the topic of this text. As such, star ratings are universal classifications of the sentiment that people actually intend to convey, whereas traditional sentiment scores tend to reflect the sentiment conveyed by the way people express themselves in natural language. Intuitively, both measures may be related to some extent, yet to the best of our knowledge, the relation between language-specific sentiment scores and universal sentiment classifications has not been previously investigated.

3 From Sentiment Scores to Star Ratings

As traditional sentiment analysis techniques are guided by the natural language used in texts, they allow for a fine-grained linguistic analysis of conveyed sentiment. In addition, they are rather robust as they take into account the actual content of a piece of natural language text, especially when involving structural and semantic aspects of content in the analysis [8, 23]. Yet, this language-dependency thwarts the cross-language comparability of sentiment thus identified.

Typically, traditional approaches are focused on assigning a sentiment score to a piece of natural language text, ranging from, e.g., -1 (negative) to 1 (positive). In order to support amplification of sentiment, such as "very good" rather than "good", sentiment scores may also range from, e.g., -1.5 (very negative) to 1.5 (very positive). Ideally, a sentiment score of for instance 0.7 would have the same meaning in both English and, e.g., Dutch, yet research has shown that this is not typically the case [3]. Therefore, in order to enable cross-language sentiment analysis while enjoying the benefits of traditional language-specific sentiment analysis approaches, a mapping from language-specific scores of conveyed sentiment to universal classifications of intended sentiment is of paramount importance.

In our current endeavors, we assume a five-star rating scale to be a universal classification method for an author's intended sentiment, i.e., consensus exists with respect to the meaning of each out of five classes. These classes are defined on an ordinal scale, i.e., a piece of text that is assigned five stars is considered to be more positive than a piece of text that belongs to the class of documents with four stars. Additionally, we assume that higher language-specific sentiment scores are associated with star ratings, which we model as a monotonically increasing step function. As such, we assume texts with, e.g., four stars to have higher sentiment scores than texts belonging to the three-star class.

Given these assumptions, we can construct language-specific sentiment maps for translating language-specific sentiment scores into universal star ratings. In each mapping, we consider five star segments, where we define a star segment as a set of sentiment-carrying natural language texts that have the same number of stars assigned to them. These five star segments are separated by a total of four boundaries, the position of which is based on the sentiment scores associated with the texts in each segment.

An intuitive sentiment map is depicted in Fig. 1. One could expect the one-star and five-star classes to be representing the extreme negative and positive cases, respectively, i.e., covering respective sentiment scores below -1 and above 1. The class of documents associated with three stars would intuitively be centered around a sentiment score of 0, indicating a more or less neutral sentiment. The classes of two-star and four-star texts would then cover the remaining ranges of negative and positive scores, respectively, in order to represent the rather negative and positive natural language texts, respectively. Many alternative mappings may exist for, e.g., different domains or languages. Mappings may for instance be skewed towards positive or negative sentiment scores or the boundaries may be unequally spread across the full range of sentiment scores.

The challenge is to find an optimal set of boundaries for each considered language in order to enable cross-language sentiment analysis by mapping language-specific sentiment scores, reflecting the sentiment conveyed by the way people express themselves in natural language, to universal star ratings, reflecting the intended sentiment. The goal of such an optimization process is to minimize the total costs c_b associated with a given set of boundaries b. We define these costs as the sum of the number of misclassifications $\varepsilon_i(b)$ in each individual sentiment class $i \in \{1,\ldots,5\}$, given the set of boundaries b, i.e.,

$$c_b = \sum_{i=1}^{5} \varepsilon_i(b). \tag{1}$$

This optimization process, yielding a set of boundaries associated with the least possible number of misclassifications, is subject to the constraint that the boundaries must be non-overlapping and ordered, while being larger than the sentiment score lower bound s_l and smaller than the sentiment score upper bound s_u, i.e.,

$$s_l < b_1 < b_2 < b_3 < b_4 < s_u. \tag{2}$$

Finding an optimal set of boundaries is not a trivial task, as many combinations exist and the boundaries are moreover interdependent. Once an arbitrary boundary is set,

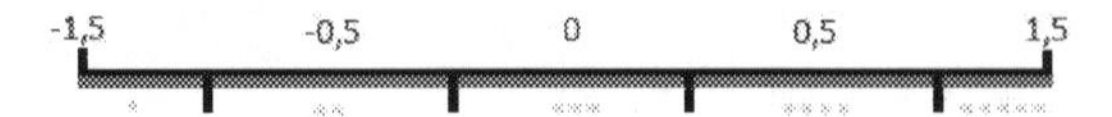

Fig. 1 Intuitive mapping from sentiment conveyed by natural language to universal star ratings

it affects the possible locations of the other boundaries. Furthermore, classes may not be perfectly separable in the sole dimension of sentiment scores.

For example, let us consider the separation problem presented in Fig. 2, where documents in *Segment A* need to be separated from those in *Segment B* by means of boundary *B*. The two segments however exhibit some overlap, which prevents the segments from being perfectly separable. Yet, some solutions are better than others in this scenario. For instance, in the intersection of *Segment A* and *Segment B*, boundary B_1 would result in all five documents from *Segment A* being erroneously classified as *Segment B* documents, one *Segment B* document being classified as a *Segment A* document, and only three documents being classified correctly in *Segment B*. Boundary B_2 on the other hand would yield only three misclassifications in *Segment A*, one misclassification in *Segment B*, and two and three correct classifications in *Segment A* and *Segment B*, respectively.

Many algorithms can be used in order to cope with such issues. One may want to consider using a greedy algorithm in order to construct a set of boundaries. Alternatively, heuristic or randomized optimization techniques like genetic algorithms may be applied in order to explore the multitude of possible solutions. Finally, if the size of the data set allows, a brute force approach can be applied in order to assess all possible boundary sets at a certain level of granularity.

By using our proposed method, the sentiment conveyed by people's utterances of opinions in natural language can first be accurately analyzed by means of state-of-the-art tools tailored to the language of these texts. The sentiment scores thus obtained can subsequently be transformed into universal star ratings by means of language-specific sentiment maps. We can thus make language-specific sentiment scores comparable across languages by mapping these scores to universal classifications of intended sentiment.

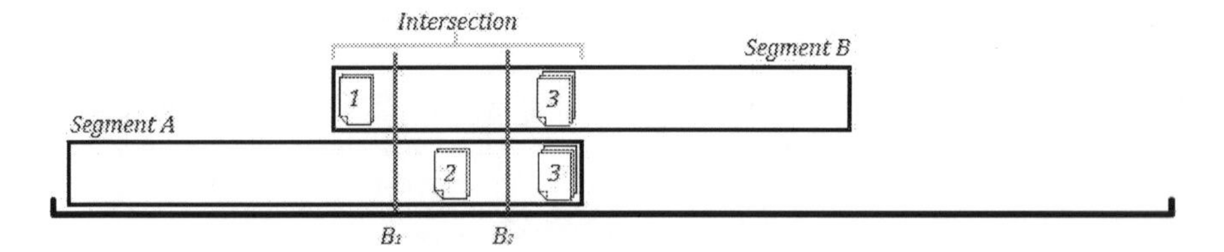

Fig. 2 Separating two segments of documents from one another by means of tentative boundaries B_1 and B_2. In the intersection of both segments, both *Segment A* and *Segment B* contain three documents with an equally high sentiment score. *Segment A* contains two additional documents with a lower score, whereas *Segment B* contains another document with an even lower score.

4 Evaluation

Our proposed approach of making language-specific sentiment scores comparable across languages by means of mapping these scores to universal star ratings, can be used to perform several analyses, as depicted in Fig. 3. First, the sentiment score of documents can be compared across languages (1). This has already been done in previous research endeavors, which have revealed that this type of cross-language sentiment analysis is not the most promising one, as sentiment scores as such do not appear to be directly comparable across languages [3]. A more suitable analysis is an exploration of how language-specific sentiment scores can be converted into universal star ratings (2) and how such mappings differ across languages (3). Therefore, we focus on this type of analysis in this paper. The (interpretation of) star ratings could also be compared across languages (4), yet this falls outside of the scope of our current endeavors, as we assume star ratings to be universal and comparable across languages.

In our analysis, we consider two sets of similar documents. One set consists of 1,759 short movie reviews in Dutch, crawled from various web sites [13, 14]. The other collection consists of 46,315 short movie reviews in English [16, 22]. Each review in our data sets contains a maximum amount of 100 words. Each review has been rated by its respective writer on a scale of one to five or ten stars, depending on the web site, where more stars imply a more positive verdict. We have converted all document ratings to a five-star scale by dividing all scores on a ten-star scale by two and rounding the resulting scores to the nearest integer. This process results in a data set in which, for both considered languages, the documents are approximately normally distributed over five star classes, while being slightly skewed towards the higher classes.

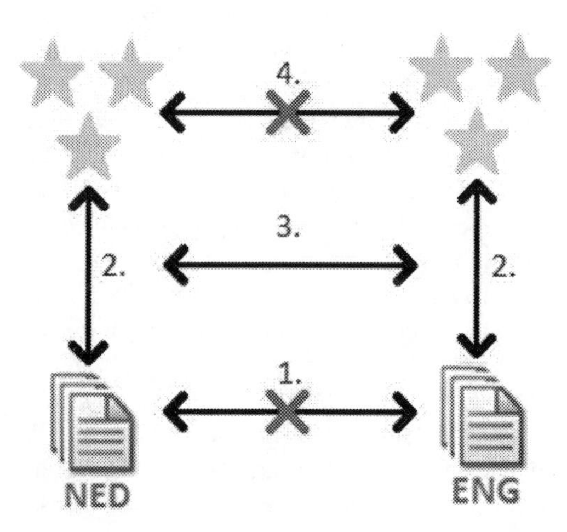

Fig. 3 Considered comparisons between Dutch (*NED*) and English (*ENG*) pieces of sentiment-carrying natural language text

Table 1 Language-specific sentiment score intervals associated with each considered number of stars for both Dutch and English

Stars	Dutch sentiment scores	English sentiment scores
1	$[-1.5,-0.5]$	$[-1.5,-0.4]$
2	$(-0.5,-0.1]$	$(-0.4,0.0]$
3	$(-0.1,0.4]$	$(0.0,0.2]$
4	$(0.4,0.9]$	$(0.2,0.4]$
5	$(0.9,1.5]$	$(0.4,1.5]$

The documents in our data set are first analyzed for the sentiment conveyed by their text by means of an existing framework for lexicon-based sentiment analysis in multiple languages, more specifically English and Dutch [3]. This framework is essentially a pipeline in which each component fulfills a specific task in analyzing the sentiment of an arbitrary document. For each supported language, this framework first prepares documents by cleaning the text – i.e., converting the text to lowercase, removing diacritics, etcetera – and performing initial linguistic analysis by identifying each word's POS as well as by distinguishing opinionated words and their modifiers from neutral words. After this preparation process, each document is scored by sum-aggregating the sentiment scores of the opinionated words, while taking into account their modifiers, if any.

Scoring each document in our data set for the sentiment conveyed by its text yields a set of 1,759 two-dimensional data points for Dutch and 46,315 similar two-dimensional data points for English, each of which represents a paired observation of a language-specific sentiment score and the associated universal star rating of intended sentiment. These data points can be used to construct a mapping between sentiment scores and star ratings for each considered language. As the size of our data set allows for it, we use a brute force approach in our current endeavors, where we assess the performance in terms of number of misclassifications for all possible combinations of boundaries, with a step size of 0.1.

The resulting sentiment score ranges per star class are reported in Table 1. These ranges are averages over all folds of our 10-fold cross-validation. The results in Table 1 indicate that sentiment maps may have different characteristics for different languages. For instance, the Dutch sentiment map appears to be more equally spread than the English sentiment map. Additionally, more than in Dutch documents, moderate sentiment scores in English documents are typically already associated with extreme star ratings. This effect hold is more apparent in positive ratings than in negative ratings.

When using the boundaries thus obtained for classifying pieces of opinionated natural language text into one out of five star categories solely based on the sentiment score conveyed by the text itself, the performance of the constructed sentiment maps turns out to differ per language as well. The 10-fold cross-validated overall classification accuracy on Dutch documents equals approximately 20%, whereas the overall classification accuracy on English documents equals about 40%. This observation suggests that the sentiment conveyed by natural language text may in some languages be a better proxy for intended sentiment than in other languages. As such, more (latent) aspects of opinionated pieces of natural language text, such as structural aspects or emoticons, may need to be taken into account when converting language-specific sentiment scores into universal star ratings in order to better facilitate cross-language sentiment analysis.

5 Conclusions and Future Work

In this paper, we have proposed to facilitate cross-language sentiment analysis by comparing the sentiment conveyed by natural language text across languages by using these language-specific sentiment scores to classify pieces of sentiment-carrying natural language text into universal star ratings. We have shown that the way natural language reveals people's sentiment tends to differ across languages, as the relation between sentiment conveyed by natural language and intended sentiment is different for the two languages considered in our current work.

The results of our initial experiments with respect to modeling this relation for each language by means of a monotone increasing step function mainly suggest that the sole dimension of language-specific sentiment scores can separate universal classes of intended sentiment from one another to a limited extent. As such, we have made first steps towards cross-language sentiment analysis through universal star ratings, yet our results warrant future research.

In future research, we consider relaxing some of our assumptions in order for the mapping between language-specific sentiment scores and universal star ratings to be more accurate. For instance, we could consider dropping the monotonicity constraint and allow for a non-linear relation between sentiment scores and star ratings. Last, more aspects of content other than the associated sentiment score, e.g., emoticons, could be used as proxy for star ratings in order to facilitate sentiment analysis across languages.

Acknowledgements. We would like to thank Teezir (`http://www.teezir.com`) for their technical support and fruitful discussions. The authors of this paper are partially supported by the Dutch national program COMMIT.

References

1. Abbasi, A., Chan, H., Salem, A.: Sentiment Analysis in Multiple Languages: Feature Selection for Opinion Classification in Web Forums. ACM Transactions on Information Systems 26(3) (2008)
2. Baccianella, S., Esuli, A., Sebastiani, F.: SentiWordNet 3.0: An Enhanced Lexical Resource for Sentiment Analysis and Opinion Mining. In: 7th Conference on International Language Resources and Evaluation (LREC 2010), pp. 2200–2204. European Language Resources Association (2010)

3. Bal, D., Bal, M., van Bunningen, A., Hogenboom, A., Hogenboom, F., Frasincar, F.: Sentiment Analysis with a Multilingual Pipeline. In: Bouguettaya, A., Hauswirth, M., Liu, L. (eds.) WISE 2011. LNCS, vol. 6997, pp. 129–142. Springer, Heidelberg (2011)
4. Bautin, M., Vijayarenu, L., Skiena, S.: International Sentiment Analysis for News and Blogs. In: 2nd International Conference on Weblogs and Social Media (ICWSM 2008), pp. 19–26. AAAI Press (2008)
5. Dau, W., Xue, G., Yang, Q., Yu, Y.: Co-Clustering Based Classification. In: 13th ACM SIGKDD International Conference on Knowledge Discovery and Data Mining (KDD 2007), pp. 210–219. Association for Computing Machinery (2007)
6. Dau, W., Xue, G., Yang, Q., Yu, Y.: Transferring Naive Bayes Classifiers for Text Classification. In: 22nd Association for the Advancement of Articifial Intelligence Conference on Artificial Intelligence (AAAI 2007), pp. 540–545. AAAI Press (2007)
7. Gliozzo, A., Strapparava, C.: Cross Language Text Categorization by Acquiring Multilingual Domain Models from Comparable Corpora. In: ACL Workshop on Building and Using Parallel Texts (ParaText 2005), pp. 9–16. Association for Computational Linguistics (2005)
8. Heerschop, B., Goossen, F., Hogenboom, A., Frasincar, F., Kaymak, U., de Jong, F.: Polarity Analysis of Texts using Discourse Structure. In: 20th ACM Conference on Information and Knowledge Management (CIKM 2011), pp. 1061–1070. Association for Computing Machinery (2011)
9. Heerschop, B., Hogenboom, A., Frasincar, F.: Sentiment Lexicon Creation from Lexical Resources. In: Abramowicz, W. (ed.) BIS 2011. LNBIP, vol. 87, pp. 185–196. Springer, Heidelberg (2011)
10. Heerschop, B., van Iterson, P., Hogenboom, A., Frasincar, F., Kaymak, U.: Analyzing Sentiment in a Large Set of Web Data while Accounting for Negation. In: 7th Atlantic Web Intelligence Conference (AWIC 2011), pp. 195–205. Springer (2011)
11. Hofman, K., Jijkoun, V.: Generating a Non-English Subjectivity Lexicon: Relations that Matter. In: 12th Conference of the European Chapter of the Association for Computational Linguistics (EACL 2009), pp. 398–405. Association for Computing Machinery (2009)
12. Jansen, B., Zhang, M., Sobel, K., Chowdury, A.: Twitter Power: Tweets as Electronic Word of Mouth. Journal of the American Society for Information Science and Technology 60(11), 2169–2188 (2009)
13. Korte Reviews: Korte Reviews (2011), `http://kortereviews.tumblr.com/`
14. Lemaire: Lemaire Film Reviews (2011), `http://www.lemairefilm.com/`
15. Melville, P., Sindhwani, V., Lawrence, R.: Social Media Analytics: Channeling the Power of the Blogosphere for Marketing Insight. In: 1st Workshop on Information in Networks, WIN 2009 (2009)
16. Metacritic: Metacritic Reviews (2011), `http://www.metacritic.com/browse/movies/title/dvd/`
17. Mihalcea, R., Banea, C., Wiebe, J.: Learning Multilingual Subjective Language via Cross-Lingual Projections. In: 45th Annual Meeting of the Association for Computational Linguistics (ACL 2007), pp. 976–983. Association for Computational Linguistics (2007)
18. Moens, M., Boiy, E.: A Machine Learning Approach to Sentiment Analysis in Multilingual Web Texts. Information Retrieval 12(5), 526–558 (2007)
19. Paltoglou, G., Thelwall, M.: A study of Information Retrieval weighting schemes for sentiment analysis. In: 48th Annual Meeting of the Association for Computational Linguistics (ACL 2010), pp. 1386–1395. Association for Computational Linguistics (2010)

20. Pang, B., Lee, L.: A Sentimental Education: Sentiment Analysis using Subjectivity Summarization based on Minimum Cuts. In: 42nd Annual Meeting of the Association for Computational Linguistics (ACL 2004), pp. 271–280. Association for Computational Linguistics (2004)
21. Pang, B., Lee, L.: Opinion Mining and Sentiment Analysis. Foundations and Trends in Information Retrieval 2(1), 1–135 (2008)
22. Short Reviews: Short Reviews (2011), `http://shortreviews.net/browse/`
23. Taboada, M., Brooke, J., Tofiloski, M., Voll, K., Stede, M.: Lexicon-Based Methods for Sentiment Analysis. Computational Linguistics 37(2), 267–307 (2011)
24. Taboada, M., Voll, K., Brooke, J.: Extracting Sentiment as a Function of Discourse Structure and Topicality. Tech. Rep. 20. Simon Fraser University (2008), `http://www.cs.sfu.ca/research/publications/techreports/#2008`
25. Wan, X.: Co-Training for Cross-Lingual Sentiment Classification. In: Joint Conference of the 47th Annual Meeting of ACL and the 4th International Join Conference on Natural Language Processing of the AFNLP (ACL 2009), pp. 235–243. Association for Computational Linguistics (2009)
26. Whitelaw, C., Garg, N., Argamon, S.: Using Appraisal Groups for Sentiment Analysis. In: 14th ACM International Conference on Information and Knowledge Management (CIKM 2005), pp. 625–631. Association for Computing Machinery (2005)
27. Wierzbicka, A.: Dictionaries vs. Encyclopedias: How to Draw the Line. In: Alternative Linguistics: Descriptive and Theoretical Modes, pp. 289–316. John Benjamins Publishing Company (1995)
28. Wierzbicka, A.: Emotion and Facial Expression: A Semantic Perspective. Culture Psychology 1(2), 227–258 (1995)

Organisational Knowledge Integration towards a Conceptual Framework

Kavoos Mohannak

School of Management, Queensland University of Technology, Brisbane, Australia
k.mohannak@qut.edu.au

Abstract. This study analyses organisational knowledge integration processes from a multi-level and systemic perspective, with particular reference to the case of Fujitsu. A conceptual framework for knowledge integration is suggested focusing on team-building capability, capturing and utilising individual tacit knowledge, and communication networks for integrating dispersed specialist knowledge required in the development of new products and services. The research highlights that knowledge integration occurring in the innovation process is a result of knowledge exposure, its distribution and embodiment and finally its transfer, which leads to innovation capability and competitive advantage in firm.

1 Introduction

In this study, knowledge integration (KI) is all the activities by which an organization identifies and utilizes external and internal knowledge. The research findings suggest that innovations occur when existing or new knowledge is integrated within organization which will result in a new product or service (Cavusgil et al. 2003). Also findings of the importance of factors across industries suggest that integration of knowledge within the companies is the most important driver for innovation (OECD 2004). Since knowledge is continuously changing and depreciating, organizations cannot possess all the required knowledge themselves. This implies that the effective transfer of external knowledge or internal creation of new knowledge is significant success factor for innovation and new product development (NPD). This process of external transfer of knowledge or internal capability development and learning which will result to knowledge integration and new product development within organization is the focal subject of this study. The conceptual research question is: How is knowledge integrated, sourced and recombined from internal and external sources for innovation and new product development?

To analyse the above question from a conceptual perspective, this paper firstly examines theoretical background of KI and it then provides some evidence of KI in Japanese firms with particular reference to the case of Fujitsu. Finally a conceptual framework will be suggested which may provide a better understanding of knowledge integration within an organisation.

L. Uden et al. (Eds.): 7th International Conference on KMO, AISC 172, pp. 81–92.
springerlink.com

2 The Theoretical Background

Knowledge integration in firms has received considerable attention in recent research (see, for example, Mohannak 2011; Brusoni et al. 2009; Zirpoli and Camuffo 2009; Becker 2003). In particular the research has highlighted the pivotal role of knowledge integration in creating and sustaining firms' innovative and competitive advantage (Kraaijenbrink et al. 2007; Grant 1996a; Kogut and Zander 1992). From the perspective of the knowledge-based theory of the firm, the main problem lies in assuring the most effective integration of individuals' specialized knowledge at the lowest attainable cost (Grant 1996a). A central claim of the knowledge based theory of the firm is that organizational capabilities depend not only on specialized knowledge held by individuals but also on an organization's ability to integrate that specialized knowledge (Galunic and Rodan 1998; Garud and Nayyar 1994; Grant 1996a; 1996b; Huang and Newell 2003; Kogut and Zander 1992; 1996; Okhuyzen and Eisenhardt 2002; Purvis et al. 2001). The knowledge-based theory thus extends existing theory on organizational differentiation and integration to include the differentiation and integration of knowledge. Stemming from the need for differentiation and integration, the theory of knowledge integration emphasizes the economic value of specialization and the effectiveness of integration. In other words, competitiveness depends on the diversity and strategic value of specialized knowledge, as well as an organization's capacity to integrate the knowledge in an effective manner. Grant (1996a) describes the integration of individuals' specialized knowledge to create value as a key capability.

Following knowledge-based theory of firm, Alavi and Tiwana (2002) has defined KI as synthesis of individuals' specialized knowledge into situation-specific systemic knowledge. This definition is based on the fact that the specialization of organization members turns organizations into distributed knowledge systems in which the range of knowledge that is required for production or innovation is dispersed over organization members (Tsoukas 1996). Therefore, organization members have to integrate dispersed bits of specialized knowledge held by individuals, i.e., to apply this dispersed knowledge in a coordinated way (Becker 2001; Grant 1996a). In this sense, knowledge integration is essentially a matter of organization, and the ability to create and exploit useful combinations is the *raison d'être* of firms (Kogut and Zander 1992; Grant 1996a).

Another definition is given by Huang and Newell (2003:167), which is defined as: "an ongoing collective process of constructing, articulating and redefining shared beliefs through the social interaction of organizational members". In fact, the emphasis on the need for communication and shared knowledge which is to be found in much product development literature is reflected in this definition. This is to say that new product development team members must be able to communicate in a manner that is meaningful. Moreover, they must be able to create new knowledge. In this way, the outcome of knowledge integration consists of "both the shared knowledge of individuals and the combined knowledge that emerges from their interaction" (Okhuysen and Eisenhardt 2002:371).

However, as emphasized by Huang and Newell (2003) it is crucial to recognize that cross functional knowledge integration within the context of a project team is not limited to a focus on the dynamics occurring within the team boundary. It is equally vital to understand the dynamics of knowledge integration beyond the team boundary, in particular in relation to knowledge integration within or outside the firm and with all stakeholder groups. In this view, knowledge creation, sharing and transfer constitute an important component of knowledge integration. Indeed, facilitating the combination of knowledge elements relies on the ability to create shared agreements across different expertise (Nonaka 1994). Therefore, for the purpose of this paper knowledge integration is defined as all activities by which an organization identifies and utilizes internal and external knowledge including creating, transferring, sharing and maintaining information and knowledge.

Knowledge integration, therefore, is a fundamental process by which firms gain the benefits of internal and external knowledge, create competitive advantage and develop capability. However, characteristics of new technologies and learning processes are such that organization members have to specialize in order to acquire a high level of expertise. Given the large amount of relevant knowledge that are available in many fields and the limitations of human information processing, individuals have to share and integrate their knowledge. While efforts aimed at knowledge sharing can constitute important mechanisms for knowledge integration, other organizational mechanisms such as codification of knowledge through procedures, instructions, and code books (Grant 1996a), knowledge platforms (Purvis et al. 2001), communication networks, knowledge integrators, communities of practice, teams (Grandori 2001), rules, directives, routines, group problem solving, decision making (Grant 1996a), tacit experience accumulation, articulation and codification (Zollo and Winter 2002) all facilitate knowledge integration.

As Gittell (2002) emphasized organizational routines in particular cause predictable patterns of collective behavior, and hence are appropriate integration mechanisms in stable environments (Gittell 2002). Routines also support complex, simultaneous and varied sequences of interactions among agents (Becker 2004; Grant 1996a). Grandori (2001) suggests that the tacit and unobservable nature of judgment and action generates epistemic complexity. Such complexities can be captured in tacit organizational routines which in turn allow for task partitioning and for specialization among organizational members (Prencipe and Tell 2001).

In contrast to these internalizing features of learning through experience accumulation and routines, literature also has emphasized the need for firms to acquire knowledge from external sources. External knowledge, generally, can be traded in labor or intellectual property markets. It tends to be rather technical and explicit, which makes it relatively easy to acquire, be it through internal training or simply by 'hiring' a specialist in the market. External knowledge does not lead to differentiation, although it may be essential for any given firm because a certain level of this type of knowledge is indispensable for competitive survival. In contrast, internal knowledge is idiosyncratic and typically related to a particular firm and refers to and is embedded in its particular organizational context. It acts as a sort of organizational glue that holds the organization together, giving it cohesiveness and sense of unity. It is therefore more valuable inside the organization than in the market, and is less subjected to imitation. Hence it constitutes a critical source of sustainable competitive advantage.

It is, therefore, through the internal development or external acquisition that an organization is able to get both the range and the quality of expertise, which is required for complex production and innovation processes. Project teams, for example, generate knowledge internally and often seek knowledge from external sources. Team members must combine their complementary, yet separately held knowledge into a new knowledge set. In order for a project team to be productive, they must have a deep knowledge of their own disciplines and an appreciation for the relevance and importance of their teammates' knowledge. All this external and internal knowledge must be integrated into team responses (Anand, Clark and Zellmer-Bruhn 2003). New product development and innovation requires the use of a multitude of skills and expertise, as well as the accumulated knowledge of the organization in order to maximize the performance of the new product. The integration of all this accumulated knowledge into the business processes used by these skilled and experienced employees has great potential to improve the new products themselves. It has been suggested that it is the degree of integration of dispersed and distributed knowledge that helps explain differences in the product development performance of different firms and that it is the effectiveness of a firm's knowledge integration that distinguishes it from its competitors (Carlile 2002).

As Grant (1996a) mentioned the integration of knowledge may be viewed as a hierarchy which is not one of authority and control, as in the traditional concept of an administrative hierarchy, but is a hierarchy of integration. In this view, organizational knowledge is treated at different levels of integration, which can be seen as an elaboration of knowledge distinctions that simply focus on different levels of analysis (Hedlund 1994; Kogut and Zander 1992). At the base of the hierarchy is the specialized knowledge held by individual organizational members which deal with specialized tasks, while at the very top lies cross-functional and new product development capability. The purpose of the multi-level integration process is to create innovation that is firm-specific. However, the crucial extension of this hierarchy is the explicit recognition of the need for specialized knowledge at the different levels. Firms may also need to combine various specialized knowledge from different disciplines such as electronics, biology, computer science, etc. This specialized knowledge might be held by individuals within the firm or can be acquired through external sources.

In this paper, the internal and external dimension of the knowledge integration is emphasized, which matches the concept of hierarchy of integration (Grant 1996a) especially considering the nature of differentiated specialized knowledge. As discussed it should be noted that in relation to different knowledge integration processes defined here – namely creation, sharing/transfer and use, knowledge integration may occur at different levels of interaction – individual, group, organizational, inter-organizational. The focus of this paper is to explore how firms respond to knowledge integration needs. In particular how firms exploit potential synergies among various internal and external knowledge sources and create competitive advantage. Knowledge exists in firms and networks, but how firms execute KI processes and whether knowledge integration performed by individuals, teams, or by firms – or whether KI is something that happens at the network level? The proposition here is that KI addresses technical, strategic and operational challenges at the various levels. To illustrate this point next section will look at KI in Fujitsu.

3 KI in Fujitsu: Towards a Human Centric Networked Society

In order to analyze the process of knowledge integration at Fujitsu Corporation, this study combined on-site interviews with analysis of company internal documents such as annual reports, web pages and internal publications. The interviews were conducted in 2010 in Japan. Senior managers at the Fujitsu headquarters and middle managers who were directly involved in the new product or system development project were interviewed. Findings suggest that, KI in Fujitsu is not limited at individual, team or organisational levels. More importantly, at technological level, Fujitsu is moving toward a Human Centric computing environment, where ICT could provide tailored and precise services wherever and whenever people needed those services (Yoshikawa and Sasaki, 2010). In this way, Fujitsu through its slogan "Shaping Tomorrow with You" is shifting the paradigm from system centric to human centric solutions. Fujitsu realised that through evolution of social activities that are supported by technological innovation in information and communication technologies there are numerous real-life fields in which the application of ICT can be further leveraged especially through application of cloud computing.

Furthermore, in Fujitsu new knowledge, whether from *inside* or *outside,* fuels innovative breakthroughs and sharing of knowledge is not purely a matter of multifunctional teams. The extensive R&D activity makes it possible to use invented technologies in new or unexpected industries and further build up the competitive edge. For example, technology fusion and integration in Fujitsu, builds from knowledge from different industries and technologies with a multi-technology basis instead of reliance on a single technology. In this regard, building integrated knowledge capital platforms from accumulated experience and expertise works advantageously in responding to technological change and resolving new issues which require speed and creativity.

In this regard, attempts are being made to achieve innovation in field operations and collaborate with service providers by sharing data in addition to the shared use of service components. For example, Fujitsu has been working on field solutions that allow a detailed understanding of field situations by introducing wireless and sensing technologies to the fields where IT technologies have not traditionally been in place, such as agriculture and healthcare, for collecting, analysing and sharing field data, namely real-time filed management, and proposed optimization of management resources, improvement of eco-friendliness and sensor solutions (Takahashi et al. 2010). To achieve this, Fujitsu is conducting R&D in integrating new technologies in areas such as Cloud computing, network technology and Smartphone evolution (Abe and Shibata 2009). Fujitsu's strength lies on its command over all of these relevant technologies and Fujitsu believes integration of these technologies will provide significant technological revolution with varied social and human centric applications.

Fujitsu intends to take advantage of the major social changes, increased business opportunities, and other such substantial changes that can be brought about by sensing the various kinds of information and acquiring knowledge from the environment that surrounds individuals. This information then can be provided to the Cloud environment via a network and converting the immense amount of collected knowledge into new value (Yoshikawa and Sasaki 2010). As a result it would be possible feeding it back to the individual and to the business environment that surrounds individuals. The question is what technology and infrastructure is needed to achieve this? Fujitsu believes it is important to place people at the centre while taking a strategic and scenario-based approach to R&D by focusing on changing events and then developing technologies and products.

For this purpose, Fujitsu has adopted a series of strategic and technological decisions to integrate the world of ICT and the real world. The aim is to be able to analyse massive volumes of sensor and web data and proactively deliver necessary services whenever are needed (Lida 2010). In this way, Fujitsu is striving towards a world where people, society and IT systems are in harmony with each other. For example, technologies can be developed to detect human movements through acceleration sensors embedded in mobile phones. By analysing human movements then it would be possible to provide health support or sports diagnostic services. Another example of human centric application would be, for instance, visualization of power consumption and environmental sensing in order to optimise the power usage based on comprehension of behavioural patterns. Fujitsu is currently working on several such applications with emphasis on integrating technologies (such as sensors, mobile devices, human interfaces, mining, ergonomics, etc.) that would merge the real world with the world of ICT and leverage knowledge and innovation (see Lida 2010).

Fujitsu's R&D strategy in this relation currently focusing on three themes: 1) large-boned themes, which consist of core research projects on important themes for the future technology of Fujitsu Group with medium- to long-term development; 2) business strategic themes, which focuses on strategic business projects with commitment from internal business segments for commercialisation with short- to medium-term technology development; and 3) seeds-oriented themes, with emphasis on new research areas for growth of future emerging technology seeds, which basically these projects are for long-term technology development (Murano 2010). According to Yoshikawa and Sasaki (2010), to enable a human centric networked society Fujitsu's laboratories have adopted several important policy initiatives including:

- Roadmap-based R&D activities related to business while looking forward ten years into the future.
- Open innovation activities utilising cooperation between industry and academia.
- Business incubation activities aimed at opening new markets.
- Strategic public relations activities for mass media, investors, and analysts, and
- Cultivation of personnel.

In summary, it seems that Fujitsu will continue to innovate by placing importance on technology and continuing to create new value. Through systemic technology and knowledge integration it will continue contributing to building a human-centric networked society in which a new social infrastructure brings people into harmony with computers and enable ICT to leverage knowledge and expertise. For this purpose, Fujitsu needs to integrate knowledge by forming networked team within the firm, while also absorbing the knowledge of external partners via integration with other firms. This internal and external knowledge integration for the Fujitsu case forms the base of the 'knowledge integration,' which integrates knowledge at various layers of the company.

4 Towards a Conceptual Framework for KI

Based on the above discussions, a framework for knowledge integration in R&D organizations will be proposed. The framework will assist in conceptual understanding of the issues related to integrating knowledge from internal and external sources in R&D firms. As emphasized before, integration capability plays an important role in acquiring and exploiting the knowledge from internal and external sources. The paper argues that knowledge integration can be characterized as having a multi-layered structure with an external (i.e., outside the firm) or internal (i.e., within the firm) orientation. Furthermore as emphasized in a R&D firm extent of the individual specialized knowledge, team–building capability, social networks, and internal/external organizational climate affect capability, which in turn will affect the creation of new products and services. Before discussing the conceptual framework, it is worth reviewing some of the current relevant frameworks.

Kodama (2005), for example, have proposed a framework in this relation. He bases his framework on the notion of strategic communities (SC) and maintains that these strategic communities are horizontally and vertically integrated within Japanese firms. These horizontally and vertically integrated SC networks, Kodama (2005) argues, promotes the external and internal integration of knowledge among corporations including external customers and internal-layers of management, so that the corporation can provide products and services that match market needs. He referred to this networking as external and internal integration capability. In this model heterogeneous knowledge from inside and outside the firm, which arising from dynamic changes to vertical and horizontal corporate boundaries, delivers two new insights regarding 'new knowledge creation': (i) the vertical value chain model distinctive to Japanese firms realizing new products, services and business models, and (ii) the co-evolution model realizing new win-win business models. Kodama investigates this integration at macro-level within Japanese firms; however he does not elaborate on the role of the individuals, teams or knowledge management systems within the knowledge integration process.

On the other hand Andreu and Sieber (2005) argue that different firms need different knowledge integration systems. They have identified different knowledge integration systems needed at the corporate level, which is determined by the type of knowledge the firm wishes to integrate. Using three classifications of knowledge (explicit vs. tacit, collective vs. individual, external and internal) these authors emphasize that different combinations, impacts overall feasibility ('integration trajectories') of knowledge integration among organizations or business units.

This study builds on the previous literature and takes the knowledge integration capability as the key capability in a dynamic environment and as a starting point of departure (see among others Grover and Davenport 2001; Probst, Raub, and Romhardt 2000; Lu, Wang and Mao 2007). This study suggest the process of integrating knowledge in R&D firms comprised of various activities that are involved in the identification, selection, acquisition, development, exploitation and protection of technologies. These activities are needed to maintain a stream of products and services to the market. In fact, R&D firms deal with all aspects of integrating technological issues into business decision making and new product development process. Furthermore, it is emphasized that knowledge integration is a multifunctional field, requiring inputs from both commercial and technical functions in the firm. Therefore effective knowledge integration requires establishing appropriate knowledge flows between core business processes and between commercial and technological requirements in the firm.

The conceptualization of knowledge integration capabilities assumes that there are both "macro" and "micro" organizational mechanisms designed to address the problem of knowledge integration in new product or process development. Therefore, this study considers a knowledge integration system that can be described by internal or external orientation. Internally-oriented knowledge systems rely primarily on private knowledge sources (both personal contacts and documents) inside the firm. In contrast, an externally-oriented knowledge system relies on company's collaborative agreements (such as consortia, alliances or partnerships) as well as employees external private networks with people at other companies or when R&D staff in internal knowledge systems access external knowledge and information.

Therefore, organizational mechanisms for effective knowledge integration should address four components: 1) team-building capability; 2) integration of individual specialized knowledge that are sources of technical and commercial information; 3) knowledge integration through communication networks within and outside the organization; and 4) technology/knowledge system integration. As

Fig. 1 illustrates, the technology systems overlap in producing the organizational capabilities in which technical staff solve problems and create new technology. Firms rely on the interaction between their organizational mechanisms and their employees' activities involving problem solving and experimentation, facilitated by their ability to import and integrate knowledge. As explained the components of the knowledge integration model can be described as internal or external according to their orientation to firm-based rules or external markets, respectively, in determining how work is organized, skills are learned, and the new technology is integrated within the new product or processes. For example, companies creating new products in an industry with short product generations find themselves relentlessly combining new internal knowledge with external knowledge to keep pace with the industry.

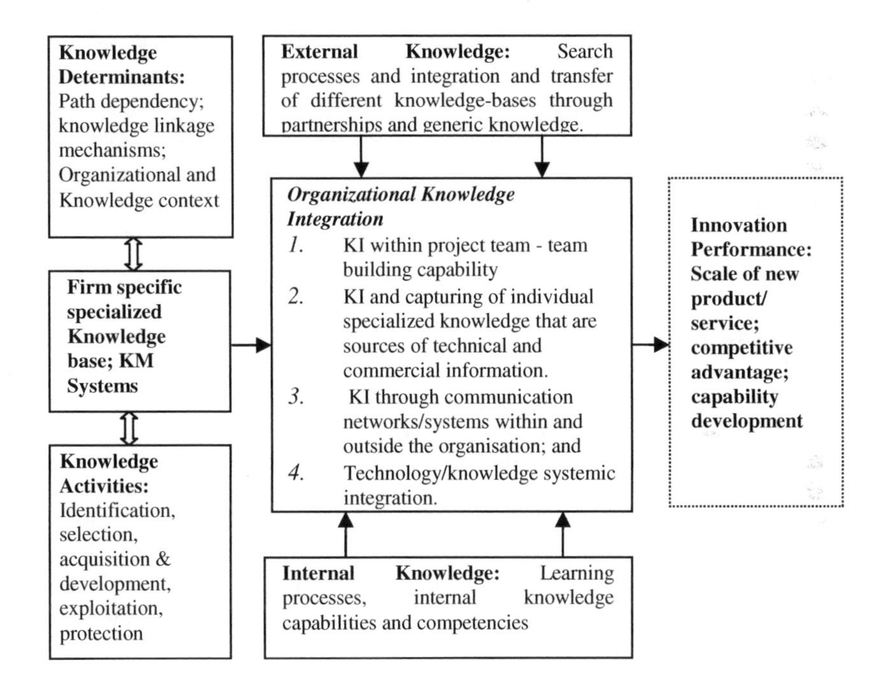

Fig. 1 The relationship between knowledge activity and knowledge integration

Hence managing the integration of knowledge do not operate in isolation, and are generally not managed as separate 'core' business processes. The various activities that constitute the knowledge integration processes tend to be distributed within business processes (for instance, technology selection decisions are made during business strategy and new product development). In fact, formal interventions and routines that focus on the improvement of the group process are also a potential way to achieve superior knowledge integration (Eisenhardt and Okhuysen 2002). In R&D firms specialized knowledge are more widespread and ideas should be used with enthusiasm, thus organization must be able to integrate them through mechanisms such as direction and organizational routines (Grant 1996a). Additionally firms that can harness outside ideas to advance their own businesses while leveraging their internal ideas outside their current operations will likely thrive in this new era of open innovation (Chesbrough 2003).

As indicated in the Fig. 1, the technology and knowledge base of the firm, represents the technological knowledge, competencies and capabilities that support the development and delivery of competitive products and services, and may include other organizational infrastructures such as KM systems. Knowledge integration activities including identification, selection, acquisition, development, exploitation and protection, operate on the technology and knowledge base, which combine to support the generation and exploitation of the firm's technology base.

This framework provides an example of how the issue of knowledge integration may be formulated into a holistic and systemic perspective by including some internal and external factors in the process of new product or process development. Emphases have been put on the context within which knowledge integration occurs within units (e.g., an individual, a group, or an organization), and the relationships between units, and properties of the knowledge itself. The overall aim is to support understanding of how technological and commercial knowledge combine to support strategy, innovation and operational processes in the firm, in the context of both the internal and external environment.

5 Concluding Remarks

As this study demonstrated, knowledge integration takes place in different levels, namely individual level, team level and systemic level. The research highlights that knowledge integration occurring in the innovation process is a result of knowledge exposure, its distribution and embodiment and finally its transfer, which leads to innovation capability and competitive advantage in firm. Knowledge management provides platforms, tools and processes to ensure integration of an organization's knowledge base. Through knowledge management structures and systems such as taxonomies, data mining, expert systems, etc., knowledge management can ensure the integration of the knowledge base. This enables staff members to have an integrated view of what knowledge is available, where it can be accessed, and also what the gaps in the knowledge base are. This is extremely important in the innovation process to ensure that knowledge as resource is utilised to its maximum to ensure that knowledge is not recreated in the innovation process.

Furthermore, this paper suggested a conceptual framework emphasising specifically team-building capability, capturing and utilising individual tacit knowledge, and communication networks for integrating dispersed specialist knowledge. Knowledge integration, as discussed, is a fundamental process by which firms gain the benefits of internal and external knowledge, create competitive advantage and develop capability. It is through internal development or external acquisition that an organization is able to get both the range and the quality of expertise, which is required for complex production and innovation processes.

References

Abe, T., Shibatam, T.: Expanding Software and Services for the Cloud Computing Era, Corporate Presentation (2009), http://www.fujitsu.com (accessed December 2010)

Alavi, M., Tiwana, A.: Knowledge integration in virtual teams: The potential role of KMS. Journal of the American Society for Information Science and Technology 53(12), 1029–1037 (2002)

Anand, V., Clark, M., Zellmer-Bruhn, A.: Team knowledge structures: Matching task to information environment. Journal of Managerial Issues 15, 15–31 (2003)

Andreu, R., Sieber, S.: Knowledge integration across organizations: How different types of knowledge suggest different integration trajectories. Knowledge and Process Management 12(3), 153–160 (2005)

Becker, M.C.: Managing dispersed knowledge. Journal of Management Studies 38, 1037–1051 (2001)

Becker, M.C.: Organizing new product development: Knowledge hollowing-out and knowledge integration – The FIAT Auto case. International Journal of Operations & Production Management 23(9), 1033–1061 (2003)

Becker, M.C.: Organizational Routines: A Review of the Literature. Industrial and Corporate Change 13(4), 643–677 (2004)

Brusoni, S., et al.: Strategic dynamics in industry architectures and the challenges of knowledge integration. European Management Review 6, 209–216 (2009)

Carlile, P.R.: A Pragmatic View of Knowledge and Boundaries: Boundary Objects in New Product Development. Organization Science 13(4), 442–455 (2002)

Cavusgil, S.T., Calantone, R.J., Zhao, Y.: Tacit knowledge transfer and firm innovation capability. Journal of Business and Industrial Marketing 18(1), 6–21 (2003)

Chesbrough, H.W.: The era of innovation. MIT Sloan Management Review 44(3), 35–41 (2003)

Eisenhardt, K.M., Okhuysen, G.A.: Integrating knowledge in groups: How formal interventions enable flexibility. Organization Science 13(4), 370–386 (2002)

Galunic, D.C., Rodan, S.: Resource combinations in the firm: knowledge structures and the potential for Schumpeterian innovation. Strategic Management Journal 19, 1193–1201 (1998)

Garud, R., Nayyar, P.: Transformative capacity: Continual structuring by inter-temporal technology transfer. Strategic Management Journal 15, 365–385 (1994)

Gittell, J.H.: Coordinating Mechanisms in Care Provider Groups: Relational Coordination as a Mediator and Input Uncertainty as a Moderator of Performance Effects. Management Science 48(11), 1408–1426 (2002)

Grandori, A.: Neither Hierarchy nor Identity: Knowledge-governance Mechanisms and the Theory of the Firm. Journal of Management and Governance 5(3), 381–399 (2001)

Grant, R.M.: Prospering in dynamically-competitive environments: Organizational capability asknowledge integration. Organization Science 7(4), 375–387 (1996a)

Grant, R.M.: Toward a knowledge-based theory of the firms. Strategic Management Journal 17, 109–122 (1996b)

Grover, V., Davenport, T.H.: General perspectives on knowledge management: Fosteringa research agenda. Journal of Management Information Systems 18(1), 5–21 (2001)

Hedlund, G.: A model of knowledge management and the N-form Corporation. Strategic Management Journal 15, 73–90 (1994)

Huang, J., Newell, S.: Knowledge integration processes and dynamics within the context of cross-functional projects. International Journal of Project Management 21, 167–176 (2003)

Kodama, M.: Knowledge creation through networked strategic communities: case studies on new product development in Japanese companies. Long Range Planning 38, 27–49 (2005)

Kogut, B., Zander, U.: Knowledge of the firm, combinative capabilities, and the replication oftechnology. Organization Science 3(3), 383–397 (1992)

Kraaijenbrink, J., et al.: Towards a kernel theory of external knowledge integration for high-tech firms: Exploring a failed theory test. Technological Forecasting & Social Change 74, 1215–1233 (2007)

Lida, I.: Fujitsu's Human Centric Computing R&D Initiatives, Corporate Presentation (2010), http://www.fujitsu.com (accessed December 2010)

Lu, I.Y., Wang, C.H., Mao, C.J.: Technology innovation and knowledge management in the high-tech industry. International Journal of Technology Management 39(1/2), 3–19 (2007)

Mohannak, K.: Knowledge integration within Japanese firms: the Fujitsu way. Journal of Knowledge Management Practice 12(2), 1–15 (2011)

Murano, K.: Fujitsu Laboratories' R&D Strategies, Corporate Presentation (2010), http://www.fujitsu.com (accessed December 2010)

Nonaka, I.: A dynamic theory of organizational knowledge creation. Organization Science 5(1), 14–37 (1994)

OECD. The significance of knowledge management in business sector. OECD, Policy Brief, Paris (2004)

Okhuyzen, G.A., Eisenhardt, K.M.: Integrating knowledge in groups. Organization Science 13, 370–386 (2002)

Prencipe, A., Tell, F.: Inter-project Learning: Processes and Outcomes of Knowledge Codification in Project-based Firms. Research Policy 30(9), 1373–1394 (2001)

Probst, G., Raub, S., Romhardt, K.: Managing Knowledge: Building Blocks for Success. John Wiley and Sons Ltd., Chichester (2000)

Purvis, R.L., Sambamurthy, V., Zmud, R.W.: The assimilation of knowledge platforms in organizations: an empirical investigation. Organization Science 12, 117–135 (2001)

Takahashi, E., et al.: Evolution of network society and technological innovation. Fujitsu Science Technology Journal 45(4), 331–338 (2009)

Tsoukas, H.: The Firm as a Distributed Knowledge System: A Constructionist Approach. Strategic Management Journal 17, 11–25 (1996)

Yang, J.: Knowledge integration and innovation: Securing new product advantage in high technology industry. Journal of High Technology Management Research 16, 121–135 (2005)

Yoshikawa, S., Sasaki, S.: R&D strategy of Fujitsu laboratories – Toward a Human-Centric Networked Society. Fujitsu Science Technology Journal 46(1), 3–11 (2010)

Zirpoli, F., Camuffo, A.: Product architecture, inter-firm vertical coordination and knowledge partitioning in the auto industry. European Management Review 6, 250–264 (2009)

Zollo, M., Winter, S.G.: Deliberate Learning and the Evolution of Dynamic Capabilities. Organization Science 13(3), 339–351 (2002)

Business Model Innovation in Complex Service Systems: Pioneering Approaches from the UK Defence Industry

Rich Morales and Dharm Kapletia

University of Cambridge, Institute for Manufacturing, 7 Charles Babbage Road, Cambridge, CB3 0FS, United Kingdom
rrm35@cam.acu.uk, Kapletia@gmail.com

Abstract. Manufacturers and operators of complex service systems are increasingly focused on customer-centric strategies. Examples include solution-based contracts, which provide more holistic approaches closely linking design, manufacture, use and reuse functions within a firm, or across a network of firms and suppliers, to deliver tailored value. Solutions deliver broader benefits that exceed the rewards of traditional transactional service delivery. Trends from the defence industry illustrate how innovative business models are applied in complex service systems to adapt and apply the knowledge resident in the firm and external networks. This paper seeks to share insights into understanding collaborative service approaches as firms adapt to changing market forces by retooling their priorities, focusing their resources, and adopting strategies driving new business models.

Keywords: Defence, Business Model, Innovation, Services.

1 Introduction

This study presents empirical data from the UK Ministry of Defence (MoD) which illustrates how innovative business models are being applied to deliver new forms of value through collaboration amongst suppliers and customers.

In their examination of customer solutions providers, McKinsey consultants illustrate how shifting from a product/manufacturing focus to a customer focus typically requires larger scale commercial and technical integration, as well as higher customization to individual customer needs (Johansson, et al 2003). By focusing on the customer's value chain, suppliers identify where they can best contribute to the customer's business (Slywotzky & Morrison, 1998). To achieve this end, many suppliers of complex systems are adopting customer centric attributes, which include adopting a customer relationship management culture, gaining a deeper knowledge of the customer's business, and initiating engagement based on customer problems/opportunities (Galbraith, 2002).

L. Uden et al. (Eds.): 7th International Conference on KMO, AISC 172, pp. 93–103.
springerlink.com

A 2008 IBM survey found that nearly all of the 1100 surveyed corporate CEOs reported the need to transform their 'business models', however few believed they had the knowledge required to do make changes (IBM, 2008). Whilst suppliers often adopt the rhetoric of being customer centric, it is argued that customer centric qualities are rarely achieved in practice (Galbraith, 2002; Shah, et al 2006). Despite recent efforts in the academic literature, there has been limited attention on how customer-centric business model innovation is taking place in industrial complex systems markets, prompting calls for further empirical research (Jacob & Ulaga, 2008; Kujala, et al 2010).

Section 2 of this paper examines the concept of business model innovation and how this applies in complex systems environments. Section 3 presents empirical data from the MoD which illustrates how innovative business models are being applied to deliver new forms of value for suppliers and customers of defence systems. Section 4 summarizes findings from two in depth case studies. Finally, Section 5 presents a discussion of findings and a conclusion.

2 Literature Review

2.1 Business Model Innovation

What is a business model and how does this theoretical concept translate into practice? While Amit & Zott (2001) declared that "a business model depicts the content, structure, and governance of transactions designed so as to create value through the exploitation of business opportunities", Casadesus et al (2010) described a business model as "a reflection of the firm's realized strategy". Each of these views captures elements of the phenomena of business model innovation; however, Osterwalder & Pigneur (2011, p 15) provide the most useful definition for further analysis, "A business model describes the rationale of how an organisation creates, delivers, and captures value and serves as a blueprint for a strategy to be implemented through organizational structures, processes, and systems". This paper adopts Osterwalder's (2005) nine distinctive business model elements as a framework for analysis as shown in Figure 1.

David Teece argues that business models 'have considerable significance but are poorly understood – frequently mentioned but rarely analysed' (Teece, 2010, p172). Similarly, in business-to-government contexts, scholars and practitioners alike identify the pressing need for business model innovation study (Miles & Trott, 2011). Kaplan & Porter cite dysfunction in US health care and describe missed opportunities, underutilized resources, and misguided business models (Kaplan & Porter, 2011). Innovation through changing business models allows firms to develop a more completive view of their organization, customers, suppliers and the environment in which the firm operates.

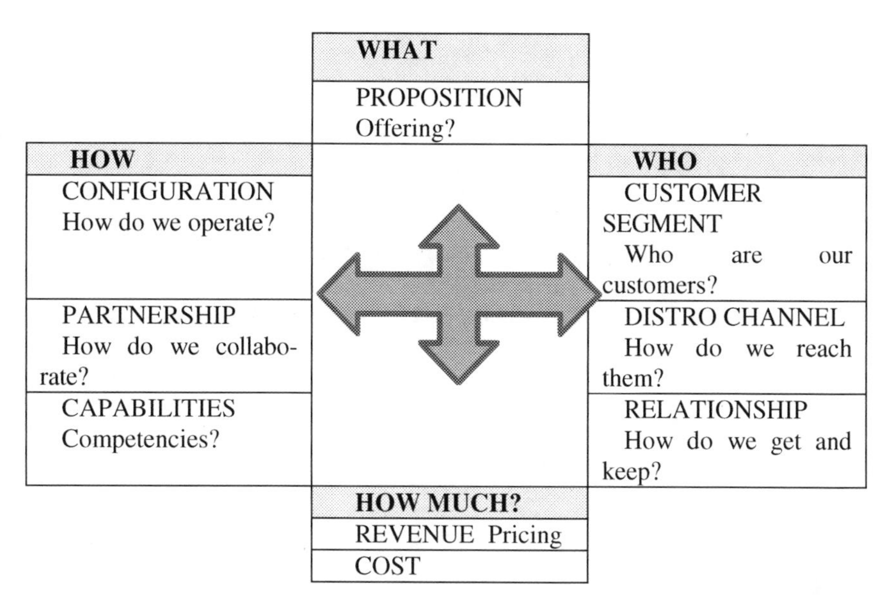

Fig. 1 Business Model Elements

Generically, this paper looks at firms engaged in the provision of complex service systems. Specifically, it examines service innovation in a defence industry context and draws on business model frameworks to capture insights.

2.2 Complex Engineering and Service Systems Industries

In complex systems industries, like the defence sector, the provision of customer solutions is an example of a fundamental departure from traditional supply and support business models (Galbraith, 2002; Davies, 2004; Kujala, et al 2010). Unlike selling a product, a customer solution is based on a value proposition, and is realised through fluid service agreements that are designed to improve the customer's operations. Customer solutions are often output or outcome based and difficult to imitate, thus providing a source of sustainable competitive advantage. Notable suppliers of customer solutions include General Electric, IBM, Rolls-Royce and Siemens (Slywotzky & Morrison, 1998; Cerasale & Stone, 2004; Wucherer, 2006).

Customers of complex systems are subject to increasing operational pressures such as the rising complexity of technology, unforeseen costs, obsolescence, poor governance decision-making and tighter budget constraints (HM Treasury, 2007). Many customers are reforming their complete system acquisition practices to address these issues (Robertson & Haynes, 2010). Unlike outsourcing, customers are now seeking to work more effectively alongside external partners/ suppliers to improve the organisation of their large scale industrial integration activities, as well as the utility and performance of core services to end users (Pew & Mavor, 2007; Whitehead, 2009).

2.3 Research Gap

Although the extant literature has usefully highlighted the case for business model innovation, as well as differences in philosophy between traditional offerings and new collaborative business models, there remains much scope for exploring how innovation is taking place at the practical level. As argued by Baines, et al (2009, p 12) service design and management is a relatively new area of study, and "yet to be explored are the detailed practices and processes needed to deliver integrated products and services."

3 Study of the UK Defence Industry

Support efforts between industry and the UK Ministry of Defence (MOD) provide an ideal context to study the on-going transformation from a manufacturing to a service based enterprise and more specifically the adoption of a customer focused business models. The UK has the fourth largest defence budget in the world of which an estimated £18 billion is spent on defence in manufacturing and service (Secretary of State for Defence, 2012, p 7).

The United Kingdom's Defence Industrial Strategy (DIS) authored in 2005 was a critical catalyst in shaping a new paradigm for UK defence acquisition. Three objectives from the UK Industrial Strategy have implications for this paper: (1) a shift in defence acquisition, away from design and manufacture of leaps in capability and upgradable platforms, toward a new paradigm focused on in-the-field operational performance, (2) an emphasis on through life capability management, characterized by modularity and sustainability thinking, and (3) longer more assured revenue/expense streams based on long term support and development (Secretary of State for Defence, 2005).

In early 2012, the UK Ministry of Defence published its latest policy paper on defence acquisition - 'National Security Through Technology', which supplements the Defence Industrial Strategy 2005 and the Defence Technology Strategy 2006 (Secretary of State for Defence, 2012). The report concedes that that neither SMEs or MoD are qualified to manage complex system solutions as SMEs "lack the capability or capacity to deliver a complete platform or weapon system, particularly where this demands complex integration, high-volume or capital-intense manufacturing" and MoD "lacks the resources and skills needed to manage the task and the associated risks, which can be considerable" (Secretary of State for Defence, 2012, p. 60). Consequently defence industry firms play the crucial role of prime contractors to deliver and support weapons or systems and manage the associated multi-organizational networks required to deliver solutions.

In response to these policy shifts and the changing economics of manufacturing as firms increasingly transition to service provision, defence firms have engaged in several innovative partnerships in support of UK military forces. These ventures are not routine collaborations; they are complete organizational transformations where structures, operating procedures, workforces, and service delivery systems are redesigned to implement a customer centric business approach.

3.1 Methodology

Empirical insights provided by company literature, reports, and industry analysis provided a means to frame the issues and opportunities associated with service business model innovation in defence. Defence firm perspectives were captured at the enterprise, operating business, and functional level. Once illustrative projects were identified that best captured innovations in contracting collaboration within the focal firm (BAE Systems), the authors engaged with key project programme managers, designers, engineers, production leaders, and customer service representatives. Semi structured interviews with key leaders involved in the design, delivery, and management of service varied in length from 30 minutes to over two hours. Twelve in-depth interviews were held with industry supplier and government (uniformed and civil MOD) staff, working on two major MOD complex service projects.

The first project focused on 'availability contracting' of Royal Air Force (RAF) Tornado aircraft, which is a mature programme established nearly 10 years ago. BAE Air Solutions is the lead in the Availability Transformation Tornado Aircraft Contract (ATTAC), a Tornado fighter jet support programme that provides a context to retrospectively examine the role of business model changes on a successful transition to availability contracting. The theory behind this approach yields stable service provision and predictable cost for MOD and revenue for industry.

In contrast to the RAF case, the second Royal Navy (RN) example is a more recent endeavour (2007) which examines support across a series of platforms both on ships, submarines and on the waterfront as part of the Warship Support Modernisation Initiative (WSMI). Like the RAF, the Navy's intent is to maximize platform availability while minimizing through life support costs. Both cases are summarized in Table 1.

Table 1 UK MoD projects examined in this study

Defence Project	End User Customer	Prime Supplier(s)	Contract features
ATTAC	Air Force (RAF)	BAE	Aircraft availability support
		Rolls Royce	Depot-level support and maint.
			Monitor system health, costs
WSMI	Navy	BAE,	Ship availability and capability
	(RN)	Thales,	Managing ship engineering
		Babcock	Monitor system health, costs
			Synch design, production, support

4 Findings

4.1 Value Proposition

In their examination of strategy in the firm, Kaplan and Norton suggest that satisfying customers is the source of sustainable value creation (Kaplan, 2004). Satisfaction in service business models is anchored in identifying a value proposition by considering benefit to customers, service differentiation, and a strong insight into customer needs and requirements. This customer centric view of a value proposition, as described by Anderson (2006), requires in-depth knowledge of the market and future trends, an understanding of the firm's capability to deliver compared to alternative solutions, and strong customer relationships.

4.1.1 Service Provision to the UK Royal Air Force

ATTAC was established as a long term availability contract which transitioned responsibility for service and support from the Royal Air Force to BAE Systems. The military pays for a specified level of aircraft availability. The level of support is extensive and includes routine and scheduled maintenance, management of spare parts, and detailed collection and sharing of crucial flight and mechanical information for the Royal Air Force. These services are provided at a fixed negotiated cost where BAE has strong financial incentives to meet or exceed availability goals. A fundamental mind-set change from the traditional sale of spares and payment for maintenance following the sale of an asset, this new model was best summarized by a service managers as being, "A reasonable profit many times, not a big fast buck once".

4.1.2 Service Provision to the UK Royal Navy

The Royal Navy in Portsmouth has relinquished ownership of various functions in the logistics arena including warehouses and related inventory and management, facilities upkeep, catering, and dockside operations. Waterfront operations, led by BAE Systems, have moved away from government provided services to contractors who serve as systems integrators across a constellation of sub-contractors in-tent on improving efficiency, applying commercial expertise, maximizing commercial technology, and lowering costs for the customer.

The value proposition in the maritime case is similar to that of the air domain as industry seeks to assume responsibility for military maintenance and support; however, the Warship Modernisation Initiative goes further than delivering availability contracts for ships and includes elements of capability contracting. The Ministry of Defence Acquisition Operating Framework describes capability as the enduring ability to generate a prescribed outcome of effect (MOD, 2009). Applied to industry, this definition translates into the ability for a platform or system to deliver a specific requirement.

What is unique about this approach is that ownership of the asset no longer rests with the customer as their interests are met not by a product or platform, but an outcome. The Royal Navy in Portsmouth has relinquished ownership of various functions in the logistics arena including warehouses and related inventory and management, facilities upkeep, catering, and dockside operations. Expanding the value proposition for the customer through a mix of capability and availability contracting model is an innovative move forward that presents the opportunity to learn from both its early successes and future challenges.

4.2 Value Configuration through Partnerships

Both cases examined demonstrate specific ways in which firm's chose to configure to deliver value. In each case the firm organized people and processes differ-ently, yet delivered successful service.

4.2.1 Integrated Teams at BAE Air Solutions

Various initiatives that increase integration across boundaries were taking place across the air domain. First, five hundred uniformed RAF work at RAF Marham as part of ATTAC. Many airmen work for civilians; conversely, civilians work for MOD. The organizational structure is very unique and not common in the UK military. Second, the use of customer-supplier 'Integrated Logistics Operation Cen-tres' (ILOCs) to collocate procurement staff, has improved overall contract responsiveness. In many instances, the extent to which service performance addressed customer needs was tied to collaborative individual interpersonal relationships and a joint-service ownership mindset. What appears to have been one of the most important factors, according to the programme manager, was the project team's ability to "coordinate, collaborate, and communicate".

When reflecting on the programme, an industry leader commented that "ATTAC is not a partnership with the RAF, it's a marriage". To underscore this point, the word trust was frequently used to describe the partnership between BAE and the RAF. Nearly everyone who participated in interviews, telephone follow ups, or presented briefings mentioned the importance of mutual trust. The Programme Manager empathized that any success they enjoyed was because of the partner-ship, "It's all about the relationships".

The senior Warrant Officer at BAE Air in Marham was particularly candid in sharing his views on the role of service and industry culture in building partnerships. He shared that the friction continues, but it is largely managed by strong leadership on both sides. Most of the friction is cultural. He said, one partner works for profit, the other to serve the country. While they are one team, their objectives are different. The RAF still doesn't quite understand BAE, BAE still doesn't quite understand the RAF, even after five years. "Organizations are different. Different is good".

4.2.2 Broad Partnerships at Portsmouth Naval Base

Unlike the organisational structure at RAF Marham, the Royal Navy's integration into BAE was not extensive (a few seconded Naval Officers being the exception). Nonetheless, the Royal Navy and BAE did uniquely configure to support service delivery. Instead of fully integrating, the approach adopted was a partnership that included customers, suppliers, stakeholders, and placed BAE Systems in the role of the primary integrator.

A visible signal that the partners involved in the maritime undertaking was the name adopted by consortium of client, firm, and suppliers – Team Portsmouth. Interviews revealed a sense of shared responsibility for both success and failure that did not serve to isolate either Prime or Customer and thus created a healthy ethos focused on problem solving versus blame.

This diverse coalition of partners provided the organization more flexibility to grow and adapt to future requirements more easily by adding more partners or refocusing and thus avoided retooling any single organization. Second, a collaborative partnership allowed competitors to participate as contributors to specific aspects of the overall naval base initiative and still keep portions of their business separate from their competitors. Finally, standing partnerships increased the speed at which decisions were made and decreased time spent in negotiations.

Importantly, partners agreed to provide stability at the top management and senior decision making levels to mitigate the knowledge drain associated with frequent personnel moves. Personnel turbulence at MOD and BAE resulted in lengthy periods of rebuilding organizational understanding and delays establishing bonds of trust. A senior manager at Portsmouth commented, "You have a customer for two or three years, then we get a different person, with a whole set of different ideas. This is not the way to run a business".

Finally, like the RAF ATTAC project, WSMI is part of a broader effort by BAE Systems at Portsmouth to support various ships that are also under separate availability contracts with BAE. Therefore, decisions that could negatively impact one element of the contract could well benefit the firm in other ways if support was improved by to that ship by reallocating resources to meet a timely or urgent need. Networked complex systems often require a different management mind-set capable of seeing a bigger picture beyond the boundaries of their business unit.

4.3 Value Delivery by Focusing the Firm's Capabilities

A firm's capabilities represent its core competency and thus provide a source of competitive advantage. As detailed earlier in this paper defence firms are increasingly redefining their value as solutions providers, moving away from traditional production models. In both cases, our observations led us to recognize the difficulty inherent in aligning corporate, business and organizational objectives and cultures, and thus delivering value. A summary the factors influencing the delivery of service is summarized in Table 2.

Table 2 Factors Influencing Service Design and Delivery

Business BAE & Partners	Customer MOD	User Front Line
Profit	Cost/Value	Cost Insensitive
Reputation,	Political Trade-offs,	Effectiveness,
Business Development	Balance Forces	Force Structures
Business Culture	Political Culture	Military Culture

Given these often divergent interests, the capability of a firm to design and manage a service delivery system to address these tensions is a prime source of competitive advantage as it works to deliver a solution that all stakeholders support.

4.3.1 Managing Contract Support to the Royal Air Force and Royal Navy

Challenges to service improvement through collaboration in the RN mirror those experienced nearly a decade earlier in the RAF. Three obstacles stand out as recurring themes in interviews with key leaders: MOD personnel turbulence, changing requirements driven by either combat missions or political and policy shifts, and skill set imbalance in a workforce accustomed to a manufacturing and now adjusting to a service and support focus.

Industry has responded to mitigate these obstacles and build more positive outcomes. BAE Military Air Solutions contracts are deliberately vague to provide flexibility to jointly prioritize effort. These cooperative themes were echoed on the maritime sector in the establishment and management of Key Performance Indicators (KPIs) developed to sustain and improve service I both cases. KPIs were broadly defined and in practice reflected achievement in all areas measured. KPIs were not frequently changed or "tightened" as was the case in more competitive contractual arrangements seen elsewhere in the firm. This arrangement was beneficial as it allowed the team to see when something was off course, yet gave them the flexibility to problem solve and avoid conflicts over variances in performance parameters. While it could be argued that this soft KPI arrangement is precisely one of the hazards of overly close collaboration, both firm leaders and customers pointed out that the programme leadership frequently challenge the status quo to deliver more "output for less money while fostering a culture of dialogue and knowledge transfer, not by using KPIs to create counterproductive competitiveness amongst the team".

Both the ATTAC and WSMI projects kept information flowing with the customer. At Portsmouth "Intelligent Customer Meetings" held three times per week encouraged frank discussions amongst stakeholders iron out both short run and longer term issues. For instance, a senior leader on the team described the type of scenario ordinarily discussed and solved at these meetings as follows:

"If someone in fleet engineering phones up and books a crane, it's covered by the plant budget. So the cost is not actually against that specific project. The commander says, 'I don't care how much the crane is costing, I want the carne there' despite its only carrying out one lift a day. Consequently that underutilized crane costs £3,000 a day to do one lift. This is a problem." Culturally, a military commander gets what he or she needs to accomplish the mission. Partnerships require more flexibility than declaring that the 'customer is always right'. They aren't. Consequently, having a uniformed officer seconded to industry, or an industry leader imbedded in a military organization, helps both sides understand the complexities of certain issues.

Air and maritime domain cases revalidated the importance of leadership, trust, and proper organizational alignment.

5 Conclusions

In the last decade, service innovation has not only taken root at firms like BAE Systems and other large scale manufacturing multinational firms, it has altered how these firms do business. This paper has identified innovative areas of collaboration based on the notion of value co-creation for the mutual benefit of members of a broader complex service network.

One area for future research is generalizability of specific elements of a customer solution based business model across a firm, industry, or a completely different context. In the defence, what may have worked for BAE and MOD in the maritime domain, may not work in the ground arena.

Four recurring themes stand out in interviews: Personnel turbulence, changing requirements driven by either combat missions or policy shifts, the importance of trust in collaborative networks, and a workforce skill set imbalance in an industry accustomed to a manufacturing and now adjusting to a service and support focus.

Successful collaborative teams were characterized by frequent interaction, dependence on one another, and delivering results and sharing credit (or sharing blame and the responsibility to fix). The means by which trust and intense knowledge of the customer's problems and the firm's capabilities are translated into value is through building robust partnerships that share information, measure system performance, and are well managed and led.

These cases illustrate that service programmes are difficult to manage and require new approaches to service design and leadership. Expanding the value proposition for the customer through a mix of capability and availability contracting model is an innovative move forward.

References

Amit, R., Zott, C.: Value Creation in E-Business. Strategic Management Journal 22(6-7), 493–520 (2001)

Anderson, J., Narus, J.A., Rossum, W.V.: Customer Value Propositions in Business Markets. Harvard Business Review 84(3), 90–99 (2006)

Baines, T.S., Lightfoot, H.W., et al.: State-of-the-art in product-service systems. Proceedings Institution of Mechanical Engineers, Engineering Manufacture 221(10), 1543–1552 (2007)

Casadesus-Masanell, R., Ricart, J.: From Strategy to Business Models and onto Tactics. Long Range Planning 43(2-3), 195–215 (2010)

Cerasale, M., Stone, M.: Business Solutions on Demand. Kogan Page, London (2004)

Chesbrough, H.: Business model innovation: it's not just about technology anymore. Strategy and Leadership 35(6), 12–17 (2007)

Davies, A.: Moving base into high value integrated solutions: a value stream approach. Industrial and Corporate Change 13(5), 727–756 (2004)

Galbraith, J.R.: Organizing to Deliver Solutions. Organizational Dynamics 31(2), 194–207 (2002)

Treasury, H.M.: Transforming government procurement. HM Treasury, London (2007)

IBM: Convergent thinking among the C-suite: Why integration and collaboration spell big opportunity for CIOs, p. 4 (2008)

Jacob, F., Ulaga, W.: The transition from product to service in business markets: An agenda for academic inquiry. Industrial Marketing Management 37, 247–253 (2008)

Johansson, J., Krishnamurthy, C., Schlissberg, H.: Solving the solutions problem. McKinsey Quarterly 3, 116–125 (2003)

Kaplan, R., Porter, M.: The Big Idea: How to Solve the Cost Crisis in Health Care. Harvard Business Review, 47–64 (September 2011)

Kujala, S., Artto, K., Aaltonen, P., Turkulainen, V.: Business models in project-based firms - Towards a typology of solution-specific business models. International Journal of Project Management 28, 96–106 (2010)

Miles, E., Trott, W.: Collaborative working: A series of personal perspectives on government effectiveness. Institute for Government, London (2011)

MOD Acquisition Operating Framework (2009), http://www.aof.mod.uk/

Ostenwalder, A., Pigneur, Y., Tucci, C.: Clarifying business models: origins, present and future of the concept. Communications of the Association for Information Systems, 2–38 (May 15, 2005)

Ostenwalder, A., Pigneur, Y.: Business Model Generation: A Handbook for Visionaries, Game Changers, and Challengers (2011)

Pew, R., Mavor, A.: Human-system integration in the system development process: a new look. National Academy Press (2007)

Robertson, D., Haynes, D.: MoD 'plays game' with industry to foil treasury. The Times, 6–8 (December 15, 2010)

Secretary of State for Defence, Defence Industrial Strategy white paper, report presented to UK parliament (2005)

Secretary of State for Defence, National Security Through Technology: Technology, Equipment, and Support for UK Defence and Security. Ministry of Defence, London (2012)

Shah, D., Rust, R., Parasuraman, A., Staelin, R., Day, G.: The path to customer centricity. Journal of Service Research 9(2), 113–124 (2006)

Slywotzky, A., Morrison, D.: The profit zone: how strategic business design will lead you to tomorrow's profit. John Wiley & Sons, Chichester (1998)

Teece, D.: Business Models, Business Strategy and Innovation. Long Range Planning 43(2-3), 172–194 (2010)

Whitehead, N.: Business transformed to deliver what the customer wants. BAE Systems News (4), 4 (2009)

Wucherer, K.: Business partnering - a driving force for innovation. Industrial Marketing Management 35, 91–102 (2006)

The Role of Trust in Effective Knowledge Capture for Project Initiation

Marja Naaranoja[1] and Lorna Uden[2]

[1] University of Vaasa, Vaasa (Sweden)
marja.naaranoja@uwasa.fi
[2] Staffordshire University, Birmingham (United Kingdom)
L.Uden@staffs.ac.uk

Abstract. The challenge in service project management is gathering business requirements from stakeholders. Requirements are often vague because it is difficult for customers to articulate their needs before they see the end product. This is especially difficult when different stakeholders are involved. Only when projects are built on trust can they work. This paper studies the importance of trust in the initiation of university project.

Keywords: Trust, knowledge transfer, project, project initiation, stakeholders.

1 Introduction

Not all projects are successful. Many fail, especially at the start of the project during requirements elicitation. The project can be unsuccessful even if it meets the timetable and budget. The project can fail if the customers are not satisfied with the projects. An elicitation requirement for project is a critical part of project management.

Requirements-specification in projects is volatile, resulting in project scope and focus evolving considerably during the course of a project. Realising requirements for project development is often impractical prior to commencing design activities.

Requirements elicitation is a key to the success of the development of non-trivial IT systems. Effective and efficient requirements elicitation is absolutely essential if projects are to meet the expectations of their customers and users, and are to be delivered on time and within budget (Al-Rawas & Easterbrook 1996). Goguen and Linde (1993) have provided a comprehensive survey of techniques for requirements elicitation, focusing on how these techniques can deal with the social aspects of this activity. The requirements elicitation problem is fundamentally social and, thus, unsolvable if we use methods that are based entirely around individual cognition and ignore organisational requirements.

Organisational requirements are those requirements that are captured when a system is being viewed in a social context rather than from a purely technical,

L. Uden et al. (Eds.): 7th International Conference on KMO, AISC 172, pp. 105–115.
springerlink.com

administrative or procedural view of the functions to be performed. Sources of such requirements could be power structures, roles, obligations, responsibilities, control and autonomy issues, values and ethics (Avison & Wood-Harper 1990). These types of requirements are so much embedded in organisational structure and policies that often they cannot easily be directly observed or articulated.

Most established techniques, however, do not adequately address the critical organisational and 'softer', people-related issues of software systems. This is particularly important when the project involves different stakeholders.
From our experience involving a renovation project management case study, we found that requirements elicitation for a project must be based on trust. The project failed because of lack of trust among and between the stakeholders. Trust is fundamental to the success of project. This paper begins with a review of trust followed by a case study discussing the failure of the project, the renovation of a university. The reasons for the failure are then presented. Suggestions for building trust are discussed. The paper concludes with suggestions for further research.

2 Trust

Requirements elicitation requires building relationships and trust among the project stakeholders. When trust is absent, the requirements elicitation process will take longer, be incomplete, and will generally become an unpleasant experience for all concerned. Although building relationships takes time and effort, it can actually shorten project time and result in improved project performance.

"Trust is a psychological state comprising the intention to accept vulnerability based upon positive expectations of the intentions or behaviour of another" (Rousseau, Sitkin, Burt & Camerer, 1998). Several factors determine the trustworthiness of an individual. According to of Sako (1992), several authors have adopted a taxonomy of trust dimensions that involves three components: ethical, technical and behavioural. Mayer and others (1950) defined trust as the simultaneous belief in a business partner's integrity, reliability and ability to care. Trust happens when there is a simultaneous belief in these three elements.

There are two definitions of trust that can be found in management literature. The first one defines trust as predictability, and one that emphasizes the role of goodwill (Hardy et al. 1998:64). Trust as predictability, points out that trust ensures that you can count on knowledge produced in a specific context by specific people or as Bachmann (2001:342) emphasizes, "trust reduces uncertainty in that it allows for specific (rather than arbitrary) assumptions about other social actors' future behaviour." The second definition points out that trust is more than predictability. It also includes goodwill, the existence of common values that can be translated into common goals (Hardy et al.1998:68). However considering trust in terms of both predictability and goodwill presupposes that predictability arises from shared meaning, while goodwill arises from the participation of all partners in the communication process whereby this shared meaning is created (ibid.:69). Thus, trust can be considered a continuous process of sense-making and negotiating that bridges heterogeneous groups.

Trust is unlikely to emerge spontaneously and, so, we need to learn how to create trust between actors with very different goals and values (Hardy et al. 1998:65). Consequently, trust is a performed achievement of a concerted and highly heterogeneous effort with actors, artefacts and other externalized knowledge representations. Trust embodies an expectation that those we work with will not take advantage of one another or exploit situations that benefit one at the expense of others. When trust is low, we usually build formal protections into these relationships. When trust is high, we are more likely to develop informal but well-understood ways of working together. Often the social context demands a formal framework for collaboration.

There are different types of trust (Dawes 2003):

- Deterrence-based trust or calculus-based trust: Results from a fear of the consequences of behaving in an untrustworthy matter. Calculus process is where a trustor calculates the costs and/or rewards for another party to be opportunistic. In the prediction process, the trustor forecasts another party's behaviour from historical data. The individual recognizes the consequences for not doing what he or she said he or she would do. Calculus-based trust rests on information-based rational decisions about the organization or person to be trusted. For example, you might decide to trust your doctor based first on his or her professional credentials and public reputation. Your trust may be challenged or reinforced as you acquire more information through personal interactions regarding your medical care.
- Identity-based trust is based on familiarity and repeated interactions among the participants. Identity-based trust also emerges from joint membership in a profession, a team, a work group, or a social group. Members of a sports team often develop deep trust in one another based on long term association and frequent interactions. Players come to know their teammates well and expect them to respond to different game situations in predictable, accepted ways.
- Institution-based trust rests on social structures and norms, such as laws and contracts, which define and limit acceptable behavior. When two companies enter into a partnership, they draw up formal contracts that specify rights and obligations, and these contracts conform to a body of accepted contract law. Even if the partners are not entirely sure about all the details of one another's operations, they know they can rely on the contract and its legal underpinnings to help the relationship work. Professional trust may rest on any or all three types of trust.
- Swift trust: The time available does not allow trust to be built up in the normal way, and team members simply assume that the other team members embrace similar values to their own (Jarvenpaa & Shaw 1998).
- Knowledge-based trust: The individuals trust each other, as they know each other sufficiently well to be able to predict each other's behavior and have shared experiences. The trustor uses information about how the trustee has carried out tasks in the past to predict future action (Jarvenpaa & Shaw 1998).
- Transferred trust: May occur when the trustor knows and trusts a person or institution that recommends the trustee. This is a form of swift trust (Jarvenpaa & Shaw, 1998).

- Psychological or behavioral trust: A predisposition towards having confidence that others will carry through on their obligations (Warrington, Algrab & Caldwell, 2000).
- Technological trust: A belief that technologies will perform reliably and will not be used for untoward purposes (Chiravuri & Nazareth, 2001).
- Organizational trust: The belief that an organization will carry through on its obligations (Cummings & Bromiley, 1996).
- Situational trust: Dependence on cues and clues in the immediate social environment when deciding whether to trust another group, organization or institution (Karake-Shalhoub, 2002).
- Interpersonal trust: An expectation that others will behave in a predictableway, and a willingness to be vulnerable during the trust relation (Dibben, 2000).

Shapiro and others (1992) identified three consecutive stages through which trust develops: calculus-based, knowledge-based and identification-based. In the beginning, calculus-based trust exists when neither party is familiar with the other. The knowledge-based stage is entered when information flow increases and behaviour becomes more predictable, therefore adding more mutuality to the relationship. Finally, trust develops into the information-based stage. By that time, there should be complete empathy within the relationship and a full understanding of each other's needs, wants and intentions (Ashleigh, Connell & Klein, 2003).

Knowledge sharing requires the following types of trust: benevolence-based trust and competence-based trust. Benevolence-based trust is when which an individual will not intentionally harm another when given the opportunity to do so. However, another type of trust is competence-based trust. Competence-based trust describes a relationship in which an individual believes that another person is knowledgeable about a given subject area. Knowledge exchange was more effective when the knowledge recipient viewed the knowledge source as being both benevolent and competent. It is it is possible for effective knowledge sharing to occur in both strong-tie and weak-tie relationships as long as competence- and benevolence-based trust exist between the two parties.

2.1 Different Types of Knowledge Require Different Forms of Trust

Researchers (Abrams et al. 2002) have showed that competence-based trust had a major impact on knowledge transfers involving highly tacit knowledge. Serrat (2009) argues that high-trust environments correlate positively with high degrees of personnel involvement, commitment, and organizational success. Decided advantages include increased value; accelerated growth; market and societal trust; reputation and recognizable brands; effortless communication; enhanced innovation; positive, transparent relationships with personnel and other stakeholders; improved collaboration and partnering; fully aligned systems and structures; heightened loyalty; powerful contributions of discretionary energy; strong innovation, engagement, confidence, and loyalty; better execution; increased adaptability; and robust retention and replenishment of knowledge workers. Nothing is as relevant as the ubiquitous impact of high trust.

According to Fukuyama (1995), trust is the expectation that arises within a community of regular, honest and cooperative behavior, based on commonly shared norms, on the part of the other members of that community.

How to build trust?

Abrams and others (2002) suggested several ways to build trust:

- Create a common understanding of how the business works
  The development of a common context or common understanding among the stakeholders is a useful way. Several of the factors are important in building benevolence- and competence-based trust, such as shared language and goals, relate to the importance of building a shared view of how the project can be achieved. Another issue is that of measurement. Creating this common understanding can make it easier for stakeholders to focus on mutually held goals and values, and reduce the amount of time and effort spent on individual issues and motivations.
- Demonstrate trust-building behaviors
  This can be achieved by modeling and recognition of trust-building behaviors, such as receptivity and discretion. Employing active listening skills and encouraging stakeholders to voice their concerns in an atmosphere where their issues will not be improperly disclosed can build trust between the various parties.
- Bring people together
  It is important to consider how to bring people to know each other.
- We can create both physical and virtual spaces where people can easily interact with one another. It possible to leverage tools, such as collaborative spaces and instant messaging, to make it easier for stakeholders to *communicate* with one another.

3 Project Front End

According to Hughes (1991), "Every project goes through similar steps in its evolution in terms of stages of work. The stages vary in their intensity or importance depending upon the project." The common stages of a project are:

- Demonstrating the need
  - Conception of need
  - Outline feasibility
  - Substantive feasibility study and outline financial authority
- Pre - construction stage
- Construction stage
- Post completion stage

A critical activity in the beginning of a project is to ensure that decisions and choices serve the best interests of the stakeholders and fulfill their long-term strategic objectives. The front end decisions embodied in a product concept define and guide the subsequent development activities later in the innovation process.

Decisions may fail if product concepts become "moving targets" (Wheelwright & Clark, 1992), if product initiatives are cancelled half-way through (Englund & Graham, 1999; Khurana & Rosenthal, 1997), if senior managers do not communicate their strategic goals and expectations, or if the strategies are too abstract to give any direction to front-end activities (Smith and Reinertsen, 1998).

4 Case Study on Requirements Specification for a University Service Renovation

4.1 Methodology

The aim of the case study is to describe the role of trust in the initiation of the project. The case material was gathered by observing the process. The other author was involved in this process. The observed stages were first written as a story. The involvement of stakeholders was also studied during each stage.

The university campus project is presented in section in sections 4.2 and 4.3. Sections 4.4 and 4.5 identified the kinds of trust that play important role in the failure of the initiation of the project.

University campus

University A is located on two campuses. One is in the centre of town where half of the students are studying in three buildings. In the seaside campus there are the laboratories and building 1, where the other half of the students study. In the seaside campus there are two other universities nearby (See figure 1). Seaside campus is located near to the middle of the town - only 20 minutes walk from the town campus. Different facility owners own all the seaside campus buildings. The University A buildings are owned by a facility management company– the laboratory is owned by the same company and by two another facility owners.

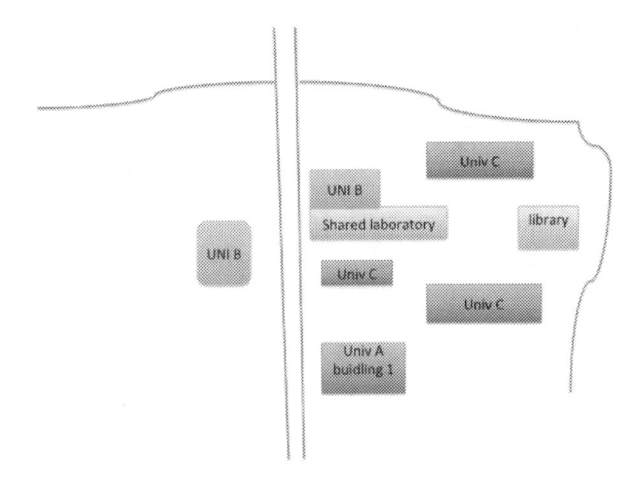

Fig. 1 Simplified map of the facilities at the seaside campus. The laboratory and library building are shared facilities.

4.2 Proposals for Renovation

The seaside building 1 of the University A was in very bad condition. There were fire risks in the building, such that the authorities considered banning the use of the facilities. Some of the personnel were not able to teach in every classroom. This was known for over five years before taking any actions. Proposal for renovation *(phase 1).*

The first university renovation proposal was written in 2007. It was turned down because the university did not use the formal process defined by the facility owner.

Detailed need specification (phases 2 and 3)
This resulted in a second need-specification for Building 1. A person who was employed by the university was responsible of writing the detailed need report. The needs were studied by forming different kinds of groups: education group, administration group, in addition two separate groups of students were asked to define their needs. It was achieved through interviews with 80 staff s and 60 students. The project also involved a survey of the state of the building where possible technical problems were described. The project was presented to the facility owner representatives during the spring of 2008 in order to write the formal need specification.

The outcome was that because the analyses contained only the functions of the Building 1, the analysis was not able to give a sufficient overall view for decision-making. The facility managers complained that the analyses were too detailed and there was a need to write a shorter description. At this stage, the facility owner representatives decided not to be involved though their involvement was requested. They wanted the university to take charge of the formal need specification to include all facilities of the university. The specification also had to include other plans of the seaside campus.

To speed up the process, the heads of departments in university A were to answer questions. The heads were given the option whether to ask the opinions of their employees. Some did and some did not. The formal need specification contains the wishes of each department.

In January 2010 a meeting was held with the facility owner. To our surprise they again said that the need specification produced was not broad enough. According to them we need to consider the seaside campus of all the universities together. The University A cancelled the process of writing the need specification of the whole university facilities.

Current situation
As one can see, the project never took off during the four years of expressing the need to renovate. Recently the facility owners have formed a new company that owns two university facilities. Meanwhile the condition of the Building 1 is becoming worse.

4.3 Analysis of the Failure to Start the Project

The actors in the project were not communicating adequately about what was required for the launch document and it appears that the ground rules changed at various times, hence hindering the appropriate knowledge capture and management. The lacking communication resulted in a lack of a common goal. The conflicting interests were not openly discussed between university and facility management.

We can speculate several reasons why the rules of need specification changed – the facility managers are also able to give reasons. In this paper we focus on one reason: ***trust.***

In the beginning and during the whole process the facility owners trusted that the University would be able to write the formal need specification document in such a way that it can be accepted. However, every time when the facility owner read the new document they learned they had not given enough instructions regarding what kind of information they need for decision making. The parties used the need specification as an information broker to transfer the knowledge from the university to the facility manager. The process was frustrating for the university representatives since the process was too slow for them and the university representatives at the same time lost benevolence-based trust, since it felt like the facility management representatives did not want to help.

Facility managers were not able to trust that all the important viewpoints were in the report and they were not able to make decision for the renovation. Thus there was a need to add new viewpoints into the presented need specification. This process was a learning process for both the facility and university representatives. Both parties were able to learn during the process..

The next section discusses how trust and knowledge transfer are linked together in the university project.

4.4 How Lack of Trust Explains Why Renovation Failed to Start

From our research, we found the following problems:

- Failure to make decisions
- Lack of trust on the information content of the need specification
- Lack of a common goal
- Conflicting interests from stakeholders.

In this paper we focus on trust viewpoint. We can argue that if the facility management representatives had believed in the message of the need specification they had been forced to make a "Go" decision in the university renovation project.

Project front end decisions

The decision makers did not want to start the project officially since all the facts related to the needs were not known. The universities plan to collaborate with each other and there are a lot of changes in the air since the amount of young generation is decreasing every year. The lacking strategic goals of the facilities management were not communicated, which made it easy to postpone the decision making.

Distrust
The most common reason for stakeholder caution and concern is distrust, caused by one or more of the following:

- Fear that the Project will, involve risk for them;
- Fear that the end product will cost them dearly;
- Fear of making decision in case the project will fail

Trust usually takes time to develop. We may initially trust or not trust those involved on our projects, based on past experience, personal filters, culture (organizational, geographical, and otherwise), and a wide variety of factors that can influence our judgments. There was no time to build trust. The parties communicated by using the need specification as a media to transfer the knowledge, and the viewpoints were learned only inside the university.

What is the role of trust in needs specification involving different stakeholders? We believe that project needs specification is embedded in a social context. Participation in the relevant social context is essential for the successful needs specification. The very existence of a relationship between participants in the project requires the presence of trust. Trust within and between members of the group can assist the dissemination of knowledge in group discussion. However, it is also important for us to recognize that issues of control and power also shape social interaction and, therefore, influence the processes of discussion. This kind of trust was visible inside the university.

Recommendations for requirements elicitation
It is important that trust can be built before the start of the project.
1. Co creation of value
One approach is to adopt the co creation of value from service dominant logic (Vargo & Lusch 2004). This can be done by discussing the content of the context with the different stakeholders involved.

2. Building of trust
Trust building can, for example, be supported by a facilitating session, in which a facilitator enables all key stakeholders to articulate their requirements in a formal way. This approach has many advantages, including using the synergy of the group to build relationships and trust.

Complexities arise when different stakeholders have different requirements that must be reconciled and finalized. Without having a clear picture of the political landscape, stakeholders will struggle, and the project is likely to take longer than expected.

5 Conclusion

University facility renovation project has several stakeholders. The facility users and owners may have different values. Co-creation of values is important with all the stakeholders. It is not enough that the facility owner observes the situation and rules what kind of need specification is broad enough and detailed as an outside assessor.

The strategic goals of the university project need to be clear enough so that the decision makers are able to take make decision to start the project.

Trust is important element in every project. Sometimes initial trust needs to be established. In other cases, unexamined initial trust was tested and needed to be rebuilt, often in the context of joint problem solving. Once earned, trust is still fragile. It needs to be nurtured through continuous attention and demonstrated in ongoing interactions (Abrams et al. 2002).

References

Abrams, L., Cross, B., Lesser, E., Levin, D.Z.: IBM Institute for Knowledge-Based Organizations, Trust and knowledge sharing: A critical combination (2002), http://www-935.ibm.com/services/in/igs/pdf/g510-1693-00-cpov-trust-and-knowledge-sharing.pdf (accessed June 30, 2012)

Al-Rawas, A., Easterbrook, S.: Communications Problems in Requirements Engineering: A Field Study. In: Proceedings of he First Westminster Conference on Professional Awareness in Software Engineering, February 1-2. The Royal Society, London (1996)

Ashleigh, M., Connell, C., Klein, J.H.: Trust and knowledge transfer: An explanatory framework for identifying relationships within a community of practice. In: Proceedings of the EIASM Second Workshop on Trust within and Between Organizations, October 23-24, Amsterdam (2003)

Avison, D.E., Wood-Harper, A.T.: Multiview: An Exploration in Information Systems Development. McGraw-Hill, Maidenhead (1990)

Bachmann, R.: Trust, Power and Control in Trans-Organizational Relations. Organization Studies – An International Multidisciplinary Journal Devoted to the Study of Organizations, Organizing, and the Organized in and Between Societies (22), 1–3 (2001)

Chiravuri, A., Nazareth, D.: Consumer trust in electronic commerce: Alternative framework using technology acceptance. In: Proceedings of the Seventh Americas Conference on Information Systems, pp. 781–784 (2001)

Cummings, L.L., Bromiley, P.: The organizational trust inventory (OTI): Development and validation. In: Kramer, R.M., Tyler, T.R. (eds.) Trust in organizations: Frontiers of Theory and Research. Sage Publications, Thousand Oaks (1996)

Dawes, S.: New Models of Collaboration A Guide for Managers. The Role of Trust in New Models of Collaboration Center for Technology in Government (2003), http://www.ctg.albany.edu/publications/online/new_models/essays/trust.pdf (accessed June 20, 2011)

Dibben, M.R.: Exploring international trust in the entrepreneurial venture. Macmillan, London (2000)

Englund, R.L., Graham, R.I.: From experience: linking projects to strategy. Journal of Product Innovation Management 16, 52–64 (1999)

Fukuyama, F.: Trust: The social virtues and the creation of prosperity. Penguin Books, London (1995)

Goguen, J.A., Linde, C.: Techniques for requirements elicitation. In: Proceedings of the First IEEE International Symposium on Requirements Engineering (RE 1993), pp. 152–164 (1993)

Hardy, C., Phillips, N., Lawrence, T.: Distinguishing trust and power in interorganizational relations: Forms andfacades of trust. In: Lane, C., Bachmann, R. (eds.) Trust within and Between Organizations: Conceptual Issues and Empirical Applications, pp. 64–87 (1998)

Hughes, W.: Modelling the construction process using plans of work. In: Proceedings of an International Conference - Construction Project Modelling and Productivity, CIB W65, Dubrov (1991)

Jarvenpaa, S.L., Shaw, T.R.: Global virtual teams: Integrating models of trust. In: Siober, P., Griese, J. (eds.) Organizational Virtualness, pp. 35–51. Simowa, Bern (1998)

Karake-Shalhoub, Z.: Trust and Loyalty in Electronics Commerce: An Agency Theory Perspective. Quorum Books, Westport (2002)

Khurana, A., Rosenthal, S.R.: Towards holistic "front ends" in new product development. Journal of Product Innovation Management 15, 57–74 (1998)

Levin, D.Z., Cross, R., Abrams, L.C., Lesser, E.L.: Trust and knowledge sharing: A critical combination. IBM Institute for Knowledge-Based Organizations (2002)

Mayer, R.C., Davis, J.H., Schoorman, F.D.: An integrative model of organizational trust. Academy of Management Review 20, 709–734 (1995)

Rousseau, D.M., Sitkin, S.B., Burt, R.S., Camerer, C.: Not so Different After All: a Cross-Discipline View of Trust. Academy of Management Review 23(3), 393–404 (1998)

Sako, M.: Prices, Quality and Trust: Inter-firm Relations in Britain and Japan. Cambridge University Press, Cambridge (1992)

Serrat, O.: Knowledge Solutions, Building Trust in the Workplace 157 (August 2009), http://www.adb.org/Documents/Information/Knowledge-Solutions/Building-Trust-in-the-Workplace.pdf (accessed June 20, 2012)

Shapiro, D., Sheppard, B.H., Cheraskin, L.: Business on a handshake. Negotiation Journal 8, 365–377 (1992)

Smith, P.G., Reinertsen, D.G.: Developing Products in half the time – New rules, new tools, 2nd edn. John Wiley & Sons, Inc. (1998)

Vargo, S.L., Lusch, R.F.: Evolving to aNewDominant Logic for Marketing. Journal of Marketing 68, 1–17 (2004)

Warrington, T.B., Algrab, N.J., Caldwell, H.M.: Building trust to develop competitive advantage in e-business relationships. Competitive Review 10(2), 160–168 (2000)

Wheelwright, S.C., Clark, K.B.: Revolutionizing product development: quantum leaps in speed, efficiency, and quality. Maxwell Macmillan Canada Inc., New York (1992)

TechnoStress in the 21st Century; Does It Still Exist and How Does It Affect Knowledge Management and Other Information Systems Initiatives

Richard J. Self[1] and Conrad Aquilina[2]

[1] University of Derby, Derby (United Kingdom)
r.j.self@derby.ac.uk
[2] Management Information Systems Officer, Government of Malta, Malta (Malta)
conrad.aquilina@gov.mt

Abstract. This paper critically evaluates the impact and consequences of TechnoPhobia and TechnoStress (Rosen and Weil 1995) on public and corporate ICT policy makers. This research is set in the context of government ministries in Malta and performed by Aquilina in 2010/2011, as part of his MSc Dissertation. It is of particular significance in indicating how little has changed in the perceptions of ICT users surveyed as to their levels of acceptance and trust or distrust in ICT in their work and recreation, using the CARS and GATS scales of Rosen and Weil. It intimates that the problem may have got worse. The fact that 56% of the respondents in Malta showed some degree of technophobia should be of particular concern to policy makers. The consequences of Technophobia need to be incorporated into the policy making forum to ensure more effective ICT systems are developed.

Keywords: Techno-phobia, techno-stress, ICT Policy Government Corporate Public Acceptance CARS GATS CTS.

1 Introduction

(Rosen and Weil, 1995) showed that between 33% and 50% of teachers, in a range of surveys between 1987 and 1989, exhibited some degree of technophobia. A range of research over the intervening years has identified some of the potential causes for technophobia as often being related to levels of experience and support.

In our daily lives, we continue to hear comments from those around us that “the system failed again” or something like that, that indicates that we generally do not have a high degree of confidence in our use of ICT or the systems themselves. We are not surprised when a credit card transaction over the internet is failed by one of the third-party security checking systems and then try it next day and it works perfectly. Most people do not use the in-built help systems in PC packages but

L. Uden et al. (Eds.): 7th International Conference on KMO, AISC 172, pp. 117–127.
springerlink.com

prefer to "ask a friend". We are pleasantly surprised when voice recognition systems work and totally unsurprised when they do not (Wilson, 2012). But, still we feel some level of stress as a result.

The EU states "The ICT PSP (Policy Support Programme) aims at stimulating smart sustainable and inclusive growth by accelerating the wider uptake and best use of innovative digital technologies and content by citizens, governments and businesses" (EU, 2007). Thus, a very clear statement of the perceived benefits of ICT in the widest possible terms for all is made. It also, clearly, implicitly assumes that there is a high degree of acceptance by many policy makers that ICT is beneficial and accepted and will lead to growth in the economy.

Technology based knowledge management systems are seen, by many, as the holy grail for ensuring that the crucial knowledge held by the staff in an organisation is captured, codified and preserved for the benefit of the organisation as productivity continues to increase and drive down the numbers of staff employed. Knowledge management is also at the heart of many government portals for the public to use, to gain knowledge, information and data that would otherwise be printed.

It is in this context, of the wide-scale policy for ever increasing dependence on ICT, that Aquilina (for his MSc Project) investigated the levels of technophobia exhibited by staff within a Maltese government ministry. Malta is one of the leading countries in the world for the development of e-government. Public Service employees in Malta have been trained in technology so that they can meet today's challenges and provide a faster, more reliable service to the public. Employees are expected to use a wide range of bespoke software referred to as e-government systems to interact with the public, that allows, amongst other supposed benefits, for quicker means of communications that helps to reduce queues and thus offer efficient service. Public Service employees can thus be regarded as super-users.

Thus, it is clearly of considerable importance to understand whether the levels of technophobia have decreased since the original research of Rosen and Weil, as it could be expected that the world should have come to terms with the provision of and use of ICT over the following 17 years. The consequences of technophobia are raised levels of stress felt by individuals during their interactions with IT which can be measured as adverse physiological impact on the cardiovascular system (Wilson 2012).

2 Literature Review

2.1 Introduction

Fear or dislike of computers and related technology, is called Technophobia and it is an issue that affects many people, many of whom don't even know it (Rosen and Weil, 1992). This may sound out of place in a world where technology overdose is considered normal and healthy, yet it is real, and although we are just beginning to comprehend the effects of such phobia, history shows us that man-kind has always reacted to technology, whether in a positive manner via adoption and use, or in negative ways such as anxiety, avoidance, incompetence, fear, stress, negative attitudes and cognitions and so on so forth (Brosnan, 1998).

Questions start to arise like, do people want technology or is it imposed on them in some way or another? Do people want technology to be part of their lives and believe that if this was so it would improve it or are we mistakenly assuming that just because people use technology, in some form or another, they are happy with it? How many of us suffer from technophobia, is this wide-spread? What are Governments doing vis-à-vis Technology, why are they promoting it and how will this decision affect us?

All these questions revolve around 2 items, technology and the human experience around it. Information Technology can be simply defined as "the modern handling of information by electronic means" (Ugwo et al, 2000). The Human Experience can be defined as when people interact with other people, an object or a situation that can happen at any point in time.

The impact of technophobia and its sibling, TechnoStress (Weil and Rosen, 1997), needs to be investigated due to the pervasive impact of ICT throughout society. There are dangers that parts of the population may become disenfranchised if they are unable or unwilling to engage with technology.

Research by (Wang et al., 2008) showed that over 75% of managers in a survey felt that technology, instead of alleviating their job and job related stress, has actually increased stress and their workloads as it not only affected usability but also availability.

2.2 Innovation and Fear of Change

Research into the take-up of innovation has shown that most populations normally are composed of approximately 10-15% of early adopters (the enthusiasts for novelty and change) about 50-60% of Hesitant "Prove Its" (who can be persuaded that something new might be worthwhile after all) and something like 30-40% of Resisters (who want nothing to do with novelty or new technology however beneficial it might be to their life) (Weil and Rosen, 1997).

In the workplace, this fear of innovation needs to be overlaid with the classic stages of the change and response curve which operates during the introduction of new processes and technology. (Bocij et al., 2003) show the sequence of shock, denial, emotional turmoil, fear, anger, guilt, grief, acceptance and letting go, new ideas, search for meaning and finally, it is hoped, integration into a productive process.

The consequences of this compound problem generates considerable stress for the workforce, which will adversely affect the organisation being able to achieve the intended changes and benefits in the planned timescale, with effective use of the new systems and procedures.

2.3 Consequence of ICT Project Failures

The Standish Group Chaos reports have, over the years, shown just how fragile the ICT project delivery process is, with a fairly consistent level of 65% to 70% of all ICT projects failing to be delivered on time, to budget and with contracted for functionality and in the 2009 report, the failure rates were higher than for a decade (Standish Group 1994, 1995, 1999, 2005, 2009). The lack of functionality in the delivered but not fully successful projects will generate additional stresses in the staff, due to the lack of crucial functionality to meet the customers needs.

In addition to all this, we are all used to random, unexpected failures of the technology that we rely on to conduct our work, our play and our other activities, credit cards are inexplicably rejected in on-line transactions, automated phone systems do not recognise the words we speak, servers crash, systems stop in the middle of a transaction, the computer systems do not understand what we want (Wilson, 2012) (Weil and Rosen, 1997) with the consequence that, often, our stress levels rise (Wilson, 2012).

2.4 Understanding User Needs

There are also dangers that significant effort may be wasted in the un-used or little used systems. The corporate world has, to some extent, learned part of its lesson from the Standish Group Chaos reports, which have demonstrated the critical value of looking to the users to define the system (Standish Group, 1994, 1995, 1999, 2005, 2009).

Jacob Neilsen continues to demonstrate the critical importance of understanding what the users need and want through his regular analyses published on his www.useit.com website. However, it is still the case that most new corporate systems are less than perfect and fail to satisfy the users. This is nowhere more important than in the technologically orientated approaches to knowledge management.

It is therefore of concern that governments and many other types of organisations are continuing the drive for the development of systems which seem not to be used or are little valued by their intended users.

2.5 Public Services

Governments world-wide are driving for replacing people with web-portals for the intermediation of knowledge between citizens and the public services. Local councils in the UK are reducing the number of planning request notices in the local newspapers and replacing these with the web-portal. Again in the UK, all local authorities are required to publish all details of purchases over a value of about £500 on their web portals in the alleged interests of transparency. In reality, these actions which are suggested to place much data and knowledge into publicly available formats and accessible modes, actually result in burying information in the oceans of available data in plain sight.

It also results in many citizens finding it very difficult to actually contact the relevant staff to gain answers to their questions and needs, particularly those citizens who find the newer ICT based processes difficult to come to terms with. This is a form of disenfranchisement.

Studies, such as (Kolsaker and Lee-Kelley, 2008) cited in (Gauld *et al.,* 2010) show low use of e-gov services in Britain, finding that 22.1% of citizens had downloaded government information, 7.1% downloaded forms and only 4.8% returned completed forms. (Gauld *et al.,* 2010) through 435 telephone interviews in Australia and 498 in New Zealand found that the majority of respondents are reluctant to use e-gov services with less then half never even visited an e-gov website. The majority of respondents (70% in New Zealand) by far still preferred to deal with government through non-digital means.

2.6 Education

For years we have been told that our children and students are the digital natives, the Millenial generation, Generation Y and Z. We are living, according to (Dede, 2005) in the era of students who are at home with Web 2.0 technologies and the many and varied forms of electronic devices for communications. It is claimed, by Dede, that these new, multi-media technologies through "mediated immersion" offer many new ways of meeting the needs of students whose learning styles do not benefit from the classic classroom style of teaching and learning, including the strengthening of "communal learning involving... with knowledge spread across a community and a context as well as within an individual".

However, it is found that even though most of the children passing through the UK educational system have attended ICT classes, often gaining European Computer Driving Licence (ECDL) and Computer Literacy and Information Technology (CLAIT) qualifications, they present as undergraduates at University with very low levels of ICT literacy (Self and Goh, 2009) with remedial classes being found to be of vital importance. They have little idea of blogging or using wikis, they have little idea of creating successful search strategies to find academic sources of information.

2.7 Consequences for Knowledge Management

In the light of the above and the continued developments of new and improved systems for the delivery of services and the modern drive for the use of ICT in enabling knowledge management for the learning organisation, it is important to determine whether the proportions of the population who are either techno-phobic or techno-stressed has changed since the original work of Rosen and Weil.

In order to evaluate whether there have been any changes as a result of the rapidly growing use of PCs and the internet during the first decade of the 21st century, Aquilina undertook a project during his MSc programme at the University of Derby to measure the levels of techno-phobia and techno-stress amongst a group of Maltese government employees, who use ICT to carry out their daily jobs.

3 Research Methodology

3.1 Research Strategy

The research strategy chosen was a quantitative approach in order to ensure that a relatively wide sample of ICT users could be evaluated, even though it is clear that the research would be measuring perceptions, using the Likert scale which, conveniently results in statistically analysable quantitative data.

3.2 Research Instrument

The questionnaire was divided into 4 parts.

The first part collected demographic data and also a "Yes" or "No" answer to the question of whether the respondents had a Maltese Electronic Identification (e-id) to access E-gov services.

The other three parts were the Technophobia tests created and validated by (Dr Rosen and Dr Weil, 1992). These are:

- questions related to Computer Anxiety named Computer Anxiety Rating Scale (CARS)
- questions related to Computer thoughts named Computer Thoughts Survey (CTS)
- questions related to General Attitudes towards Computers named General Attitudes Towards Computer Scale (GATCS).

3.3 Research Instrument Design

CARS contains 20 likert scale questions scored from 1 to 5 with 1 indicating a response of "not at all" and 5 reflecting a response of "very much". CARS contains items created to assess technological anxiety that includes anxiety about the machines, their role in society, computer programming, use (including consumer use) and issues with computers and technology. Scoring is divided into three parts: Interactive Computer Learning Anxiety, Consumer Technology Anxiety and Observational Computer Learning Anxiety with the higher the score the more computer anxious the subject is.

CTS was modeled after the CARS with a similar 20 likert scale questions that ranks from "not at all" to "very often". CTS reflects how often the respondent had each thought (depending on the question) when working or thinking of technology and assesses directly computer enjoyment. CTS differs from CARS as 11 questions are phrased negatively while the other 9 items are phrased positively. Negative items were reversed-scored to yield a summated score with higher numbers reflecting more positive cognitions and lower numbers reflecting more negative cognitions.

GATCS is also modeled with a 20 likert scale questions that ranges from "Strongly Agree" to "Strongly Disagree". GATCS contains items created to assess attitudes towards computer in education, in employment and in health care. It also assesses attitudes towards computer control, inequity in computer ability, solving societal problems using computer and computers and future jobs. Similar to the CTS, GATCS also was formed with 10 items stated negatively while the others positively. After reverse scoring the 10 negative items, the GATCS higher scores will yield more positive attitudes towards computers and technology while lower scores reflect more negative values.

The scores for each of the three scales can then be combined, according to the (Rosen and Weil, 1992) procedure to give an indication of the levels of TechnoPhobia and TechnoStress felt by the respondents. This results in each dimension or instrument (i.e. CARS, CTS and GATCS) being divided into 3 possible scoring: No Technophobia, Low Technophobia and Moderate/High Technophobia.

Any subject who scores in the Moderate/High Technophobia Group on any one measure is considered to possess moderate or high technophobia. Any subject who scores in the No Technophobia Group on all measures is considered to have no technophobia. Any subject who scores in the Low Technophobia Group on one or more measures, but does not score in the Moderate/High Technophobia Group on any measure is considered to have low technophobia (Rosen and Weil, 1992).

3.3.1 Research Tool Validation

The Questionnaire was developed originally by (Rosen and Weil, 1992) following 14 studies using research best practices with thousands of university students, elementary and secondary school teachers, Business people and second school students. This included a study of 3,392 students in 38 universities from 23 countries (Weil and Rosen, 1995). These instruments have been used by many other authors (Shashaani, 1993), (Heinssen Jr et al., 1987), (North and Noyes, 2002), (McIlroy et al., 2001) validating Dr Rosen and Dr Weil's work.

It was, therefore, appropriate to use these scales to evaluate the situation within the Maltese environment.

3.3.2 Research Procedure and Sample

The survey was mainly conducted using an on-line questionnaire, however, a back-up of using a paper based form was also produced.

The questionnaire was sent to 250 Government of Malta employees, out of which 116 (46.4%) submitted the completed form back. Out of these 116 respondents 96.6% (112) of respondents used the online questionnaire while only 3.4% (4) of respondents used manual forms.

10 people where asked to participate in an interview, but only 4 (40%) accepted such invitation.

Gender	Age	Position Level
Male	54	Operational Staff
Female	30	Senior Management
Female	33	Senior Management
Male	55	Middle Management

It is noticeable that in this self-selected group of respondents three of the four respondents are of management level, which became important in interpreting the interviewee perspectives.

4 Findings

4.1 Demographics

In terms of age ranges, the 116 respondents were relatively evenly spread across the age range from 20 to 60.

Age	Count	Percent
<= 20	1	0.9%
21 to 30	31	27.4%
31 to 40	32	28.3%
41 to 50	21	18.6%
51 to 60	24	21.2%
Over 60	4	3.5%

In terms of gender, 55% of all respondents were female while 45% were male. Out of all respondents 64% are currently registered and own an E-id to access the Maltese e-government website while 36% do not.

4.2 Key Findings

The most important findings are as follows:-

1. 58.6% of respondents suffer from some degree of technophobia.
2. Out of these 58.6% technophobic respondents, 19% suffered from moderate to high technophobia while 39% suffered from Low Technophobia in any of the three dimensions.
3. From all of respondents, 72 (64%) had an E-id and 41 (56.9%) of these E-id holders where technophobic.
4. out of a total of 58.6% respondents, 25% where male while 33% where female.
5. 56% of all male respondents suffered from Technophobia on any of the 3 dimensions
6. 59% of all female respondents suffered from Technophobia on any of the 3 dimensions

4.3 Discussion

Several issues are raised by this analysis, the most significant of which is that the passage of 17 years has done little to change the levels of technophobia and

Technostress, even with the endemic levels of ICT usage in both work and private life.

Indeed, given that the current sample use computers during most of the workday, compared to many of the school teachers in the original work by Rosen and Weil, who often did not need to use computers during their teaching, it might be considered very surprising that the levels have been sustained at such high levels. It is therefore extremely important to reflect on this fact and the implications for the providers of ICT based systems in all areas of life.

Most people are unaware that they are technophobic, or if they are, they are reluctant to admit it (Rosen and Weil 1995) and this does not benefit the workplace as if they do not admit it, they would not try to improve the situation and that can lead to aggravating even further their phobia. This was clearly seen during the interview process where it was noted that all respondents discussed the subject with a firm believe that they were not technophobic yet some were very defensive in their answers and showed signs of uneasiness when confronted with diverse questions. In addition, they were all positive about the benefits to their organisation of ever greater provision of the e-government systems.

Supposing that these levels of technophobia and technostress should be confirmed in the educational sector, the implications for the provision of technology enhanced learning are also very significant in terms of both the speed and the extent of the development of the delivery platforms for learning institutions.

Weil and Rosen (1977) suggests that much can be done by carefully constructed and delivered education and training programmes.

5 Conclusions and Consequences

With technophobia and technostress being perceived to be one of the most challenging issues in Information Systems Research today (Sami and Pangannaiah 2006), Governments and other organisations around the world have a significant task on their hands, as in one way or another technophobia and technostress affects them. With Governments, like other Service Providers implementing a diverse range of technologies to allow citizens/customers to produce and consume services electronically (Gilbert et al., 2003), the last thing that they want to hear about, is a phobia that unfortunately affects use and the adoption of such technologies.

Moreover, **if** a phobia that is affecting a large proportion of people, as for the Maltese Public Sector or students in higher education, can be generalised to the majority, this could explain circumstances such as: slower performance of certain employees while using technologies, low reliance and usage on computers and related technologies to users and students getting anxious, nervous or stressed to the extent of producing less accurate performances with a subsequent increase in levels of computer anxiety and negative attitudes (Rosen and Maguire, 1990, cited in Brosnan and Lee, 1998).

Even worse, is the clear picture Aquilina received from the interviews performed, where some people felt that this was something from the past, that people had overcome technophobia and use computers and related technology regularly and willingly. The research suggests that this perception is a denial of reality.

These results bring to the fore just how much technophobia can be a problem if it is unnoticed. Without the realization of the problem, the problem will never be tackled, let alone resolved.

However, there also remains the question as to the usability of the technology that is being delivered today and its design congruence (or lack thereof) with human thought and behaviour patterns.

Given the rapidly increasing use of knowledge management and all the other technology based systems world-wide, unless both **training**, as advocated by Rosen and Weil, and **design issues**, as advocated by Neilsen, are incorporated into the design, implementation and change management plans, it is very likely that the projects will continue to not deliver any significant benefits to the organisations commissioning them and they will become shelf-ware.

It is clear that far larger survey is necessary, spanning countries and types of organisations and age ranges to fully comprehend the levels and implications of technostress.

References

Bocij, P., Greasley, A., Hickie, S.: Business Information Systems: Technology. Development and Management for the E-Business. Prentice-Hall (2008)

Brosnan, M.: Technophobia: The psychological impact of information technology. Routledge (1998)

Dede, C.: Planning for Neo-millennial Learning Styles. EDUCAUSE Quarterly 1, 7–12 (2005); EU. About ICT Policy Support Programme (2007), http://ec.europa.eu/information_society/activities/ict_psp/about/index_en.htm (accessed February 12, 2012)

Gauld, R., Goldfinch, S., Horsburgh, S.: Do they want it? Do they use it? The 'Demand-Side' of e-Government in Australia and New Zealand. Government Information Quarterly 27(2), 177–186 (2010)

Gilbert, D., Lee-Kelley, L., Barton, M.: Technophobia, gender influences and consumer decision-making for technology-related products. European Journal of Innovation Management 6(4), 253–263 (2003)

Heinssen Jr, R.K., Glass, C., Knight, L.A.: Assessing computer anxiety: Development and validation of the Computer Anxiety Rating Scale. Computers in Human Behavior 3(1), 49–59 (1987)

McIlroy, D., Bunting, B., Tiernet, K., Gordon, M.: The relation of gender and background experience to self-reported computing anxieties and cognitions. Computers in Human Behavior 17(1), 21–33 (2001)

North, A.S., Noyes, J.M.: Gender influences on children's computer attitudes and cognitions. Computers in Human Behavior 18(2), 135–150 (2002)

Rosen, L., Weil, M.: Measuring Technophobia. In: A Manual for the Administration and Scoring of the Computer Anxiety Rating Scale (Form C), the Computer Thoughts Survey (Form C) and the General Attitudes Toward Computers Scale (Form C). California State University (1992)

Rosen, L., Weil, M.: Computer availability, computer experience andtechnophobia among public school teachers. Computers in Human Behavior 11(1), 9–31 (1995)

Sami, L.K., Pangannaiah, N.B.: "Technostress" A literature survey on the effect of information technology on library users. Library Review 55(7), 429–439 (2006)

Self, R., Goh, W.W.: Addressing The Challenges Of Low ICT Literacy In UK University Students, e-Learning Africa, Dakar (2009)

Shashaani, L.: Gender-based differences in attitudes toward computers. Computers & Education 20(2), 169–181 (1993)

Standish Group, The Chaos Report, The Standish Group International Inc. (1994)

Standish Group, Unfinished Voyages, The Standish Group International Inc. (1995)

Standish Group, Chaos: A Recipe for Success, The Standish Group International Inc. (1999)

Standish Group, Chaos Rising, The Standish Group International Inc. (2005)

Standish Group. Chaos Newsroom 2009, The Standish Group International Inc. (2009), http://www1.standishgroup.com/newsroom/chaos_2009.php (accessed February 19, 2012)

Ugwo, L.O., Oyebisi, T.O., Ilori, M.O., Adagunodo, E.R.: Organisational impact of information technology on the banking and insurance sector in Nigeria. Technovation 20(12), 711–721 (2000)

Wang, K., Shu, Q., Tu, Q.: Technostress under different organizational environments: An empirical investigation. Computers in Human Behavior 24(6), 3002–3013 (2008)

Weil, M., Rosen, L.: The psychological impact of technology from a global perspective: A study of technological sophistication and technophobia in university students from twenty-three countries. Computers in Human Behavior 11(1), 95–133 (1995)

Weil, M., Rosen, L.: TechnoStress: Coping with Technology @work @home @play. John Wiley and Sons (1997)

Wilson, R.: On Hold, Channel 4 TV, January 12 (2012), http://www.channel4.com/programmes/richard-wilson-on-hold/4od (accessed February 12, 2012)

Knowledge Elicitation Using Activity Theory and Delphi Technique for Supervision of Projects

Sanath Sukumaran[1], Akmal Rahim[1], and Kanchana Chandran[2]

[1] School of Computing, Taylor's University, Petaling Jaya, Malaysia
sanath.sukumaran@taylors.edu.my, akmalrahim@gmail.com
[2] Bangsar, Kuala Lumpur, Malaysia
kanch168@gmail.com

Abstract. Even though many Knowledge Management initiatives are already in operation, most do not have the right approach when it comes to capturing tacit knowledge. This paper brings to light how social, cultural and organizational paradigms can be infused within a KM initiative to capture tacit and new knowledge. This research paper also brings to light a holistic approach to investigate, analyze, probe and document an activity system for a KM initiative. This case study based investigation uses Delphi techniques to elicit responses from experts from the lens of Activity Theory (AT). It probes into supervision of projects as a case on point to demonstrate how tacit knowledge could be elicited from a team of supervisors.

Keywords: Activity Theory, Delphi Technique, Knowledge Management.

1 Introduction

Great philosophers, leading authors and ancients scriptures dwelling on the subject of knowledge have somewhat appeared to have reached a common consensus. They have all viewed knowledge being linked to (effective) action and application. If such knowledge were to only reside within the realms of human minds and not made explicit, then the whole discipline of Knowledge Management (KM) shall cease to exist. Herein lies the challenge.

Capturing the tacit knowledge from individuals in a way that can be leveraged by the company is perhaps one of the biggest challenges in KM. Arguably; effective knowledge elicitation is a precursor to any KM initiative. Although managers generally have understood the importance of tacit knowledge and KM, most are still skeptical if the KM initiative they are embarking on may bring the desired results especially when investments in technology and resources are involved (Hurley & Green, 2005). Hence, at the heart of the whole discussion still lies the perennial challenge which is to effectively capture tacit and new knowledge.

L. Uden et al. (Eds.): 7th International Conference on KMO, AISC 172, pp. 129–139.
springerlink.com

2 Problem Statement

Implementing effective methods to counteract impediments in knowledge transfer may not always be plausible. For example, it may be too much to expect that contributors describe a knowledge artifact, including the factors that associate the strategy with the original context, and how the strategy should change when applied to different contexts (Szulanski, G., 1996).

Although in theory there are sufficient amount of literature espousing the need for holistic knowledge capture, in actual practice, knowledge capture has not been that holistic (Akhavan et al., 2005). Although many research findings have pointed out social, cultural and organisational issues as impediments towards successful KM implementation, current frameworks are ill-suited towards alleviating these hurdles.

3 Literature Review

Working with leading companies and government organizations, the IBM Institute for Knowledge-Based Organizations has identified a number of important roadblocks that organizations typically face when implementing knowledge management programs. These roadblocks are (Fontain and Lesser, 2002):

- Failure to align KM efforts with the organization's strategic objectives;
- Creation of repositories without addressing the need to manage content and context;
- Failure to understand and connect KM into individuals' daily work activities

Although the aforementioned roadblocks are not exhaustive, they represent issues that can hinder the effectiveness of a KM effort, costing organizations time, money, resources and perhaps, most importantly their ability to affect meaningful business results (Akhavan et al., 2005).

KM approaches may fail when they are designed without input from all stakeholders. This happens when systems analysis and design ignores the community processes and organizational culture (Lee et al., 2002). This is very likely to increase resistance in adoption of the approach and considered a failure factor in the development of any systematic approach.

From the old adage; *Tell me and I'll forget. Show me, and I may not remember. Involve me, and I'll understand.* Much like the old adage, Activity Theory (AT) is a way of examining human practices and also the study of the progression of human activity within a relevant context (Engeström, 1999). It is also a theoretical framework for clarifying human activity and illustrates human point of view, interest and interaction. People are embedded in a physical and socio-cultural context and their behavior cannot be understood outside of it. Human activities are dynamic and if at all, a theory is used to understand and analyse human activities and contexts, it must be capable of the same (dynamic). That being said, AT is a clarifying tool and not a predictive theory but this is sufficient to unravel intricacy of human behavior, cultural sensitivities and organizational context (Kaptelinin et ak., 1995). Using AT one is able to explain why things are done the way it's done, highlight deviations from the norms (human activity) and explicating best practices, all of which would contribute to the body of knowledge. In this context, it is evident that AT has much to offer to the realm of KM.

4 Case Study

4.1 Background

Programming project is a module where a student undertakes a software development project independently for duration of 14 weeks. Students who enroll themselves for this module typically do not attend lectures but engages in weekly consultations with a nominated supervisor. The deliverable of this subject includes a software prototype and documentation. Supervision of projects is inherently different from that of lecturing. A competent lecturer may not necessarily make good supervisors and vice versa. It has been observed that experienced supervisors play a key role in bringing out the best out of their students.

In order to ensure students' achieve his or her full potential, it is of paramount importance that a supervisor must not only embody sufficient technical prowess in the area that he or she is supervising but must possess adequate people-skills. There is also no such thing as 'one size fits all'. An approach which might benefit one student may not necessarily benefit another. Likewise, an approach that benefitted one student in the past may not necessarily benefit him in another context or project. It is also interesting to note that no two supervisors may use the same approach in supervision. Yet both may potentially achieve a good outcome respectively. What is even more challenging is that despite effective supervision, a student may not necessarily do well since 90% of a student's success rests on him and not the supervisor. Each supervisor has his or her own style of supervision. Good and consistent supervision provides a good practice guideline for the students who are undertaking programming projects to ensure consistency of approach for supervisors and students across the school. The aforementioned challenges and complexities often demand the best from supervisors alike. For new and budding supervisors, these demands often weigh them down consequently affecting the quality of project supervision.

The study was conducted at Taylor's University, Malaysia involving 10 experienced undergraduate supervisors but restricted to only lecturers from the School of Computing.

4.2 Aim of the Case Study

The aim of this case study based investigation is to elicit best practices with regards to supervision of programming project. The outcome of the exercise would be a deployment of a KM system although this is beyond the scope of the case study. This research however, shall delve on the gathering of tacit knowledge required for the KM system. It is hoped that the exercise can shed better insights on the intricacies and methods employed by supervisors in differing contexts. Programming project is a subject that requires students to develop a software artifact as a solution to a problem as part of their submission. Students will have to provide a documentation outlining their effort and methods employed. There is no examination for this subject and students are largely on their own with periodic contact with a nominated supervisor.

4.3 Modus Operandi

The Delphi technique is a systematic method of collecting opinions from a group of experts through a series of questionnaires. The interactive approach espoused in Delphi allows KM experts to reconsider their judgments in the light of feedback from peers. The process also gives KM experts more time to think through their ideas before committing themselves to them, leading to a better quality of response. The anonymity of the approach enables experts to express their opinions freely, without institutional loyalties or peer pressure groups getting in the way.

The key functions are identified and should be fulfilled in terms of the supervisory process. The researcher is aware that within the supervisory process, lies a unique and specific "style" that is not standardized and subjective. The "style" belongs to the supervisor's own way of supervising a student (supervisee).

All current knowledge elicitation frameworks have strengths and weaknesses but very few provide useful support for the critical - people-related (or softer) organizational behavioral issues crucial to the successful implementation of any new process or system. In this context, AT would appear to have much to offer, incorporating, as it does, notions of intention, history, mediation, motivation, understanding, communication, culture and context.

The Delphi technique was used as mentioned above involving a 10 senior supervisors (experts) from the School of Computing, Taylor's University. The projects being supervised are limited to ICT projects which are also typically the research area of the supervisor. The Delphi technique involves a set of common questions to elicit responses aimed to devise criteria and elements required for effective supervision on the part of undergraduate supervisors.

4.4 Analysis of the Activity System Using Activity Theory

A significant attempt to operationalize theoretical concepts of AT was undertaken by (Uden, 2001). Uden in her paper used AT as a descriptive tool for understanding what is happening within an activity system and also a practical tool for guiding the requirements elicitation process. The following ten-step model of Activity Theory into requirements elicitation has been adapted from (Uden, 2001).

An activity is transformation process driven by people's needs. The process transforms objects into the outcomes that satisfy those needs. Subject is the active component that carries out the activity is the subject. Tools are element used in transformation. Meanwhile communities are the sets of subjects related directly or indirectly to the same objects. The relationships between subject and communities are determined by rules, for instance laws or social conventions. Where else the relationship between community and objects are determined by the division of labors.

Clarifying the purpose of the activity system (Step 1)
This stage is the most important stage in the process because the main motive and expectations about the outcome is being anticipated by the stakeholders. Based on (Uden, 2001), this step has a dual purpose:

- Firstly, to understand the context within which activities occur. What are the supervisors concerns when it comes to the supervision of the students in programming a project?
- Secondly, to reach a thorough understanding of the motivations for the activity being modeled and any interpretations of perceived contradictions. Each supervisor has his or her own way of supervision and this kind of supervision may be differing and tacit in nature which can either be a problem or a boon.

From the above two purposes, getting an input from the stakeholders of the system is crucial. The stakeholders are the lecturers (supervisors). It is important to ensure that the activity system is meaningful from their input on how to understand the relevant context within the activities. Each supervisor has his or her own set of motives, habits and rules which are not standardized when it comes to supervision. The goal of this activity is to make the supervision more consistent and robust and to highlight the best practices that can be formulated upon the elicitation of tacit knowledge from the supervisors. Inconsistencies where detected has its place in AT provided such inconsistencies (contradictions) are brought to light.

The activities embedded within a typical supervision are as follows:

1. Motivation – supervisor's style of motivating the student to perform better;
2. Consultation / Communication – communication style between the student and the supervisor;
3. Management – administrative duties of the supervisor;
4. Guidance – guiding supervisees on the right way of work or practice in terms of time management and technical skills;

Analyze the activity system (Step 2)
This is where the author starts to define the mechanism of activity in detail. From this case study, the author can analyze the activity system based on specified *subjects, objects, communities, rules, division of labour* and *tools*. The details are listed below:

Activity: The activity of interest in the case study is the supervision process.

Object: The purpose of this activity is to ensure successful supervision of project.

Subjects: Supervisors who are in charge of handling students enrolled for Programming Project.

Tools: Supervisors' approach, email, Blackboard®, documents and guidelines.

Community: Supervisors, Students

Rules: Log sheets, course outline, examiners' remarks

Division of labour: None (Project carried out independently)

Produce an activity system of the application (Step 3)

This step entails analyzing information of the activity system collected from the preceding stage before enabling the researcher to obtain basic knowledge of the case study and to produce the activity system. The purpose of this step is to map the activity system into the Engestrom's model from the viewpoint of the supervisor as shown below:

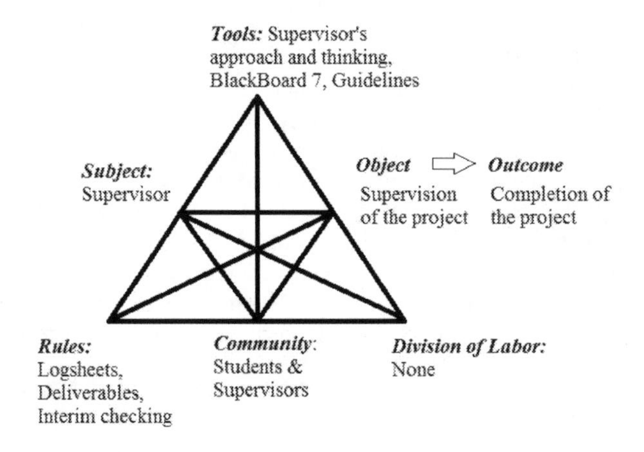

Fig. 1 Activity system from a view point of the supervisor

Analyze the activity structure (Step 4)

The activities can be broken down into operations to analyze and understand the context within which it was set for. As the levels go deeper, the interaction process between the supervisor and the student can be understood more clearly as seen below;

Table 1 Activity structure for discussing topic with student

Activity	Actions	Operations
Discuss topic with student	Choose the student's topic	1. Student choose the lecturer
		2. Listen to the student
		3. Evaluate the topic
		4. Give advice to the student
		5. Approve the student's topic
	Choose the topic of interest	1. Student choose the lecturer
		2. Student inquires about available topic
		3. Give the topic
		4. Mutual agreement with student about topic
		5. Approve the student's topic
	Reject the topic	1. Student choose the lecturer
		2. Listen to the student
		3. Evaluate the topic
		4. Topic inappropriate
		5. Reject the topic

Analyze tools and mediators (Step 5)
From the analysis of the activity structure, the author can obtain information regarding tools and mediators. The subject, object and community do not co-relate directly with each other. The interaction between subject, object and community is mediated by tools. Decomposition of one such activity is shown below:

Discuss topic with student

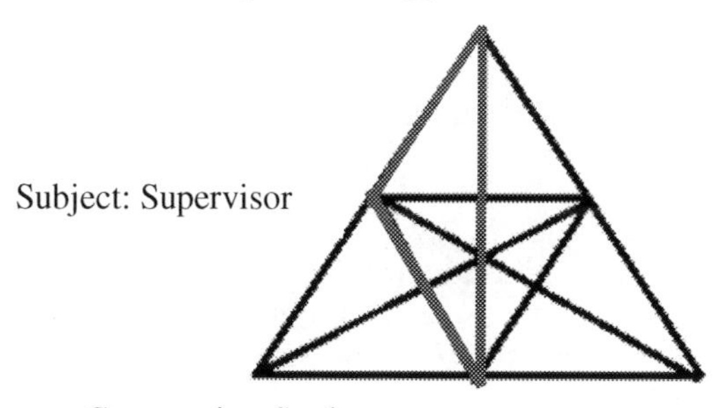

Fig. 2 Activity system for subject-tool-community

Referring to the above figure, a student can discuss a topic with a supervisor using the supervisor's approach or email the supervisor, both of which act as mediators. A supervisor's approach and method (tool of thinking) can be used as a mediator to conduct the activity because each supervisor has his or her method of handling the student. Besides that, the usage of email is also possible in case the supervisor or the student is not free. The tool of thinking is basically a tacit knowledge that is concealed within the supervisor's approach of supervision.

Decompose the situation's activity system (Step 6)
Activity notation can be used to assist the development of narrowing down the state of the larger activity into smaller manageable units. One such decomposition of an activity is shown below:

Table 2 One of the activities that being decompose into an activity system

Actors(Doers)	Mediators	Objective(Purpose)
Subject: Supervisor	Tools: Supervisor method, email (directive tools)	Approving the topic

Generate questions for each activity notation (Step 7)
Questions that are specific to a particular combination within the activity notation and sub-activity triangle are generated. The questions can be general or specific to a particular situation (Uden, 2001). Getting feedback from the experts is crucial. The challenge here is handling the students consistently and effectively. In order to achieve this, tacit knowledge from the experts (supervisors) must be acquired. Capturing the tacit knowledge from the experts can be done via the Delphi method. The Delphi Technique was originally conceived as a way to obtain the opinion of experts without necessarily bringing them together face to face (Stuter, 2003).

In this step, a series of questions will be asked to the experts. Delphi is an expert survey in two or more 'rounds' in which in the second and later rounds of the survey the results of the previous round are given as feedback. Therefore, the experts response from the second round can influence the response of other colleagues' opinions (Cuhls, 2003). The process of Delphi mindfully uses the AT notion in obtaining human related information allowing social-organization processes to be accounted.

Conduct detailed investigation (Step 8)
This step is to conduct detailed investigation based on the questions and interview that have been generated in the previous stage. The main goal here is to discuss and bring to light conflicting issues within the activity notation. Some of which are listed below;

- How does other automated or non automated tools help create better supervision (subject –tool – object)
- How does interim deliverables schedule can help supervisor to supervise the student better (subject –rule – object)

- How do can help the student to perform better (community-rule-object)
- How does student response to the methods of supervising from their supervisor? (community-tool-object)

Analyze the context (Step 9)
Based from the questions posed to the panel of experts, the context can be discovered by analyzing the answers provided after several rounds of Delphi interviews. Several contexts were unearthed from this exercise and each analyzed as follows:

- Supervisors' ways of discussing the topic of the project with the student
- Encouragement regarding new idea about the project
- Estimating student's ability
- Strategies to aid students in avoiding delay progress
- Usage of tools among supervisor
- Overall approach in supervision by the supervisors

Discussion on each of the above could not be highlighted here due to the elaborate nature of discussion.

Analyze activity system dynamics (Step 10)
In this final stage, all the data that has been gathered and analysed and interpreted in terms of AT's view on contradictions. Engeström (1987) emphasized the importance of contradictions in understanding how an activity system works. In this case study of supervision of Programming Project, each supervisor has his or her own view and this personal view will eventually lead to a contradiction of opinions. By identifying the tensions and interactions between the elements of an activity system, it is possible to reconstruct the system in its concrete diversity and richness, and therefore explain and foresee its development (Engeström, 1999).

Several types of contradictions are identified from the case study which forms the basis of our knowledge elicitation exercise as stated below:

- Rule node: Log sheets

Log sheets serve as mechanism for the students to document the progress of their project. The usages of log sheets are effective if students are honest and discipline. The supervisor evaluate the logs and authorizes next course of action.

- Rule Node: Deliverables

Interim progress checking is effective to guide students in avoiding delayed progress. Even though most of the experts agree with this method, to perform this method practically requires a lot of time and effort from the supervisor. Interim checking can be done weekly or once in two weeks depending on the supervisor. The supervisor might be engaged in other academic activities leaving him or her unable to perform interim progress checking.

- Tool node: Thinking tools (Approach)

Most supervisors prefer to have casual supervision which refers to providing greater freedom to supervisees to carry out their course of research independently without exercising excessive demands on them. This is true since most supervisors generally supervise students who are academically above than average, i.e. discipline and self-motivated. It was seen from the interviews that supervisor's approach may change if they tend to supervise students that are not motivated and academically weak. In such cases, supervisors generally employ a more regimental approach in supervision instead of a casual one mentioned earlier.

- Subject node: Supervisor way of evaluating student's capabilities.

Some supervisors believe that attitude and the willingness of the student is an important aspect to ensure that students can embark on a project. Other supervisors feel that a student's ability to undertake a project depends on one's academic performance. These two perceived opinions contradict but forms an important tacit input to the way a supervisor engages with his supervisees.

5 Conclusion

The increasing importance of knowledge for organizational success is hardly questionable nowadays resulting in significant attention directed towards elicitation of knowledge. KM frameworks should be realigned from a best practice to a best fit approach. This means knowledge elicitation framework should first provide a mechanism to investigate and understand an organization's context, and then, based on the context, provide insight into the most suitable and holistic approach to KM. The investigative method employed in this case study through the lens of Activity Theory (AT) exemplifies that in order to harvest tacit knowledge, KM practitioners must leverage on situations and context which may give rise to different meaning and consequently new knowledge. By using both AT and Delphi technique, deep analysis was performed, situations and contexts probed into which was otherwise not possible.

The main challenge the researchers faced was that while the study was only restricted to project supervision within ICT domain, large number disciplines within ICT has rendered some questions relatively insignificant when applied in some disciplines and vice versa. This issue was palpable in the study since each supervisor was involved in projects with distinct research areas from the other which proved to be an important determinant influencing the nature of supervision. This has resulted in a criterion or element deemed important for one supervisor appears insignificant to another. This has led to the limitation of the research because while the knowledge elicited using Delphi technique has shed interesting and deep insights into the nature of project supervision from experts, one cannot still generalise the findings with regards to criteria and elements required for projects supervision due to the reasons mentioned above.

On a positive note, the authors have shown that it is possible to elicit knowledge seamlessly taking into account social, cultural and organizational paradigms and its dynamics through Engeström's AT model of activity system. The ability of AT to find contradictions amongst opinions and thoughts is truly remarkable providing authors with valuable insights and new knowledge. A combination of AT and Delphi technique fused in a KM initiative as shown in this paper has paved the way and laid the foundation for the construction of a new knowledge elicitation framework which is potentially more holistic and grounded.

6 Further Research

Further research is needed to develop a full-fledged knowledge elicitation framework building upon the tenets discussed in this paper. Subsequently, researchers could further twig the framework so that it can be generalized with carefully designed plug-points to enable seamless knowledge elicitation practices across domains. Conversely, researchers can also consider applying the same steps as outlined in this case study to other KM initiatives.

References

Akhavan, P., Jafari, M., Fathian, M.: Exploring Failure-Factors of Implementing Knowledge Management Systems in Organizations. Journal of Knowledge Management Practice (May 2005)

Baskerville, R., Dulipovici, A.: The theoretical foundations of knowledge management. Department of Computer Information Systems, Georgia State University, Atlanta (2006)

Benbya, H.: Knowledge Management Systems Implementation: Lessons from the Silicon Valley. Chandos Publishing, Oxford (2008)

Davenport, T.H., Prusak, L.: Working Knowledge: How Organizations Manage what they Know. Harvard Business School Press, Boston (1998)

Engeström, Y.: Learning by Expanding: An Activity Theoretical Approach to Developmental Research. Orienta-Konsultit, Helsinki (1987)

Engeström, Y.: Activity theory and individual and social transformation. In: Engeström, Y., Miettinen, R., Punamaki, R. (eds.) Perspectives on Activity Theory. Cambridge University Press, Cambridge (1999)

Fontain, M., Lesser, E.: Challenges In Managing Organizational Knowledge. IBM Institute For Knowledge-Based Organization Publication (2002)

Hurley, T.A., Green, C.W.: Knowledge Management and the Nonprofit Industry: A Within and Between Approach. Journal of Knowledge Management Practice (2005)

Kaptelinin, V., Kuutti, K., Bannon, L.: Activity Theory: Basic Concepts and Applications. In: Blumenthal, B., Gornostaev, J., Unger, C. (eds.) EWHCI 1995. LNCS, vol. 1015, Springer, Heidelberg (1995)

Lee, C.C., Egbu, C., Boyd, D., Xiao, H., Chinyo, E.: Knowledge Management for Small Medium Enterprise: Capturing and Communicating Learning and Experiences. Knowledge Management (2002)

Rodriguez, H.: Activity theory and Cognitive Sciences. Perspectives on Activity Theory (1998), http://www.nada.kth.se/~henrry/papers/ActivityTheory.html (retrieved)

Stuter, L.M.: The Delphi technique as a forecasting tool: issues and analysis. International Journal of Forecasting (1996), doi:10.1016/S0169-2070(99)00018-7

Uden, L.: Activity Theory for Requirements Engineering. In: Argentina Symposium on Software Engineering (ASSE), September 10-11. Buenos Aires. University of Buenos Aires, Argentina (2001)

Uden, L., Sukumaran, S., Chandran, K.: Requirements Elicitation for Knowledge Management Systems using Activity Theory (2011)

Customer Knowledge in Value Creation for Software Engineering Process

Anne-Maria Aho[1] and Lorna Uden[2]

[1] Seinäjoki University of Applied Sciences, Seinäjoki, Finland
anne-maria.aho@seamk.fi
[2] Staffordshire University, Stafford, United Kingdom
l.uden@staffs.ac.uk

Abstract. The aim of this paper is to explain how we can achieve and integrate customer knowledge into the software engineering process in multi-disciplined product development, through developing dynamic capabilities. According to service dominant logic, knowledge assets are key drivers in creating competitive advantage for organizations. This research will provide significant new information about the role of customer knowledge in value creation for the software engineering process in the machinery industry. As the result of the research, a model for acquisition and use of customer knowledge in value creation of the software engineering process is created and guidelines produced.

Keywords: customer knowledge, dynamic capabilities, software engineering.

1 Introduction

The design and development of embedded software for machinery engineering is not trivial. Firstly, the role of software engineering in the machinery industry is changing. Software products are increasingly becoming more important. Traditionally, the product's competitive edge in terms of mechanical engineering is based on the characteristics of the physical product. Today, it is increasingly em-bedded in software-based products. Software production in a machinery company has a strategic role in the company's competitive advantage. It has shifted from a complementary role to the focus of the product development in machinery companies.

Secondly, according to machinery industry reports [15], the following drivers of change are influencing the strategic position and value creation in the machinery industry: changing customer needs, changing competitive position, including price competition and the requirements of focusing, and technological changes and environmental issues. Today, customers require detailed and comprehensive solutions. The production of customized products requires accurate knowledge about (and from) the customer. This is especially important in the value creation for software engineering. We have developed a value creation model in software engineering for a multi-national machinery company [1, 2].

L. Uden et al. (Eds.): 7th International Conference on KMO, AISC 172, pp. 141–152.
springerlink.com

The methodology involves a case study using action research. In data collection, several methods were used: interviews, workshops, modeling of information flows and processes, inquiries and observation. The data was collected over a seven-year period in a machinery company using longitudinal empirical research.

Our study is based on a multi-national group of machinery companies. The company group operates several manufacturing plants in Finland, one in Europe and sales subsidiaries in Europe and Northern America. The aim of the software engineering process is to implement common continuous product development and customer-oriented product development projects. The most typical product sold to customers is the standard product complemented with customer-specific features. This leads to the demand for new standards and exact customer knowledge. The use of customer knowledge in the value creation of the software engineering process requires acquisition and integration of different knowledge assets in value creation through dynamic capabilities. Dynamic capabilities refer to the firm's willingness to re-develop processes and routines in order to develop and renew the resources of the organization [18, 26, 27, 28].

From our research, we have identified problems with customer knowledge in the different phases of software engineering process: requirements, specifications, design, re-design and testing. The requirement phase is informal and very often critical customer knowledge is missing. Inadequate knowledge produces uncertainty for both customer and salesman. The specifications are often unclear, resulting from inappropriate knowledge about features required by customers. Better definitions of interfaces are required. The boundary between the phases of design and re-design influences the quality of the software engineering process in multidiscipline product development processes. There are problems with communication between mechanical and electrical engineering processes. Typically, the communication is unidirectional and few. There is the risk that important customer knowledge does not reach the software engineers. Because proper software testing functions are missing, software bugs often appear during physical machine testing at the factory or with customers. There are no inspection methods or metrics in use. Customer information, for instance customer claims, are also not properly documented. It would seem that currently the use of customer knowledge in software development is insufficient.

This research seeks to answer how we can improve value creation of the software engineering process and what kind of development actions are needed in order to create and integrate customer knowledge in multidisciplinary product development in the machinery industry.

This paper begins with a brief review of the value creation model. This is followed by brief reviews of the customer knowledge and dynamic capabilities. Subsequent sections discuss how to use dynamic capabilities in order to capture customer knowledge in value creation in software engineering process. This is followed by a brief evaluation of the model of the integration of customer knowledge in software engineering. The paper concludes with suggestions for further research.

2 Value Creation Model

Based on the action research cycles during seven years the model of value creation has been developed (See Figure 1). The model describes the value creation process through dynamic capabilities from balanced scorecard perspectives. It is divided into three dimensions as follows: core, three-level layers and environmental perspectives.

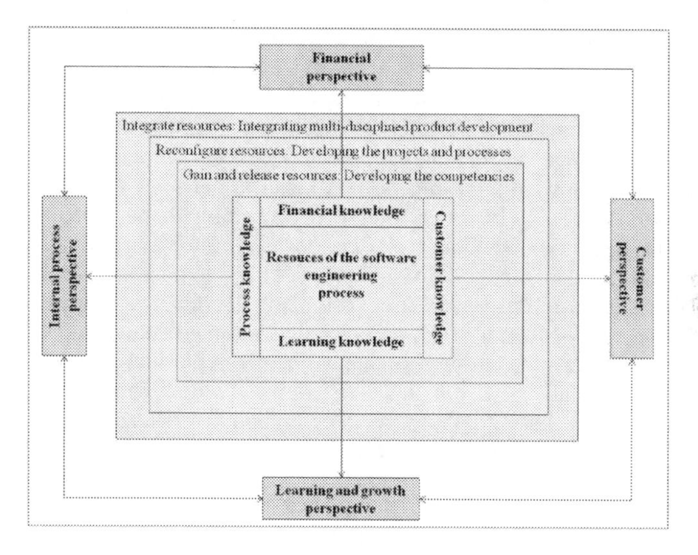

Fig. 1 The model of value creation in software engineering

Firstly, the core of the model consists of the resources of the software engineering process. In order to confront the requirements of the changing environment, the competences of different perspectives are required. Secondly, the layers around the core describe the dynamic capabilities, which are needed to sustain competitive advantage. The development of dynamic capabilities in this model is mainly based on the information flows [11, 18].

In order to create and sustain competitive advantage, the software engineering process must renew and develop its organizational capabilities. The inner layer of the model consists of the gain and release of resources, which are needed to confront the demands of a changing environment. It is important that the resources of the software engineering process are assessed and new resources are acquired and other resources are released continuously. The middle layer of the model describes reconfiguration of resources, which is based on the need to renew and develop the process and working practices all the time.

The outer layer illustrates the integration of resources. The integration comprises the integration between processes, i.e. inside and outside the software engineering process, between the other assets of the firm. Thirdly, the environmental perspectives are described as follows: customer, financial, internal process and learning and growth.

3 Customer Knowledge

Knowledge is a key competitive factor in the global economy. It is now generally accepted that the customer is a key strategic resource in any company's success. It is important to manage customer knowledge (CK), in order to support the generation of better new products and services to secure customer loyalty and to obtain the synergy of collaboration between companies and customers [12].

According to Sofianti et al. [24] Customer Knowledge Management (CKM) is the strategic process by which cutting edge companies emancipate their customers from passive recipients of products and services, to empowerment as knowledge partners. CKM is about gaining, sharing, and expanding the knowledge residing in customers, to both customer and corporate benefit.

CKM is described as an ongoing process of generating, disseminating and using customer knowledge within an organization and between an organization and its customers [24]. In order for the organization to be more efficient and effective in delivering products or services to customers, and thus creating customer de-light, knowledge of customers will have to be managed to ensure that the services organizations provide are those that will address customer needs [21].

According to Al-Shammari [3] successful Customer Knowledge Management (CKM) requires transformation of organizations from product–centric to customer centric, from vertical to network structure, from individualistic to collective work, from hoarding to a sharing culture, from ceremonial to results-oriented practices, from functional to process work orientation, and from centralized to distributed computing. CKM should no longer be a tool like traditional Customer Relationship Management, (CRM), but be a strategic process that captures, creates and integrates customers dynamically [23]. It is important to identify what types of capabilities might facilitate CKM activities. Customer integration is regarded as a vital method to improve innovation performance.

Customer knowledge management (CKM) is a combination of knowledge management (KM) and, customer relationship management (CRM) principles. [9] The objectives of CKM is gaining, sharing and expanding knowledge of (inside) the customer. This implies individual or group experiences in applications, competitor´s behaviour, possible solutions, etc. CRM aims to achieve knowledge about the customer. This means the use of the customer's database. In KM the perspective is internal and it is focusing on sharing knowledge about customers among employees. In the view of business objectives KM is aiming to add efficiency and speed, CRM is intending to nurture the customer base and maintain customers. The objectives of CKM are to collaborate with customers and co-create value. In order to implement CKM we must create favourable conditions to develop it. That means understanding of customer value should be integrated in the software engineering process. In this paper we have produced a model based on dynamic capabilities.

4 Dynamic Capabilities

The dynamic capability perspective is used to develop CKM in the software engineering process. The dynamic capabilities approach is based on the work of

Teece *et al.* [25, 6]. According to Teece *et al.* [27], dynamic capabilities are, "The firm's ability to integrate, build, and reconfigure internal and external competences to address rapidly changing environments". Dynamic capabilities refer to the firm's willingness to re-develop processes and routines in order to develop and renew the resources of the organization [18, 26, 27, 28]. In dynamic markets this ability becomes increasingly important. It is insufficient to have strong resources and organizational capabilities, with continuous developing and renewing of its resources and organizational capabilities also required. [27, 28]. According to Eisenhardt and Martin [14] dynamic capabilities are "The firm´s processes that use resources – specifically the processes to integrate, reconfigure, gain and release resources – to match and even create market change. Dynamic capabilities thus are the organisational and strategic routines by which firms achieve new resource confirmations as markets emerge, collide, split, evolve and die."

Cepeda and Vera [11] studied the knowledge management processes behind the development and utilization of dynamic capabilities and their impact on operational capabilities. In contrast, Nielsen [18] described dynamic capabilities through knowledge management in his research. According to him, dynamic capabilities can be seen to be composed of concrete and well-known knowledge management activities. Both Cepeda and Vera [11] and Nielsen [18] integrated the dynamic capabilities approach and knowledge management approach.

Anand et al. [5] presented a framework of continuous improvement infrastructure, which is derived from the dynamic capabilities perspective and its underlying Zollo´s and Winter´s [29] theory of organisational learning. They studied the content of continuous improvement strategies and identified infrastructure decision areas that are essential for continuous improvement initiatives.

Barrales-Molina et al. [7] empirically studied the influence of managerial perceptions of the environment in the creation of dynamic capabilities. In their paper, they identified three dimensions of competitive environment: dynamism, complexity, and munificence. Moreover, they used Zollo´s and Winter´s [29] theoretical model in order to explain the creation of dynamic capabilities. According to Barrales-Molina et al. [7] the managerial perception of the environment has a significant role in the processes of dynamic capabilities.

Drnevichi´s and Kriauciunas´ [13] contribution focuses on the conditions under which ordinary and dynamic capabilities contribute to higher relative firm performance. They brought out the positive and negative contributions of capabilities to relative firm performance. Further, they examined the effects of environmental dynamism and the degree of capability heterogeneity. They found that environmental dynamism had negative effects on the contribution of ordinary capabilities and a positive effect on the contribution of dynamic capabilities to relative firm performance. According to Drnevichi and Kriauciunas [13], the direct effects of capabilities are stronger with a process-level performance measure, while environmental dynamism and heterogeneity have a stronger influence with firm-level measures.

Khavul et al. [13] concentrated on the development of customer-focused dynamic capabilities, which are needed because the globalization of innovation calls for entrepreneurial new ventures from emerging economies. Shamsie et al.

[22] approached dynamic capabilities in a framework of two complementary processes. Firms can develop existing capabilities in products and markets in which they have experienced recent success, or they can also purposefully focus on other products and markets in which they are going to build capabilities. Shamsie et al. [22] studied these two processes within project-based industries. Based on the study, they identified replication and renewal as two types of strategies that firms adopt in order to add a dynamic component to their capabilities.

Barreto [8] concludes based on past research of dynamic capabilities and proposes a new definition of dynamic capabilities that accommodates old and new suggestions within the field and also attempts to overcome some of their limitations, as follows: "A dynamic capability is the firm's potential to systematically solve problems, formed by its propensity to sense opportunities and threats, to make timely and market-oriented decisions, and to change its resource base." [8]

Dynamic capabilities emphasize management capabilities and inimitable combinations of resources that cut across all functions, including R&D, product and process development, manufacturing, human resources and organizational learning. They focus on the firm's ability to face rapidly changing environments, in order to create and re-new resources, and change the resources mix [4, 10, 27]. Dynamic capabilities are important because knowledge flows from one capability to another, through the reconfiguration of organizational capabilities, leading to new knowledge that enables the firm to create superior customer value.

According to Landroguez et al. [17], a firm's external and internal organizational capabilities are of vital importance for increasing the value created for the customer. Because of increased customer importance, a firm should focus on improving those capabilities which view the customer as its key component, in order to maximize the value created for them. The authors proposed three capabilities that they believe are mainly related to customers: 'market orientation' (MO), 'knowledge management' (KM) and 'customer relationship management' (CRM). Landroguez et al. [17] suggested that combination and integration of those external and internal organizational capabilities (MO, KM and CRM) in the form of a dynamic capability could be used in the creation of customer value and in management in general.

5 CKM in Software Engineering for Machinery Engineering

This section describes a case study involving the development of CKM for an international machinery company in Finland. The context of this study is presented in Figure 2. The creation and sustaining of competitive advantage require customer knowledge asset as a key driver. The borders between the software engineering process and the customer impose certain barriers in acquisition and integration process of customer knowledge. In order to solve this challenge the development process of dynamic capabilities is applied in this study.

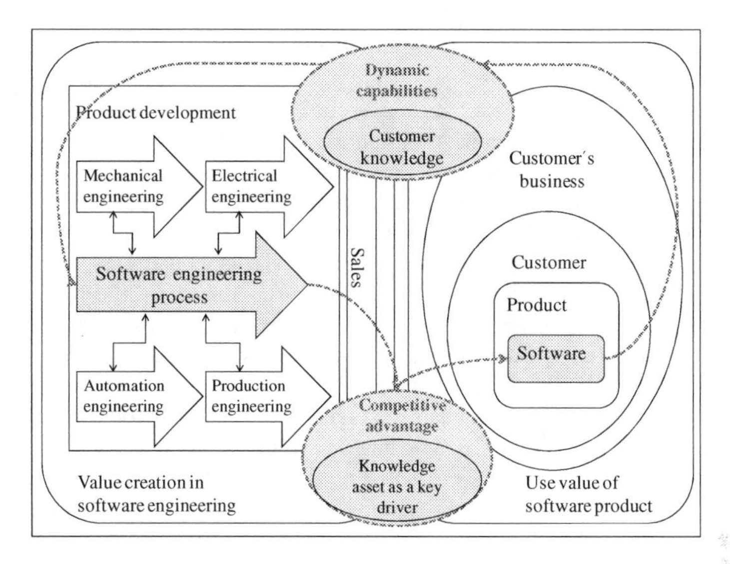

Fig. 2 The context of the research

The study gives a deep understanding of the role of customer knowledge in value creation of software engineering process, which will be achieved by analysis of information flows and processes in multidisciplinary product development. In this case study the acquisition and use of customer knowledge in value creation of the software engineering process is created, implemented and evaluated. In order to start the CKM process, customer understanding was developed by developing dynamic capabilities. The development process of dynamic capabilities necessitates development methods which are easily integrated in common routines in order to create commitment to the new value creation.

The developed model aims to extend the perspective of software engineering in order to move from understanding of one's own process towards understanding of the customer's business operations. The customer has an essential role in the value creation as in Bowman and Ambrosini´s [10], Norman´s [19] and O´Cass and Ngo´s [20] models.

In order to create and sustain competitive advantage, the software engineering process must renew and develop its organizational capabilities. The role of customer knowledge is described according to Eisenhard and Martin´s [14] definition of dynamic capabilities is used as a basis for the classification: integrate, reconfigure, gain and release resources (Cf. Figure 1).

Regarding customer knowledge, the following integration activities were developed: sales policy, project management, internal communication, and development of testing process and release policy of software product. The reconfiguration of resources is considered in the form of the development of project management facilities and project work to acquire and integrate customer knowledge. The reconfiguration of resources is based on the need to renew and develop the process and working practices continuously.

The gaining and releasing consist of the development of the customer knowledge. It can be stated that the resources of the software engineering process should be assessed and new resources acquired and other resources released continuously. While there is a need to gain new kinds of knowledge, the attitudes toward the new knowledge areas are important. Based on this study, it can be seen that creating possibilities to participate and express personal opinions are key elements in successful development processes. So-called learning cases were planned in order to gain customer knowledge. The customer knowledge of the software engineering process was developed by conducting three business plan projects as learning cases including the following themes: product description, markets and customer needs, sales and marketing, product development, production process and business concept.

After completing the single business plans, the workshop for the staff of the whole department was organized. The business plans are used as training material. In the workshop the plans were presented by the person who created the plan and after that the plans were discussed with the whole department. The aim was to disseminate knowledge throughout the staff. In the same context, the success of the process was evaluated by group discussion and personal interviews. The evaluation focused on the awareness of customer needs and benefits as well as the awareness of cost and risks. In addition to these, the process itself and consequent personal learning were evaluated.

The development of dynamic capabilities requires participants´ commitment to the development process, which requires customized developing methods and the use of small steps over a long time period. A positive organizational climate supports the development process. We can compare the created model to CRM, MO and KM as presented in Table 1.

Table 1 The comparison of the created model to CRM, MO and KM

	CKM by DC	CRM	MO	KM
Aim	Responding to changing customer needs	Identifying of customer needs and satisfaction	Satisfying customers by being better than competitors	Achieving and using customer knowledge
Process	Mainly internal	External	External	Internal
Focus	Employees and customers	Customers	Market	Employees

6 Evaluation

The created model is implemented during seven-year research period and it is currently in use. The evaluation of the CKM model is accomplished according to classification of dynamic capabilities, as follows: 1) integration of resources, 2) reconfiguration of resources, and 3) gain and release of the resources.

Firstly, by resource integration customer knowledge is integrated into the software engineering process. Through these development activities the quality of software requirements and design improved. The process development enables software engineers to gain, share and expand the customer knowledge. By using these development activities software engineers have constant interaction with customers providing information flows within product development and sales. In implementing process improvements the method of small steps was used in order to manage the change and create commitment to new practices.

Secondly, with the resource reconfiguration we can ensure the functionality of the project work and its ability to include customer knowledge and react to changes in customer needs. In this model CKM is seen as an ongoing process of using customer knowledge within software projects and between the organization and its customers. Project leaders are responsible for the reconfiguration of project resources. By using customized project development methods, commitment was created among software engineers.

Thirdly, the gaining and releasing of resources focuses on the development of software engineers´ customer knowledge. The aim was to add understanding about customer value by using business plans as learning methods. The tools developed for business plans are currently in use and the planned learning objectives were realized. Again, the choice of development actions has great importance, because the goal is to develop customer knowledge, which has traditionally not been a natural part of the engineering knowledge. Software engineers´ commitment and motivation played a pivotal role in this development phase.

From pure evaluation, we can argue that the created model is effective and useful. The company is satisfied with the results obtained using the model. The developed model is useful for companies where software engineering is part of the multi-disciplinary product development.

The model provides guidelines for the development process of CKM in software engineering in the multidisciplinary product development process. It gives us a model for the development of CKM by using dynamic capabilities. In addition it also gives us the know how to sustain competitive advantage, which requires continuous developing and learning.

7 Conclusion

Customers play a crucial role in knowledge-creation. The recognition of customer knowledge as a key resource for firms in the current business environment confirms the need for processes that facilitate individual and collective knowledge creation, transfer and leverage. Value co-creation is associated with the opportunity to gain competitive advantage by developing unique competences, together with the appropriate organizational resources and technological capabilities, aiming at better satisfying customers' demands for personalized products, services and experiences. Knowledge and value co-creation with customers – but also with suppliers and other business partners - is becoming important for organizations.

Simply possessing valuable and rare resources and capabilities do not guarantee the development of competitive advantage or the creation of value; firms must be capable of managing them effectively. Value can be created by recombining existing resources and capabilities. It should be possible to reconfigure organizational capabilities to enable the firm to be continually creating value, and this is where dynamic capabilities come into play. Because of increased customer importance, a firm's external and internal organizational capabilities are of vital importance for increasing the value created for the customer. A firm should focus on improving those capabilities which view the customer as its key component, in order to maximize the value created for them

In order to achieve and sustain competitive advantage, software engineering requires value creation, which responds to current and future environmental requirements. Understanding of customer value and taking it into account in software engineering are important. The development of customer knowledge in order to confront changing environments helps to create and sustain competitive advantage. As the role of software engineering in industry changes towards being a creator of competitive advantage, it is increasingly important to consider the value creation from strategic perspectives. Value creation must be such that it creates and maintains continuous development and learning, enabling the use and deployment of a company's resources. Dynamic capabilities play a pivotal role in value creation in software engineering, where continuous integration, active reconfiguration, gain and releasing of resources are required.

In future research, the customer perspective may have a more important role in the embedded software engineering process. It would be interesting to examine the methods by which customer understanding is achieved. One option is to use social media.

Acknowledgments. This paper is a product of the FUTIS (Future of Industrial Services) research project. The financial support of the Finnish Funding Agency for Technology and Innovation (Tekes), FIMECC and the companies involved is gratefully acknowledged.

References

1. Aho, A.-M.: Value creation of software engineering in the machinery industry – A case study approach. University of Vaasa. Department of Production. Doctoral thesis (2011)
2. Aho, A.-M., Uden, L.: Value creation for software engineering in product development. Building Capabilities for Sustainable Global Business: Balancing Corporate Success & Social Good. In: The 12th International Conference of the Society for Global Business and Economic Development (SGBED), Singapore (July 2011)
3. Al-Shammari, M.: Customer Knowledge Management: People, Processes and Technology. Information Science Reference, Hershey (2009)
4. Ambrosini, V., Bowman, C.: What are dynamic capabilities and are they a useful construct in strategic management? International Journal of Management Reviews 11(1), 29–49 (2009)
5. Anand, G., Ward, P.T., Tatikonda, M.V., Schilling, D.A.: Dynamic capabilities through continuous improvement infrastructure. Journal of Operations Management 27, 444–461 (2009)

6. Augier, M., Teece, D.J.: Dynamic capabilities and multinational enterprise: Penrosean insights and omissions. Management International Review 47(2), 175–192 (2007)
7. Barrales-Molina, V., Benitez-Amado, J., Perez-Arostegui, M.N.: Managerial perceptions of the competitive environment and dynamic capabilities generation. Industrial Management & Data Systems 110(9), 1355–1384 (2010)
8. Barreto, I.: Dynamic capabilities: A review of past research and an agenda for the future. Journal of Management 36(1), 256–280 (2010)
9. Belkahla, W., Triki, A.: Customer knowledge enabled innovation capability: proposing a measurement scale. Journal of Knowledge Management 18(4), 649–674 (2011)
10. Bowman, C., Ambrosini, V.: Value Creation Versus Value Capture: Towards a Coherent Definition of Value in Strategy. British Journal of Management 11(1), 1–15 (2000)
11. Cepeda, G., Vera, D.: Dynamic capabilities and operational capabilities: A knowledge management perspective. Journal of Business Research 60(5), 426–437 (2007)
12. Chen, Y.-H., Su, C.-T.: A Kano-CKM Model for Customer Knowledge Discovery. Total Quality Management & Business Excellence 17(5), 589–608 (2006)
13. Drnevichi, P., Kriauciunas, A.P.: Clarifying the conditions and limits of the contributions of ordinary and dynamic capabilities to relative firm performance. Strategic Management Journal 32, 254–279 (2011)
14. Eisenhardt, K.M., Martin, J.A.: Dynamic capabilities: What are they? Strategic Management Journal 21(10–11), 1105–1121 (2000)
15. Jääskeläinen, E.: Kone- ja laiteteollisuus. Toimialaraportti 7. Työ- ja elinkeinoministeriö, Helsinki (2009)
16. Khavul, S., Peterson, M., Mullens, D., Rasheed, A.A.: Going Global with Innovations from Emerging Economies: Investment in Customer Support Capabilities Pays Off. Journal of International Marketing 18(4), 22–42 (2010)
17. Landroguez, S.M., Castro, C.B., Cepeda-Carrión, G.: Creating dynamic capabilities to increase customer value. Management Decision 49(7), 1141–1159 (2011)
18. Nielsen, A.P.: Understanding dynamic capabilities through knowledge management. Journal of Knowledge Management 10(4), 59–71 (2006)
19. Normann, R.: Reframing Business: When the Map Changes the Landscape. John Wiley & Sons, New York (2001)
20. O'Cass, A., Ngo, L.V.: Examining the firm's value creation process: A managerial perspective of the firm's value offering strategy and performance. British Journal of Management (2010), doi:10.1111/j.1467-8551.2010.00694.x (cited on December 12, 2011)
21. Plessis, M., Boon, J.A.: Knowledge management in eBusiness and customer relationship management: South African case study findings. International Journal of Information Management 24, 73–86 (2004)
22. Shamsie, J., Martin, X., Miller, D.: In with the old, in with the new: Capabilities, strategies, and performance among the Hollywood studios. Strategic Management Journal 30(13), 1440–1452 (2009)
23. Smith, H.A., McKeen, J.D.: Developments in Practice XVIII – Customer Knowledge Management: Adding value for our customers. Communications of the Association for Information Systems 16, 744–755 (2005)

24. Sofianti, T.D., Suryadi, K., Govindaraju, R., Prihartono, B.: Customer Knowledge Co-creation Process in New Product Development. In: Proceedings of the World Congress on Engineering 2010, WCE 2010, London, U.K., June 30-July 2, vol. 1 (2010) ISBN: 978-988-17012-9-9. ISSN: 2078-0958 (Print); ISSN: 2078-0966 (Online)
25. Teece, D.J., Pisano, G., Shuen, A.: Firm Capabilities, Re-sources and the Concept of Strategy. In: Economic Analysis and Policy. Working Paper EAP38, University of California, Berkeley (1990)
26. Teece, D.J., Pisano, G.: The dynamic capabilities of firms: an introduction. Industrial and Corporate Change 3(3), 537–556 (1994)
27. Teece, D.J., Pisano, G., Shuen, A.: Dynamic capabilities and strategic management. Strategic Management Journal 18(7), 509–533 (1997)
28. Wheeler, B.C.: NEBIC: a dynamic capabilities theory for assessing net-enablement. Information Systems Research 13(2), 125–146 (2002)
29. Zollo, M., Winter, S.: Deliberate learning and the evolution of dynamic capabilities. Organisation Science 13(3), 339–351 (2002)

The Influence of System Interface, Training Content and It Trust on ERP Learning Utilization: A Research Proposal

Chris N. Arasanmi, William Y.C. Wang, and Harminder Singh

Auckland University of Technology

Abstract. This paper examines the factors that affect the use of enterprise resource planning (ERP) systems in an e-commerce environment after they have been adopted. The model presented here extends the technology acceptance model (TAM) to examine how IT trust, training content, and the quality of the system interface influence the extent to which knowledge about using an ERP system is applied after it has been learnt during a training program. The model extends research on ERP utilization by proposing a set of variables which have not been explored in depth in studies on the use of ERP systems, especially in the context of e- commerce.

Keywords: IT trust, system interface, system utilization, training content.

1 Introduction

The advent of information systems has changed the global business environment. Companies today face the challenges of higher competition, global markets and rising customer expectations [18]. To remain competitive, as well to maintain their market share, organizations must increase their capabilities and competencies to communicate accurate information [18]. Firms facing "do or die" situations that require a radical transformation of their business processes [15] have been motivated to adopt ERP systems.

ERP systems are integrated software packages which enhance organizational performance by integrating information and business processes [11]. ERP packages enhance information flow and accessibility among units in an organization. The benefits of ERP systems include faster and more accurate information exchange, higher productivity, better-integrated supply chain links, and more coordinated business processes [10] [18]. ERP systems integrate functional units through a database for increased coordination. Empirical findings suggest that ERP implementations are costly, time-consuming and complex [2]. The purchase and implementation of an ERP system can run into the millions of dollars [18]. In spite of the benefits of ERP, there is concern about the performance of ERP systems, especially in terms of how well they meet the goals of the adopting organisations.

L. Uden et al. (Eds.): 7th International Conference on KMO, AISC 172, pp. 153–161.
springerlink.com

1.1 Problems with ERP Learning Utilization

Several ERP systems have not offered an adequate return on investment when evaluated against the cost of purchasing the system and the losses arising from the overall organizational disruption during implementation [18]. The lack of effectiveness of ERP systems in meeting the goals of the adopting firms has left many practitioners in doubt about the advantages of deploying these systems.

A reason that has been put forward for the failure of ERP systems to meet the adopting organization's goals is the limited use of the system by the organization's employees. Limited utilization or under-utilization of these systems, in terms of only using some features to complete work-related tasks, hampers and obstructs good practice in the domain of a particular task. It represents ineffective system exploration and slight or low technological proficiency [8]. Limited utilization of the system constitutes a potential barrier in achieving organizational goals [8]. Usually, ERP systems are the single largest IT investment of a firm, and significantly affect its ongoing business operations [15]. Therefore, ERP systems are instrumental to a firm's operational performance. However, this benefit can only come about if the system is utilized at an optimum capacity. This indicates the importance of skills utilisation as a determinant of ERP effectiveness.

Regular optimization of the system affects the success of the ERP system. This requires a critical mass of knowledge and skills in solving problems within the system's framework [18]. However, ERP learning utilization can be limited by many reasons, such as unrealistic and unrelated training content and a complex system interface, which may hinder learning during the implementation training. For instance, ERP learning utilization has been limited by poor training design, which makes training inadequate. Also, a lack of trust in the system could affect learning and the migration from legacy systems. Trust in I.T (information technology) is associated with the functionality of the system in carrying out the responsibilities expected of information systems in the work environment. Also, the ERP system's interface affects the learning process, user reactions and overall learning outcomes. Consistently, system interface has been identified as a factor that affects system usability; however, investigations into its impact on ERP learning outcomes at the distal and proximal level remain anecdotal. This issue is highlighted because system interface compatibility connotes the ease of use and perception of system usability during training.

This paper proposes that these factors affect learning utilization, or the application of skills learned during training to the use of the ERP system. Learning utilization refers to the use of learned skills on information technologies by organizational members [8]. Successful deployment of ERP systems depends on the skill-set of the user. Several studies acknowledge the crucial role of ERP training in the implementation success. However, investigations based on the technology acceptance model (TAM) have concentrated on the influence of the antecedents of TAM on system usage, and have ignored training-related antecedents of TAM and their influence on ERP learning utilization. This paper focuses on filling this important gap.

The remainder of the paper is organized as follows: The second section provides the literature review. The third section proposes the research model and research propositions. We conclude with a look at the contributions of the study.

2 Literature Review

2.1 System Interface

Human interaction with technology is well documented in human-computer studies and information systems literature. The impact of well-designed technology on subsequent use makes this dimension important in system acceptance adoption.

The technological dimensions of a new innovation consist of the system's quality its design features. This enhances human perception of the ease of use and usefulness of the system. The socio-technical school of thought concludes that technology designs affect the post-adoption behavior. The components of a system's design are the board, slides, menus, sound, images, capability, and its compatibility with the user. Good interfaces should possess navigation guides for users to explore the capabilities of the system. A system interface facilitates information-seeking and exploration of the system's features. Research on ERP training indicates that a user-friendly interface increases ERP usage efficacy. Also, it potentially minimizes the problem of limited system use. Research has also suggested that an easy-to-use interface increases system usability [3] [7] [19]. This is consistent with our position in this paper that a user-friendly interface is an effective supportive mechanism for learning utilization.

A user-friendly interface could be the needed impetus to increase users' motivation and encourage positive post adoption behavior. Rich technology seems to stimulate users' engagement and concentration in learning environments. Such cognitive immersion is related to the development of positive attitudes, perceived usefulness and ease of use and a greater exploratory use of the technology [13]. Utilizing the various modules of a complex software application like ERP, it is imperative to have a user friendly and easy to navigate interface. A complicated interface is likely to increase users' frustration and anxiety, and make them hesitate to use the skills they have learned. The system interface is one component of system design, and has tremendous impact on users' perception of the ease of use of a technology [6]. Koh et al. [6] indicate that system design is a key reason for ERP utilization. Similarly, Osei-Bryson and Ngwenyama [10] found a positive relationship between ERP design features and perceived ease of use and perceived usefulness.

The concept of complexity from diffusion of innovation theory (DOI) [12] refers to the degree to which an innovation is perceived to be difficult to understand, learn and operate. A complex system interface is not desirable because it de-motivates, disrupts a user's flow, and increases users' frustration. This leads to acceptance and utilization being nearly extinguished.

The concept of compatibility in DOI indicate the degree of an innovation's fit with the potential users' existing values, previous experiences and current needs. A user's positive perception of an innovation's compatibility increases positive system usage intention and behaviour [16]. Empirical evidence suggests that systems interface increases innovation acceptance in a social milieu [3] [4].

2.2 Training Content

Training content encompasses an explanation of the relevance of the training to task performance, and has been shown to influence job performance [7]. The compatibility between training and task performance indicates the link between the training intervention and its effectiveness in a particular work environment.

The extent to which training contains elements of the business processes of the firm is an important aspect of training content. In some cases, it involves the re-creation of business situations and information in the work environment during training. The goal is to learn about the new system and be able to use it in task performance [14]. Put differently, ERP adoption demonstrates business goals through the training content. Research findings have shown that ERP solutions which support a business' vision are more likely to lead to success when they are used [5].

Usually, there is a connection between training content and the skills required in the job environment. Where such a condition exists, application of the skills becomes regular and continuous. Users' hesitation or limited use of the required skills is minimized, because the learning content corresponds to the task. In an ERP setting, learners' perception of the training content serves as a person-task interaction in the person-artifact-task model [3]. Training content provides the expected components (in this case, the features and modules of the systems) and the outcomes of the training (system appropriation). Training content fills the gap between training needs and system utilization. Empirical evidence has shown that task-related content increases individual performance [7]. Lim et al. [7] argued that, the ultimate goal of a training program is to increase task ability with advanced knowledge, technology and attitude on the system. It has been emphasized that ERP training should demonstrate the use of the system, so that users can develop mental cognitions of how the system can be utilized in real task domains. The relevance of the training content and instructor's competence and the ability to impart knowledge to end-users is crucial in connecting training content to the real task environment.

2.3 IT Trust

Research in this domain identifies technology-related variables which potentially influence end-user post-training adoption behavior. Some of the technological variables of interest include technology availability, cost, security, reliability and capability, functionality and quality [21] [12]. These factors affect the attitude and perception of users towards the adopted innovation. Moreover, these technology-related factors help form IT trust among end-users. For instance, the reliability and capability of a software application shore up trust in it, which connotes dependability on software attributes.

Trust is defined as a psychological state where one is willing to depend on another and has favourable beliefs about others' attributes [17]. Extended into IT, IT trust connotes the belief that a technology will perform efficiently. Technology trust is connected with a belief in the predictability and functionality of a system to carry out the functions it is programmed with. It has been suggested that post-adoption behavior goes beyond satisfaction or success from utilizing the technology. Post-adoption behavior also includes object-specific beliefs about the particular technology [17].

Trust in IT affects the exploratory and transfer intentions of users [17]. While IT trust is popular in some sections of the literature, investigation of the influence of IT trust appears scarce in the context of this study, particularly as a predictor of post-training behavior. IT trust is an important factor that must be considered in ERP system selection. Its impact during training and in the post-training environment is tremendous. We argue that end-users' resistance to systems is due to a lack of trust in the learning and post-learning environment. We propose that this lack of IT trust will negatively affect end-users perceptions of the ease of use and usefulness of the technology. A lack of trust can be due to system errors, unreliable performances and malfunctioning features. The uncertainty associated with system capabilities highlights a gap and a motivation for the study of IT trust.

Following [17], it is reasoned that trust in IT, (in this context, trust in the ERP pack- age, in terms of its integrity, competence and benevolence [20]) would enhance its utilization. Trust in IT indicates the helpfulness, functionality, and predictability of the system, and is demonstrated in actual usage. Users will not use an ERP solution that negatively affects their performance. In fact, the literature indicates that trust is positively related to PU, PEOU, the intention to explore a system [17], willingness to use, and actual utilization [8]. The installation of expensive, unreliable or malfunctioning systems appears to negate sound and ethical organizational practice. This study provides an understanding of the impact of IT trust on system learning utilization.

2.4 Technology Acceptance Model (TAM) Constructs

Drawing from psychological theories [1], TAM argues that, the decisions to accept an innovation is based on individual reactions towards the technology. Davis argued that affective reactions determine the level of adoption and extent of use of an innovation. The core affective reactions, or beliefs, in this case are perceived ease-of-use and perceived usefulness.

Perceived ease-of-use is an individual's perception of the amount of effort required to use the new system and perceived usefulness represents a user's perception of the degree to which the system will improve job performance. Perceived usefulness focuses on the extent to which using or dissipating energy on the system will translate into increased productivity through the performance of the deployed technology. These two elements influence the attitude, perception and behavioural intention of users and the extent of system exploration.

3 Research Propositions

We extend TAM and its antecedents into the ERP training setting. In our model, systems interface, training content and IT trust are proposed as antecedents of PU and PEOU and ERP learning utilization. Most of the research on ERP adoption that used TAM focused on individuals' intention to adopt the ERP system and few studies examined ERP learning utilization. This model addresses this gap and proposes an approach to investigate ERP learning utilization in an attempt to understand ERP utilization.

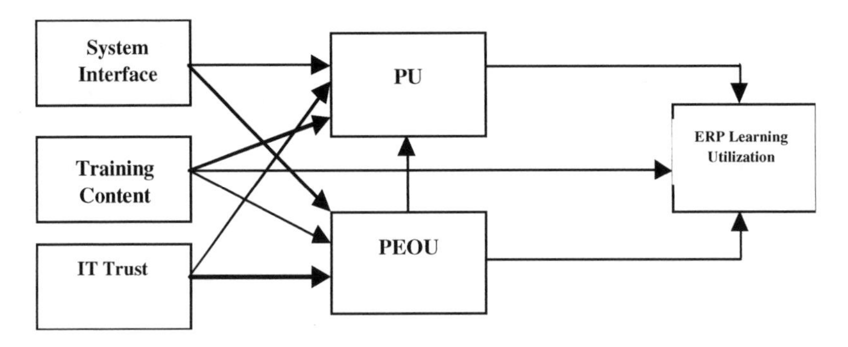

Fig. 1 ERP Learning Utilization Model

Figure 1 summarizes the research propositions. The model proposes that systems interface, training content and trust in IT influence perceived ease of use, perceived usefulness, and these two constructs subsequently affect ERP learning utilization.

Drawing on the elements of relative advantage, complexity and compatibility elements (Roger, 1995), we propose that:

Proposition 1:

The interface of an ERP system influences end-users' perception of its usefulness.

Proposition 2:

The interface of an ERP system influences end-users' perception of its ease- of-use.

Several researchers have investigated the impact of training on enterprise system effectiveness. While training affects user behavior, the role of training content on end-user reaction is unclear. The relative advantage (Roger, 1995) obtained from training is the benefit derived from the design and content of the training program. The tasks that can be carried out with an ERP system are packaged into the design of the training program. Hard skills utilisation is a person-artifact interaction that involves trialability (Roger, 1995) or experimentation of learned skills on the system. Training designs that lack a demonstration of know-how on the system (utilization) are considered failures. In line with the explanation, we propose that:

Proposition 3:

Training content influences end-users' perception of ERP usefulness.

Proposition 4:

Training content influences end-users' perception of ease-of-use of ERP systems.

Proposition 5:

Training content influences ERP system utilization.

One factor that resonates among the critical failure factor is the poor selection of the software package. End-users often report that the failure of an ERP system can be traced to a poor selection decision. This leads to hesitant behaviors during implementation. End-users dissociate themselves from a system when they assess it as being untrustworthy. Trust evaluation process leaves certain psychological conclusions with users, which mostly influence their judgment of the relative advantage or value of such system.

Trust evaluation acts in conjunction with perceived usefulness and ease of use. A lack of trust in an IT package negates the value of any relative advantage that comes about from training on the package. This indicates that lack of IT trust negatively influences perceived ease-of-use and usefulness of the package, as well as its utilization. In such a scenario, users become hesitant and resist the use of the software. However, when there is a trust in an application, the system is perceived to be useful, increasing the application of their learning on the system. We propose that:

Proposition 6:

IT trust influences end-users' perception that ERP is easy to use.

Proposition 7:

IT trust influences end-users' perception that ERP is useful.

TAM posits that perceived ease-of-use affects usage through perceived usefulness [8]. Several studies have reported significant relationships between these two constructs [2] [8] [9] [10] [17], and the direct and indirect influence of perceived ease of use on system acceptance. We recognize that the context of technology use affect users' behavioural intention and actual usage. Our proposition is based on the assumption that: a) post-implementation training satisfies the criteria of adequate training content (i.e., provides a relative advantage), enhancing the relevance and value of the training for ERP utilization; (b) IT trust (the assessment of system functionality and usefulness) is positive and (c) system interface satisfies Rogers' compatibility and trialability concepts. We propose that:

Proposition 8:

Users' perception that ERP is easy-to- use influences ERP utilization.

Proposition 9:

Users' perception that ERP is easy-to- use influences the perception that ERP is useful.

Proposition 10:

Users' perception that ERP is useful influences ERP utilization.

4 Conclusion

Enterprise resource planning systems are a strategic business intervention for achieving competitive advantage in a turbulent market place. While some of the variables in this model are known to influence system usage, the literature on their influence in relation to ERP learning utilization is lacking. This study extends research in this area by proposing relationships between systems interface, IT trust, training content, and ERP learning utilization.

The model contributes to the existing body of knowledge in the following ways. Firstly, it extends TAM constructs as learning reactions and intervening variables in the explanation of the role of system interface, training content and I.T trust on ERP learning usability. Secondly, we adapt the theory of diffusion of innovation to justify our propositions.

References

1. Ajzen, I.: The theory of Planned Behavior. Organizational Behavior and Human Decision Processes 50(2), 179–211 (1991)
2. Amoako-Gyampah, K., Salam, A.F.: An extension of the technology acceptance model in an ERP implementation environment. Information & Management 41, 731–745 (2004)
3. Choi, D.H., Kim, J., Kim, S.H.: ERP training with a web-based electronic learning system: The flow theory perspective. International Journal of Human-Computer Studies 65, 223–243 (2007)
4. Davis, F.D.: Perceived usefulness, perceived ease of use, and user acceptance of information technology. MIS Quarterly 13(3), 319–339 (1989)
5. Ifinedo, P.: Impacts of business vision, top management support, and external expertise on ERP success. Business Process Management Journal 14(4), 551–568 (2008)
6. Koh, S.C.L., Gunasekaran, A., Cooper, J.R.: The demand for training and consultancy investment in SME-specific ERP systems (2009)
7. Lim, H., Lee, S.-G., Nam, K.: Validating E-learning factors affecting training effectiveness. International Journal of Information Management 27, 22–35 (2007)
8. Lippert, S.K.: Investigating Post adoption Utilization: An Examination into the Role of Inter-organizational and Technology Trust. IEEE Transactions of Engineering Management 54(3) (2007)

9. Lippert, S.K., Forman, H.: Utilization of Information Technology: Examining Cognitive and Experiential Factors of Post-Adoption Behavior. IEEE Transactions Of Engineering Management 52(3) (2005)
10. Mouakket, S.: Extending the Technology Acceptance Model to Inves- tigate the Utilization of ERP Systems. International Journal of Enterprise Systems 6(4), 38–54 (2010)
11. Osei-Bryson, K.M., Dong, L., Ngwenyama, O.: Exploring manage- rial factors affecting ERP implementation: An investigation of the Klein-sorra model using regression splines. Information Systems Journal 18(5), 499–527 (2008)
12. Rogers, E.M.: Diffusion of Innovations. Free Press, New York (1995)
13. Scott, J.E., Walczak, S.: Cognitive engagement with a multi- media ERP training tool: Assessing computer self-efficacy and technology acceptance. Information & Management 56, 221–232 (2009)
14. Simonsen, M., Sen, M.K.: Conceptual Frameworks in Practice: Evaluating End- User Training Strategy in Organizations. Presented at the meeting of the SIGMIS, Tucson, Arizona (2004)
15. Srivardhana, T., Pawlowski, S.D.: ERP systems as an enabler of sustained business process innovation: A knowledged-based view. Strategic Information Systems 16, 51–69 (2007)
16. Taylor, S., Todd, P.A.: Understanding Information Technology Usage: A Test of Competing Models. Information Systems Research 6(2) (1995)
17. Thatcher, J.B., McKnight, D.H., Baker, E.W., Arsal, R.E., Roberts, N.H.: The Role of Trust in Post Adoption IT Exploration: An Empirical Examination of Knowledge Management Systems. IEEE Transactions of Engineering Management 58(1) (2011)
18. Umble, E.J., Haft, R.R., Umble, M.M.: Enterprise resource planning: Implementation procedures and critical success factors. European Journal of Operational Research 146, 241–257 (2003)
19. Volery, T., Lord, D.: Critical success factors in online education. The International Journal of Educational Management 14(5), 216–223 (2000)
20. Wang, W., Benbasat, I.: Integrating TAM and Trust to explain online recommendation agent adoption. Journal of the Association of Information Systems 6, 72–101 (2005)
21. Wymer, S.A., Regan, E.A.: Factors Influencing e-commerce Adoption and Use by Small and Medium Businesses. Electronic Markets 15(4), 438–453 (2007)

Technological Tools Virtual Collaborative to Support Knowledge Management in Project Management

Flor Nancy Díaz Piraquive[1], Víctor Hugo Medina García[2], and Luis Joyanes Aguilar[3]

[1] Catholic University of Colombia, Bogotá, Colombia
[2] District University "Francisco José de Caldas", Bogotá, Colombia
[3] Pontificial University of Salamanca, Madrid, Spain
fndiaz@ucatolica.edu.co, vmedina@udistrital.edu.co, joyanes@upsam.net

Abstract. This article briefly described as collaborative virtual tools contribute to the generation of knowledge in project management. First we will describe what are virtual collaborative tools, what are the benefits they bring to managers and other project team members, what their main characteristics and main categories. Thereafter there will be a recount of some of the technological tools used in project management and the impact that they cause in the organization as well as successful completion of projects. Finally, we show the evolution of technological infrastructure in the world has caused a steady movement towards these systems forcing organizations to make better use of these resources, creating virtual learning environments, communities, collaborative software and media optimized for the most out of these and generate knowledge that will be leveraged in project management to convey the experiences and other situations that can be controlled.

1 Introduction

In regard to education, it is faced with many challenges generated in response to technological changes of the last two decades. With the development of ICT have undergone major changes in the way of giving, acquire knowledge and carry out educational activities, left behind some activities such as delivery of written or printed works, documents, paper presentations, etc. It has been incorporating the use of computers, digital documents, email, chats, PDAs and other more complex giving the student the opportunity to comment, contribute and share their learning and learn collaboratively [1].

This has led to a wide variety of educational alternatives offered by technology in its different shades: E-learning, educational software, online course, e-learning, remote desktop teaching, learning objects (OA) and repositories (ROA), etc. Providing multiple opportunities to improve educational processes based on WBT.

L. Uden et al. (Eds.): 7th International Conference on KMO, AISC 172, pp. 163–174.
springerlink.com

All the tools involve collaborative learning in virtual environments, are called "Groupware Tools". These applications aim to provide skills and abilities in a group of students to be knowledge-generating activities. Thus, as intended to give a brief account of some virtual technology tools, some elements of the collaborative platform as e-learning, messaging (email, calendar, contacts), real-time group collaboration (instant messaging, chat, conferences) and social collaboration (wikis, blogs, taxonomy, ontology, folksonomy, social networks etc.), some features of such tools as virtual objects and types of classifications that describe these platforms.

The Project Management is a challenge for the Project Manager not only manage the efforts of your project team about the financial and technological assets, but lead the development and their capacity to influence through motivation, sharing experiences, best practices and lessons learned through facilitating virtual tools to create a collaborative learning environment conducive to the success of the undertaking of projects and thus to achieving the objectives of the organization. The Project Management software includes collaborative tools together with the software programming, resource allocation, communication and documentation systems should be used to help organize a complex project with a different task in a given time and with a budget previously assigned. Hence the importance of the project team know, be trained, appropriating and using collaborative virtual tools, which support and guarantee the successful implementation of projects.

2 Fundamentation

2.1 Collaborative Learning as a Means of Transmitting Knowledge in Project Management

Learning by virtual tools or collaborative learning expresses the importance of interaction with others to acquire knowledge and of the need to share goals and distribute responsibilities as desirable forms of learning.

Pfister and others [2] believe that collaborative learning is a teaching strategy through which two or more individuals interact to construct knowledge. This social process results in the generation of shared knowledge, which represents the common understanding of a group with respect to the contents of a specific domain.

In project management tools are in place to facilitate these learning processes and make a more direct approach between the methodological organization of project implementation, commissioning, human resources and documentation.

2.2 Collaborative Virtual Technology Tools

The collaborative tools could be defined as systems that allow access to certain services that facilitate users to communicate and work together. Tools are grouped mainly in Groupware and formally can be defined as "computer-based systems that support groups of people working in a common task and provide an interface to a shared environment" [3]. The groupware can be divided into four categories:

1. Communication tools: They allow communication and dialogue between members of a group, email, forums, chat.
2. Shared Applications: They make it possible to work simultaneously. Shared desktop or whiteboard, concept mapping, decision making (voting, polls) or organization of work (Calendars, Diaries).
3. Shared documentation: You can store, access and edit documents.
4. Management: Allows different levels of access to workspaces (read, modify, create new spaces, to convene meetings).

These tools are known as LMS (Learning Management Systems), CMS (Content Management Systems) or Virtual Learning Enviroments. All the features of these tools promote knowledge sharing in the environment of project management, supports their implementation more successful.

- **Content management systems:** It involves software that allows us to manage the contents of a Web automatically. In this sense, we can publish, edit, delete, grant access permissions or set also visible to the visitor modules. It consists of two elements: the content management application (CMA) and the dispensing application content (CDA).

- **Learning management systems:** E-Learning platforms or LMS (Learning Management Systems), which group management capabilities and distribution of training content, communication tools, user management, and training materials, manage access, control and monitoring the process of learning, assessments, reporting, managing communication services such as discussion forums, video conferencing, among others, very favorable elements for project management.

The LMS supports learning processes and the minimum instruction unit that manages the course itself and the CMS manages the content and the unit that handles are learning objects [4].

The following table shows the main characteristics of virtual collaborative tools regardless of the platform on which they are developed, these properties make the groupware is the main channel of interaction between project management and transmission of knowledge:

Table 1 Features of online collaborative tools. Source: Authors.

Custom installation and environment	
Instalation	Most have a wizard that guides the student through the installation.
Personalization	Templates, style sheets, templates and configuration parameters to modify the appearance and dignity, making it more students friendly.
Idioms	Mainly English and Spanish. Some tools such as Dokeos incorporate more than 20 languages in installation.
Managing users, groups and profiles	
Users, Defining Groups, Membership Groups and Profiles	Allow individual user loads and massive, some tools loaded from databases and / or flat files. Rich user information, photos, features and custom functions. Definition of Roles (Administrator, Students, Faculty, Guest). In some tools you can define the role of application (process) and / or platform. Allow to rename or belong to roles at certain times.
Web 2.0 functionality	
Features	Almost all the tools Web 2.0 have basic features like forums, blogs, news, messages, calendar, video, chat, glossaries
Course management and audit	
Management and auditing	Manage one or several courses that can be categorized. They have audit tools to manage the work, activities, tests, assessments and so on. It can import / export of courses in different formats. Manage databases or lists of questions for the test and evaluation. Management of exercises in different ways. AutoComplete, multiple choice, true or false others. Learning objects: Function as tasks, course material on which to base your knowledge of specific area. Ratings, degree of progress, level of detail, timing and reporting of student employees about the course that is performing. Instructions based on the course, indicating procedures, functions and methods.

2.3 e-Learning systems

The trend of e-learning systems has generated four key features:

- Interactivity: get a person who uses the tool to be the star of their own learning.
- Flexibility: features that allow the system meets the necessary requirements such as content and curriculum and projects.
- Standardization: the ability to use courses taken by others, in this way, the courses are available to the organization that created them and for others who meet the standard.
- Scalability: the ability of e-learning platform to work with a small or large number of users.

- **e-Learning:** The e-learning can be defined as all methodologies, learning strategies or systems that use digital technology and / or computer-mediated communication to produce, transmit, distribute and organize knowledge between individuals and community organizations [5]. E-learning is primarily an electronic medium for distance learning or virtual, where you can interact with teachers via the Internet. The user can manage the schedules, is a completely autonomous.

2.4 Learning Environment Platforms

- **Moodle:** Is an open source tool that has become very popular among educators around the world as a tool for creating online dynamic web sites for their students [6]. To be effective, must be installed on a web server somewhere, either in one of their own computers or one at a Web hosting company. Platform Moodle has features that allow students to have a display of very large scales, but can also be used for an elementary school or who do not belong to an educational community.

  Project is used as a platform to develop activities that complement the online management, through forums, wikis, databases, and other community building collaboration and team learning.

- **Dokeos:** Is an e-learning environment and an application of course content management and collaboration tool. Is free software and is licensed under the GNU GPL, is international development and collaborative [7]. He is also certified by the OSI and can be used as a content management system (CMS) for education and educators. This content includes feature to manage content distribution, calendar, training process, text chat, audio and video, test management and records kept. Until 2009, was translated into 34 languages (and several are complete) and is used by over one thousand organizations.

- **Microsoft Groove:** Allows users to have a single workspace that allows others to simultaneously view the changes. Files can be viewed, shared and discussed within the team workspace. Groove also has the ability to integrate with Microsoft software. This application is ideal for medium businesses. Microsoft Groove is a collaboration tool that allows "different people from different groups in different companies work together on projects from anywhere" [8].

- **Mambo:** Open Source (July 2006) is an application written in PHP code, based on content management systems (CMS) that allows easy creation and maintenance of web sites and portals[9]. The simplicity of mambo means that no great knowledge needed to update, maintain and customize the contents.

- **WebCT:** Is a software platform training (e-learning) for building interactive courses and training over the Internet, conducting mentoring and monitoring students [10]. This provides information concerning the time, place and date on has been visited. WebCT has plenty of communication tools, content, evaluation and study. It also allows unparalleled flexibility in customizing the presentation of an online course and the type of files that can be incorporated into the course. For this reason can be incorporated, for example, audio and video in which the student can read a text and by activating a button to hear the pronunciation of a native, and what is essential to the teaching of languages, chance to hear it many times as desired.

- **Claroline:** Is a learning platform and virtual work (eLearning and eWorking) open source and free software (open source) that enables trainers to build and manage courses online learning activities and collaboration on the Web **[11]**. Translated into 35 languages Claroline has a large community of developers.

- **Blogs:** A weblog, blog or binnacle is a website with dated notes in reverse chronological order, so that the most recent entry is the one that first appears. In the educational world are often called edublogs. "A weblog is, above all, a free form of expression, creation and sharing knowledge" [12]. As for the social aspect of these, we find two definitions:

  - Blogs are the Internet service for personal publishing's most famous web today, which has put millions of people to write and share experiences, hobbies and interests.
  - Blogs are a media group that promotes the creation and use of original and truthful information, and causes, with much efficiency, social and personal reflection on the issues of individuals, groups and humanity.

2.5 Tools to Tag and Share on the Web

These tools are characterized because they can publish content without having to install any software and are complemented with the tools to organize and rebuild the same content. The trend of using virtual learning tools can be defined in a very colloquial and "Rip, Mix and Learn" and applies to information "down" daily to learn…videos, files, images, documents.

The way you interact on the Internet based on a single form: Words. Because this form has been created ambiguity and heterogeneity in the data. As a result, partial solutions were created as methods of editing, assigning tags and folksonomy, syndication and content organization and resources on the network.

- **Folksonomy:** Is a system of categorization based on the aggregation by adding non-hierarchical and overlapping labeling of individual community members or service users. According to the definition postulated, we can say that the taxonomy does not require its components are connected by a specific type of relationship, that is, simply requires that its components are organized [13]. Unlike traditional taxonomic categorization is not part of previous ontologies. Folksonomies are distinguished from other organizational systems documentary classification because they are collaborative, not a documentary expert who assigns the classification, are people who find or add information on the web and descriptors assigned according to their wisdom and interest [14].

- **Contents syndication:** Ability to navigate the browser by subscribing to a particular website and can be informed at all times when you update the page without entering the web, through feed readers. It's like subscribing to a page and automatically receives information through this subscription. Automatically update the content of blogs, wikis, forums, websites, etc. most visited without having to remember each of the URL. The requirements: The site must allow to distribute their content through one or more syndication formats or content syndicators: RSS, RDF, Atom, etc. (XML-based languages and designed to distribute content).

- **e-Portfolio:** Personal collection of user information that describes and documents the achievements of that person, in which there are records, achievements and work done among other materials. The versatility of weblogs has made their use as such are used to descend and blogs (or part thereof) as eportfolio. The digital portfolio is used to "combine the technological tools in order to gather evidence to enable monitoring and evaluation of the learning process of students" [15].

- **Wikis:** A collaborative website carried out by the collective work of many authors. It is similar to a blog in structure and logic, but allows anyone to add, edit and delete content, even if they were created by other authors. It is linked to concepts of collaboration and community.

We are beginning to live in an age where the cooperative will have a relevant participation Wiki. Not only has captivated education but also in the business world, as some companies are replacing groupware applications for business wiki solutions that are simple and inexpensive. It allows: Interact and collaborate dynamically with team project, share ideas, create applications, propose definitions, lines of work for certain purposes, recreate or make glossaries, dictionaries, textbooks, manuals, information repositories.

- **Virtual classroom:** They are an area of blended learning. They are platforms that are built on: education and information technology. Usually many of them are based on a constructivist approach (there are some who copied the classical model of behaviorism). These tools integrate many of the tools described in previous slides.

3 Questioning

What are the benefits of technology tools virtual collaborative knowledge management?

What is the role of collaborative Virtual Tools Project Management?

4 Methodology

To build the item identification virtual collaboration tools that support knowledge management in project management, we used the process of identifying information, which was located, selected, collected and synthesized. This process began with the formulation of the questions that guided the search in the documentary collection of this study, they were identified and selected sources, then the strategy defined search and retrieval of documents, conducted the analysis and evaluation of information and finally the publication was carried out both of the tools to support the management and the results obtained by using the same.

5 Experimentation

What are the benefits of technology tools virtual collaborative knowledge management?

"The learning environment seeks to foster cooperative and collaborative spaces which give the individual skills and group discussion from the students, the time to explore new concepts, each one being responsible both for their own learning and that of other group members" [16]. The collaborative tools facilitate interaction among participants in the project, making contributions enhance its development smoothly, accurately and with optimal resource utilization. It also allows each member feel free to make their contributions as their experiences and mistakes are exploited to complement each interaction in the management.

Several learning theories can be applied to these types of environments, including those of Piaget, Vygotsky [17] and Dewey. The environments of collaborative learning prepare students for: Actively participate in the collective, integrate into a group, give help to others and ask when required, make information available to others their individual strengths, accept the views of other, understand the needs of others, discover solutions that benefit everyone, establish meaningful contact with communities with different cultures, to compare their activities and beliefs with those of others, establish goals, tasks, resources, roles, etc., accept the reasonable critic of others, yield to evidence or argument weight, recognize the foreign credits, negotiate language and methods and develop interpersonal skills.

For virtual tools to support collaborative project management, there must be commitment from management is, as it must lead in organizing the creation of enabling environments for interaction among participants who interact with projects.

Virtual technology tools provide techniques to capture, organize, store knowledge. Providing flexibility and interactivity. Flexibility is also related to the ease of access, as this can be done from anywhere. The project team members can choose the route to follow to build their own learning, according to the needs and characteristics and can choose speed over that way: go slower in the sections where you find major problems, or will move faster in certain points at which finds more easily. This allows for more customization in the process, and results in increased motivation.

On the other hand, collaborative virtual tools is through a process of synergy in which 'whole is greater than the sum of its parts', so that collaborative learning has the potential to produce higher learning gains to learning alone.

There is a very important element about virtual education and communication. This must occur to meet certain requirements to ensure their effectiveness, such as it is frequent and fast, and to promote and energize the team work [18]. The ideas put forward by the group must be reasoned and debated constructively evolve by learning together. Therefore, the concept of virtual community emerges as a necessary step for collaborative learning. Learning communities are the vehicle through which learning is accomplished virtually the benefit of the project. Within the virtual tools, learning communities are the vehicle through which collaborative learning is achieved, which in project management becomes very important.

Project managers, thinking of increasing awareness of all team members, have made companies or organizations allow the implementation of a new organizational scheme, based on the culture of knowledge and relied primarily on the use of information technologies and communications with collaborative learning approaches within the new corporate management paradigm.

What is the role of collaborative Virtual Tools Project Management?

Virtual technology tools provide techniques for capturing, organizing and storing knowledge, fundamental objectives in project management. Allow those involved in the project access to information and tools from anywhere at any time. Thus able to adapt to the needs of the disadvantages of distance equipment and / or time occurring at the time of managing or develop a project. For example Microsoft Dynamics GP for Petrotesting generated significant benefits are reflected in the budget execution process, the entire management of the accounting process, the adaptability to the industry of hydrocarbon projects in the quality of information by keeping it updated in reducing time and opportunity to deliver real-time reporting [19].

Case to highlight is the "Sanitas", which manages to improve the effectiveness, efficiency and quality of project management, thanks to a change in methodology that was used in Enterprise Project Management and his concept of earned value and earned value. Are very satisfied with the results because they have gotten better training, communication with internal and external users, greater customer satisfaction and greater flexibility in decision-making. They also highlight the importance of knowing at all times the value it brings each project. The interesting thing is that earned value deviations in cost and time are measured in real terms. At all times the Department of Systems Sanitas know how their projects contribute value to the organization [20].

Decision making does not depend on geographical variables, organizational or time. That is, the tools allow the generation of collaborative virtual spaces supported rapid and effective communication in the documentation and analysis of the project variables. Organizations like Applies Chile, with project management solutions, process and technology; the "Araucana" with social solutions through portfolio management; Synapsis with corporate project management through the PMO OnDemand; Vizagroup with joint projects and outsourcing through the PMO (project Management Office), Sherwin Williams collaborative environments for project management, among others, have increased assertiveness in providing services and in making appropriate decisions, since virtual tools have enabling collaborative knowledge sharing of successful practices, control of risks and problems, optimal resource management, excellent managing expectations of stakeholders or involved with the project, monitoring the same in all its variables, stages and processes.

It is worth noting that the virtual tools foster collaborative learning, as they build knowledge collectively, through discussion, experience and argument, and this is also an important potential of the project development environments.

These tools allow you to redefine structures, resources, methodologies and learning processes in the development of projects, since they are easily exchanged experiences and views focused on the process of production and productivity of the development team.

Collaborative learning involves peer exchange, peer interaction, and interchangeability of roles, so that different members of a project or project groups can play different roles at different times, depending on the needs.

Collaboration to create new knowledge. Different people working together for long periods to achieve common goals: such as technological projects in which the sharing of tasks and work plans generate satisfactory results in the completion of stages of development and specific activities.

In the case of projects is essential to use tools to manage information repositories and managers have versions for members of the development team can focus on the generation of knowledge and effective use of learning tools that will to advance in the assigned activities.

Collaborative virtual tools in technology projects, provide key elements for your organization and implementation, they help define strategic elements as: Objectives (the nature of the project must be real, sustainable and measurable), calendar of activities (must have a program of activities or work plan), demand resources (requires skills, knowledge, capital and manpower in various areas of an organization or community), organizational structure (has roles and responsibilities, for example, Project manager, project leader, sponsor, customers, etc.), monitoring and information system (at least one manual or automated system to record documents and information related to the project).

6 Conclusions

The application of collaborative virtual tools in knowledge management projects can be concluded that:

- These virtual collaborative tools have a high impact on project management. Depending on the effectiveness and the virtues of the software can generate greater competitiveness in the market.
- By leveraging virtual collaborative tools that support the generation of knowledge, empowers the implementation of projects, as these are used to share best practices, to show the mistakes and learn from them, promote ways to solve problems, train the team project when required or encourage participation and knowledge transfer.
- In general the virtual tools are presented as a training strategy, participation and sharing of knowledge, which can take advantage of project managers to innovate in methods, techniques or models of project management.

- The groupware is to become the main technological tool that enables enterprises to manage and share knowledge. Current requirements that demand higher productivity to organizations and reducing technology costs to knowledge becoming an invaluable resource for organizations.
- The implementation of project management technologies in business is promoting a cultural change, because the information is shared among all employees and ownership around it generates new knowledge. This, in turn, is reflected on best practices and the development, implementation and collaboration on future activities and processes. The collaborative management technologies have been introduced in companies not only as a tool to solve problems with employees, suppliers or customers, but have fostered an environment of collaboration between all stakeholders, to share solutions, continuous improvement in time real and more and better attention to both internal and external customers.

References

1. Myriam, B.A.: Contrucción Colaborativa de Objetos de Aprendizaje. In: En II Conferencia Latinoamericana de Objetos de Aprendizaje, LACLO 2007, Community of Learning Objects, L.A. (2007)
2. Wessner, M.P.: Using Learning Protocols to Structure Computer Supported Cooperative Learning. In: World Conference on Educational Multimedia Hypermedia & Telecommunications, Seattle Washington (1999)
3. Dave, C.: Groupnetsware, Workflow and Intranets. In: Reengineering the Enterprise with Collaborative Software. Digital Press (1998)
4. Diaz Anton, G.: Hacia una ontología sobre LMS. In: Proceeding VII Jornadas Internacionales de las Ciencias Computacionales, Colima Mexico (2005)
5. Bernandez, M.L.: Diseño, Produccion e Implementacion de E-Learning. Bloomington, Indiana (2007)
6. http://moodle.org (n.d.). (retrieved Noviembre 25, 2009)
7. http://www.dokeos.com/ (n.d.) (retrieved Noviembre 20, 2009)
8. Jones, D.: Groove: Virtual Office 3.1: Get Your Groove On. RedmondMag.com (2005), http://redmondmag.com/reviews/article.asp?EditorialsID=511 (retrieved)
9. http://www.mambohispano.org/ (n.d.) (retrieved Noviembre 25, 2009)
10. http://newwebct.uprm.edu:8900/webct/public/home.pl (n.d.) (retrieved November 23, 2009)
11. http://www.claroline.net/index.php?lang=es (n.d.) (retrieved Noviembre 29, 2009)
12. Blogs, I. E.: (Diciembre 14, 2007), http://tadiguea.wordpress.com/2007/12/14/weblogs-wiki-redes-socialesweb-20/ (retrieved Noviembre 27, 2009)
13. Díaz, F.N., Joyanes, L., Medina, V.: Taxonomía, ontología y folksonomía, Enero-Junio. Enero-Junio Universidad y Empresa, Universidad del Rosario 8, #16, 242–261 (2009)

14. Mathes, A.: Folksonomies-Cooperative Classification and Communication Through Metadata (2004)
15. Cabero, J.: e-Actividades: un referente básico para la formación en Internet. Alcalá de Guadaira (Sevilla), EDuforma (2006)
16. Gutiérrez, M.A.: Educación Virtual: Encuentro Formativo en el Ciberespacio. UNAB, Bucaramanga (2001)
17. Gómez, L.: Computadores en la Nueva Visión Educativa. Escuela de ingeniería (2000)
18. Guitert, M.: El trabajo Cooperativo en entornos Virtuales (2002)
19. Petrotesting. Caso de éxito. Pionera en el control de proyectos con herramientas virtuales. Consultado (February 03, 2012), http://www.acis.org.co/fileadmin/Revista_109/cuatro.pdf
20. Sanitas. Gestión de Proyectos por valor ganado. Casos de éxito de Microsoft. Soluciones de colaboración. Consultado (January 02, 2012), http://www.microsoft.com/spain/enterprise/casos-exito/detalle-casos-de-exito.aspx?ContenidoID=20100304001

Promoting Knowledge Sharing and Knowledge Management in Organisations Using Innovative Tools

Aravind Kumaresan[1] and Dario Liberona[2]

[1] Analyst Programmer, Financial Times, One Southwark Bridge, London, SE1 9HL.UK
aravind.kumaresan@ft.com
[2] University of Santiago, Chile
Dario.Liberona@Usach.cl

Abstract. Businesses have recognized that knowledge constitutes a valuable intangible asset, and currently they have access to an extensive pool of knowledge, from the skills and experience of the workforce and also their understanding of customers' needs. The challenge's for the businesses are not just processing these big data but also dealing with company's cultural issues that allow knowledge sharing and information exchange. Sharing of knowledge constitutes a major challenge in the field of knowledge management because some employees tend to resist sharing what they know with the rest of the organisation. It takes the right environment to create an effective knowledge sharing program and for such an environment to be nurtured, organisations need to look inwards at the type of culture they promote before investing in knowledge sharing tools. The purpose of this study was to analyse the various factors affecting the knowledge sharing process in organisations and the use of innovative social business tools to promote knowledge sharing in organisations. A global survey was conducted for this study relating to the willingness to share knowledge, and the tools available for sharing knowledge. Also, a practical case study was conducted on a UK based company.

Keywords: Knowledge sharing tools, Social business tools, knowledge management.

1 Introduction

We live in a digital age, everything is transforming into its digital form. Traditional print newspapers are transforming into digital news. Print books are transforming into e-books etc. Along with the tremendous growth in the digital space, the willingness to share the digital information is also on the increase amongst people. According to the Facebook, the world's most popular social networking site, more than 5 billion pieces of content in the format of web links, news, blog posts, notes, photo albums, etc. are shared each week. The statistics given by Facebook on content sharing is beyond the belief and attention grabbing.

L. Uden et al. (Eds.): 7th International Conference on KMO, AISC 172, pp. 175–184.
springerlink.com

As Francis Bacon said "knowledge is power. Knowledge can be used for learning new things, solving problems, creating core competences, and initiating new situations for both individual and organisations now and in the future. The shared and processed knowledge over the time would serve as a strategic competitive advantage for the organisations. Knowledge sharing is considered to be a key component and also a substantial barrier in achieving an effective knowledge management. (Chen et Al., 2009) says there are various impediments identified for knowledge sharing in organisations like negative attitude toward knowledge sharing, organisation's culture, inadequate organisations support, lack of trust, lack of motivation. These identified factors are not related to any technological factors as they are general issues effecting knowledge sharing and it may impede any organisation trying to promote knowledge sharing using web based tools. According to (Hendriks,1999), technological factor that may obstruct the knowledge sharing process could be the tool itself. The inability of the tool to provide a proper platform for knowledge sharing could have a significant impact on the success of knowledge management. The three main objectives of this paper are:

- To analyse the various factors affecting the knowledge sharing process in organisations.
- The use of innovative social network context tool to promote knowledge sharing in organisations. The case study on Neo tool used by Pearson group will be discussed in detail.
- To make recommendations in overcoming the factors affecting in knowledge sharing and for further enhancement of the Neo tool discussed in the case study.

2 Importance of Knowledge Sharing and Knowledge Management in Businesses

Although knowledge is considered to be a vital asset of an organisation, most of the organisations lack in implementing a successful strategic process for managing it. One of the main reasons for the failure in knowledge management can be pointed out to the knowledge extraction phase. Knowledge sharing is a key component and a stepping stone in knowledge management. According to (Tsui et al., 2006), Knowledge sharing can be defined as a process of exchanging knowledge (skills, experience and understanding) among individuals, a community and within an organisation.

Various organisations have realised the importance of knowledge management and knowledge sharing process, a fine example would be a research conducted by (KPMG, 2000), to find out the willingness of its organisational units across Europe and the Unites States to implement knowledge sharing programmes. The survey results revealed that about 81% of the organisational units have either adopted knowledge sharing programmes or they are willing to adopt one. Knowledge sharing is very important and it's vital in eliminating the key person dependency. When the individuals leave their job they take away the valuable knowledge that they had learned from the organisation. In the year 2007 Chuck Law and Eric Ngai conducted an extensive first-hand research on 134 firms to determine the importance of knowledge sharing activity in various fields of industries including manufacturing, retail and wholesale. The results from the research emphatically reveal that knowledge sharing, learning methods and the workers behavior are powerfully correlated in getting the business procedures and production better.

3 Difficulties of Knowledge Sharing in Organisations

One of the hardest challenges of knowledge management is that of convincing and making people to share their knowledge. According to (Hansen et al., 1999), organisations normally give emphasis to either the personalisation or the codification strategy. On the other hand, irrespective of which knowledge management system is executed individuals will accumulate the knowledge they hold. (Husted and Michailova 2002). Absence of trust in fellow employees is another major challenge in knowledge sharing. (Szulanski, 1995). With the past researches of (Blau, 1964), (Wasko and Faraj, 2005), (Kankanhalli et al., 2005), (Bartol and Srivastaa, 2002), (Bock et al., 2005) has determined numerous possible motivation of knowledge sharing behavior and the three most potential motivation of online knowledge sharing can be reputation, mutual benefit and rewards. According to (Paul, 1999), individuals should have a positive thought towards sharing knowledge. Individuals fear that their shared knowledge might be stolen and used by others. This negative thought has to be removed and trust and security should be provided to the individuals in order to promote knowledge sharing. Based on the above literature review, the following hypothesis will be tested in this paper:

- H1- The positive perception of an individual's willingness to share knowledge is an influencing factor for knowledge sharing.
- H2- The positive support and encouragement given by an organisation promotes knowledge sharing.
- H3- The available technology platform for individuals is not adequate to share their knowledge.
- H4- The main influencing factor for knowledge sharing is the motivational aspect which comprises of the Incentives in the form of bonus, performance appraisals and awards.
- H5- Adapting to social networking model based tools, portals and software tools has a positive influence on knowledge sharing in organisations.

4 Research Methodology

The online web based questionnaire method was used to collect the primary data. The questionnaire was divided into two sections. Section A contains the demographic information of the respondents. Section B contains questions relating to this study. The questionnaire was sent totally to 864 members all over the world belonging to various fields of industries including education, manufacturing, retail, software services, financial services, newspaper; and the questionnaires was sent to various level of employees such as managerial, executives, administrative, software engineers and general staffs. Also, to identify the difficulties in knowledge sharing for a specific UK based global newspaper firm, The Financial Times is included in the survey. The survey was kept open from October 2011 to Feb 2012 to gain more responses. A response rate of 69.33% (599 responses) was received from various countries including United Kingdom, India, United States, Canada, Colombia, Chile, Ecuador, Germany, Finland, Spain, Japan, Malaysia, Spain, Panama, Argentina and Philippines. A total of 15 responses came from The Financial Times employees.

5 Hypothesis Testing Using Descriptive Analysis

We have used the descriptive analysis method to test the 5 hypothesis specified above. Hypothesis 1 argues that the positive perception of an individual's willingness to share knowledge is an influencing factor for knowledge sharing. 576 respondents, which is 96% of respondents, have shown their strong interest in willingness to share knowledge. The majority of the 4% negative responses to willingness to share the knowledge are from Asian and South American countries. 544 respondents, which constitutes to 90.82% of respondents have considered knowledge sharing very important to them. 98% of respondents have replied learning and growth opportunities when asked for their perception on knowledge sharing. Also, 82% of respondents have accepted that knowledge sharing leads to innovation. The above positive statistics on willingness towards knowledge sharing fully supports the Hypothesis 1. Hypothesis 2 argues that positive support and encouragement given by an organisation promotes knowledge sharing. 32% of respondents either agree or strongly agree their organisation support for knowledge sharing. 60% of respondents either disagree or strongly disagree their organisations support for knowledge sharing. Amongst the 15 respondents from the Financial Times, 13 of them either strongly agree or agree for the support for knowledge sharing from their organisation. Hypothesis 3 argues that the available technology platform for individuals is not adequate to share their knowledge. 70% of the respondents either disagree or strongly disagree the available tools provided by their organisations for knowledge sharing. Interestingly in a contradicting way about 86.66% of Financial Times employees has either agree or strongly agree the available tools provided by their organisation for knowledge sharing is sufficient for the purpose. Hypothesis 4 argues the main influencing factor for knowledge sharing is the motivational aspect which comprises of the Incentives in the form of bonus, performance appraisals and awards. Below are the responses that we have received in our survey relating to the motivational factors for knowledge sharing. Individuals were asked to specify the various motivational factors affecting the knowledge sharing. The answers were then interpreted to the following suitable categories.

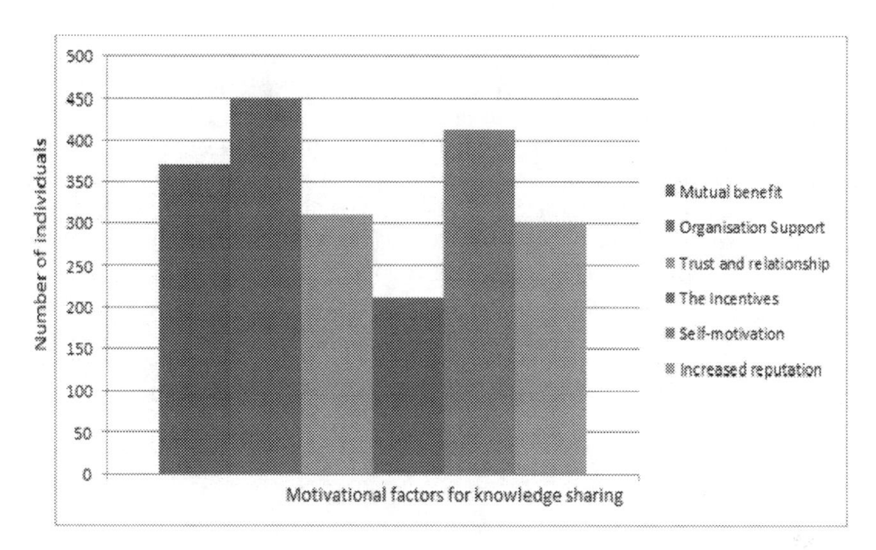

Fig. 1 Motivational factors for knowledge sharing

From the above figure 1 more than the 70% of respondents highly rated the organisation support and self-motivation for the main motivational factors for knowledge sharing. Only 20% of respondents rated highly the incentives in the form of bonus awards, performance appraisals as a motivational factor for knowledge sharing. This result re-iterates the (Gammerlgaard, 2007) findings on use of incentives for knowledge sharing in organisations. According to the Jens Gammerlgaard findings the extrinsically motivated incentives such as bonuses or promotions, are not most effective at motivating knowledge sharing. Instead, employees favour intrinsically-motivated incentives, such as colleagues' acknowledgement and respect, improved reputation, and the possibility of professional or personal development. Thus, the results are not supporting the hypothesis 4.

Hypothesis 5 argues that adapting to social networking model based tools, portals and software tools have a positive influence on knowledge sharing in organisations. 73% of respondents either agree or strongly agree the usefulness of web portals, social networking tools for knowledge sharing in organisations. 86.66% of Financial Times employees who are already using the social networking context based tool NEO in their organisation have replied positively for the use of social networking tools for knowledge sharing. More evidence can be derived further for Hypothesis 2, 3 and 5 from the case study of the knowledge sharing tool provided by Financial Times to its employees.

6 Tools for Knowledge Sharing in Organisation

(Dennis et al., 1999), focused their analysis on e-collaboration systems commonly called group decision support systems, which normally are applied to support group meetings in the same place and at the same time for knowledge sharing. These systems reduced the wastage of meeting time, especially in the important sessions that involve knowledge sharing activities like brainstorming and decision-making tasks. It has been said by (Qureshi and Keen, 2005) that the electronic communication and collaboration tools have made it quite possible to extract intelligence and knowledge through time and space. (Jeed, 2008) argues that using Web 2.0 tools or social software inside organisations enhances collaboration, knowledge sharing and innovation.

Below are the responses that we have received in our survey relating to the tools currently used by the organisations for knowledge sharing. Individuals were asked to specify the various tools their organisation is using to promote knowledge sharing. The answers were then interpreted to the following suitable categories.

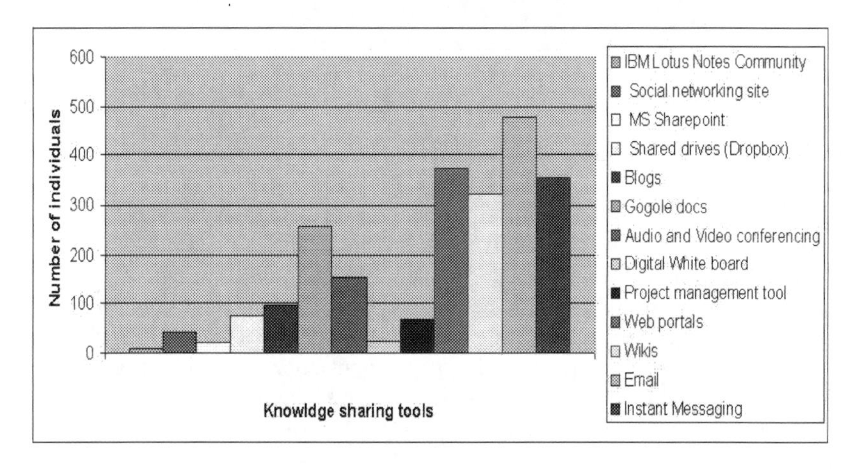

Fig. 2 Knowledge sharing tools in organisations

From the figure 2 Email, web portals, wikis, instant messaging were appeared to be most preferred methods for knowledge sharing in the organisations. But the survey finding on satisfaction on the tools provided by their company clearly opposes the organisations view. 426 respondents among the 552 were not satisfied with the tool provided by the organisation for knowledge sharing. The number clearly illustrates the lack of available tools ability in promoting knowledge sharing in organisations.

7 Knowledge Sharing Tools Used at the Financial Times

The Financial times, is an international business newspaper. Apart from the print newspaper business, Financial Times invests heavily in the online newspaper channel FT.com. The Financial times uses various tools for knowledge sharing and management. These various tools are used in a various methods across different departments. These various tools and methods create the centralised knowledge in the organisation, which is difficult to process when information is needed. The developer community with in the organisation uses heavily wiki pages for capturing the explicit and permanent knowledge. In order to promote the sharing of tacit knowledge the company often arranges a discussion sessions or presentation sessions. The company also uses the tools like email, intranet web portals, instant messaging, shared network drive, Google docs, Audio and video conferencing etc for knowledge sharing. In mid of 2011, Pearson plc, a parent company of The Financial Times launched an innovative collaboration tool NEO to promote the knowledge sharing across its three operating divisions: The Penguin Group, Pearson Education and Group. In the current years, Pearson has acquired a lot of companies emerging in the new markets, widening its international reach and changing the shape of its workforce. With new companies on the stream and an increasing need for its different brands to work together, Pearson needed a tool that would help its people collaborate and communicate effectively in a group. Pearson's digital products and services are an essential part of its offering to customers, and it was very precise in providing its employees with a high technology-driven collaboration tool. The NEO works on the principle of social business model and it promises to easily facilitate the process of knowledge sharing.

8 Case Study on Neo Tool

Pearson is an international media company with world-leading businesses in education, business information and consumer publishing. All that Pearson were in need of was an easy uncomplicated and an effortless approach to all major resources, tools, and knowledge across the organisation. Pearson expects its employees to be in a position to seek for expert skills in distinguishing areas without having an admittance to know the names of those individuals. This can be explained more in a case where a production editor in one business unit had a cover that required tweaks and the designers were not available. The editor was able to find substitute designers via Neo and meet the deadlines. Pearson is of the opinion that it is essential to help the acquired companies make them a part of the parent company to share content and resources and bring on board the wider conversation with regards to Pearson's market strategy. Now a day's company to company communications have been taken over with centralised blogs, discussions and posts on the Neo. The clarity and the clearness of activities on Neo drive a number of alliances between the organisations. For Example: A new program was developed for a new certification of UK's vocational business organisations. The employees of Pearson in Australia watched the posts about the new certification and immediately apprehend its suitability to their market, and established an equivalent and a similar program from content the UK shared with them. Neo has enlightened with the content sharing across a lot of functions, inclusive of conference development. Pearson initially tested this Neo tool with about 700 users and with its success it grew to about 5000 users before the tool Neo, went live. Six months after the launch of Neo, it was used by 27,000 users. By late 2011, almost 75% of Pearson employees and all the employees from their acquired companies have also engaged themselves into this community. Karen Gettman, Vice President and Director for Community and Collaboration said they were very proud with the achievements Neo has made within a short span of time. He also said that a lot of changes have started in the way different businesses work.

9 Features and Evaluation of Neo Tool

Neo tool allows the users to know the other employees on what they are doing. It is kind of status messages on Facebook and like a Tweet on a twitter. The simple update messages are very powerful in connecting the employees. The my profile space is to have the up to date information about the users like their expertise, business group, biography, languages known etc. The information entered in my profile space will be indexed in the search engine and in turn it's used when someone does a search for any specific expertise or skills. The my stuff area, where the users can able to have their photo albums, videos, blogs posts, documents etc. Also from the my stuff area the users can able to follow specific communities or projects across the businesses. The connections features are to connect to the various contacts across the businesses via the people directory. The connection represents the same way the LinkedIn, a popular professional network, works. There is also private messaging service, which is similar to that of email services. The Pearson heavily promotes the Neo tool to the users by means of bill board advertising with in the companies, arranging for professional photo shoots for profile pictures. Recently the senior management started to post the various companies' performance announcements in the Neo space rather than sending out the bulk traditional email. Of course, they still do send out the traditional bulk email for announcements, but instead of composing the whole message in their email they now attach a Neo page link to read further. Also, most of the intranet based content websites like HR policies; communications etc were all moved into Neo communities. The Neo tool in general is a mixed breed of Wikipedia, LinkedIn, Twitter and Facebook. Few of the team members at Financial Times who heavily uses the Neo tool feel it fits the purpose for documentation and other organisational social networking needs. Hypothesis 2, 3 and 5 are further evident from the case study.

10 Recommendations for Improvements

Although Neo tool has made a significant progress for the whole company in bringing the social context in to their business, to promote the knowledge sharing and management, there are still quite a lot of opportunities to further improve the system. Some of the teams are still using the traditional systems for their documentation, shared files in network etc. The knowledge is still floating in different systems. The company have not enforced the groups to strictly use the Neo tool. The features of Neo tool is much overloaded. Imagine the features of Facebook, Twitter, LinkedIn and Wikipedia all in one place and with out any privilege for users to customize the required features. There is a feature overload in Neo and the company has to remove the unused features by the users. The points are given for users based on their activities but many of the users are still unaware of how this points system works. The company has to provide more regular workshops on the tool for the users.

11 Conclusion

In today's modern world change in communication methods are so rapid that the businesses are forced to innovate new tools for their employees in order to share and reuse the existing knowledge to compete other organisations. The global survey and responses from the Financial Times employees clearly rates the organisational support and the individuals willingness to share knowledge are the major influence for knowledge sharing in organisations. Also, it is becoming more clear that the employee's are not relying on bonus schemes to promote knowledge sharing instead they prefer more support from organisation and enhanced reputation for knowledge sharing. An interesting 70% of respondents disagreeing in ease of use of their existing tools to share knowledge is evident that many organisations are still struggling to bridge the gap and unable to easily facilitate their employees in knowledge sharing process. From the case study of Neo tool used by Pearson group, it is becoming more evident that this kind of social business tools could be a right step in promoting knowledge sharing in organisations. Although these social business tools could facilitate the ease of use for the users for knowledge sharing, but the real success will only be achieved when there is a realisation from the businesses to achieve their knowledge sharing and knowledge management goals.

Acknowledgments. We would like to express special thanks to Dr. Lorna Uden, Staffordshire University for her reviews and thoughtful comments.

References

Bajwa, S.D., Lewis, F.L., Pervan, G.: Adoption of collaboration information technologies in Australian and US organizations: a comparative study. In: Paper presented at the Proceedings of the 36th Hawaii International Conference on System Sciences, United States, Hawaii (2003)

Bock, G.W., Zmud, R.W., Kim, Y.G., Lee, J.N.: Behavioral Intention Formation in Knowledge Sharing: Examining the Roles of Extrinsic Motivators, Social-Psychological Forces, and Organizational Climate. MIS Quarterly 29(1), 87–111 (2005)

Chen, I.Y.L., Chen, N.-S., Kinshuk: Examining the Factors Influencing Participants' Knowledge Sharing Behavior in Virtual Learning Communities. Educational Technology & Society 12(1), 134–148 (2009), http://www.ifets.info/journals/12_1/11.pdf (accessed January 20, 2012)

Dennis, A.R., Hayes, G.S., Daniels, R.M.: Business process modeling with group support systems. Journal of Management Information Systems 15(4), 115–142 (1999)

Gammelgaard, J.: Journal of Knowledge Management Practice. Why Not Use Incentives To Encourage Knowledge Sharing? 8 (2007), http://www.tlainc.com/articl127.htm (accessed February 12, 2012)

Hansen, M., Nohria, N., Tierney, T.: What's your strategy for managing knowledge? Harvard Business Review 77, 106–116 (1999)

Hair Jr., J.F., Anderson, R.E., Tatham, R.L., Black, W.C.: Multivariate Data Analysis, 5th edn. Prentice Hall, Upper Saddle Rive (1998)

Hendriks, P.: Knowledge and Process Management. Why Share Knowledge? The Influence of ICT on the Motivation for Knowledge Sharing 6(2), 91–100 (1999), http://mapule276883.pbworks.com/f/Why%20share%20Knowledge.pdf (accessed January 15, 2012)

Husted, K., Michailova, S.: Knowledge Sharing in Russian Companies with Western Participation 1. Knowledge Sharing in Russian Companies with Western Participation 1 (2002), http://www.knowledgeboard.com/download/1708/knowledge sharing_Russia.pdf (accessed March 02, 2012)

KPMG, Knowledge Management Research Report 2000. Knowledge Management Research Report 2000, 7–23 (2000), http://www.providersedge.com/docs/km_articles/KPMG_KM_Research_Report_2000.pdf (accessed February 10, 2012)

Keyes, J.: Identifying the Barriers to Knowledge Sharing in Knowledge Intensive Organizations 69(5A), 15–37 (2008), http://www.newarttech.com/KnowledgeSharing.pdf (accessed February 10, 2012)

Law, C., Ngai, E.: An empirical study of the effects of knowledge sharing and learning behaviors on firm performance 34(4) (2008), http://www.mendeley.com/research/an-empirical-study-of-the-effects-of-knowledge-sharing-and-learning-behaviors-on-firm-performance (accessed February 13, 2012)

Liao, S.H.: Employee relationship and knowledge sharing: a case study of a Taiwanese finance and securities firm 2, 34 (2004), http://www.semgrid.net/Citation-Before-2006.1/+Knowledge%20sharing.pdf (accessed January 18, 2012); Nunnally, J.C.: Psychometric theory, 2nd ed. McGraw-Hill, New York (1978)

Paul: Assessing the Role of Culture in Knowledge Sharing (1999), http://www2.warwick.ac.uk/fac/soc/wbs/conf/olkc/archive/oklc5/papers/d-3_hendriks.pdf (accessed March 1, 2012)

Szulanski, G.: Exploring internal stickiness: Impediments to the transfer of best practice within the firm. Strategic Management Journal 17(winter S), 27–43 (1996)

Tsui, L. et al.: A handbook on knowledge sharing: Strategies and recommendations for researchers, policymakers, and service providers. Community-University Partnership for the Study of Children, Youth, and Families: Edmonton, Alberta (2006), http://www.ice-rci.org/research_ops/Knowledge_Sharing_Handbook.pdf (accessed February 2, 2012)

Qureshi, S., Vogel, D.: Information Technology Application in Emerging Economies. In: A Monograph of Symposium (2005), http://www.hicss.hawaii.edu/Reports/40ITApplication.pdf (accessed February 16, 2012)

Knowledge Management Model Applied to a Complex System: Development of Software Factories

Víctor Hugo Medina García, Ticsiana Lorena Carrillo, and Sandro Javier Bolaños Castro

District University "Francisco José de Caldas", Bogotá, Colombia
vmedina@udistrital.edu.co, aiscit@yahoo.com, sbolanos@udistrital.edu.co

Abstract. This article shows the connection and interrelation between complexity theory and knowledge management. It is planned to find mechanisms to communicate and share knowledge efficiently, improving knowledge management in a complex system such as the development software factories. As a result, it propose a model of knowledge management, taking into account notions of complexity theory applied to a development system software factories, where normal development involves different forms of complexity, which must be managed properly to help solve. Of course, the results of the application can be seen in the medium term best practices and lessons learned.

1 Introduction

First of all, because we need to know why you get to apply two complex issues by them selves, and each complex. The first concept, knowledge management, frames the modern world, knowledge as a source of power and the best partner when developing organizational strategies. But if the organizations have the knowledge, what is the reason now to enter the word management?

Then bring to mind, the organizational level, where technical skills are required, especially social processes-methods. What happens if there is no support in the generation and transmission of knowledge? If every time a new oficial or programmer enters, they have to try to develop or invent the wheel again to learn how to do things? Obviously the company or the project itself will have an extension in time to execute tasks and therefore increase costs and a decrease in performance.

On the other hand, the term complexity exists and has existed forever, even comes embedded with the man with the world and nature itself. Thousands of simple processes are complex to our eyes and at the same time fascinating.

The complexity and knowledge come side by side according to the hypothesis of Seth Lloyd [1]: "Everything that is worth under-standing complex systems about can be understood in terms of how it processes information", the complexity

L. Uden et al. (Eds.): 7th International Conference on KMO, AISC 172, pp. 185–195.
springerlink.com

can be taken into terms of processing information that action involving, namely knowledge.

Therefore, you can take advantage of the concepts and the different applications of complexity theory to propose a model of knowledge management to facilitate the development of a complex system such as software factories.

2 Conceptualization

As the basis of this research, described below most relevant concepts that affect the understanding of the proposed model.

2.1 Knowledge Management

In order to develop or try to implement knowledge management, it is necessary to have different ideas about it, for example, it can be explained in terms of knowledge transfer: The Management or Knowledge Management is a concept used in companies, which seek to transfer knowledge existing experience and employees, so that it can be used as a resource for others in the organization. Knowledge management aims to make available to each employee the information you need at the right time so that their activity is effective [2].

You can also set based on the satisfaction of needs: "The process of continually managing knowledge of all kinds to meet present and future needs, to identify and exploit resources of knowledge that exist and are acquired to develop new opportunities" [3].

Other definitions are focused on intellectual capital: "The set of processes and systems that allow an organization's intellectual capital to increase significantly through the management of their problem-solving capabilities efficiently (in the least about of time possible), with the ultimate goal of generating sustainable competitive advantage in time" [4]. And the generation of value through it: "Creating value from intangible assets of the organization" [5].

Just as ideas are aimed at competitiveness: Management planned and ongoing activities and processes to enhance knowledge and increase competitiveness through better use of resources of creation of individual and collective knowledge [6].

“The European KM Framework is designed to promote a common European understanding of KM, show the value of the emerging KM approach and help organizations towards its successful implementation. The Framework is based on empirical research and practical experience in this field from all over Europe and the rest of the world. The Framework addresses the most relevant elements of a KM approach and aims to serve as a point of inspiration and as a reference basis for all types of organizations aiming to improve their performance through dealing with knowledge in a better way” [6]. This framework is outlined in Fig. 2.

But ultimately, all knowledge management concepts point to the way it obtains, classifies, distributes, transmits and stores the knowledge: It is a concept used in companies, which aims to transfer knowledge and experience existing within the employees, so that it can be used as a resource for others in the organization.

Knowledge management aims to make available to each employee the information you need at the right time so that their activity is effective [2].

Based on the above notion, it can be established that the objectives of knowledge management are:

- Develop an outreach strategy for organizational development, acquisition and application of knowledge.
- Implement strategies to knowledge.
- Promote continuous improvement of business processes, emphasizing the generation and use of knowledge.
- Monitor and evaluate the achievements through the application of knowledge.
- Reduce costs associated with the repetition of errors.
- Reduce cycle times in developing new products.

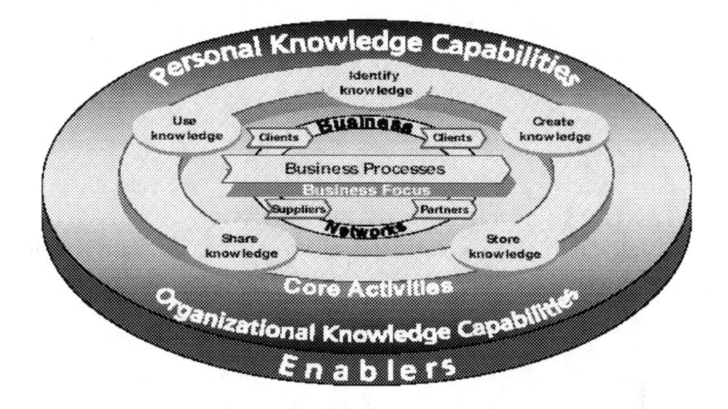

Fig. 1 Knowledge Management Framework: A European Perspective. Source: [6].

And it can be defined as a supporting technology: A "Knowledge Management System" is a platform for information and communications technologies (ICT) that supports the KM processes in the organization. It focuses on creating, collecting, organizing and disseminating "knowledge" of an organization, rather than information or data [7].

2.2 Complexity Theory

Many events and facts that are inherent in the complexity theory: the theory of relativity of Albert Einstein, the beginning of uncertainty or indeterminacy, fractals, artificial intelligence, chaos theory, agents, emergency, self-organization [1], which makes finding a single definition almost impossible. It is usually associated with complexity theory to the study of complex systems.

A system is complex when it consists of many parts that are interconnected so intrinsic [8]. A system has dynamic complexity where cause and effect are subtle, over time [9]. A system is complex when it consists of a group of related units (subsystems) for which the degree and nature of the relationships is imperfectly known [10].

2.3 *Software Factories*

Now a day, software systems that can be developed in a short time are needed, in other words, reduce the time lag of specifications to implementation. You should also be able to recycle a set of procedures, processes, designs, strategies, alternatives and experiences to improve the software development process. Also, statistically, 64% of IT projects are considered failures either because they are made within the time / budget plan or because they are within the projected range, all this confirms the need to automate the software production process through production lines, similar to the auto assembly process.

For the next decade the demand for software will increase significantly, promoted by the global economic changes, new technological applications, the rise of mobile devices, internet, and artificial intelligence. In other words it is not possible that the software developments meet this demand, then how do you get the ability to meet the above demand? The possible solution is not to increase the number of developers, because it could bring more problems than advantages. What should be done is radically change the methods of software development so that the people involved do not perform repetitive tasks.

Another aspect that adds to this scenario is the constant evolution of technology, which adds complexity to both the project and process and the product, while coexisting with the evolution of the domain problem [11]. This increases the demand on the development of solutions, forcing software development companies to look for options that allow them to have agile and efficient processes, reduce consumption of resources and maintaining high quality levels [12].

Today, you can still consider software development as a traditional process based on the skills and knowledge of the engineers involved. Each new product required by the software development company is a new solution, unique and customized to the costumer.

The industries have increased their capacity from the craft, where all products are created from scratch by individuals or small teams, to the industry, where a wide range of product variants are quickly assembled from reusable components created by multiple vendors or the same industry.

With the passage of time the customers needed a larger number of solutions, improved quality, reduced time on development, therefore it is necessary to change the product solution approach and focus on the process as a central source of the product. To move towards the process of product development software, factories appear to improve productivity, through a knowledge bank, components, tools and processes used to grant customers greater quality, flexibility, lower price and delivery time [13]. By reusing, leverage can take advantage of the skills of several developers on multiple projects at the same time, this has been considered as the main focus for significant improvements in productivity and quality of software engineering [14], [15]. The development focuses on the use of existing modules.

A software factory can be defined as "a software product line that configures extensible tools, processes and content using a software factory template based on a software factory schema to automate the development and maintenance of variants of an archetypical product through adaptation, installation and configuration of component-based framework" [16] (Fig. 2). Other more practical ways are defined as "a product line that configures extensible development tools like Microsoft Visual Studio Team System (VSTS) with packaged content and guidance, carefully designed to build specific types of applications."

Fig. 2 Software factory: Product. Source: Authors.

This type of project can be conducted within the framework of an experimental design [17], using the methodology MSF for agile Software Development at Microsoft [18], [19], it supports iterative development and refinement learning continuous. The product definition, development and testing occur in incremental interactions (Fig. 3). The main principle is to add interactive functionality that improves the quality of not only software but also the way it builds [20]. You can choose this methodology for spiral development and ease of integration with Visual Studio.net

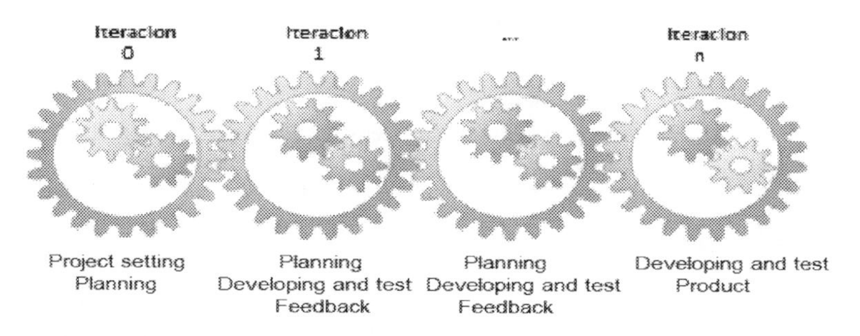

Fig. 3 Cycles and interactions. Source: Microsoft Developer Network.

Each interaction has the following phases: forecasting, planning, development, stabilization, and implementation.

3 Model

In a society that seeks tirelessly methods that facilitate the processes to produce their goods and services, arises the urgent need to implement and develop knowledge transfer processes, using tools that are fast, reliable and economical.

In trying to implement a knowledge management model to a complex system, the model may take the qualities of complex systems and become too complex. Also, if the basis is the characteristics of complex systems such as emergency, self-organization and chaos can be an organized model in an emergent way (Fig. 4).

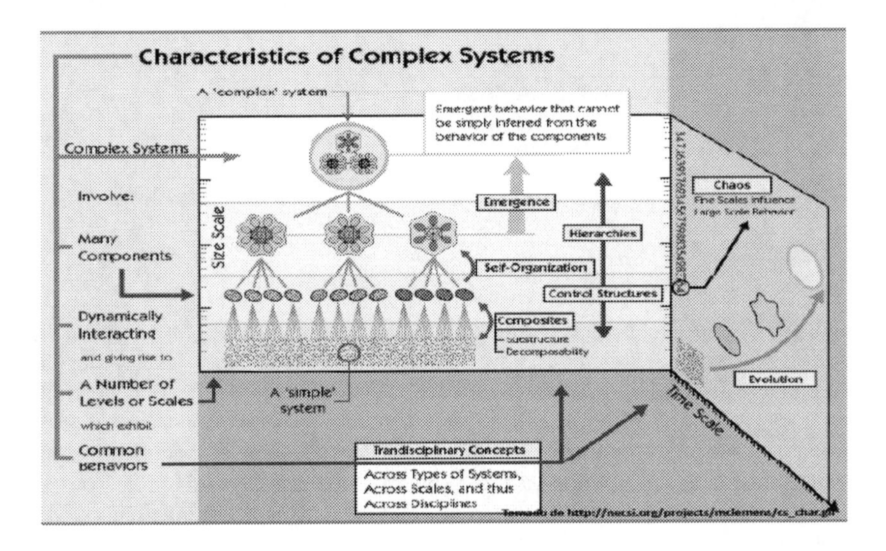

Fig. 4. Characteristics of Complex Systems. Source: Institute of Complex Systems of England.

The best way to support a software factory development is through a process of continuous learning and refinement, which can be achieved through the qualities of a complex system, then in building components could begin implementing each of these concepts:

3.1 Organization

For the development of each software component it is necessary to organize the working group structures that interact and change according to the immediate goal, a reason that you might want to distribute, prioritize work groups or contract according to the abilities of people, the number of hours / person, or type of product to be obtained (Fig. 5). With this changing distribution of knowledge flows get really dynamic and strong, because knowledge has several directions and reaches all individuals naturally.

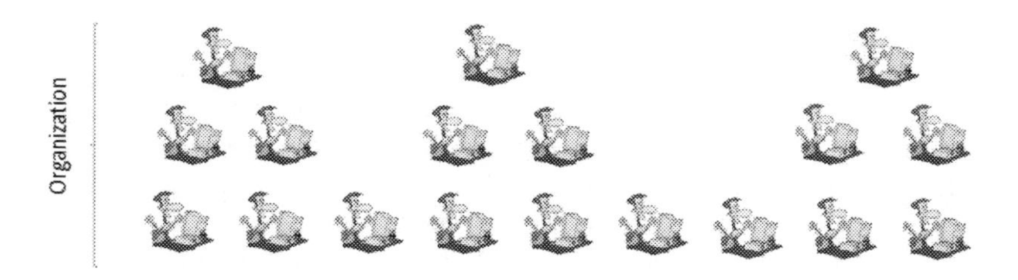

Fig. 5 Forms of organization. Source: Authors.

3.2 Interaction

In addition to the interaction that is achieved between each of the people doing the project, you get other types of interactions such as those generated through conversations with customers, managers, colleagues and even people outside the project and events from the outside world (Fig. 6), which makes the system an open system in which the inspiration to solve difficulties may stem from the same system as external events. The distribution of information increases the capacity of people to solve problems.

Fig. 6 Types of interaction. Source: Authors.

3.3 Feedback

Each of the events generated and interactions that commonly occur in the organization produce a feedback that supports all processes in software development (Fig. 7).

Fig. 7 Events and feedback. Source: Authors.

3.4 Adaptation

The system, based on feedback obtained as a response adaptation making decisions and providing a better link from modifications or changes in methods, processes and environment. External and internal events and the feedback, function as incentives to achieve harmony between the current form of development of components and the best way of taking into account all the environmental variables involved in the society (Fig. 8). Knowledge flows equally fit, allowing them to have the knowledge about why the change and then a log can be consulted to understand the decisions made. The events, feedback, skills and existing methods cannot adapt to causes that improve a new process.

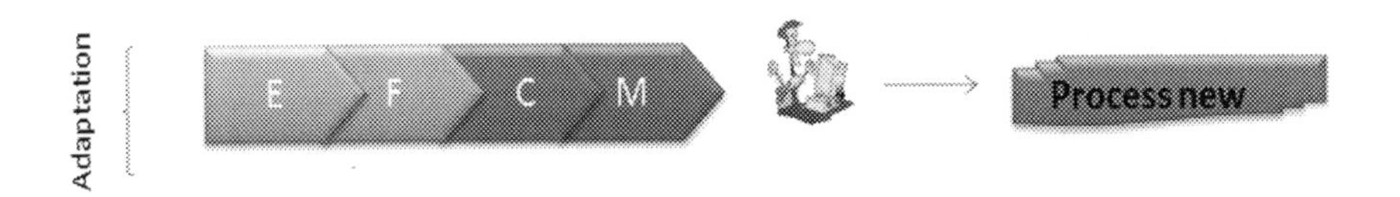

Fig. 8 Process of adaptation. Source: Authors.

3.5 Evolution

All changes generated by the organization, interaction, feedback and adaptation forged with the passing time optimizing the system itself and knowledge management processes, achieving a continuous improvement that leads to evolution (Fig. 9).

Fig. 9 Continuous Improvement. Source: Authors.

3.6 Stage and Packages

Each of the processes seen above apply knowledge management to the development of the packages or components, and this whole process can be repeated over and over again in the development of the package (Fig. 10), where by different stages are formed in the end and increase more precisely in evolution.

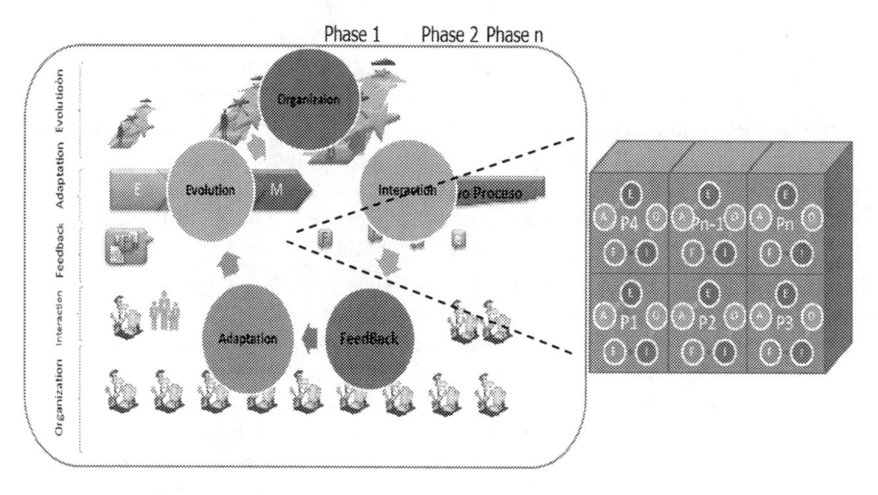

Fig. 10 Process for generating packages and components. Source: Authors.

3.7 Establishment of the Model

In reviewing each of the above concepts it is clear that the development of a complex system such as the software factory, these concepts apply in the development of a whole (Software Product) while an item (Package). When you begin to develop a package, the organization, interaction, feedback, adaptation and evolution adhere to this process, and this occurs again when the packages are grouped into one function, allowing the system to reach the point of becoming complex (Fig. 11).

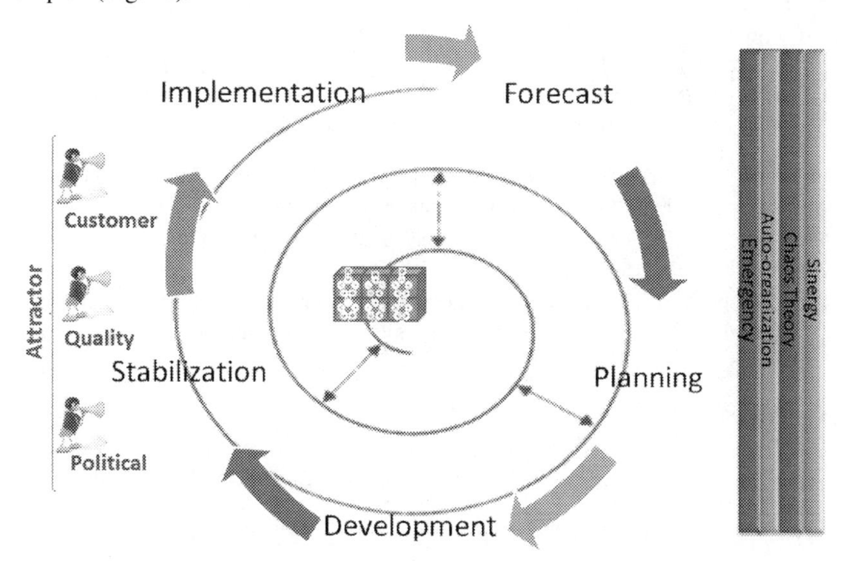

Fig. 11 Model of knowledge management applied to a complex system: Development of software factories. Source: Authors.

Once you take each of the packages, they are implemented to the methodology MSF for agile Software Development at Microsoft under a spiral approach, which includes forescasting, planning, stabilization and implementation. Several packages together represent functionality and this in turn united with others go to join the software product, when all this happens it is notable that there is a self-organization based on events and feedback obtained prior to which can respond to any stimulus. As the functionality of each particular package it can be converted into a goal based on the community which may be called *Emergency* and therefore to obtain each function or procedure separately, will never give the same value as if having the entire project, because 4 + 4 = 5 (Synergy).

But even when you can see the entire structure, there may be some *attractors* like *customers, the quality and politics*, forms and methods with which the organization works, which constitute the firm basis and return the project horizon. But keep in mind it is clear that because the methodological approach is a spiral, any decision taken in the early stages affects the latest and therefore the failure or success of the project, so you can see reflected in this model the theory of chaos.

The resulting model shows that knowledge management is implemented in complex systems because to become complex it must obtain, sort, distribute, transmit and store information making the most of the organization, interaction, feedback, adaptation and evolution (Fig. 11).

4 Conclusions

Both knowledge management and software factories are complex systems, which makes the development of a knowledge management model for a complex system, also be complex. Applying the concepts of complexity to a software development process shows that knowledge flows can be dynamic, open and distributable.

The core complex concepts applied to a management model of the complex system of knowledge, development of software factory are:

- The sum of the parts is greater than the whole. It is a set or combination of things or parts forming a complex or unitary whole. Synergy.
- It is composed of several elements that together can generate a certain behavior. Self-organization emergency.
- The present iterative and incremental development that is contained, can lead to wrong decisions in the first interaction system-wide impact. Chaos theory.

References

1. Lloyd, S., http://en.wikipedia.org/wiki/Seth_Lloyd (Consultado Noviembre de 2010)
2. La gestión del conocimiento [En línea], http://www.gestion-conocimiento.com/ (Consultado Octubre de 2, 2010)

3. Paul, Q., Paul, L., Geoff, J.: Knowledge Management: a Strategic Agenda: Long Range Planning 30(3), a385–a391 (1997)
4. Carrión, J.: Nuevos modelos en Internet para gestionar el talento Conjugación eficiente de las competencias estratégicas, técnicas y conductuales, con el entorno y la organización, para el logro de un desempeño superior. y el conocimiento (2001)
5. Sveiby, Erik, K.: Capital Intelectual: La nueva riqueza de las empresas. Cómo medir y gestionar los activos intangibles para crear valor. Gestión (2000)
6. EKMF. The European Knowledge Management Forum (2004), `http://www.knowledgeboard.com`
7. Medina, V.: Sistemas de Gestión del Conocimiento. Notas de Clase. Universidad Distrital Francisco José de Caldas, Bogotá (2010)
8. Moses, J.: Complexity and Flexibility. Professor of Computer Science and Engineering. MIT/ESD (2001)
9. Senge, P.: The Fifth Discipline. Senior Lecturer, Organization Studies. MIT/ESD (1995)
10. Sussman, J.: The New Transportation Faculty. Professor of Civil and Environmental Engineering. MIT/ESD (2007)
11. Greenfield, J.: Software Factories (2005)
12. Pohl, K., Bockle, G., van der Linden, F.: Software Product Line Engineering. Springer (2005)
13. Ivica, C., Magnus, L.: A Case Study: Demands on Component-based Development. In: ACM International Conference on Software Engineering: Proceedings of the 22nd International Conference on Software Engineering (2000)
14. Jacky, E., France, G.: Reuse and Variability in Large Software Applications. In: Foundations of Software Engineering: Proceedings of the 10th European Software Engineering Conference Held Jointly with 13th. ACM, Lisbon (2005)
15. Lisboa, B.L., Leandro, M.N., de Almeida, E.S., de Lemos Meira, S.R.: A Case Study in Software Product Lines: An Educational Experience. Federal University of Pernambuco and C.E.S.A.R, Recife, PE, Brazil (2004)
16. Short, Greenfield, J., Keith: Software Factories-Assembling Applications with Patterns, Models, Frameworks, and Tools. Wiley Publishin. (2004)
17. Kleppe, A., Warmer, J., Bast, W.: MDA Explained The Model Driven Architect: Practice and Promise. Addison-Wesley (2003)
18. Pressman, R.: Ingenieria de Software un enfoque práctico. McGraw-Hill (2006)
19. Lano, K.: Advanced System Design with Java, UML and MDA. Elsevier (2005)
20. Warmer, J., Kleppe, A.: The Object Constraint Language, getting your models ready for MDA. A. Wesley (2003)

A New Metric to Estimate Project Development Time: Process Points

Charles Felipe Oliveira Viegas[1] and Maria Beatriz Felgar de Toledo[2]

[1] Charles, Instituto de Computação, IC, UNICAMP, 13084-971, Campinas-SP, Brasil
charles.viegas@gmail.com
[2] Maria Beatriz, Instituto de Computação, IC, UNICAMP, 13084-971, Campinas-SP, Brasil
beatriz@ic.unicamp.br

Abstract. In this article, a new metrics to estimate project development time is proposed. The approach is based on business process models. The estimates are divided into four categories: process, screen, integration and service. They are quantified in process points. The project must be modeled as a business process, the four estimates must be calculated and added to obtain the final project complexity. It is a faster method to evaluate the design of complex software systems.

Keywords: Project design, Complex Software Systems, Business Process, BPM, Estimation Metric.

1 Introduction

The dynamism of the nineteenth century has made the search for methods that solve problems quickly more intense in many areas of modern society. In computing, specifically in the area of information technology, this quest is no different. From the stage of elaboration of a project to its completion, there are several stages that have been studied and analyzed in order to obtain a final product with more quality quickly.

In this context, the problem to be addressed in this paper seeks to speed up the preparation of a project proposal estimating its total time in a practical and fast way. This estimation is based on business process models. Sharp and McDermott (2001) [1] define a business process as a structured flow of activities that supports business objectives. Other elements of a process model besides activities are gateways and roles responsible for the activity execution. These are the main elements considered in the proposed estimate. The estimates are divided into four categories: process, screen, integration and service. They are quantified in process points (PP). A process point is the unit that measures the complexity of a process by deriving the size of the process to be automated.

For the time estimate to be feasible, it is necessary that the process documentation is specified as a flow of activities. The four estimates mentioned above are

L. Uden et al. (Eds.): 7th International Conference on KMO, AISC 172, pp. 197–208.
springerlink.com

calculated, and, at the end, the measurement obtained from each category is added to obtain the total project time.

This new approach based on process points is faster and simpler than other previous methods to estimate project development complexity. As software design involves a large number of tasks and many people working collaboratively, it can benefit from this more efficient approach.

This paper is divided into the following sections. Section 2 presents previous approaches that deal with the mentioned problem. Section 3 presents the new approach and Section 4 concludes the paper.

2 Background

The problem of estimating the design of software systems has been studied since a few decades with the aim of developing metrics of project quantification that can estimate the effort and time for the development of programs and systems. One of the oldest forms of estimation is counting of the number of lines of source code. In the literature, there are several ways to count lines of code (Albrecht and Gaffney, 1983 [2]). Some of these forms may only consider or count the number of statements, variables or lines of code. As it counting the number of lines, this metric is also considered a physical measure.

The physical measure is becoming obsolete as it is unable to measure a system or program considering what is really important for this software. It has been replaced by the use of the so-called functional measures. The functional measures are intended to measure the functionality provided by the software thus enabling the use of estimates at the beginning of a project, a fact unlikely to happen if a physical metrics was used. Function Points, Use Case Points and Process Points are examples of functional metrics.

The function point metrics (Dunn, 2002 [3]) estimates the size of a system or the time spent to develop it according to the number and complexity of its functional requirements, from the point of view of the user. This metrics is independent of the computer language, development methodology, technology or capability of the development group.

The points per use case metrics (Kusumoto *et al.*, 2004 [4]) allows measuring software design from the use case. In addition to the use cases, this technique also considers environmental factors and factors of technical complexity to obtain the estimate. Environmental factors determine the level of experience and efficiency of professionals related to the project. Technical factors help determine how difficult it is to carry out this project.

The process point metrics is described in this paper. Given a process model containing the flow of business activities and their actors, the estimate of the time to complete a project can be made. We call PP (process point) the unit that measures the complexity of a process by deriving the size of the process to be automated.

3 An Approach Based on Process Points

The proposed project time estimate is based on the business process model.

Firstly, the process is specified as an activities flow. The estimates are divided into four categories: process, screen, integration and service. They are quantified in process points (PP). A process point is the unit that measures the complexity of a process by deriving the size of the process to be automated.

Afterwards, the four estimates mentioned above are calculated, and, at the end, the measurement obtained from each category is added to obtain the total project time. The four categories are closely related so that from the process category the quantification of the other categories are derived.

The categories are described below.

3.1 Process

A process consists of a flow of activities that models a business system. The process model is composed of elements such as activities, gateways, roles and exception handling (OMG, 2009 [5]). These are the elements considered in the approach proposed in this paper.

The quantification process is performed as an analysis of the characteristics of the process domain. Moreover, these features serve as inputs to measure and estimate the other categories of a project.

The *gateway* is an element that deals with a response of yes-or-not. It usually occurs when a particular flow branches into more than one way or when there is a junction of several branches. For purposes of counting, it will be given a weight equal to "1", considering that it is an item with low complexity and does not generate a great deal of construction.

The *automatic activity* results in the execution of a computer program within the workflow. For purposes of counting, it will be given a weight equal to "2", considering that this is an item of medium complexity, as it involves program calls inside and outside the workflow. The integration with external programs is handled in the "Integration" category(in section 3.3).

Human activity results in the interaction with the end user via friendly screens. These screens can be configured in the tool or coded in an appropriate programming language, as described in the "Screens" category(section 3.2). For purposes of counting, it will be receive a weight equal to "3", considering that the activity of this nature involves more configuration variables that an automatic activity.

The *role* attributes activities to a person, group of persons, or a computer system. It will receive a weight equal to "5", because this item requires a considerable complexity requiring differential treatment to messages, access, interfaces and data exchange.

Exception handling deals with unexpected situations that may occur during process execution. For the purpose of counting, it will have a weight equal to "1" for each automatic or human activity.

The table 1 shows the weights assigned to the items in process analysis.

Table 1 Weights assigned to the items in process analysis

Process	Complexity	PP
Gateways	Low	1
Automatic Activity	Medium	2
Human Activity	Medium	3
Role	High	5
Exception Handling	Low	1

3.2 Screen

The screen category is the most important visual channel for interaction and communication between the business process and the user. The screen can be evaluated by visual factors as color, font type, the screen size or structural factors such as the number of fields and buttons. To make estimates of a screen is only considering te structural factor of a screen because the moment to estimate a screen are those screen elements that define the information that needs to be displayed. Screen estimates only consider the structural factors as they define the information to be displayed.

The screens are divided into two groups according to how they are implemented: configured or coded. Configured screens are generated dynamically through the use of native resources of the workflow and / or BPM tool. The coded screens, in turn, are created by a Web programming language compatible with the used tool. Each group of screen has a specific weight distribution, which varies with the level of complexity, as shown in Table 2.

Table 2 Weight versus complexity given to configured and coded screens

Complexity	Weight of Configured Screen	Weight of Coded Screen
Very Simple	1	30
Simple	2	60
Medium	6	85
High	10	110
Very High	14	135

The difference between the weights associated with the complexity of each group is partly because the encoded screen requires a greater effort in terms of life cycle of software development.

The quantification of screens depends on the amount of information at the time of counting. Therefore, regardless of the type of the screen, its quantification can be divided into:

- An unknown set of screens and / or not specified and;
- A set of known screens and / or specified, that is, screens that show at least a layout sketch.

The unknown or unspecified screens are assigned weight “1” to every human activity existing in the process flow. Next, the way of construction of the screen on the tool (configured or coded), as shown in Table 3, is considered. It is observed that the distribution of complexity is based on the historical of screen development in previous projects.

Table 3 Distribution of the complexity of screens

Complexity	Tax of Configured Screens	Tax of Coded Screens
Very Simple	30%	5%
Simple	25%	35%
Medium	20%	40%
High	15%	15%
Very High	10%	5%

The classification of the level of complexity when screens are known and specified depends on the amount of screen elements. These are the elements considered: fields, buttons, combos, radio and tables. Each element will receive weight 1 except for table that will receive 1 for each column. Thus, a screen with two fields, one button, a combo, a radio and a table of three columns will have a total of 8 elements.

Table 4 is an evolution of Table 2 and incorporates the number of elements of the screen to the complexity and the respective weights assigned to configured and coded screens.

Table 4 Number of elements versus complexity versus the weight given to the screen

Number of elements in the screen	Complexity	Weight of Configured Screen	Weight of Coded Screen
$x \leq 5$	Very Simple	1	30
$5 < x \leq 10$	Simple	2	60
$10 < x \leq 30$	Medium	6	85
$30 < x \leq 60$	High	10	110
$x > 60$	Very High	14	135

3.3 Integration

The integration category includes a communication process with other business systems to facilitate the process operations. In this category, information systems external to the process are obtained automatically without user intervention. The estimate integration is based in the difficulty of communication between the process and other system.

The integrations always consider the client and server sides. The client side includes the creation of programs with the goal of sending and receiving data from legacy systems. The server side represents outsourcing of the functionality of legacy systems that are accessible through interoperable standards. In Khalaf *et al.* (2006) [6] there are examples of integrations. Table 5 defines the levels of complexity for the integration of client and server sides.

Table 5 Definition of levels of complexity for the integrations

Complexity	Client
Very Simple	Native clientes of the workflow and/or BPM tool that can communicate with legacy systems.
Simple	Native clients of the workflow and/or BPM tool that can communicate with legacy systems such as Web services and EJB.
Medium	Clients in programming languages compatible with the workflow tool that uses native connectors of platforms such as JMS, XML e CSV.
High	Clients in programming languages compatible with the workflow and/or BPM tool that do not use native connectors of the used platform, but communicate with legacy systems of lower platforms such as DLL, CMD e OLE.
Very High	Clients in programming languages compatible with the workflow and/or BPM tool that do not use native connectors of the used platform and communicate with legacy systems with high platform or middleware such as EntireX.

Complexity	Server
Very Simple	Server code of low or medium difficulty level written in commercial programming languages such as Java and .NET
Simple	Server code of low or medium difficulty level written in commercial programming languages such as Java and .NET that do not use APIs such as Web services and EJB.
Medium	Server code in commercial programming languages such as Java and .NET that use proprietary API.
High	Server code in programming languages with either infrequent commercial use or from discontinued platforms running in lower platforms.
Very High	Server code in programming languages with either infrequent commercial use or from discontinued platforms running in high platforms.

The counting of integrations varies with the level of information in the process. When there is insufficient information, it is necessary to count the automatic activities that will become future integrations with external systems.

For each automatic activity two integrations will be considered: one for the client side and one for the server side. At the end of the count, the distribution of complexity for both sides is applied, as shown in Table 6.

When there is information or integrations are specified, it is enough to map the integration according to its level of complexity, as described in Table 5.

Table 6 Distribution of integration complexity for unspecified integrations

Complexity	Client	Server
Very Simple	80%	10%
Simple	20%	35%
Medium	0%	35%
High	0%	15%
Very High	0%	5%

Next, the weight distribution is applied according to the complexity level for client and server sides, as shown in Table 7.

Table 7 Distribution of weight in relation to the complexity of integration

Complexity	Client Weight	Server Weight
Very Simple	2	30
Simple	4	60
Medium	8	85
High	12	110
Very High	16	135

3.4 Service

The service category is responsible for estimating the systemś functional requirements which include the actions performed by a user. These are used in the process functionality as services to provide reusability and maintainability of the system. The estimated service depends of the level of complexity of the functionality involved.

Authors such as MacKenzie *et al.* (2006) [7] define service as a mechanism to enable access to one or more capabilities provided through a pre-defined interface. A service is consistent with the descriptions of restrictions and policies. Erl (2007) [8] defines services as standalone software programs with different characteristics that allow the alignment of strategic objectives with service-oriented computing.

Services may be process flow or logic operations. The flow services are operations to execute the state machine of process flow and logic operations services include manipulation or data analysis. Typically, flow services are actions that involve the verbs receive, send and inform. The services of logical operations are actions to calculate, verify, update, simulate, among others.

Table 8 shows the definition of complexity associated to services.

The counting of services also varies with the level of information in the process in relation to automatic activities. When there is only basic information, services are distributed according to the distribution of complexity presented in Table 9. When there is more concrete information, services are mapped according to their level of complexity, as shown in Table 8.

Table 8 Complexity of services

Complexity	Service Types
Very Simple	Native services of the workflow tool that, by means of configurations, facilitate development in Java.
Simples	Native services of the workflow tool that, by means of configurations, facilitate development of Web services and EJB.
Medium	Services in programming languages compatible with the workflow tool that use native connectors of the used platform. Examples are JMS, XML e CSV.
High	Services in programming languages compatible with the workflow tool that do not use native connectors of the used platform, but communicate with legacy systems of lower platform such as DLL, CMD e OLE.
Very High	Services in programming languages compatible with the workflow tool that do not use native connectors of the used platform and communicate with legacy systems of high-level platform or middleware such as EntireX.

Table 9 Distribution of the complexity of services

Complexity	Flow Operations	Logic Operations
Very Simple	100%	80%
Simple	0%	20%
Medium	0%	0%
High	0%	0%
Very High	0%	0%

Table 10 Relationship between the distribution of complexity versus weight

Complexity	Flow Operations	Logic Operations
Very Simple	2	30
Simple	4	60
Medium	8	85
High	12	110
Very High	16	135

Next, weight distribution is applied varying with the level of complexity, as shown in Table 10.

It is important to highlight that each automatic activity can originate a service or an integration, depending on the action to be performed by this activity.

3.5 Example

Suppose that, from a business process model, we have the following information about the process:

- (i) Gateways = 5;
- (ii) Automatic activities = 15;
- (iii) Human activities = 34;
- (iv) Manual activities = 5;
- (v) Roles = 7;
- (v) Exception handling = 15.

About the 15 automatic activities, we have the following information:

- (i) 10 integrations with external systems;
- (ii) 5 services, 2 flow operations and 3 logic operations;
- (iii) 15 services, 2 flow operations and 3 logic operations;15 exception handling (at least one for each activity).

Considering that there are 34 human activities, the minimum number of screens are 34. All of them are coded screens. Among these, 14 are not specified and 20 are known. For those 20 specified screens:

- (i) 10 screens have at most 5 elements;
- (ii) 7 screens have 12 elements;
- (iii) 3 screens have 31 elements.

For this scenario, table 11 has the estimate for each category.

For the screens, firstly, it is necessary to obtain the number of screens of each complexity and, then, the process point is applied. Table 12 shows the estimate for screens. The total number of screens is known screens plus unknown screens for a given complexity.

Table 11 Example of the estimated size of the process category

Process	Amount	Weight	PP
Gateways	5	1	5
Automatic Activity	15	2	30
Human Activity	34	3	102
Manual Activity	5	1	5
Role	7	5	35
Exception Handling	15	1	15
		Total	192

Table 12 Example of the estimated size of the screen category

Complexity	Amount Known Set of Screens	Amount Unknown Set of Screens	Weight of Screen	Coded PP
Very Simple	10	0.7	30	321
Simple	0	4.9	60	294
Medium	7	5.6	85	1071
High	3	2.1	110	561
Very High	0	0.7	135	94.5
			Total	2341.5

The process has 10 integrations with external systems that are non-specified integrations composed as Client (C) side integration and Server (S) side integration. Table 13 shows the estimate for integration.

Table 13 Example of the estimated size of the integration category

Complexity	Amount C	Weight C	PP C	Amount S	Weight S	PP S
Very Simple	2	8	16	30	1	30
Simple	4	2	8	60	3.5	210
Medium	8	0	0	85	3.5	297.5
High	12	0	0	110	1.5	165
Very High	16	0	0	135	0.5	67.5
		Total	24		Total	770

There are 5 services, 2 flow operations (FO) and 3 logic operations (LO). Table 14 has estimates for service.

Table 15 shows that the estimate for this scenario was 3.439,5 PP. Analyzing the application of process points in concluded projects, we may say that 1 PP corresponds to a productivity that varied from 2 HH/PP in the first iterations to 1.25 HH/PP when the project was finishing.

Table 14 Example of the estimated size of the service category

Complexity	Amount AF	Weight AF	PP AF	Amount OP	Weight OP	PP OP
Very Simple	2	2	4	30	2.4	72
Simple	4	0	0	60	0.6	36
Medium	8	0	0	85	0	0
High	12	0	0	110	0	0
Very High	16	0	0	135	0	0
		Total	4		Total	108

Table 15 Estimate For This Scenario

Category	Table	PP
Process	Table 11	192
Screen	Table 12	2341.5
Integration	Table 13	794
Service	Table 14	112
	Total	3439.5

4 Conclusions

This paper proposes a new method to estimate project development time. This estimation is based on business process models. The estimates are divided into four categories: process, screen, integration and service. They are quantified in process points. A process point is the unit that measures the complexity of a process by deriving the size of the process to be automated.

This new approach based on process points is faster and simpler than other previous methods to estimate project development complexity. As software design involves a large number of tasks and many people working collaboratively, it can benefit from this more efficient approach.

As future work we intend to further refine each category, we want to study and perform the estimation of productivity of the team that participated in the project and we intend to calculate the total hours spent by the team involved in the project and the time spent in each stage of a project.

References

1. Sharp, A., McDermott, P.: Workflow Modeling. Artech House, Boston (2001)
2. Albrecht, A.J., Gaffney, J.: Software Function, Source Lines of Code, and Development Effort Prediction: A Software Science Validation. IEEE Transactions on Software Engineering 6 (1983)
3. Dunn, A.: Function Points, Functional Requirements, Functional Documentation. In: IFPUG Annual Conference (2002)

4. Kusumoto, S., Matukawa, F., Inoue, K., Hanabusa, S., Maegawa, Y.: Estimating effort by use case points: method, tool and case study. In: Proceedings of 10th International Symposium on Software Metrics, vol. 2, pp. 292–299 (2004)
5. Business Process Modeling Notation. OMG (2009), http://www.omg.org/docs/formal/09-01-03.pdfofstandarddocument
6. Khalaf, R., Keller, A., Leymann, F.: Business processes for Web Services: Principles and applications. IBM System Journal (2006), http://www.research.ibm.com/journal/sj/452/khalaf.html
7. MacKenzie, C.M., Laskey, K., McCabe, F., Brown, P.F., Metz, R.: Reference Model for Service Oriented Architecture 1.0. OASIS (2006), http://docs.oasis-open.org/soa-rm/v1.0/ofstandarddocument
8. Erl, T.: SOA - Principles of Service Design. Prentice Hall International (2007) ISBN: 978-0132344821

Intellectual Assets and Knowledge Engineering Method: A Contribution

Alfonso Perez Gama and Andrey Ali Alvarez Gaitán

Fundación Educación Superior San José, Bogota, Colombia
japerezg@ieee.org, desarrollotecnologico@fessanjose.edu.co

Abstract. The Higher Education Institutions, HEI are the natural space where the ethos of knowledge can be developed and the Intellectual Capital, IK can be accumulated. It's no secret the inadequacy of the Management Information Systems, MIS to incorporate knowledge. In fact is shared by many authors that well over half the market value is explained by the IK in companies of the Third Millennium. Traditional systems do not account for the IK and only estimate the book values vs. market values when selling focused on negotiation processes. On the other hand, there are already quite a few financial models of the knowledge economy that considers the IK. For the universities is a critical issue in the way of the Competitiveness of Higher Education. Our initiative EIDOS constituted unconventional ideas that have not had an audience in the context of Colombia. Now beginning to break through what an open space is required with basic units in the enterprise architecture of the HEI focused on innovation; This was stated by the Government Plan 2010-2014, in this country as an national strategy. These new architectures must be defined and supported in order to facilitate the stream of information and knowledge with new sources of generation, especially those where the IK did not exist. It begins to think that IK is the basis of sustainability and financial governance. Knowledge Engineering (KE) maturity level is inderway. As background it should be mentioned that this research originated to facilitate the use of development methodologies, information systems and evolve them to the incorporation of layers of knowledge for an Intelligent Management Information IMIS, to the various organizational and technological changes that occur in building new systems to meet customer requirements in improvement of processes, new demands of companies in the Knowledge Society and basically make the intelligent performance of the firm. Our research effort represents ongoing work with several University projects and unpublished graduate works and also consulting works for institutions in Colombia. In the FESSANJOSE the model has evolved to our proposal of social innovation Leontief Model that incorporate the knowledge sector. The KE methodology is shortly presented in this paper has been an attempt for reducing the complexity involved: is part of that effort and has not obeyed to a large-scale project with good funding. They have been serene developments that we want to share with the academic community. We describe an attempt to formalize a KE methodology in 8 steps which are presented in detail, the mix of ideas coming from both Engineering and

L. Uden et al. (Eds.): 7th International Conference on KMO, AISC 172, pp. 209–218.
springerlink.com

Management we tried to integrate in this construct. The full method can be read in the 2nd chapter titled "Analytical Models for Tertiary Education by Propaedeutic Cycles Applying Knowledge Engineering and Knowledge Management".

Keywords: Knowledge Engineering, Intelligent Management Information Systems, Intellectual Capital, Knowledge Base.

1 Backgrounds

We have faced the problem of HE in Colombia with the use of the Propaedeutic Cycles, which deals many serious problems of student dropout and mainly the lack of competitiveness of this sector. The foregoing led us to rethink the matrix of input / output in terms of curricular knowledge and competences to be cultivated in students to submit papers in several international conferences. In recent years we have redefined the architecture of the Leontief model by introducing the sector of knowledge and applied to our university, FESSANJOSE.

The Internationalization strategy in our Institution, FESSANJOSE seeks further movement of resources, including knowledge assets. It is also intended to promote quality trans-boundary HE, developing programs and giving them more competitiveness as an worldwide dimension, promoting international qualifications, language skills and multicultural understanding in the process as well as through the high training of lecturers for intellectual assets knowledge transfer competitiveness with global vision. The process of internationalization for HEI results in mobility and circulation of people, knowledge, goods and services including pedagogical, ideas and values cultivated for abroad institutions.

The HEI are the natural space where the ethos of knowledge can be developed and the IK can be accumulated. It's no secret the inadequacy of the Management Information Systems, MIS to incorporate knowledge. In fact is shared by many authors that well over half the market value is explained by the IK in companies of the III Millennium. Traditional systems do not account for the IK and only estimate the book values vs. market values when selling focused on negotiation processes. On the other hand, there are already quite a few financial models of the knowledge economy that considers the IK. For the universities is a critical issue in the way of the competitiveness of HE.

Now beginning to break through what an open space is required with basic units in the enterprise architecture of the HEI focused on innovation; this was stated as a national strategy. These new architectures must be defined and supported in order to facilitate the stream of information and knowledge with new sources of generation, especially those where the IK did not exist. It begins to think that IK is the basis of sustainability and financial governance.

Our work has been developed with academic rigor, disciplined and continued: the documentation of the project is a key part of the methodology we use to formalize and enrich our research ideas and innovation. They are subjected to the contrasting, in international meetings, world class, and obviously by systematizing the feedback provided by international pairs of very high scientific level.

1.1 Proposals for an Intellectual Assets and Knowledge Engineering Methodology

KE is considered as the systematization of intelligence and relies on the work of organization, representation and knowledge management as a substitute for human reasoning and reasonable basis to encapsulate solutions to hard problems and difficult tasks, also serves to identify the layers of knowledge of an information system. In itself, the KE, is the technology for building expert systems and intelligent agents. Clearly the maturity of the KE requires a disciplined work and treatment of problems, analysis and solutions, supported by knowledge.

1.2 Some Implications of Intelligent Engineering Solution

We found that extensive knowledge in the subject or study target, informal or explicit is one of the consequences. It is required the support for heuristics analysis or for the same creativity. Finally the main competences required are: the ability to infer knowledge, the ability to reason or symbolic processing and the ability to apply an appropriate logic of the problem.

In e-Commerce, the increase in business intelligence (BI) is impressive, supported by ES (expert systems) and intelligent systems. An ES is seeking a satisfactory solution, good enough for the job, although not optimal. The output of an ES depends on the amount of knowledge and depth of analysis is made on the knowledge of experts.

2 Engineering of the Knowledge

This construct refers to the techniques of obtaining knowledge, representing, building and using it appropriately to assemble and explain reasoning of the knowledge based system, creating a product which tries to solve the involved problem. The following defined steps were indentified: The following figure describes graphically the proposed methodology of KE. The first 2 phases is related to the problem Definition and Determination.

1. Cognitive Planning: defining the system (structure and components), model (behavior), detailed specification of the problem, selecting methods and techniques, user requirements and interface, identifying the data types and objects and solution quality.
2. Learning: Acquisition of knowledge or machine learning process: acquisition of knowledge required by the system, in a systematic or automatic or with help from the experts. This knowledge can be specific when the problem domain

and solution procedures are the same, or may be general knowledge, i.e. knowledge across an enterprise.

The subsequent 3 phases are related to the development of the intelligent system:

Ontology, representation and organization of knowledge: how it is distributed, classified, organized, developed the map of knowledge and is represented by some formalism or structured information or objects in the KB, to enable computer use.

3. Knowledge Acquisition Methods
4. Problem Solving Design alternatives and selection of the solution.
5. The final 3 phase are related to the system construction.
6. Prototype Construction.

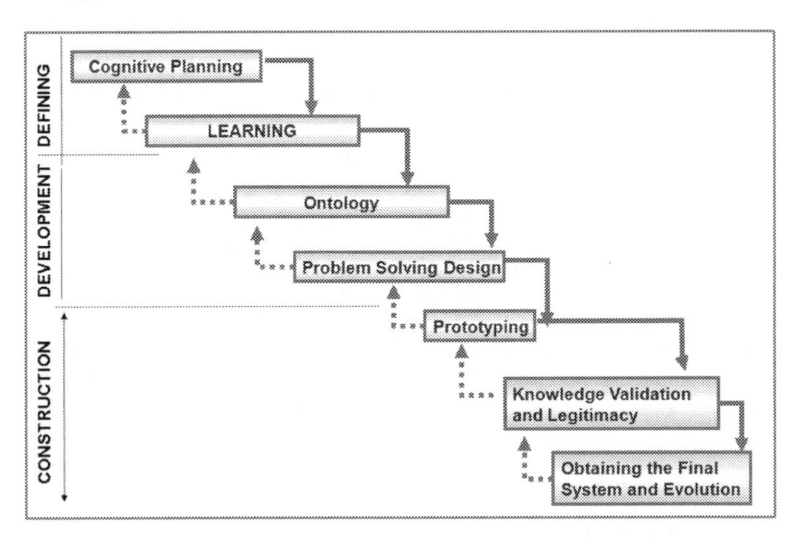

7. Verification, validation, contrasting and legitimacy: the methodology used successfully for the problem? Verify sources of knowledge. Legitimate and check that the knowledge is represented in the KB, is correct. To demonstrate that the solution and conclusions are expected. The inference is based on the design of software, which allows the computer to infer on the knowledge represented in the BC. This includes: Explanation and justification
8. Obtaining the Final System and Evolution.

1. KE Cognitive Planning

We begin by establishing the commitment of senior management and stakeholders and also the involvement of the group of domain experts. Preliminary meetings are held with this group, which reviews the key issues to consider in connection with the layers of knowledge to add. Planning is also made as to form and means to acquire such skills.

1.1 System Determination.
1.2 Determination of the subsystem to incorporate knowledge
1.3 Determination of the specific module:
1.4 Determination of Intellectual Assets.

1.1 Intellectual Assets

There is no consensus on the definition of what an industry knowledge is nor what is an intellectual assets and how can be quantified. One question is related to determining the processes of knowledge within a company, enabling design of a new enterprise architecture as a basis for developing competitive strategies in the Third Millennium.

Based upon strategic information (Mission, Vision, Strategies) and supported by the organization with its current information system, process engineering and academic institutions is built as shown bellow:

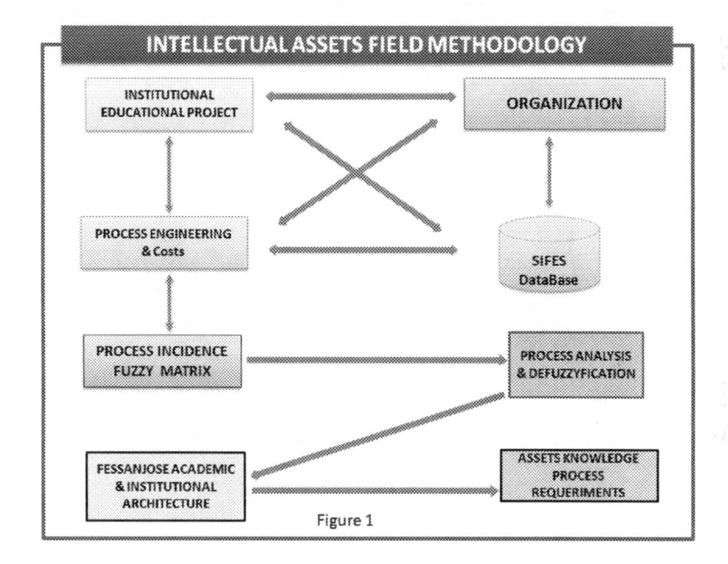

Figure 1

The incidence matrix is inferred by the process flow analysis: determining the interactions between the processes in terms of purchases (inputs) and sales (outputs); This is powered by a survey of process owners. The result is presented below in Figure 2.

This cell may contain other qualitative values (fuzzy logic or ambiguity). Each value represents the incidence between the subject and skills.
$[\![a]\!]\_(i,j)]$ = {yes, not} (black, white), or$[a_(i,j)]$ = {null, medium, high} (white, red, and blue for a dashboard chart)..

	Incidence Matrix of Institutional & Academic Processes														Academic Programs						
	Management	Knowledge process	Accountancy	Treasury	Human Resources	Admissions	General Services	Information Systems	Inventories	University Welfare	Labs	Extension Service	Library	Research	Curricular	Lecturers	Students	Permanence & Desercion	Alumni & Graduates	Research & Social	Virtual Education
Management	2	2	1	2	1	2	1	2	2	1	1	1	1	1		1	1		1	2	2
Knowledge process	2	2			1				2						2	2	2		2	2	2
Accountancy	2	1	2	2	1	2	1	2	2	1	1	1	2			1	1		1		
Treasury	2	1	2	2	1	2	1	2	2							1	1		1	1	1
Human Resources	2			2	2	1	1	2	1	1			1			2					1
Admissions	2			2	1	2		1	1												
General Services	2				2	2	2	2	2												
Information Systems	2	2	2	2	2	2	1	2	1	1	1	1	1		2	2	2	1	2	1	1
Inventories	1		1	1	2		2	1	2		2		2							1	1
University Welfare	1				1	1		1	1	2		2				1	2	2	2	2	2
Labs	1		1		1	1	2	2	2		2		1	1	2	2	2	1	2	1	1
Extension Service	1	2			1	1	1	2			2				1	1	1	2	2	2	1
Library	1	2	1	1			1	2	2		2		2	1	1	2	2		2	2	2
Research	1	2						1					2	2	2	2	2	1	1	1	1
Academic Programs: Curricular		2			2	2	1	2	1	2	2	2	2	2	2	2	2	2	2	2	2
Lecturers	1	1	1		2	1		2	2	2	2	2	2	2	2	2	2	2	2	2	1
Students	1					1	1	2		2	1	2	2	1	2	2	2	2	1	1	1
Permanence & Desercion	1						1	2		2		2			1	1	1	2	2		1
Alumni & Graduates	1	1						1		2		2	2	2	1	1	1	1	2	2	1
Research & Social	2	2	1	1	1			2	1	1		2		2	1	1		1	2	2	1
Virtual Education	2	1	1	1	2	2	2	1	1	2	2	2	2	1	2	2	2	2		1	2
Costs	g_1	g_2	g_3	g_4	g_5	g_6	g_7	g_8	g_9	g_{10}	g_{11}	g_{12}	g_{13}	g_{14}	g_{15}	g_{16}	g_{17}	g_{18}	g_{19}	g_{20}	g_{21}

Figure 1: Incidence Matrix

Process quantification from the INCIDENCE MATRIX:

The contribution of the j process is obtained by the following matrix:

$$\sum_{i} a_{i,j} g_j = d_j \qquad (1)$$

Where g_j is the corresponding cost d_j is the input contribution of the j[a] process

The process production i^a is as follows:

$$\sum_{j} a_{i,j} g_j = b_i \qquad (2)$$

Where b_i is the i^a process production mentioned, expressed for each rows.

Knowledge has played the second role in connection to other factors of production like land, capital and labor within any company. A few years ago that knowledge as an Intellectual asset was brought into the mainstream of modern organizational theory. It is now very clear that the primary business of organizations (specially the universities) is to sell knowledge based products and that competitiveness hinges on the effective management of all organizational intellectual assets. The internationalization inside the university is a process leading to the Knowledge Society.

Phase Outcomes: The recognition of the classes of knowledge; The Identification of the layers of knowledge of the company (knowledge architecture) and knowledge processes; The determination of the intellectual assets; The specification of the methodology, tools and techniques; A Portfolio of the intelligent subsystems; The Intellectual capital resources: e.g. Knowledge bases, Study Plans; Scheduling and Time table; Quality Assurance Plan

2 Knowledge Acquisition Phase: Learning

Taking the problem and its modular decomposition, followed by the conceptualization of relevant knowledge, must define its scope. When knowledge has been collected, must be analyzed, coded and documented so that these activities take place according to the chosen acquisition technique. It also should be prioritized to represent in the KB. Later must be formalized to determine the method of acquisition. The acquisition stage can be difficult to the extent that knowledge is extracted directly from human experts. Once the execution is carried out, it should be the implementation, which includes programming and coding of knowledge in the computer, designing a prototype that allows refining and contrasting the results. Finally, the knowledge engineer test at the KB by means of examples (cases) and compares the results with experts to examine the validity of knowledge.

Phase Outcomes: Knowledge Recovery; Knowledge Bases; Repositories; Conceptual Map of the Knowledge or Syllabus; Knowledge Structures

3 Knowledge Representation Phase: Ontology

The representation of knowledge involves the consistent use of mathematical logic and information structures. The KE theories fall into two categories: mechanisms and content. Ontology contents theories about classes of objects, their properties and their interrelationships, which allow the specification in the domain of knowledge.

Phase Outcomes: Knowledge Bases with fully reorganization; Internal and External Networks; Portfolio Capabilities: Assimilation, Strategic Technology and Innovation.

4 Knowledge Acquisition Methods

Working with KE involves a transdisciplinary team, in which the engineer is the intermediary between the KB and the Experts, refining and representing the KB. There are at least three types of methods for the acquisition of knowledge such as: manual, semi-computer based and automatic: artificial or machine learning.

Phase Outcomes: Assessed and appropriate method; Determination of appropriate computational resources; Tutorial Online Help; Determination of alternative sources of knowledge

5 Problem Solving Solutions Design

The design of responses to problems or needs raised to be knowledge based is presented in several steps which are described as follows:

5.1 Defining of Potential Solutions
5.2 Determining the knowledge related to the relevant parts of the module
5.3 Determining of Knowledge Related for Problem Solving
5.4 Tasks Decomposition
5.5 Identifying of Production Rules
5.6 Identifying of meta-rules and frameworks

Phase Outcome: Selected Method for solving the problems identified; Problem solution domains; Meta-cognition; Specifications accomplished.

6 Prototyping

The first objective is to build a small prototype, for which selects a subset of the KB and carries to the KE tool, which must be done quickly. The result is a prototype which can quickly verify the implementation and testing and verifying ideas. In this sub-step is the representation of knowledge with the tool by means of a prototype, since this technique to identify weaknesses and strengths of the model developed, by which you can refine the results to get quality. A prototype is also a good way to test the concepts before investing in a larger program. With the use of a shell can quickly assemble a small prototype to determine if you are on the right track. It allows for demonstrations, as its assessment will also be important in determining the quality of the result.

6.1 Graphical Representation of Knowledge Layers
6.2 Inference Proof

Phase Outcome: Verified operational prototype and with quality assurance; Documentation of software testing and prototype; Effective and proven outcomes; Prototype of Dual-Use: as a Professional Tool and for Powered Training Tutorial.

7 Knowledge Validation and Legitimacy

A used specific case to test the quality of the KB is acceptable, verifying that the source of knowledge is accurate.

7.1 Validation and contrasting of results with the Experts
7.2 Refinement
7.3 Assessment
7.3.1 Verifying and Validation:

Phase Outcome: Correct source of knowledge; Methodology used properly; The solutions checked and conclusions as expected; Refined knowledge System.

8 Obtaining the Final System and Evolution

One advantage of rule-based systems is that they are modular, so can build subdivisions of large systems and then test them step by step, which is made possible with prototypes.

One approach is using the evolutionary prototyping, where a prototype is enhanced until obtaining the final system, which can be built gradually by adding pieces, parts or modular components. This is a cyclical process, which is an advantage that can improve the result and get the best approach to user requirements and needs of knowledge of IMIS.

If each subsection is evaluated and approved separately, the final system will most likely work the first time.

In this sub-step is performed integration of subsystems that is, if KB is created different or heterogeneous in terms of sources, we must take into account the need to create interfaces between them, allowing optimal operation and communication between the system and users.

Phase Outcome: Integrated, Robust intelligent System; Populated Knowledge Bases; Institutional Learning System; Intellectual and Knowledge Assets; Institutional Memory; Tutorial Intelligent Engineering Knowledge.

3 Conclusions and Future Work

This proposal provides methods that reduce the complexity of knowledge management which are required to enterprises of the Third Millennium. The formalization of accounting for assets and liabilities of IK with endpoints is required and it is also required the reengineering of business processes to identify which of them generate value and advantage competitive and decide which of them have to be outsourced but fundamentally require investments in building capacities of knowledge within the company. We are aware that more empirical work is required complementary and many theoretical efforts. The research has resulted in a tool that has allowed us to quantify our goals, to make viable and to face the serious problems of the Colombia Higher Education Sector for which we have innovated the Inter-industry model I/O W Leontief incorporating the knowledge as a sub-sector of the economy, by the identification of a model architecture. The world has become more complex, enterprise application development requires new engineering approaches and architectural approaches. Similarly, more specialized cognitive abilities around the working groups. It takes audacity KE alternatives, systems architecture and design, taking into accounts the systemic qualities and cost-benefit to the organization.

Acknowledgments. To "Fundacion de Educación Superior FESSANJOSE", to Lecturers and Officers, especially to SAN JOSE EIDOS INTERNATIONAL RESEARCH TEAM including Dr Carlos F. Pareja, Dr. Guillermo Hoyos, Byron Alfonso Pérez G., Leyini Parra, José Manuel Perilla, Roger Londoño, Luis Jefferson Arjona, to all of them a debt of gratitude. Special mention of our Graduate Students: Hada Jessica Perez G, Raúl Alexander Alonso Marroquín, Javier Humberto García Torres, Alcira Fernandez F, Dr. Oscar Vanegas and Professor Octavio Ramirez Rojas from the Remington University at Medellin. In particular, our gratitude to Ministry of National Education for the support received under the Agreement 755, 2011: Dr Diana Marcela Duran, Jorge Hernan Franco-Gallego and Sandra Patricia Sanabria.

References

New Research on Knowledge Management Models and Methods. Hou, H.-T. (ed.) Intech Open Books, ISBN 978-953-51-0190-1 2012, http://www.intechopen.com/books/new-research-on-knowledge-management-models-and-methods

Gama, A.P., Tommich, M.Q.: The International Accreditation of the Engineering by Propaedeutical Cycles: An Imperative For Colombia. In: EDUCON 2012 IEEE Global Engineering Conference, Marrocco (April 2012)

Gairín, J., Rodríguez-Gómez, D., Armengol, C.: Agents and Processes in Knowledge Creation and Management in Educational Organisations. In: Hou, H.-T. (ed.) New Research on Knowledge Management Models and Methods, Department of Applied Pedagogy, Universitat Autonòma de Barcelona; Spain, ch. 15. Intech Open Books (2012)

Campbell, H.M.: The Liberation of Intellectual Capital Through the Natural Evolution of Knowledge Management Systems. In: Hou, H.-T. (ed.) New Research on Knowledge Management Models and Methods, ch. 19. Intech Open Books, Vaal University of Technology, South Africa

Rhaods, C.J.: Managing Technology Information Overload: Which Sources of Knowledge are Best. IEEE Engineering Management Review 38(1) (March 2010)

Chandrasekaram, R., et al.: What are Ontologies and Why Do We Need Them. IEEE Intelligent Systems 14(1) (1999)

Fang, L., et al.: A Decision Support System for Interactive Decision Making - Part I and Part II. IEEE Systems Man and Cybernetics (February 2003)

Rouse, Will: Need to Know –Information, Knowledge and decision Making. IEEE Systems Man and Cybernetics (November 2002)

Octavio, R.R.: La entidad Capital Intelectual y financiero; vol. I & vol. II. temaspara la reflexión y el accionarestratégico y operativo

Gama, A.P., Gómez, G.H., Gutiérrez, B.A.P.: Knowledge for the Society: Social Inclusion to the Superior Education High Competetivity: A Knowledge-Based Governability Model for the Higher Education in Colombia by Propaedeutic Cycles. In: IEEE EDUCON 2011 Annual Global Conference Engineering Education – Amman Jordan, April 4-6 (2011) ISBN : 978-1-61284-641-5

Gama, A.P., Gomez, G.H., Mena, A.M., Gutiérrez, B.A.P.: A Knowledge Based Analytical Model of Propaedeutic Cycles: Linking Media Education to Higher Education in Colombia. In: IEEE EDUCON 2010 Annual Global Engineering Education Conference, Madrid España (April 2010); Conference Book ISBN 978-84-96737-70-9. CD ROM IEEE CATALOG CFP10 EDU-CDR ISBN 978-1-4244-6569-9

The Effect of Connectivism Practices on Organizational Learning in Taiwan's Computer Industry

C. Rosa Yeh* and Bakary Singhateh

Graduate Institute of International Human Resource Development,
National Taiwan Normal University, Taiwan, ROC

Abstract. Technology has altered the way we learn and work. This study hopes to help business leaders and corporations recognize the crucial role of these societal-changing technologies that link people to information in the digital age. This study explored the effect of technology on organizational learning from the perspective of connectivism. Practices of connectivism studied include social software technologies and knowledge management practices. Quantitative survey questionnaires were sent to 301 companies in the computer industry across Taiwan, resulting in 80 valid responses. Hierarchical regression was used to test study hypotheses. Hypotheses on the direct effects among innovative corporate culture, practices of connectivism and organizational learning were supported. Additionally, companies that were younger or in more remote locations were found to have higher motivation to innovate, learn and adopt new technologies formally.

Keywords: Innovative corporate culture, Social software technologies, Knowledge management practices, Connectivism, Organizational learning.

1 Introduction

Technology has altered the way we learn and work in the digital age. Prevalent theories of learning and knowing (behaviorism, cognitivism, and constructivism) have not kept pace with societal and technological progresses. Connectivism is therefore seen as an alternative theory of learning that addresses the inadequacies of current theoretical models (Siemens, 2005). Connectivism is a new theory of learning that recognizes technology has impacted society and that thoughts on how we work and learn are shifting. Connectivism posits that "know-how" is now replaced by "know-where" the information/knowledge is, hence proposing a more flexible and networked learning environment.

As knowledge creation is pertinent to today's knowledge-intensive and highly globalized workforce, it is critical that knowledge reaches the people who need it

* Corresponding author.

L. Uden et al. (Eds.): 7th International Conference on KMO, AISC 172, pp. 219–229.
springerlink.com

to effectively do their jobs. Therefore this paper posits that if the knowledge workers within the workplace have to maintain a sustained competitive growth, then they have to be part of a "wired workforce", a connected working environment where information flow, sharing, collaboration, networking, and exchange of knowledge is not restricted. Towards this endeavor, it is not enough to have only social software technologies in place, but also prudent that knowledge management practices are also utilized to ensure information or knowledge is disseminated in a timely fashion to all organizational functions that need it. For this purpose, a terminology called "practices of connectivism" is deployed in this study to envelop social software technologies and knowledge management practices.

For these practices to impact positively on organizational learning, it must be supported by the organization's corporate culture. Organizations are therefore challenged to develop corporate cultures that exploit the opportunities brought about by web 2.0 and it's accompanying new technologies. Globally today, companies are faced with unprecedented environmental changes, and as such speedy and effective adaptation is required; therefore these organizations must have to learn in order to survive. Hence, how organizations learn in the realm of massive information and abundance of knowledge becomes a crucial factor. This paper therefore investigates the impact an innovative corporate culture will have on practices of connectivism, as this may ultimately impact on organizational learning in companies.

2 Literature Review

2.1 Organizational Learning

The term organizational learning has been viewed from different perspectives by researchers (Jones & Hendry, 1992; Garvin, 1993; Argyris & Schon, 1996). Furthermore, various studies have attributed organizational learning to different organizational phenomena, such as management innovation (Stata, 1989), increasing an organization's capacity to take effective action (Kim, 1993), and reducing defensive routines (Argyris & Schon, 1996).

A review of literature provides an insight into the various definitions of organizational learning. Many studies defined organizational learning from a process view, in which members of an organization detect errors or anomalies and correct it (Argyris, 1977), knowledge about action outcome relationships between the organization and the environment is developed (Daft & Weick, 1984), actions can be improved through better knowledge and understanding (Fiol & Lyles, 1985), performance can be maintained or improved based on experience (Nevis, et al., 1995, p.73). Other studies defined organizational learning subtly different as facilitating its members and continuously transforming itself (Pedler, Burgoyne & Boydell, 1991), or creating, acquiring, and transferring knowledge, and modifying its behavior to reflect new knowledge and insights (Garvin, 1993). To sum up, Hurley & Hult (1998) broadly defined organizational learning as an "organization-wide activity of creating and using knowledge to enhance competitive advantage." This includes obtaining and sharing information about customer needs, market changes and competitors.

2.2 *Innovative Corporate Culture*

West and Farr (1990) define innovation as "the intentional introduction and application within a role, group or organization of ideas, processes, products or procedures, new to the relevant unit of adoption, designed to significantly benefit the individual, the group, organization or wider society".

A critical part of innovativeness is the cultural openness to innovation (Zaltman et al., 1973). According to Kono (1990), corporate culture is the concurrent values, decision making methods, or thinking model of corporate members. The basic elements of culture influence innovation through socialization and basic values, assumptions and beliefs (Tesluk et al., 1997) that become the guide for behaviors.

In line with the arguments above, Dobni (2008) defines innovative corporate culture as the degree to which the organization has a formally established architecture within their business model to develop and sustain innovation. This architecture is communicated through vision, goals, objectives, and operationalized through the business model and business processes. It includes the organization's ability to execute value-added ideas, thus proactively co-align systems and processes with changes in the competitive environment.

From their review of organizational learning and learning organization literatures, Gatignon et al. (2002), and Chiva (2004) were able to develop a group of factors that facilitate organizational learning. Five underlying dimensions were arrived at: experimentation, risk taking, interaction with the external environment, dialogue and participative decision making. These dimensions that facilitate organizational learning share a lot of characteristics with organizations that have an innovative corporate culture. Thus:

Hypothesis 1: Innovative corporate culture has a positively effect on organizational learning.

2.3 *Practices of Connectivism*

Connectivism is a theoretical framework for understanding learning in the digital age. According to Siemens (2005), previous theories were developed in a time when information development was slow as compared to the digital era of today when the flow of information is fast and fluid. In connectivism, learning is considered as a process of informal information exchange, organized into networks and supported with electronic tools such as social networks like LinkedIn, MySpace, Facebook etc. and other social media such as Blogs, Wikis, and YouTube etc.

To further characterize connectivism, Siemens (2005) stated that it "addresses the challenges that many corporations face in knowledge management activities. Knowledge that resides in a database needs to be connected with the right people in the right context in order to be classified as learning. A real challenge for any learning theory is to actuate known knowledge at the point of application.

Some knowledge management practices include Communities of Practice (CoP) and e-Learning. Downes (2007) described what CoP can do, which includes facilitating collaboration, answering specific questions via subject matter expert, filtering out incorrect information by peer group, capturing institutional knowledge and re-use it, preventing re-inventing wheels by sharing knowledge, and sharing successful best practices. On the other hand, "e-learning is the acquisition and use of knowledge distributed and facilitated primarily by electronic means" (Wentling, et al., 2000). This form of learning currently depends on networks and computers but will likely evolve into systems consisting of a variety of channels (e.g., wireless, satellite), and technologies (e.g., cellular phones, PDA's) as they are developed and adopted.

Practices of connectivism is thus defined by this study as "those practices that use technology both as enabler or enhancer for informal and social learning in the workplace, through the use of social software technologies and knowledge management practices, in order to foster sharing, collaboration, and exchange of information and knowledge".

An important function of social networking and communities of practice is the provision of a collaborative learning environment within the organization, in which problems encountered are collectively solved and solutions are shared among peers, bridging the gap between procedures and practice (Orlikowski, 2002). The above literatures seem to support a relationship between social networking and organizational learning. Thus:

Hypothesis 2: Practices of connectivism have a positive effect on organizational learning.

Jaworski et al., (2000) stated that an innovative culture is more likely to be adaptive, adopts new ideas, accepts new business ideas and technological breakthroughs, and takes aggressive competitive moves. Therefore, an innovative company is most likely to support practices of connectivism and not act as a barrier. Thus:

Hypothesis 3: Innovative corporate culture has a positive effect on the practices of connectivism.

Furthermore, according to Glynn, Lant and Milliken (1994) "newer perspectives on learning focus on the more emergent nature of learning; implying that information to be learnt is constructed through the ongoing interactions among organization members" (p. 55). As the literature shows, it is reasonable to predict that a company with innovative culture will seek to elevate its organizational learning through the use of technological advances in connecting people and information. Thus:

Hypothesis 4: Practices of connectivism have a mediating effect on the relationship between innovative corporate culture and organizational learning.

3 Research Method

3.1 Sample

This study uses the Taiwanese computer industry as a sample. As the world's computer component manufacturer, Taiwan has been a country leading in technology use. As of March, 2011, there are 16,147,000 internet users and 8,970,880 Facebook users, a 70% and 38.9% respective penetration rate of Taiwan's 23,071,779 population (Internet World Stats, 2011). The high penetration rate of both internet and Facebook usage implies minimal restrictions in technology usage which provides a valid context for this research.

For the purpose of this study, the sample comprised of three business types within Taiwan's computer industry; software development, manufacturing industries, and systems integration; totally, 301 companies are sampled within the industry using the Taipei Computer Association (TCA) Members Directory and Computer Suppliers (2010) that contains the most comprehensive listing of computer companies available in Taiwan with address details, telephone and e-mail contacts provided. The company location in this study covers north, south, and central Taiwan.

Out of a sample of 301 companies, there were 80 valid responses, which equates to a response rate of 27%. Personal level demographic information shows 86% of the informants hold an executive or managerial position in their company, 65% are male, 49% have a master's degree, and 53% have more than 5 years of seniority with their company. Company level demographics reveal 66% of companies belonging to the small and medium segment with employees less than 200, 56% with tenures of more than 10 years, and 65% locating in northern Taiwan. Business types of responding companies include software 44%, manufacturing 20%, and system integration 29%.

3.2 Survey Instrument

Data was collected using a survey questionnaire with 49 items. The questionnaire was divided into four parts: The first part measures innovative corporate culture with two dimensions and fourteen items coded as ICO1-ICO14, adopted from Dobni (2008). Only the innovation intension and innovation implementation context, out of Dobni's original four dimensions, were adopted.

The second part measures practices of connectivism. A new instrument was developed from a survey of the literature to measure practices of connectivism in the organizational context. Accordingly, these practices were divided into social networking technologies, social media technologies, and the knowledge management practices that includes Communities of Practices (CoPs) and E-learning. These three dimensions contain nine items coded as SNT1-3, SMT1-3, and KMP1-3. The instrument mainly measured the degree to which these companies were utilizing the practices of connectivism.

The organizational learning construct adopted measures from Panayides (2007) and Pedler et al. (1991), and was characterized by the presence of an intra-organizational culture that values learning manifested by top-management commitment towards learning, a shared vision, open-mindedness towards change and intra-organizational sharing of knowledge, as well as learning of external competitive information. Thus, the organizational learning construct had five dimensions and 18 items coded as CL1-4, SV1-5, OM1-3, IKS1-4, and IEE1-3.

In addition, there were also 8 demographic variables, including four company demographics (business type, employee number, company tenure, and location) and four personal data (position, gender, education and seniority).

3.3 Reliability and Validity Analysis

To ensure the validity of all the measures, an exploratory factor analysis was carried out. The resulting factor structure cumulatively explained 56.47%, 75.96%, and 75.96% of the variances of innovative corporate culture, practices of connectivism, and organizational learning, respectively. In particular, two new dimensions, Openness to Ideas and Management Effectiveness, emerged in addition to the two dimensions in the original innovative corporate culture measurement model, as seen in the top part of Table 3.1. Furthermore, the analysis on the practices of connectivism measure had resulted in 2 dimensions as opposed to the original model of three, as shown in the middle part of Table 1. The organizational learning measure was shown in the last part of Table 1.

Table 1 Exploratory factor analysis results

Dimensions	Items	Factor Loadings	Dimensions	Items	Factor Loadings
Innovative Corporate Culture					
Innovation Intention	ICO1	.732	Innovation Implementation	ICO13	.735
	ICO6	.731		ICO14	.717
	ICO5	.636		ICO12	.649
	ICO4	.626		ICO9	.574
Openness to Ideas	ICO11	.812	Management Effectiveness	ICO2	.747
	ICO7	.771		ICO8	.721
	ICO10	.621		ICO3	.414
Connectivism Practices					
Social Media Technologies	SNT1	.927	Knowledge Management Practices	KMP2	.917
	SNT2	.874		KMP3	.875
	SNT3	.869		KMP1	.865
Organizational Learning					
Open mindedness	OM2	.801	Commitment towards Learning	CL1	.791
	OM1	.767		SV1	.650
	OM3	.762		CL4	.643
	SV5	.626		CL2	.543
	SV4	.525		SV3	.527
Interaction with the External Environment	IEE1	.786	Learning from the Past	CL3	.764
	IEE2	.783		IKS1	.647
	IEE3	.669		IKS2	.546
Intra-Organizational Knowledge Sharing	IKS3	.768			
	IKS4	.667			

Reliability Analysis was carried out after the exploratory factor analysis. The three variables show acceptable Cronbach's Alpha at .80 for Innovative Corporate Culture, .84 for Practices of Connectivism, and .88 for Organizational Learning.

4 Findings and Discussions

4.1 Correlation Analysis

As shown in the correlation analysis in Table 2, company location is found to be correlated with all three major variables in the positive direction. Since location is coded from business center of north as 1 to the furthest east as 4, this seems to suggest that those companies located further away from the business center are better at cultivating innovative culture ($p<0.05$), using connectivism practices ($p<0.05$), and promoting organizational learning ($p<0.01$). Company size and tenure are found to both negatively relate to all three major variables. Significant values are found between tenure and innovative culture at -.308 ($p<0.01$), tenure and organizational learning at -.317 ($p<0.01$), as well as employee number and connectivism practices at -.243 ($p<0.05$).

Table 2 Correlation Analysis Results

Variables	Mean	S.D	1	2	3	4	5	6	7
1. Business Type	2.00	1.02	-						
2. Employee Number	2.98	1.74	-.079	-					
3. Company Tenure	3.80	1.22	.174	.405**	-				
4. Location	1.51	.80	-.128	-.131	-.291**	-			
5. Culture	3.84	.42	.060	-.218	-.308**	.242*	(.795)		
6. Practice	3.76	.69	.116	-.243*	-.114	.220*	.488**	(.838)	
7. Org. Learning	3.91	.45	.125	-.155	-.317**	.291**	.669**	.618**	(.883)

*p<0.05 **p<0.01 ( ): Cronbach's α

4.2 Hypothesis Testing

Hierarchical regression analysis was conducted to test all four hypotheses of the study. Table 3 summarizes the result of these tests. In model 1, only the company demographics were entered in the regression equation with organizational learning as the dependent variable. Innovative corporate culture was subsequently entered into the same equation as shown in Model 2, which shows a significant R2 change of .30 and a positive effect of innovative corporate culture ($\beta = .59$, $p<0.001$) on organizational learning. The first hypothesis, H1: Innovative corporate culture will have an effect on organizational learning, is therefore supported. When combined with the demographics, innovative corporate culture explains 49% of the variances in organizational learning.

Further regression analyses have been conducted at the dimensional level to show which dimensions of innovative corporate culture have stronger effects on the organizational learning. As seen in model 3, except the dimension of innovation intention, the other three cultural dimensions do show significant effect on organizational learning. In particular, innovation implementation seems to have the most significant effect on organizational learning with (β = .30, p<0.001).

Similar to the first hypothesis, for H2: Practices of connectivism will have an effect on organizational learning, model 4 shows a significant R2 change from model 1 when practices of connectivism was entered into the equation. It also shows that practices of connectivism does have a strongly significant effect (β = .57, p<0.001) on organizational learning in the positive direction. Therefore, hypothesis 2 is also supported in this study.

The dimensional level analysis, as seen in model 5, shows that both social media technologies (β = .27, p<0.01) and knowledge management mechanisms (β = .43, p<0.001) have significant impact on organizational learning. In particular, social media being a new practice in business today implies that companies in Taiwan's computer industry are quite advanced in utilizing social media tools to enhance organizational learning.

Table 3 Effect of Innovative Corporate Culture and Culture Dimensions on Organizational Learning

	Organizational Learning						Practices of Connectivism		
Model	1	2	3	4	5	6	7	8	9
Business Type	.20	.13	.12	.13	.13	.10	.12	.07	.04
Employee Number	.00	.06	.05	.13	.12	.12	-.210	-.17	-.12
Company Tenure	-.28	-.14	-.14	-.29	-.30	-.19	-.01	.12	.09
Location	.23	.13	.12	.11	.12	.07	.21	.13	.14
Innovative Corporate Culture		.59***				.42***		.45***	
Innovation Intention			.06						.17
Innovation Implementation			.30***						.28**
Openness to Ideas			.20*						.06
Management Effectiveness			.25**						.11
Practices of Connectivism				.57***		.39***			
Social Media Technologies					.27**				
Knowledge Management Practices					.43***				
R^2	.18	.49	.51	.47	.48	.60	.11	.28	.30
Adj. R^2	.14	.45	.46	.44	.44	.56	.06	.23	.22
F	4.23*	14.29 ***	9.43 ***	13.59 ***	11.37 ***	18.34***	2.3	5.92 ***	3.90 ***
N	80	80	80	80	80	80	80	80	80

Standardized regression coefficients are shown, *p<0.05 **p<0.01 ***p<0.001.

For H3: Innovative corporate culture will have an effect on the practices of connectivism, as shown in model 8, the regression achieved a significant F value, a significantly higher R2 than Model 7, and a positive effect of innovative corporate culture ($\beta = .45$, $p<0.001$) on practices of connectivism. The dimensional level analysis, as shown in model 9, only innovation implementation has a significant impact on practices of connectivism ($\beta = .28$, $p<0.01$). This means that in the case of businesses in Taiwan's computer industry, only those who pay special attention to the implementation aspect of innovative culture will likely to adopt practices of connectivism.

Of special interest to this study is the testing of the fourth hypothesis, H4: Practices of connectivism will have a mediating effect on the relationship between innovative corporate culture and organizational learning. Based on Baron & Kenny (1986) procedure for testing mediation, as seen in model 6, although the Beta value of innovative corporate culture dropped from 0.59 in model 2 to .42 in model 6, it still remained significant, rejecting the hypothesis of a full mediation. Practices of connectivism have more a direct effect on organizational learning.

5 Conclusions and Implications

This study is the first to explore connectivism practices and its impact on organizational learning. The findings of this study have clearly shown that these practices have a direct significantly positive effect on organizational learning. The study findings also showed that an innovative corporate culture does have a significant positive effect on the connectivism practices. Findings of this study showed that companies in Taiwan's computer industry are yet to fully utilize social networking tools; this may be because some of these tools are just emerging in the workplace, and that these companies are not yet aware of their potential impact on organizational learning.

One theoretic and two practical implications are derived from this study. The theory of connectivism has been popularized by scholars such as Siemens (2005) and Downes (2007, 2010) as a response to the rapid development and wide-spread use of internet technologies. However, it has also been criticized as to whether it qualifies as a theory due to a lack of empirical research to prove the claims. This research was founded on the basis of connectivism theory and has successfully tested the relationship empirically between practices of connectivism and organizational learning in one industry segment. Thus the study serves as one small step towards the validation of connectivism as a useful theory in modern society.

The first practical implication for business leaders is that they can no longer ignore society-changing social software technologies such as MySpace, Facebook, blogs, wikis, IM, and YouTube since many companies still forbid employees to utilize some social software technologies whilst at work. This is contrary to the findings of this study, as practices of connectivism have a direct significantly positive effect on organizational learning. Another implication is that business leaders should experiment with an innovative corporate culture type, which creates opportunities for business leaders to be creative and proactive. Employees should see themselves as part of a 'wired workforce', where sharing of information and knowledge should be nurtured, and where collaborative problem-solving skills are encouraged.

Since the research is the first of its kind to study connectivism in the computer industry, many limitations were encountered. Firstly, small sample size (80 responses, 26.6%) and unequal distribution of the responses (65% of north, 20% of central, 13.8% of south, and 1.3% of east Taiwan) indeed caused limitations to generalize the findings to certain locations. Second, because of the nature of the topic this study had attempted to address and the difficulties in data collection, this research had relied on a single informant at each sample company as the source of data. The use of single informant has raised some concerns as the major source of common method variance.

Two future research directions can be also suggested. First, as an attempt to operationalize connectivism, this study used a list of technologies which share the common feature of linking people to information. Although it was empirically proven to predict organizational learning, the new measure for practices of connectivism probably needs better refinement in future research because of the speed of technology development. Second, some company demographic variables such as tenure and location were found to be related to the degree of innovativeness, connectivism adoption, and organizational learning. These were not hypothesized in this research but can potentially produce interesting results in future research as to the reasons why companies in these specific demographics have higher motivation to innovate, learn and adopt new technologies formally.

References

Argyris, C.: Double-Loop learning in organizations. Harvard Business Review 55(5), 115–125 (1977)

Argyris, C., Schon, D.A.: Organizational Learning II: Theory, Method, and Practice. Addison-Wesley, Reading (1996)

Baron, R.M., Kenny, D.A.: The moderator-mediator variable distinction in social psychological research: Conceptual, strategic, and statistical considerations. Journal of Personality and Social Psychology 51, 1173–1182 (1986)

Chiva, R.: The facilitating factors for organizational learning in the ceramic sector. Human Resource Development International 7(2), 233–249 (2004)

Daft, R.L., Weick, K.E.: Toward a model of organizations as interpretation systems. Academy of Management Review 9(2), 254–295 (1984)

Dobni, C.B.: Measuring Innovation Culture in Organizations: The Development and Validation of a Generalized Innovation Culture Construct Using Exploratory Factor Analysis. European Journal of Innovation Management 11(4), 539–559 (2008)

Downes, S.: Introduction to connective knowledge. In: Hug, T. (ed.) Proceedings of the International Conference on Media, Knowledge & Education - Exploring New Spaces, Relations and Dynamics in Digital Media Ecologies, June 25-26 (2007)

Downes, S.: Learning networks and connective knowledge. In: Yang, H.H., Yuen, S.C. (eds.) Collective Intelligence and E-Learning 2.0: Implications of Web-Based Communities and Networking. IGI Global (2010)

Ernst, D.: Inter-organizational knowledge outsourcing: What permits small Taiwanese firms to compete in the computer industry? Asia Pacific Journal of Management 17(2), 23 (2000)

Fiol, C.M., Lyles, M.A.: Organizational learning. Academy of Management Review 10, 803–813 (1985)

Garvin, D.: Building and learning organization. Harvard Business Review 71(4), 78–91 (1993)

Gatignon, H., Tushman, M.L., Smith, W., Anderson, P.: A structural approach to assessing innovation: Construct development of innovation locus, type and characteristics. Management Science 48(9), 1103–1122 (2002)

Glynn, M.A., Lant, T.K., Milliken, F.J.: Mapping learning processes in organizations: A multi-level framework linking learning and organizing. In: Meindl, J., Porac, J., Stubbart, C. (eds.) Advances in Managerial Cognition and Organizational Information Processing, vol. 5, pp. 43–83. JAI Press, Greenwich (1994)

Hurley, R.F., Hult, G.T.M.: Innovation, market orientation, and organizational learning: an integration and empirical examination. Journal of Marketing 62, 42–54 (1998)

Internet World Stats. Asia Marketing Research, Internet Usage, Population Statistics and Facebook Information (2011), http://www.internetworldstats.com/asia.htm#tw (retrieved June 10, 2011)

Jaworski, B.J., Kohli, A.K., Sahay, A.: Market-driven versus driving markets. Journal of the Academy of Marketing Science 28(1), 45–54 (2000)

Jones, A., Hendry, C.: The learning organization: A review of the literature and practice. Warwick Business School, University of Warwick, The HRD Partnership (1992)

Kim, D.H.: The link between individual and organizational learning. Sloan Management Review 35(1), 37–50 (1993)

Kono, T.: Changing a company's strategy and culture. Long Range Planning 27(5), 85–97 (1990)

Levitt, B., March, J.G.: Organizational Learning. Annual Review of Sociology 14, 319–340 (1988)

Nevis, E., DiBella, A., Gould, J.: Understanding organizations as learning systems. Sloan Management Review 36(2), 73–85 (1995)

Nonaka, I., Takeuchi, H.: The Knowledge-Creating Company: How Japanese Companies Create the Dynamics of Innovation. Oxford University Press, New York (1995)

Orlikowski, W.J.: Knowing in practice: enacting a collective capability in distributed organizing. Organizational Science 13(3), 249–273 (2002)

Panayides, P.M.: Effects of Organizational Learning in Third-party Logistics. Journal of Business Logistics 28(2), 133–158 (2007)

Pedler, M., Burgoyne, J., Boydell, T.: The Learning Company. McGraw Hill Book Company, N.Y. (1991)

Siemens, G.: Connectivism: A Learning Theory for the Digital Age. International Journal of Instructional Technology and Distance Learning 2(1) (2005), http://www.itdl.org/Journal/Jan_05/article01.htm (retrieved December 12, 2009)

Stata, R.: Organizational learning - the key to management innovation. Sloan Management Review 30(3), 63–82 (1989)

Tesluk, P.E., Faar, J.L., Klien, S.R.: Influences of organizational culture and climate on individual creativity. The Journal of Creative Behavior 31(1), 21–41 (1997)

Wentling, T.L., Waight, C., Gallaher, J., La Fleur, J., Wang, C., Kanfer, A.: eLearning – A review of literature (2000), http://learning.ncsa.uiuc.edu/papers/elearnlit.pdf (retrieved December 14, 2009)

West, M.A., Farr, J.L.: Innovation at work. In: West, M.A., Farr, J.L. (eds.) Innovation and Creativity at Work. Psychological and Organizational Strategies, pp. 3–13. Wiley, Chichester (1990)

Zaltman, G., Duncan, R., Holbek, J.: Innovations and Organizations. Wiley, New York (1973)

The Impact of a Special Interaction of Managerial Practices and Organizational Resources on Knowledge Creation

Jader Zelaya-Zamora[1], Dai Senoo[1], Kan-Ichiro Suzuki[2], and Lasmin[2]

[1] Tokyo Institute of Technology, Japan
zelaya.j.aa@m.titech.ac.jp,
senoo.d.aa@m.titech.ac.jp
[2] Ritsumeikan Asia Pacific University, Japan
k1suzuki@apu.ac.jp,
la10@apu.ac.jp

Abstract. This study analyzes how an apparent paradoxical combination of managerial practices and organizational resources affects the knowledge-creation capability of organizations. The propositions developed are theoretically supported and for the greatest part empirically validated, thus transcending the conventional understanding and generating insightful perspectives and practical implications for both academicians and practitioners.

1 Introduction

There is now a plethora of empirical evidence confirming that effective knowledge management indeed leads to superior organizational performance in terms of efficiency, innovation and competitiveness. As a result, nowadays almost everyone involved in organizational sciences already recognizes the importance of knowledge management. This recognition and general agreement, however, is good as long as we do not forget to regularly test the basic assumptions.

This study is a challenge to one of the most classical assumptions in management: the belief that exploration is merely for the future, and exploitation for the present. Successful organizations normally engage in exploration activities to ensure the achievement of future-oriented goals, and in the same fashion they carry out exploitation activities to meet short/mid term objectives. However, this reality has led many to assume that there is only a fit between long term and exploration, on one hand, and short term and exploitation, on the other.

Testing such assumption from a knowledge management perspective is of crucial importance. A reason is obvious: if our suppositions are flawed, the resulting conclusions and decisions will tend to be imperfect as well. What is more, the authors of this study found in the literature a tendency among

L. Uden et al. (Eds.): 7th International Conference on KMO, AISC 172, pp. 231–242.
springerlink.com

knowledge management researchers to forget that managerial practices and organizational resources are like the two wings of a bird: both must function and interact in order to produce a synergistic impulse to achieve organizational goals, particularly related to knowledge creation.

Therefore, the specific purpose of this study is to make a unique contribution by elucidating the interaction between two key organizational variables, namely, (1) long and short term managerial practices, and (2) knowledge-exploitation and knowledge-exploration resources. The effect of this interaction on organizational knowledge creation is assessed using actual data from more than one hundred research and development departments of Japanese companies. The findings, discussed along the way, are insightful and have tangible implications for both academicians and practitioners interested in the field of knowledge management.

2 Concepts and Hypotheses

An organization is an identifiable social entity pursuing several objectives through the coordination of its members and objects (Hunt, 1972, p. 4). Setting objectives and establishing coordination is essentially the task of managers or leaders. Other members, objects, and their relations and particularities, serve mainly as organizational resources. Both the intrinsic purpose for and the manner in which managers decide to utilize specific resources originate managerial practices. Behind the veils of higher-order variables and capricious groupings and labels, these two, managerial practices and organizational resources, can be considered as the most elemental components of any organization.

Even in a basic production facility, for example a farm that grows vegetables, the productive capacity is largely dependent on the quality and quantity of both the farmer's decisions (e.g. irrigation and fertilization method, timing of plantation and harvest, etc.), and the resources used (e.g. water, equipment, etc.). From a knowledge management perspective, the productive capacity of an organization can be thought of as its knowledge-creation capability.

Correspondingly, by means of a well-designed Delphi process, Holsapple and Joshi (2000) identified three broad factors that influence knowledge management in organizations: managerial influences, resource influences, and environmental influences. Then, controlling for environmental factors, it is primarily the combination of managerial practices and organizational resources that determine the capacity of any firm to achieve its knowledge management goals.

2.1 Managerial Practices and Organizational Resources

At the core of all managerial action is the task of not only establishing but also accomplishing both short term and long term objectives. While long term managerial practices are characterized by leaders' emphasis on future-oriented activities, short term managerial practices are characterized by leaders' focus on more present-oriented activities. The former are visionary and most apt to transform, the latter are consensus oriented and apt to solve current problems (Thoms, 2004).

Over the years, organizational resources have been categorized using several interesting labels such as Assets and Capabilities(Wade & Hulland, 2004), Schema and Content Resources(Holsapple & Joshi, 2001), Knowledge Assets (Nonaka, et al., 2000), Tangible and Intangible Resources(Barney, 1991), Physical, Human and Organizational Resources(Pee & Kankanhalli, 2009), Employee Knowledge and Embedded Knowledge(Leonard-Barton, 1995), among many others. However, none of these classifications seems to capture a dialectical reality in organizations: resources of any kind are used at all times for exploration and/or exploitation undertakings. Leaders must seek a well balanced condition of knowledge-exploration resources and knowledge-exploitation resources in order to guarantee a robust system of knowledge creation (March, 1991; Crossan & Hurst, 2003).

In organizations, exploration has to do with flexibility, search and discovery, and experimentation. On the other hand, exploitation is more related to efficiency, selection and execution, and implementation(March, 1991). Therefore, two representative exploration resources are the so called Knowledge Diversity and Inter-unit Communications Enablers; and two exploitation resources are Common Knowledge and Intellectual Property.

2.2 *Hypotheses*

When reviewing the knowledge management literature, it is evident that the influence of managerial practices and organizational resources on knowledge creation is typically studied independently, as shown in Fig. 1 (a). For example, the effect of only managerial practices (e.g. leadership style, top management knowledge values, rewards and incentives, supervisory control, visioning, centralization and formalization, etc.) on knowledge-creation-related variables is examined in many studies (Hirunyawipada et al., 2010; Hsu, 2008; Kankanhally et al., 2005; Kulkarni et al., 2006-7; King & Marks, 2008; Kim & Lee, 2005; Lee & Choi, 2003). Likewise, the impact of only organizational resources (e.g. common knowledge, knowledge diversity, information technology, intellectual property, communication enablers, motivation, improvisation, etc.) on knowledge-creation-related variables is analyzed abundantly (Kogut & Zander, 1992; Hirunyawipada et al., 2010; Grant, 1996; Chen et al., 2009; Namvar et al., 2010; Senoo et al., 2007; Zsulanski, 1996; Vera & Crossan, 2005; Lee & Choi, 2003).

We argue that there is a synergistic interactional effect of these two variables on the knowledge-creation capability of organizations, as shown in Fig. 1 (b). And so far, this important combinatory effect has been neglected in most of past studies. As reasoned above, managerial practices and organizational resources are the fundamental factors in any organization. Like the two wheels of a bicycle or the two wings of a bird, normally both must interact and work in harmony in order to produce a synergistic impulse to achieve organizational goals. From this perspective, it is reasonable to conceive that some managerial practices in combination with organizational resources can originate a more favorable condition for knowledge creation. Then, our first hypothesis is:

H_1: Managerial practices and organizational resources interact to produce a significant impact on the knowledge-creation capability of organizations.

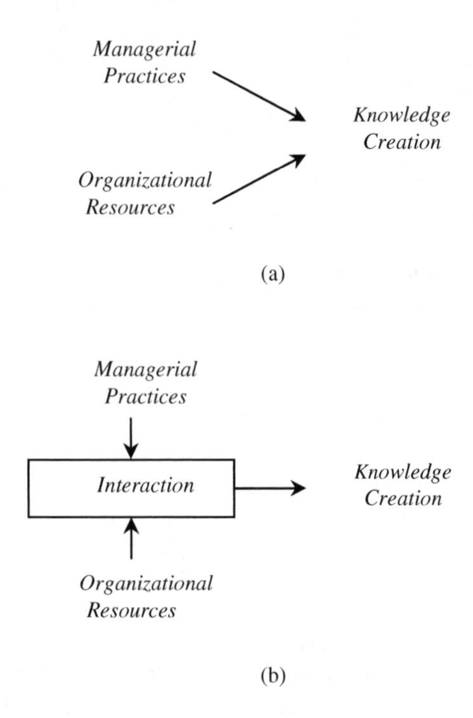

Fig. 1 (a) Typical framework to study the influence of managerial practices and organizational resources on knowledge creation; (b) This study's model: the interaction of managerial practices and organizational resources originates some influencing patterns on knowledge creation

Now, the key to successful knowledge creation is the ability of organizations to manage seemingly paradoxical variables (Nonaka et al., 2000, p. 7). Order and chaos, tacit and explicit knowledge, theory X and theory Y, centralization and decentralization, flexibility and bureaucracy, stability and change are just some examples of organizational conditions that may seem antithetical, but they are actually complementary; and managing its coexistence has a powerful positive effect in organizational success(Nonaka I. , 1994; Peters & Waterman, 2006; Farjoun, 2010).

Furthermore, nowadays it is of paramount importance for managers to go beyond conventional thinking. For instance, to learn things is important, but to unlearn things is also necessary; to improve old products is important, but to develop completely new products is also necessary; to lead a group in a consistent, agreeable manner is important (as it allows organizational members to perform their tasks effective and stresslessly), but to lead them in a surprising, challenging manner is also necessary (as it awakes them from inertia and complacency). These conventionality-breaking combinations also have a great positive effect on organizational success.

As a result, combining long-term managerial practices and exploitation resources, on one hand, and short-term managerial practices together with exploration resources, on the other, makes perfect sense. The interaction of these pairs of apparent paradoxical organizational variables, which also transcends the conventional wisdom, has the potential to produce a positive effect on knowledge creation. Therefore, we hypothesize that:

H_2: Long-term managerial practices have a superior impact on the knowledge-creation capability of organizations when combined with exploitation resources, rather than with exploration resources.

H_3: Short-term managerial practices have a superior impact on the knowledge-creation capability of organizations when combined with exploration resources, rather than with exploitation resources.

3 Methodology

The following paragraphs present specifics about the measurement of the variables and the sample used in the study.

3.1 Measures

Muti-item, five-point Likert scales were used to compose a survey questionnaire. Based on a systemic review, Zelaya-Zamora & Senoo (2011) identified six fundamental knowledge creation accelerators. These catalyzing factors were used to generate a knowledge-creation capability score. Similarly, the other variables were operationalized guided by the theoretical basis mentioned previously. Table 1 shows more details, including some examples of questionnaire items used to measure all the research constructs in this study.

Table 1 Measurement details of constructs

Constructs	Indicators	Example of Questionnaire Items
Knowledge-Creation Capability	Absorptive Capacity	*The organization is aware of the outside and tries to use and absorb technologies more than competitors.*
	SECI Balance	*Researchers share their thoughts, worries and feelings in their interaction with colleagues.*
	Inter-unit Ties	*How often do regular meetings with other departments take place?*
	External Ties	*There is technical assistance from, or research collaboration with, universities.*
	Members' Commitment	*Researchers work very hard for the success of the organization.*
	Cooperation and Trust	*Employees help each other when in need.*
Long-term Managerial Practices	Goal Orientation	*The leaders set and put emphasis on very high goals.*
	Vision Orientation	*The leaders always highlight the corporate philosophy and vision.*
Short-term Managerial Practices	Consensus Orientation	*The leaders are of the type oriented toward consensus and mutual agreement with organizational members.*
	Problem-Solving Orientation	*The leaders are of the type oriented to finding solution to day-to-day problems.*
Knowledge-Exploitation Resources	Common Knowledge	*Researchers use very similar jargons and terminologies at work.*
	Intellectual Property	*The company owns a large number of intellectual property rights compared with competitors.*
Knowledge-Exploration Resources	Knowledge Diversity	*The educational background of researchers is diverse.*
	Inter-unit Communication Enablers	*There are common places for spontaneous communication at work –such as smoking rooms, coffee corners, etc.*

Here "researchers" mean the members of the R&D departments surveyed.

3.2 Sample

The survey was conducted in two formats: paper-based and Internet-based. The questionnaire and a special cover letter in Japanese were sent to the leaders of research and development organizations of around 700 companies registered in the Ministry of Economy, Trade and Industry of Japan. A survey research center was hired to manage the communication between the authors of this study and the respondents of the questionnaire. At the end, a total of 125 valid responses were obtained from organizations in different industries: machinery (85), chemical (18), pharmaceutical (11) and others (11).

4 Analysis and Results

Several statistical analyses were performed in order to examine the data collected and to test the hypotheses. The specifics are provided next.

4.1 Reliability and Validity Analysis

The content validity of all the indicators has already been established by previous studies. Three of the constructs (knowledge-creation capability, exploitation resources, and exploration resources) were measured using formative scales. Naturally, these types of scales do not require neither internal consistency nor convergent validity confirmation, only discriminant validity (Hair et al., 1998; Chin, 1998). However, as their constituent indicators are reflective constructs (like long-term and short-term managerial practices), all scales' internal consistency, convergent validity and discriminant validity were examined. Due to space constrains, detailed results of reliability (internal consistency) and validity tests are not provided here, but the Cronbach's alpha scores, the item-to-total correlation scores, and the factor loadings are all higher than the accepted values, that is: 0.7, 0.4 and 0.5, respectively (Kline, 1999; Field, 2005; Comrey, 1979). These results indicate that the scales used are reliable and valid.

4.2 Analysis of Variance

As the study design resulted in one continuous dependent variable (knowledge-creation capability) and two categorical independent variables (long/short managerial practices and exploitation/exploration organizational resources), a two-way ANOVA was performed. This approach is appropriate for testing the kind of bivariate fit hypothesized in this study (Venkatraman, 1989). The dependent variable, knowledge-creation capability, was normally distributed for the groups formed by the combination of the managerial practices and organizational resources as assessed by the Shapiro-Wilk test. There was homogeneity of variance between the groups as assessed by the Levene's test for equality of error variances. Table 2 shows the core results of the analysis of variance computed in SPSS 17.

Table 2 Results of the analysis of variance (Dependent Variable: Knowledge-Creation Capability)

Source	Mean Square	F	Sig.
Corrected Model	.884	6.852	.001
Intercept	458.819	3554.770	.000
Managerial Practices (MP)	1.854	14.361	.000
Resources (RES)	.323	2.506	.120
MP*RES	.382	2.959	.092

R Squared = .291 (Adjusted R Squared = .249).

Because the sample size is relatively small, a significance level of 0.1 was adopted. This criterion is fairly common and appropriate as a means to increase statistical power(Aguinis & Harden, 2009). The Sig. column in the table above reveals that there is a statistically significant interaction between managerial practices and organizational resources (MP*RES) at the P = 0.092 level (H1: confirmed). This significant interaction may be better appreciated in the profile plot depicted in Fig. 2.

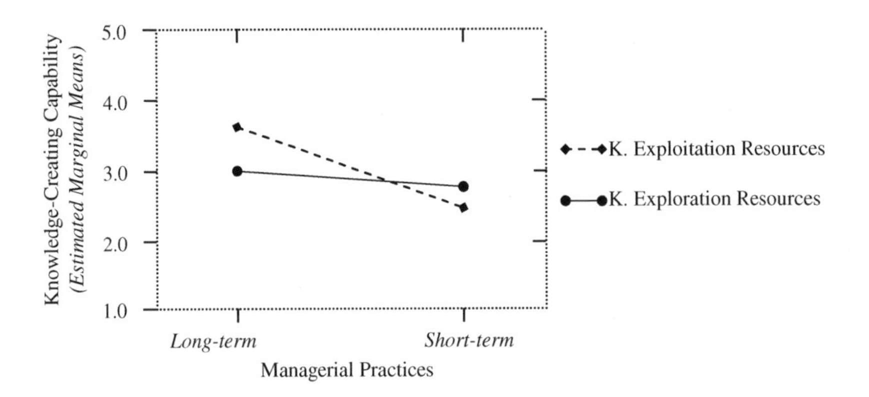

Fig. 2 The interaction effect of managerial practices and organizational resources on the knowledge-creation capability of organizations

Besides the numerical results, judging by the fact that the lines in the plot above are not parallel, it can be easily observed that there is a significant interaction taking place between the independent variables of the study. Concretely, this means that the knowledge-creation capability of organizations is higher when long-term managerial practices are combined with exploitation resources (H2), and when short-term managerial practices are combined with exploration resources (H3). This significant interaction was further analyzed by running tests for simple main effects, in order to make sure that the mean difference in knowledge-creation capability between exploitation and exploration resources is significant for each kind of managerial practices. The results in Table 3 show that the knowledge-creation capability is statistically different between

Table 3 Results of the simple main effects analysis

(Dependent Variable: Knowledge-Creating Capability)

Managerial Practices	Mean Square	F	Sig.
Long-term	.819	6.349	.015
Short-term	.001	.008	.928

Each F tests the simple effects of *Organizational Resources* within each level combination of the other effects shown (*Managerial Practices*).

exploitation and exploration resources for long-term managerial practices at the P = 0.015 (H2: confirmed), but the knowledge-creation capability is not statistically different between exploitation and exploration resources for short-term managerial practices at the P = 0.928 (H3: not confirmed).

5 Discussion

The theory introduced in this study and the subsequent empirical validation of the interaction proposition confirm that managerial practices and knowledge resources are like the two wings of a bird, that is to say, both are necessary for knowledge creation to rise in organizations. This is a word of caution for those academicians and managers/leaders that only focus on one of the wings, neglecting the other. It is not only managerial practices that matter for knowledge creation, nor is it only organizational resources, but rather both managerial practices and resources together interact to generate a combined effect on the knowledge-creating capability of organizations. This first finding is in perfect agreement with the original knowledge creation theory which suggests that managers can intervene in the SECI process by altering the organizational context, for which knowledge assets (resources) are used (Nonaka & Takeuchi, 1995).

In particular, the theoretical foundation and the statistical significance found for the second hypothesis provide powerful evidence that managers oriented toward the long-term can confidently use knowledge-exploitation resources (such as the common knowledge among the members, and the intellectual property of the organization) as a knowledge creation strategy. Not always is it rational to consider long-term and exploitation as mutually exclusive in organizations. Although the third hypothesis was not corroborated using the data collected, the theory and the results of the statistical analysis point to the absence of any offsetting effect. This means that combining short-term managerial practices and exploration resources has an effect on the knowledge-creating capability of organizations that is at least as good as the effect of the traditional combination of short-term managerial practices and exploitation resources. These findings provide supporting empirical proof for the theoretical observation of previous research: the coexistence of organizational variables conventionally seen as incompatible is of primordial importance to enhance performance (Peters & Waterman, 2006; Farjoun, 2010; Nonaka et al., 2000).

In frank and open conversations with managers and group leaders from diverse companies in Japan, during training seminars and other settings, one of the authors of this study have heard in several occasions from these practitioners a view that is in agreement with the tenets of this study: it is no longer wise to lead organizations with a black-or-white managerial strategy. Although it may sound paradoxical and somewhat illogical for some folks, in the real world it is now necessary for organizations aiming at competitive performance to carry out apparent contradictory activities such as learning and unlearning, improving the old and developing the new, leading with consistency and leading with surprise, creating and destroying, changing and remaining stable, etc. This seems to be the new wave of leadership.

6 Conclusions

Traditionally, most research on knowledge management only examines how knowledge creation is affected by either some managerial practices, or by some organizational resources. This study is partly original in that it provides empirical evidence that these two fundamental organizational variables actually can interact to produce a combined effect on the knowledge-creation capability of organizations.

Furthermore, this study challenges an assumption that most people take for granted: that only long-term and exploration go together, as well as short-term and exploitation. A thorough analysis using actual data from 125 research and development organizations in Japan shows a unique perspective in which long-term managerial practices have a better fit with exploitation resources when their interaction effect on knowledge creation is considered.

These findings enhance the existing literature on knowledge management and provide practical implications for managers/organizational leaders. In particular, future-oriented managerial interventions (such as vision-seeking and goal-achieving practices) fit better with knowledge-exploiting resources (such as the common knowledge among employees and the intellectual property of the company). As mentioned in the discussion section above, the essence of this empirical conclusion is in line both (1) with previous theoretical propositions stating that successful organizations reconcile apparent paradoxical variables and (2) with similar actual views expressed by practitioners during training seminars and unstructured conversations. This is a significant indication of the emergence of a new managerial paradigm, which we can call, Duality Leadership.

As future research opportunity, this study's model welcomes re-evaluation, specially overcoming obvious limitations, for example by using a larger sample, a more diverse respondent base (not only Japanese R&D organizations), and an enriched taxonomy for managerial practices and resources. Although the theoretical foundation presented here is robust enough, testing the model in different scenarios will make the generalizability of the findings much more universal.

References

Aguinis, H., Harden, E.: Sample Size Rules of Thumb: Evaluating Three Common Practices. In: Lance, C., Vandenberg, R. (eds.) Statistical and Methodological Myths and Urban Legends: Doctrine, Verity and Fable in the Organizational and Social Sciences, pp. 267–286. Routledge, New York (2009)

Barney, J.B.: Firm resources and sustained competitive advantage. Journal of Management 17(1), 99–120 (1991)

Chen, D.-N., Shie, Y.-J., Liang, T.-P.: The Impact of Knowledge Diversity on Software Project Team's Performance. In: The 11th International Conference on Electronic Commerce, Taipei, pp. 222–230 (2009)

Chin, W.: Issues and Opinion on Structural Equation Modeling. Management Information Systems Quarterly 22(1), 7–16 (1998)

Comrey, A.: A First Course in Factor Analysis. Academic Press, New York (1979)

Crossan, M., Hurst, D.: Strategic renewal as improvisation: reconciling the tension between exploration and exploitation. Working Paper, Ivey Business School, Ontario (2003)
Farjoun, M.: Beyond Dualism: Stability and Change as a Duality. Academy of Management Review 35(2), 202–225 (2010)
Field, A.: Discovering Statistics Using SPSS. Sage Publications, London (2005)
Grant, R.: Prospering in Dynamically-Competitive Environments: Organizational Capability as Knowledge Integration. Organization Science 7(4), 375–387 (1996)
Hair, J., Andersen, R., Tathan, R., Black, W.: Multivariate Data Analysis. Prentice-Hall, NJ (1998)
Hirunyawipada, T., Beyerlein, M., Blankson, C.: Cross-functional integration as a knowledge transformation mechanism: Implications for new product development. Industrial Marketing Management 39, 650–660 (2010)
Holsapple, C.W., Joshi, K.D.: Organizational knowledge resources. Decision Support Systems 31, 39–54 (2001)
Holsapple, C., Joshi, K.: An investigation of factors that influence the management of knowledge in organizations. Journal of Strategic Information Systems 9, 235–261 (2000)
Hsu, I.-C.: Knowledge sharing practices as a facilitating factor for improving organizational performance through human capital: a preliminary test. Expert Systems with Applications 35, 1316–1326 (2008)
Hunt, J.W.: The restless organization. Wiley and sons Australasia Pty. Ltd., Sydney (1972)
Kankanhally, A., Tan, B., Wei, K.: Contributing Knowledge to Electronic Knowledge Repository: An Empirical Investigation. Management Information Systems Quarterly 29(1), 113–143 (2005)
Kim, S., Lee, H.: Employee Knowledge Sharing Capabilities in Public and Private Organizations: Does Organizational Context Matter? In: 38th Annual Hawaii International Conference on System Sciences, Hawaii, pp. 249–258 (2005)
King, W., Marks, J.P.: Motivating Knowledge Sharing through a Knowledge Management System. International Journal of Management Science 36(1), 131–146 (2008)
Kline, P.: The Handbook of Psychological Testing. Routledge, London (1999)
Kogut, B., Zander, U.: Knowledge of the Firm, Combinative Capabilities, and the Replication of Technology. Organization Science 3(3), 383–397 (1992)
Kulkarni, U., Ravindran, S., Freeze, R.: A Knowledge Management Success Model: Theoretical Development and Empirical Validation. Journal of Management Information Systems 23(3), 309–347 (2006-2007)
Lee, H., Choi, B.: Knowledge Management Enablers, Processes, and Organizational Performance: An Integrative View and Empirical Examination. Journal of Management Information Systems 20(1), 179–228 (2003)
Leonard-Barton, D.: Wellsprings of knowledge: building and sustaining the source of innovation. Harvard Business School Press, Boston (1995)
March, J.: Exploration and exploitation in organizational learning. Organization Science 2(1), 71–87 (1991)
Namvar, M., Fathian, M., Akhavan, P., Reza, M.: Exploring the impacts of intellectual property on intellectual capital and company performance. Management Decision 48(5), 676–697 (2010)
Nonaka, I.: A Dynamic Theory of Organizational Knowledge Creation. Organization Science 5(1), 14–37 (1994)
Nonaka, I., Takeuchi, H.: The Knowledge-Creating Company: How Japanese Companies Create the Dynamics of Innovation. Oxford University Press, New York (1995)
Nonaka, I., Toyama, R., Konno, N.: SECI, Ba and leadership: a unified model of dynamic knowledge creation. Long Range Planning 33, 5–34 (2000)

Pee, L., Kankanhalli, A.: Knowledge management capability: a resource-based comparison of public and private organizations. In: International Conference on Information Systems, Phoenix, pp. 1–19 (2009)

Peters, T., Waterman, J.R.: Managing Ambiguity and Paradox; Simultaneous Loose-Tight Properties. In: Search of Excellence: Lessons from America's Best-Run Companies, pp. 89–118, 318–325. Harper Collins Publishers, New York (2006)

Senoo, D., Magnier-Watanabe, R., Salmador, M.P.: Workplace reformation, active ba and knowledge creation: from a conceptual to a practical framework. European Journal of Innovation Management 10(3), 296–315 (2007)

Thoms, P.: Driven by time: time orientation and leadership. Greenwood Publishing Group, Inc., Westport (2004)

Venkatraman, N.: The Concept of Fit in Strategy Research: Toward Verbal and Statistical Correspondence. The Academy of Management Review 14(3), 423–444 (1989)

Vera, D., Crossan, M.: Improvisation and Innovative Performance in Teams. Organization Science 16(3), 203–224 (2005)

Wade, M., Hulland, J.: The resource-based view and information systems research: Review, extension and suggestion for future research. MIS Quaterly 28(1), 107–142 (2004)

Zelaya-Zamora, J., Senoo, D.: An approach to develop novel and practical knowledge management research. In: Sixth International KMO Conference "Knowledge Management for Sustainable Innovation", Tokyo (2011)

Zsulanski, G.: Exploring Internal Stickiness: Impediments to the Transfer of Best Practice within the Firm. Strategic Management Journal 17, 27–43 (1996)

Use of Learning Strategies of SWEBOK© Guide Proposed Knowledge Areas

Andrea Alarcón, Nataly Martinez, and Javier Sandoval

Universidad Pedagógica y Tecnológica de Colombia, Grupo de Investigación en Software (GIS), Avenida central del Norte, Tunja, Colombia
{andreacaterine.alarcon,aida.martinez,
javier.sandoval}@uptc.edu.co

Abstract. This paper gives a general vision of the knowledge areas that compound the software engineering according to the IEEE SWEBOK (Software Engineering body of knowledge) guide, and starting at that point proposed a pedagogic strategy, to be applied as a cathedra complement at the Pedagogic and Technologic University of Colombia (UPTC), due the knowledge areas proposed at SWEBOK provide an appropriate structure that adapts itself to the teaching-learning process at the Systems and Computation engineering college. The strategy that square up the educational objectives of Bloom's Taxonomy cognitive domain is designed and later evaluated by an advisor and further applied to an study group belonging to the Software Engineering study line; taking the results as a base to define the viability of this propose to be extended to another knowledge areas proffered by SWEBOK guide.

Keywords: SWEBOK, methodological strategies, Software Engineering, Knowledge Area, Bloom's Taxonomy.

1 Introduction

Within the Systems Engineering and associated programs study, the Software Engineering is a changing area that requires the knowledge acquirement in an effective and long term way; also it is necessary an update in Software Engineering related tools use, hereby and taking as starting point the postulated knowledge areas in SWEBOK (Software Engineering Body Of Knowledge), that is a proposed IEEE guide that contains topics that provides a cognitive structure that can be adapted to any teaching-learning process of the main contents, that compose the Software Engineering study line. This current study is aimed to enforce the personal and collaborative knowledge, and it proposed practical methodological strategies for Software Engineering areas, based in the appreciation of the knowledge acquisition level by a part of the sample population which take the final test, in the given complex topic selected by the population, where a practice workshop is developed with a tutor collaboration, and taking as starting point the selected topic for subsequent application in a selected group.

L. Uden et al. (Eds.): 7th International Conference on KMO, AISC 172, pp. 243–254.
springerlink.com

2 Preliminary Concepts

For better structuration of the adequate practice, and according with recent and important topics within Soft-ware Engineering area, it's important allow for some concepts like SWEBOK structure and definition and teaching-learning strategies' concepts.

2.1 *What's SWEBOK©?*

It´s a guide developed and approved by IEEE Computer Society and it was published in a formally accepted version in 2004, with the purpose of collect criteria that establish the adequate practices to accomplish a well-developed process of Software products and solutions, establishing 10 fundamental areas that must be allowed as essential components of Software Engineering.

The guide defined by IEEE, has as a main target to accomplish with the following areas:

- Define the body of knowledge required and best practices.
- Define ethical and professional standards.
- Define study plans for undergraduate, postgraduate and later studies.

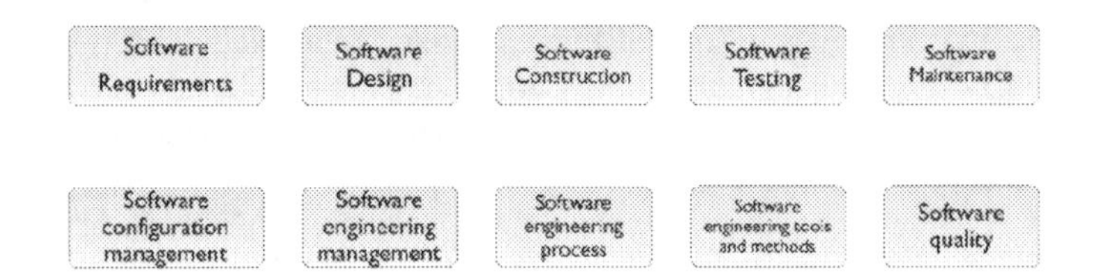

Fig. 1 Software Engineering Knowledge Areas. The SWEBOK guide document shows a 10 KA's (Knowledge Areas) division, and each one shows a subdivision of topics.

2.2 *Teaching-Learning Strategies*

The strategy refers to projection and direction; where someone projects, command and manages the operations to achieve the proposed objectives. Thus, the learning strategies are a set of cognitive operations that students implement to organize, integrate and develop information, and it can be meant as process or sequences of activities that underlie the achievement of intellect tasks and facilitates the building, permanence or transfer of information or knowledge (Campos 2000).

2.3 *Bloom´s Taxonomy*

This taxonomy let in a specific way to measure the level of the Software Engineering area students' skills accomplishing clearly to cognitive domain of Bloom's Taxonomy, defining the level of each specific topic proposed at SWEBOK within the fields shown in figure 2.

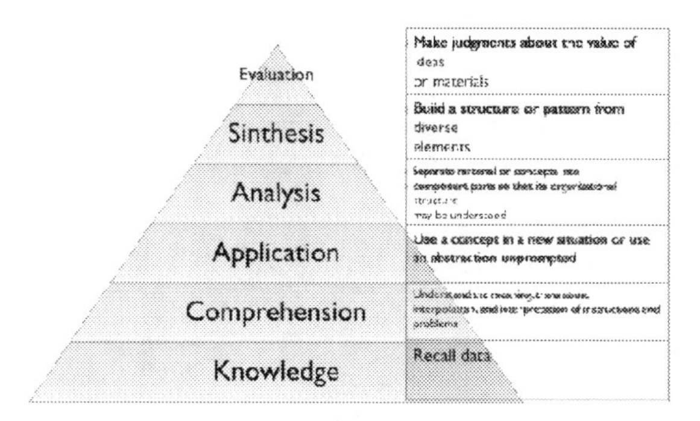

Fig. 2 Bloom's Taxonomy Cognitive Domain. It is part of the widely known classification of educational objectives of a teaching program being one of the focus domains of human being.

3 State of Art

Focusing in the educational context, it's observed that the use and implementation of the SWEBOK guide has had notable scopes, but as is shown in (Alonso et.al. 2011), the basic unities of a Software Engineering course was bound to a based reading format, so that at realize a effectiveness study to compare these learning techniques with others, like distance education or merged techniques, showing that of this form it's possible improve the student abilities, experience to develop maintain and acquire complex systems (Ardis et. al. 2011). In the international context as shown in (Lizhi et. al. 2009), the collaborative learning and the team work are fundamental components for the learning, but the use of web tools, magnifies the based instruction education paradigm without leaving behind the telecommunications infrastructure.

It has worked this guide as basis for the Software engineering enterprise progress, cases like the developed in Venezuela, where in the Universidad de los Andes de Merida, the guide was taken as a basis to develop a learning hybrid process of the software engineering based in an agile and disciplined software implementation process (Castillo 2010). Another important experience told in Costa Rica is described by Lizano Madriz and others (Madriz et. Al. 2011), where are shown, in first place, the SWEBOK guide implementation benefits and advantages, in an enterprise and academic environment, where are linked the systems and software engineering teaching of a concrete way and highlights the evaluation per competences of the model proposed by the SWEBOK guide.

On the other hand is important to mention important projects like the one developed at the *Pontificia Universidad Javeriana* (Payán 2010), which proposes a model for the establishment of good practices related to the first 4 areas proposed by SWEBOK: Requirements, Design, Construction and Testing. Undoubtedly the most valuable contribution to the integration of educational models that approach a business environment and supporting this research, is found in the project is developed at University EAFIT, developed by Raquel Anaya PhD (Anaya 2006), where propose the student oriented paradigm from the practical aspect, ie, facing the students in their learning process to a business environment, showing that this is one of the biggest challenges presented in organizations.

4 Current Knowledge Transfer Model at UPTC

Currently at the UPTC a mixed model of knowledge transfer is implemented, so we have to allocate it or adapt it to one of the four models proposed at the Laval University of Canada (Becheikh, et al., s.f.):

- *RDD (Research, Development and Dissemination) models*: this kind of models focuses on the advancement of knowledge as the most critical factor for research utilization (Hargreaves 1999).
- *The Problem-Solving Models:* Within this kind of models the knowledge users are the initiators of change, and are direct responsible of needs' identification and formulation (Neville and Warren 1986), and this process has five fundamental steps: 1) needs identification, 2) articulation of the problem, 3) search for solutions, 4) selection of the best solution, and 5) implementation of the retained solution to satisfy the need.
- *The Linkage Models:* These models integrate the two ones and highlight the implementation of mechanisms to ensure linkages between producers and users of knowledge.
- *The Social Interaction Models:* These models emphasize the knowledge utilization as a result of multiple interactions of knowledge producers and users (Huberman 2002).

In Becheikh's formulation we can determine that the model of knowledge transfer that has a better adaptation to the current state of UPTC, is the linkage model, because the transfer mechanisms used between researchers and users are very specific and exclusive according to the user' needs so that the final users -in this case the Systems and Computation Engineering students of UPTC- has a narrow perspective of the specific knowledge that they need and there is a loss of opportunities to get broader knowledge, specificly in Software Engineering study line.

As we can see in figure 3, the current model of knowledge transfer at UPTC' Systems and Computation Engineering Study Program has notorious failures regarding to the role that play the student in itself training process, where the teachers prepare their cathedra material based on the recent researches produced artifacts, but they doesn´t consider the students – or for the model, the knowledge users – feedback respect to the topics that they consider important in their transition to the enterprise environment, but is notorious that they have accomplish with the process of knowledge transfer that propose Becheikh (see figure 4).

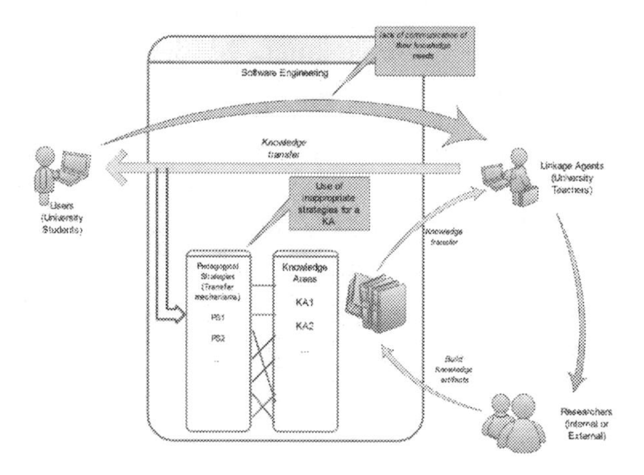

Fig. 3 Current knowledge transfer model at UPTC' Systems and Computation Engineering study program. Here we can see that additional to the lack of feedback of the students to the Linkage Agents (University Teachers), the knowledge transfer is based on pedagogical strategies that not always satisfy the knowledge assimilation needs of students.

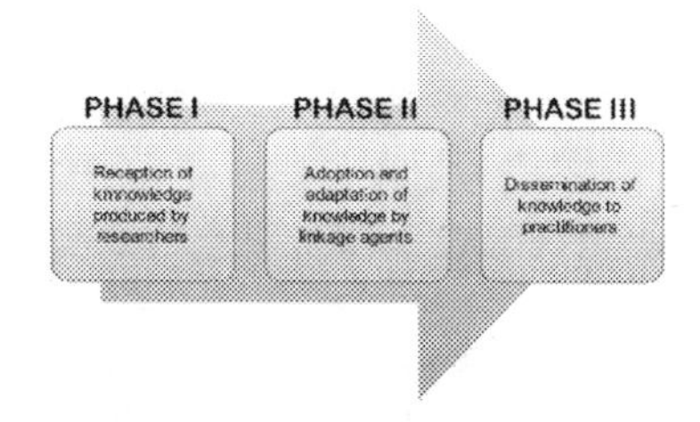

Fig. 4 Transfer of Knowledge from Linkage Agent's perspective. Of the linkage agent's role depends the success or the fail of the definition of a knowledge transfer model because they must do all the treatment to the knowledge for this be understandable.

Another important issue of this model to highlight is the fact that there are not standardized contents for the Software Engineering subjects, and each one of the teachers take the information that they consider relevant, whereby the classes turn into magisterial classes only, and there is a little use of pedagogical strategies for teaching, so that it probably has a negative impact in the training process of the student that want some variety in the knowledge transfer mechanisms not to fall into the routine.

It´s important to accentuate that in some cases inside the University, the Linkage Agents act as Researchers, but in a partial way, so some of them has this duality role that sometimes contribute to the student training process but sometimes reduce the capability of the student for belong in the course topics and learning strategies definition.

5 Proposed Knowledge Management Model

In order to improve the teaching-learning process of Software Engineering subjects at UPTC It's necessary to punctuate that there are determinants at the moment of define a knowledge transfer model, that implicates an analysis to establish the model that adapts better for the educational environment, without setting aside the organizational background that serves as transition between the educational and enterprise environments.

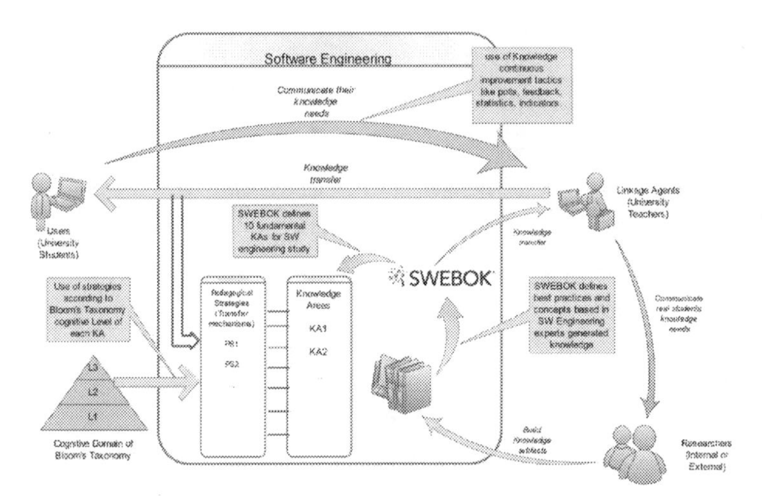

Fig. 5 Knowledge transfer model proposed with applied changes. This model pretend to cover the deficient aspects of the current knowledge management model of UPTC, improving the capability of the knowledge final users (students) to assimilate it in a better way.

Otherwise, three perspectives are taken into account to develop de model, the first concerning to the attributes and features of the knowledge, the second is the perspective of the actors involved in this knowledge transfer model, and Finally, the perspective of knowledge transfer mechanisms, that we can represent as the pedagogical strategies or resources that let a better knowledge transfer process.

5.1 Transferred Knowledge Attributes

As is known, the main source ok knowledge that is taken to define the model of knowledge transfer of Software Engineering is the SWEBOK guide, 2004 version, that define the required body of knowledge and recommended practices that could be defined in any educational curricula for undergraduate, graduate and continuing education concerning to the Software Engineering.

In the proposed knowledge transfer model, it's recommended that exists a classification of the knowledge according to de definition of Eraut of the types of knowledge that is acquired at an University classroom (i.e. Theoretical Knowledge, Methodological Knowledge, Practical skills and techniques, Generic Skills and General knowledge about the occupation) and the mapping to the knowledge that is used in the Workplace (i.e. Codified Knowledge, Skills, Knowledge Resources, Understanding, Decision-making and Judgement) with the purpose of make special emphasis in the enterprise working contexts and variables (Eraut, 2004).

5.2 Knowledge Transfer Model Actors

Later an identification of main actors at the model was made, showing 3 type of actors that adapts to the model that is proposed. But also as propose Becheikh et. al., it is recommended to view the actors including in a knowledge transfer model from two perspectives: 1) the individual, that concern to the features of an actor as a person and 2) the organizational, that points to the characteristics of the actor that evolves institutional factors.

Table 1 Determinants viewed from actor's perspective. It's taken the individual and the organizational attributes.

Actors	Individual Attributes	Organizational Attributes
Researchers (Anis et al. 2004)	- Adaptation efforts - Contextualization efforts - Dissemination efforts - Researchers' credibility	- Experience in knowledge transfer - Emphasis given to knowledge transfer
Linkage agents (Beier and Ackerman 2005)	- Professional experience - Cognitive abilities - Social capital - Personal attributes	- Organizational structure - Resources dedicated to knowledge transfer - Policies to encourage knowledge transfer
Practitioners (Hemsley-Brown 2004).	- Time allowed to acquire and adopt new knowledge - Motivation to acquire and adopt new knowledge - Ability to understand research results	- Organizational climate - Organizational culture - Organizational structure - Organizational procedures and policies - Organizational resources

5.3 Knowledge Transfer Mechanisms

For a better definition of the pedagogical strategies (transfer mechanisms) that are appropriated for each Bloom's Taxonomy Level, it's necessary an exhaustive revision of each one of these with Pedagogical support, with the purpose of determine the more optimum to be applied in each topic defined at SWEBOK that are mapped to its correspondent Bloom's Taxonomy level.

Into the lowest level of Bloom's Taxonomy "Knowledge", it is necessary to define transfer mechanism that let student to make an appropriate and complete recollection of data. The products of this type of mechanisms are listings, ideas, summaries, etc. Some of these mechanisms are Roundtable, Interview, Survey, Brainstorm or Portfolio. For the second level "Comprehension" it is necessary looking for strategies that let the student to understand problems and contribute with possible solutions. The products of this kind of strategies are generally manuscripts, diagrams, mind maps, conclusions, etc.; some strategies to apply to this Taxonomy level are conceptual map, problem-based learning, discussions, Buzz Groups, Pyramid or Snow Ball or learning by investigation. (USQUID, 2009).

For the third level "Application", it is necessary looking for strategies that let the student to apply the acquired concepts to real or hypothetic problems, or current developments. The products of these strategies are generally report redaction, corrections, fixings or complements, etc.; and its strategies are study cases, role games, application Workshops, and Projects (Marquès, 2001). For the fourth level "Analysis", it is necessary looking for strategies that let the student to separate concepts, analyze texts, interpretation of organizational structures. The products of these strategies are generally written or oral productions, etc; and their strategies are like sustentation, audiovisual material making, presentations, posters, learning contracts, puzzles, or Press Writing (Rajadell, 2001).

6 Study Case at UPTC

6.1 Selections and Development of SWEBOK Strategy and Topics for Content Adaptation

In order to test the proposed model, it is necessary to develop some diagnosis and preferences tests in SWEBOK guide proposed knowledge areas, and this underlies the implementation in study groups at Pedagogic and Technologic University of Colombia (UPTC).

Survey Application

A Virtual Diagnosis Survey was performed, in 2011 2nd semester, using Google Docs Survey Tool and Paper Surveys too, applied to the six Software Engineering area teachers at Systems and Computational Engineering graduate program of Pedagogic and Technologic University of Colombia (UPTC). The Survey investigated about knowledge, importance, complexity, ignorance, difficult, and other aspects of each one of the SWEBOK proposed Software Engineering Knowledge areas.

SWEBOK topic selection

Once the SWEBOK topics of interest diagnosis Survey is performed, the target knowledge area Software Quality is defined for this research, but for following phase corresponding to the practical workshop design and pilot test execution, it is necessary to choose a concrete topic inside the KA "Software Quality", for which 30% of IX & X semesters Systems Engineering students was consulted because they know about the given topics at UPTC concerning this KA, looking for content preference referent to "Software Quality", and based in SWEBOK guide proposed topics, that let us make a low-scale study.

Design of Pedagogical Strategy

After selection of the specific topic "Techniques of Software Quality Management" and with the UPTC cathedra Software Engineering II tutor support, we proceed to build an educational workshop, since this strategy provides the student not only theoretical but practical possibility of the conceptual framework presented, according to their level within Bloom's taxonomy, it is clear that issues of "Techniques of Software Quality Management" reach the level of Application within that taxonomy, which provides a starting point for developing the strategy that want to be implemented.

Pilot test execution to population sample

After preparing the workshop concerning software quality, a group of software study line is selected to make a pilot test, and the group chosen was the "Software Engineering II" area. To perform the test, the group is divided in two parts, in which half of the group is given a master class, and the other group half performed the workshop with the relevant explanations. Finally, both group of population, are submitted to a final assessment that defines the effectiveness of pedagogical strategies applied.

6.2 Results Analysis and Conclusions

The Survey results (Fig 6 & 7), show that the more difficult and complexity knowledge area is Software Quality, therefore indicates that it is a critical and urgent care area within the software product development.

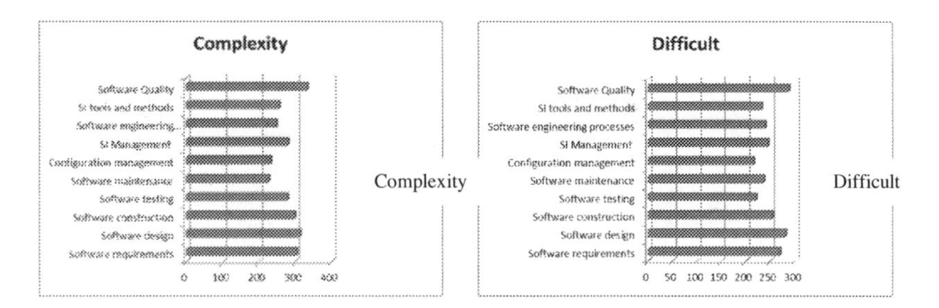

Fig. 6 & 7 Survey determined complexity and Difficult Level

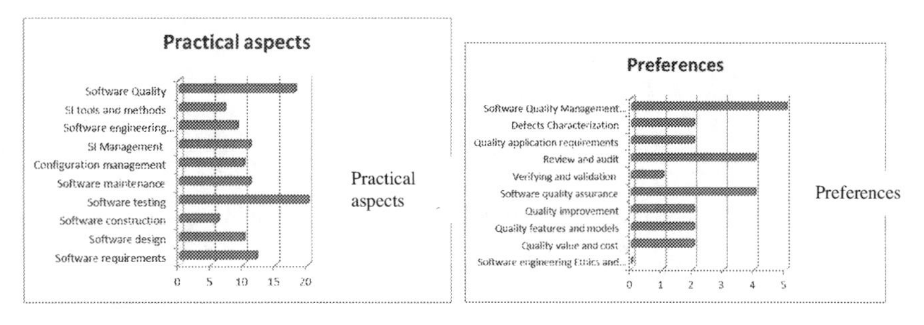

Fig. 8 & 9 KA's that require more practical aspects according to the group survey and Poll results of interest topics selection of Quality Software KA

For preferences poll in Figure 9, we see that the topic "Software Quality", selected from the results of the diagnostic test, it's considered important because it is a fundamental aspect in the software development. From the above poll we can conclude that the topic to be addressed in the teaching strategy is "Techniques of Software Quality Management" because it is one of the subjects according to respondents that produce more interest

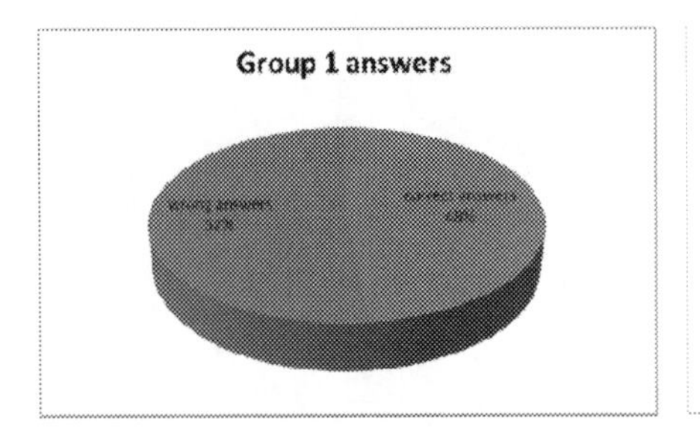

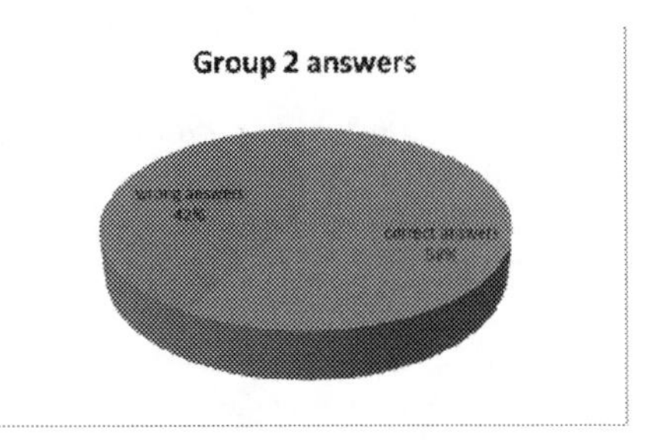

Fig. 10 & 11 Group Assessments (Master Class & Practical Workshop). it shows the percentage of questions answered correctly within the strategy of each group. The percentages indicates that the implementation and enforcement of workshops optimize learning and knowledge assimilation and that practice is essential to reinforce knowledge.

To the Learning factors Inquiry question (Figure 12), we see that for those receiving the application aspects of the workshop and practice, time and technology resources are very important when learning or assimilate knowledge. For people who received the lecture, knowledge and interest of the subject are quite relevant, therefore we conclude that the type of lecture caused a deficit of attention and interest on the topics, while practices awaken and motivate the student to acquire ever more knowledge.

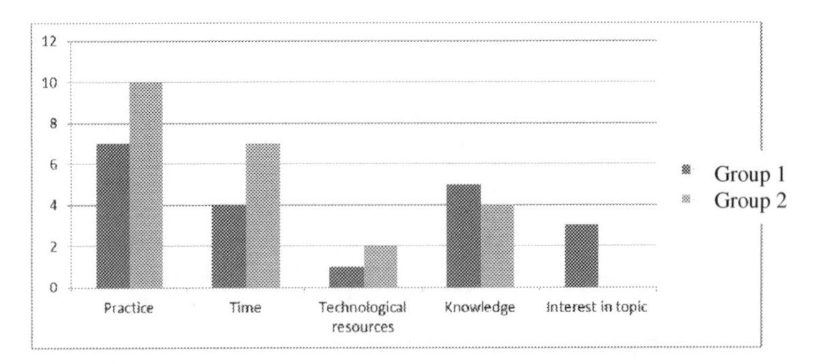

Fig. 12 Learning factors Inquiry question results. The results of inquire question to the students themselves that shows important issues that consider the two groups of the experiment (Master Class & Practical Workshop).

After of the respective test of the knowledge transfer model, we can conclude that

- SWEBOK areas are chosen for this research because each contains a cognitive structure adequate, easily adapted to the teaching-learning process, in addition to supplying all stages in the software development.
- Teamwork and participation of students, in all phases of software development, is critical to achieving high standards of quality.
- Is necessary a continuous improvement model that involve tactics that enrich the required knowledge and strategies pointing to the improvement of the teaching-learning context knowledge transfer, and this must be defined with the interest of the students in participation of curricula definition.

References

Alonso, F., et al.: How Blended Learning Reduces Underachievement in Higher Education: An Experience in Teaching Computer Sciences. IEEE Transactions on Education 54(3), 471–478 (2011), http://ieeexplore.ieee.org/stamp/stamp.jsp?arnumber=5607328

Anaya, R.: Una Visión de la enseñanza de la Ingeniería del Software como apoyo a las empresas de Software. Revista Universidad Eafit, Universidad EAFIT (2006), http://redalyc.uaemex.mx/pdf/215/21514105.pdf

Anis, M., Armstrong, S.J., et al.: The Influence of Learning Styles on Knowledge Acquisition in Public Sector Management. Educational Psychology 24(4), 549–571 (2004)

Ardis, M., et al.: Advancing Software Engineering Professional Education. IEEE Software 28(4), 58–63 (2011), http://ieeexplore.ieee.org/stamp/stamp.jsp?arnumber=5590235

Becheikh, N.Y., y otros, S.F.: How to improve knowledge transfer strategies and practices. Research in Higher Education Journal, 21

Beier, M.E., Ackerman, P.L.: Age, Ability, and the Role of Prior Knowledge on the Acquisition of New Domain Knowledge: Promising Results in a Real-World Learning Environment. Psychology and Aging 20(2), 341–355 (2005)

Campos, Y.: Estrategias de enseñanza-aprendizaje: Estrategias didácticas apoyadas en tecnología, México (2000), http://virtuami.izt.uam.mx/e-Portafolio/DocumentosApoyo/estrategiasenzaprendizaje.pdf

Castillo, A.: Conceptualización del proceso de implementación de software: perspectivas ágil y disciplinada. Revista Ciencia e Ingeniería 31 (2010), http://erevistas.saber.ula.ve/index.php/cienciaeingenieria/article/viewFile/1147/1102

Eraut, M.: Transfer of Knowledge between Education and Workplace Settings. University of Sussex (2004), http://old.mofet.macam.ac.il/iun-archive/mechkar/pdf/TransferofKnowledge.pdf

Hargreaves, D.H.: The Knowledge-Creating School. British Journal of Educational Studies 47(2), 122–144 (1999)

Hemsley-Brown, J.: Facilitating research utilization: A cross-sector review of research evidence. The International Journal of Public Sector Management 17(6/7), 534–552 (2004)

Huberman, M.A.: Moving Towards the Inevitable: the sharing of research in education. Teachers and Teaching: Theory and Practice 8(3), 257–268 (2002)

Lizhi, C., et al.: Test Case Reuse Based on Ontology. In: 15th IEEE Pacific Rim International Symposium on Dependable Computing, PRDC 2009, pp. 103–108 (2009)

Madriz, F.: SWEBOK aplicado: "Experiencias en la Cátedra de Ingeniería de Sistemas, http://www.escinf.una.ac.cr/ingenieria.sistemas/images/documentos/sweebok.pdf (accessed May 10, 2011)

Marquès, P.: La enseñanza. Buenas prácticas. La motivación. Departamento de Pedagogía Aplicada, Facultad de Educación, UAB (2001), http://peremarques.pangea.org/actodid.htm

Neville, J., Warren, B.: The Dissemination and Use of Innovative Knowledge. The Journal of Product Innovation Management 3(2), 127 (1986)

Payán, R.: Recomendaciones para desarollar software internacionalizado. Pontificia Universidad Javeriana (2010)

Puiggros, N.R.: Los Procesos formativos en el aula: Estrategias de enseñanza-aprendizaje. In: Didáctica General para Psicopedagogos, Universidad de Barcelona, Facultad de Pedagogía (2001)

USQUID. Estrategias metodológicas de enseñanza-aprendizaje. Universitat Pompeu Fabra (2009), http://www.usquidesup.upf.edu/es/estrategias-metodologicas

Outsourcing of 'On-Site' User Support – A Case Study of a European Higher Education Centre

Hilary Berger and Tom Hatton

Cardiff Metropolitan University, United Kingdom
Hberger@cardiffmet.ac.uk, hatton.tom@hotmail.ac.uk

Abstract. Outsourcing has garnered significant attention regarding the perceived benefits or repercussions. Our research investigates the specific impact of outsourcing 'on-site' IT user support provision considered by many organizations to be an ideal candidate for outsourcing. We found that the cost implications of the outsourcing exercise provided no real benefit in comparison with in-house staff provision. Notably, moreover, the exclusion of end-users in the process had significant impact on the outsourced agent. We provide insight of real world practice to inform current and further practice and propose two considerations that should be addressed aimed at improving benefits to be realized.

Keywords: Outsourcing, User Support, Service Delivery, Financial Impacts.

1 Introduction

Outsourcing has become both a defined term and a formally recognized business strategy. The prominent 1990s Eastman Kodak contract transformed the way that outsourcing was exploited. Differing interpretations have been proposed dependent upon the context and domain involved. Nonetheless the application of outsourcing is not a transitory phenomenon and continues to accelerate rapidly. Literature reports that 70% of organizations are actively engaged with some form of outsourcing (Dibbern et al. 2004; Kussmaul et al. 2003). The most influential factors in the outsourcing decision are cost savings (Johnson et al. 2008) driven by a lack of relevant in-house expertise and the desire to focus on core in-house functions (Fink and Shoeib 2003; Lacity and Willcocks 2001). However, despite reported success rates outsourcing projects continue to present challenges.

The majority of literature reflects the strategic outsourcing of IT projects, IT solutions, Business Process Outsourcing and so on rather than the outsourcing of specified IT positions. Due to the perceived lack of influence that user support has upon an organization's general direction, it is a popular area for an organization to outsource. Moreover, literature argues the positive and negative impacts and provides a warning regarding the possible negative connotations of outsourcing user support, particularly within an academic environment (Bulchand-Gidumal and Melian-Gonzalez 2009; Majchrzak et al. 2000).

L. Uden et al. (Eds.): 7th International Conference on KMO, AISC 172, pp. 255–266.
springerlink.com

Utilizing the four categories of user support tasks identified in Gingrich's (2003) framework (Beisse 2009) and the actor groups identified by KPMG (1995) it was possible to determine a similar breakdown within the case study setting and subsequently examine any realized benefits (Bulchand-Gidumal and Melian-Gonzalez 2009). Our case study concerns the European Higher Education Centre (EHEC) where the role of user support is affected by a defined group of end-users. Our research focuses primarily upon an outsourcing exercise within the academic environment where the required user support provision was outsourced to a commercial company. The decision to outsource was made by top management who purposefully excluded any end-users involvement or participation. Our aim is to provide insight into the realities of this act, and hopefully to encourage further studies within the user support arena.

The paper is structured as follows: next we explain the context of the case study, present the theoretical context, and describe the research design. This is followed by analysis and discussion of the empirical data collected and the conclusions drawn. Finally, we put forward two key considerations to enlighten current and future practice for similar environments.

2 Case Study Setting

The case study environment concerns the European Higher Education Centre (EHEC), a world-renowned international university, established by the European Member States. Based in Central Italy it provides education at post-graduate and post-doctoral levels and it supports a complimentary research initiative, the Huber Advanced Research Facility (HARF). Our study focuses specifically upon the user support provision provided to the HARF end-users i.e. approx. 175 Academic Staff (Professors, Programme Directors and Research Assistants), Administrative Staff (Director, Secretaries, Project Managers) and Fellows required to work with computing equipment, thus their computing needs must be met.

The Computing Service Department (CS) of EHEC, consisting of 20-25 staff, provides systems maintenance and user support via four *Site Offices*. Our research focuses primarily upon the user support provision provided by the *Dominion Site Office* to HARF end-users which is considered as a *First Line of Support*. A top management decision was taken to outsource the *Dominion Site Officer* position of the user support service for HARF to *TechEd*, a commercial company. There was no end-user involvement or participation in this process.

TechEd, as the Supplier, provide outsourced User Support services to the EHEC (HARF) for a €35,000 annual fee. Therefore, as the contract exists with the organization for the provision of a continuous service rather than with any individual EHEC should not be affected by individual matters that could manifest themselves in the form of "unsuitability or unavailability of the individual". Indeed, the Service Level Agreement (SLA) sets out that TechEd are obliged to provide a suitable replacement without causing any interruption to the service if, for example, the Outsourced Technician becomes unavailable then an immediate replacement must be sourced by TechEd who must also meet all related costs. A further provision in the SLA dictates that the EHEC participate in the selection process by interviewing the potential technician which is aimed at guaranteeing the suitability of a technician to the host organization. Thus, an outsourced technician does not receive an offer of employment until passing both interviews thereby meeting the specific requirements of the EHEC, namely, "*speaking English and basic technical competencies*" which they deemed necessary. Consequently an Outsourced User Support Technician (Peter) was assigned to the host organization for the stated remuneration.

From EHEC's point of view they purchased a service labeled 'Junior User Support Technician' costing €35,000 p.a. to replace the in-house position of 'Site Officer' that would otherwise have been employed internally a similar salary. The primary motivation to outsource this position was to relieve EHEC from the responsibility of providing cover for holidays/sick days of the in-house position as this obligation would now be met by TechEd. A further advantage was that EHEC's standard contract of 7.5 working hours (daily) could be increased to 8 hours a day to ensure that somebody is always present. Additionally, from a legal and professional perspective because the technician is outsourced EHEC has no duty of care beyond providing a safe environment to work.

3 Theoretical Context

In literature, one of the earliest definitions of outsourcing was provided by Lacity and Hirschheim (1993), it states that the act of outsourcing refers to the purchase of goods or services that were previously provided internally. Studies indicate that due to the popularity of methods such as economies of scale and corporate networks, outsourcing did not become a formally recognized business strategy until 1989 (Mullin 1996).

Outsourcing has become both a defined term and a formally recognized business strategy. However, prior to its definition, it is believed that outsourcing as a concept was first discussed by philosopher and political economist Adam Smith in 'The Wealth of Nations' (1776). He postulated that the greatest barriers to trade were the restrictions imposed by a lack of information, the disability to negotiate, the cost of transportation and the enforcement or honouring of contracts. These barriers were theorized to be breachable by taking a particular business activity outside of the organization i.e. outsourcing. More recently Bergkvist and Fredriksson (2008) propose three further interpretations. Firstly, outsourcing is the contracting of ISD activities by the Client with an IT-supplier regardless of the supplier location (Gonzales et al. 2006). Secondly, contracting out ISD activities with a domestic IT supplier (Hirschheim and Lacity 2000). Thirdly, the management of assets and services for financial returns over an agreed period of time (Kern and Willcocks 2002). Our case reflects aspects of the latter two interpretations.

The application of outsourcing is not a transitory phenomenon and continues to accelerate rapidly (Outsourcing Institute). Statistics show that 70% of organizations are actively engaged with some form of outsourcing i.e. business analysis, project management, IT/IS development and so on to streamline and integrate processes, and reduce operational costs (Dibbern et al. 2004; Kussmaul et al. 2003). Johnson et al. (2008) postulate that the most influential factors in the outsourcing decision are cost savings, the desire to focus on core in-house functions by outsourcing lesser functions, and a lack of relevant in-house expertise (Fink and Shoeib 2003; Lacity and Willcocks 2001). However, despite reported success rates outsourcing projects continue to present challenges.

It is important to understand that the world of IT outsourcing contains many variables and specifics which therefore makes it impossible to write a piece of literature that could be referred to as definitive. Whilst there is a substantial amount of literature available within the field of IT outsourcing, the majority of this literature refers to the strategic outsourcing of IT projects, IT solutions, Business Process Outsourcing and so on rather than the outsourcing of specified IT positions. Due to the perceived lack of influence that user support has upon an organization's general direction, it is a popular area for an organization to outsource (Bulchand-Gidumal and Melian-Gonzalez 2009). Our empirical case study aims to provide some insight into the realities of this act, and hopefully to encourage further studies within the user support arena.

Previous research studies offer different interpretations of the effects that the outsourcing user support positions can have. Smith (1776) initially stated that the outsourcing of any activity within an organization creates a dependency upon a particular third party, namely the supplier. An underlying movement during the past 20+ years within public institutions to reduce the human cost has led to the paradox of being required to perform more efficiently and at a higher standard, with less human resources (Olias de Lima 2001). Bulchand-Gidumal and Melian-Gonzalez (2009) believe that this notion has led to a trend of public organizations to outsource their IT functions, or parts thereof. They further postulate that whilst user support may appear to be an ideal function or process to outsource, due to the level of low specificity and number of service providers available, help desks are in fact becoming more and more complex (Marcella and Middleton, 1996). Moreover, literature is divided as to the overall effect that outsourcing may have in terms of positive and negative impacts. Kuh and Vesper (2001) implicitly state the positive effects, whilst other literature provides a warning regarding the possible negative connotations of outsourcing user support, particularly within an academic environment. Within such environments support is not limited to technical support as due to the presence of the students who are considered to be early adopters of technology and the professors who generally adopt technology later provisions for the supply of technical knowledge must be made (Bulchand-Gidumal and Melian-Gonzalez 2009; Majchrzak et al. 2000).

The identification of Gingrich's (2003) framework provides a typical breakdown of User Support tasks that consequently recognizes the skills and qualities required from User Support. Gingrich's framework identifies four categories of user support tasks and their frequency that require different levels of skills and qualities (depending upon complexity) in terms of organizational, technical and communicative competences. These are [1] *Infrequent Occurring tasks* that require high level of technical competence (i.e. electronic faults), [2] *Routine tasks* requiring little communication, technology or organizational competence (i.e. resetting of passwords), [3] *Business Related User Support tasks* requiring an equal level of all competences (i.e. setting up e-mail systems), and [4] *Business Project Work tasks* which utilize medium technical knowledge, but require extensive organizational and communicative competences (i.e. support of administration software). Thus, Gingrich's framework demonstrates that the skill-set of user support is far more diverse than the stereotypical perception and identifies that whilst technical skills are required, they are now less-utilized than communicational, organizational and business knowledge competencies. Users however, do not typically recognize that user support requires qualities other than technical competency (Beisse 2009).

We also recognize the need to examine awareness of the key factors affected and the issues involved in the outsourcing of user support positions. KPMG (1995) define four terms that are used when discussing the topic of IS or ICT outsourcing: *Customer* (organization being supplied with the outsourcing service), *End-User* (person who uses the outsourcing service), *Business Manager* (business sponsors of ICT within the organization) and *Supplier* (the organization providing the outsourcing service). We refer to these as 'actors'. The first and the last must always be present in any outsourcing situation, however depending on the specifics of the case being examined it may also be possible to identify the existence of the second and third actors. In context, aligned to Gingrich's framework of required competencies, we identified the 'actors' involved and determined a breakdown of typical tasks within the case study setting.

4 Research Approach

We adopted a mixed method research approach in an interpretative case study setting that involved both qualitative and quantitative inquiry utilizing an inductive research style as the research is grounded in theory (Collis and Hussey 2003; Saunders et al. 2007). In line with KPMG (1995) we identified the four actor roles relevant to our case study setting to get both the specific knowledge from key persons and further knowledge from the wider population. We conducted informal, semi-structured interviews with key personnel and a questionnaire with a group of 20 end-users in the case study setting.

Table 1 Population Details

'Actor Role'		No	Name	Job Title	Method
Customer	HARF/EHEC	1	Michael	User Support Co-ordinator	3 Interviews
Supplier	TechEd	2	Peter	Outsourced Technician	3 Interviews
			Frank	Supplier Co-ordinator	2 Interviews
BusinesManager	EHEC/HARF	1	Maria	Admin Co-ordinator	3 Interviews
End-users	HARF	20	Various	Various	Questionnaire

Table 1 above sets out the details of the participants, method of inquiry and frequency adopted for data gathering. In order to identify these groups the sampling method for both interview and questionnaire subjects needed to be non-probability based (Huberman and Miles 1994). This led directly to the use of convenience sampling and following the suggestion by respondents of additional, further respondents, an element of snowballing sampling was used (Salganik and Heckathorn 2004). Significantly each of the participants is a key representative of one of the groups of actors involved in our case study and the questionnaire participants are representative of the overall environment.

The use of interviews was a particularly strong tool within this investigation as it allowed the discovery of information that is subjective and could not be obtained by observation or the collection of statistics (DeMarrais 2004; Patton 2002; Saunders et al. 2003). Interviews were conducted in situ, on a 1:1 basis with informed consent, and lasted approximately one hour that afforded the opportunity to explore and expand in broader context any particular issues and ideas that may not have been previously acknowledged.

The questionnaire facilitated both quantitative and qualitative data collection. Consequently it was designed and structured to return a range of medias, including Boolean and full sentences (Saunders et al. 2003). Of the 20 end-users contacted a 100% response rate was achieved. We conducted a pilot questionnaire involving six participants, whose feedback and suggestions were incorporated and the questionnaire was subsequently amended (Saunders et al. 2003). Despite being a single-organization our real world case study, we believe our setting is representative of university environments throughout the European education sector that will enable findings to be transferable across other similar cases (Cresswell and Plano Clark 2011; Saunders et al. 2003; Stake 1995). Although common criticisms refer to limited breadth it is possible to generalize from one case study to theory (Yin 2003).

5 Findings and Discussion

In this section we discuss the findings on two levels. Firstly to represent those 'working on the ground' we examine the perspectives of those receiving the outsourced user support service (end-users) as well as the outsourced technician's perception (Peter). On one side this deals with the skills and qualities end-users expect from user support services, and how they perceived the outsourced provision. On the other side it presents the organizational issues that impacted upon the outsourced position. Secondly, we examine the business perspective from both the Customer's (EHEC) and the Supplier's (TechEd) positions i.e. a management view. More specifically we examine cost benefits and disadvantages of outsourcing the User Support provision against in-house provision.

5.1 End-Users' Perspectives

Our initial question to end-users was designed to ascertain whether they were aware how 'user support' was provided. We found that the majority of users

(60%) believed that their user support is provided through a mix of in-house and outsourced staff. The remaining 40% was split evenly (i.e. 20% each) between in-house and outsourced. We then asked, bearing this answer in mind, had they experienced any alternative support to that which they had indicated. Unexpectedly, we found that the majority (81.8%) identified that they had indeed experienced an alternative type of support within their working environment. This suggested that there was some confusion about end-users' understanding of the current situation. Indeed, a common response was "*We are currently going through a transition towards outsourced support*". Further investigation explains end-users perceive 'user support' to be in a transitional state based on the situation across EHEC rather than within the HARF where it is in fact already outsourced.

We next examined the nature of problems that end-users typically experienced. We aligned this to Gingrich's (2003) framework (discussed earlier) i.e. whether problems were *Technical*, *Communicational* or *Organizational*. The highest response (80%) reported that *Communicational* problems were the most problematic. Situations where end-users were unable to successfully access, carry out or complete a business action or system function (i.e. unable to make the computer do something they wanted to do). *Technical* problems accounted for 10% (i.e. issues of configuration or connection of the computer, network etc.), the remaining 10% concerned *Organizational* problems (i.e. inability to do something in line with the EHEC's principles because system is lacking). The mean response for each category is 28.07%, 42.81% and 29.12 respectively. However, when we looked at the respective standard deviations (SD) for each category, 25.70, 23.97 and 19.43, the high value indicates that the range of values provided for all categories is wide, with *Organizational Problems* having the smallest range. We suggest that such a wide range implies that the situation is influenced by the diversity of different end-users' problems that fluctuate significantly and, which due to their non-predictive nature, are prone to variation.

Having substantiated that end-users were not consulted about the outsourcing of the user support position we next examined the skills and expertise end-users felt it necessary for a User Support Technician to have. Our aim was to provide a context for comparison of the outsourced service provided with previous experience of the in-house provision. It was clear that they believed "*a good degree and experience*" was an indication of technical competence. However, we found a particular emphasis placed upon the ability to apply such competence to the end-users specific environment that is particularly relevant in the context of provision as inherently in-house technicians are familiar with the organizational context but outsourced agents are not. Significantly end-users' felt that "*knowledge of the tasks that staff carry out*" and "*exceptional knowledge of all computer program used at the Institute*" are considered key skills (Beulen 2011). End-users maintained that in-house user support personnel "*have a greater understanding of HARF activities that is lacking with outsourced staff*", and furthermore end-users believed that "*outsourcing made the user support less communicative and prompt*". Interestingly, although *availability* may not generally be considered to be a skill, 40% of the responses stated that availability is a required skill more readily provided by in-house provision.

Finally from the end-users' perspective we asked them what other attribute[s] they '*desired*' from an outsourced user support position. In other words what could be done to exceed end-user's expectations? Interestingly we found that what was '*desired*' reflected behavioural qualities rather than skills and expertise. Only 20% indicated that they felt that the skills that were desired from user support were those skills already being provided by the outsourced technician. What is noteworthy here is that 47% responded that *"humility, patience and a caring attitude*" are desired skills. This is far more characteristic of 'behaviour' than an ability to resolve a problem. End-users posit that although they want problem resolved, they want user support technicians to *'want'* to find solutions and "*not because they are simply doing it because it is their job*". There is a strong indication here that end-users would prefer a greater level of personable interaction with outsourced user support people. Indeed, one respondent chose to use the open nature of this question to 'rate' different '*desired*' qualities of user support interaction and put forward the following (out of 10): *"Availability - 9, Perseverance - 4, Expertise - 6, Experience - 6, Patience - 5"*. The suggestion here that these 'attributes' although previously present are not currently being provided through the outsourcing provision. It is clear from the above analysis that there is a bias towards in-house user support provision. The rationale behind this trend unfolds in the following section.

5.2 Outsourced Technician's Perspective - Peter

Interviews with Peter, the outsourced technician, gave us an opportunity to see the situation from the other side and respond to some of the points made above. The two stage interview process established that Peter possessed the required level of technical competence to the satisfaction of the EHEC's own criteria "*speaking English and basic technical competencies*" requisite for the outsourced Junior User Support Technician position. He was not educated to degree level as considered necessary to be regarded as technically competent by the end-users who consequently had some reservation about his technical skills.

Peter's expectation was that in addition to an initial induction period TechEd would provide a level of training to pick up any short fall in his knowledge *vis-à-vis* Microsoft or Cisco Certification. However, Peter reports that "*no training has been provided"*. He also informed that he experienced some difficulties related to in-house procedures indicating that the induction process was not sufficient. As a consequence he does not possess in depth *"knowledge of the tasks that staff carry out"* or the "*exceptional knowledge of all computer programs used"* that end-users expect. Furthermore, we rationalize that understanding and familiarity of common problems is context specific and gained through experience typically learnt over a period of time in situ.

Peter's revealed that as the outsourced User Support Technician he feels quite separated from either organization. He claims that "*this feeling is further enhanced by his experiences inside the environment as he does not get invited to in-house events*". He believes that he is perceived as an external entity rather than an internal asset. For example he is unable to access certain areas of the Intranet because as an external member of staff he does not have the necessary status and this contributes towards him "*feeling like an outsider*". He is further aware of a prior outsourcing experience that although occurred within a different department has left end-users with a negative perception of outsourcing making his job more difficult. Such pre-conceived conjecture meant that end-users assumed that Peter provided a low quality service prior to actually requesting his help. We surmise that the end-users negative attitude may be behind their expressed '*desired*' attributes of "*humility, patience and a caring attitude*". However, if 'one' is not '*en famille*', not receiving any reciprocal respect or appreciation within the setting then these characteristics are consequently all the more challenging to achieve. When we questioned the EHEC management about this we found, surprisingly, that even at management level there was an understanding that an outsourced technician cannot provide a much needed '*personal touch*'. Indeed, the Business Manager (Maria) stated that "*the 'personal touch' is not so readily available from a technician who is not employed internally*".

It is hardly surprising then that Peter admitted that he feels demotivated. In addition to the above issues he believes that his "*career opportunities are limited*" and that there is actually "*no possibility of promotion either inside the EHEC environment, or with TechEd.*" Contrarily, when questioned Frank confirmed that TechEd acknowledges that "*every person must have career opportunities and we try to accommodate this*" and qualifies that career opportunities are equally "*achievable by an outsourced technician*". However, this does not appear to have been the situation in Peter's case dampening his enthusiasm.

5.3 Business Perspective – Supplier (TechEd); Customer (EHEC/HARF)

We examine both these in tandem. On the Supplier side Frank (TechEd Co-ordinator) believes that a €35,000 fee is justified explaining that "*TechEd must meet a range of costs, of which only one cost is fully determined - the salary of the outsourced technician*". When pressed further he reveals that they pay the Outsourced Technician €14,100 p.a., and justifies that the remaining €20,900 is available to meet any additional costs. He puts forward the two main areas of potential extra cost as typically "*maintaining the continuity of the service*" and "*administrative management*". For example, in order to maintain continuity of service TechEd "*must ensure a backup technician is available*". TechEd "*assume that a 'person' can provide 210 working days a year (taking into account holidays and sick leave)*" however Frank explains that "*the service provided to the EHEC is contracted for 250 days per annum*". Consequently, TechEd must meet the costs related to "*employing another technician for 40 additional days including an induction period of a few working days*". In terms of the administrative management costs these are not extensive, and are primarily related to the formation of the contract and recruitment but the sum of both these provisions falls short of the remaining €20,900 balance that indicates a substantial profit.

From a Customers perspective, Maria (Business Manager-HARF) and Michael (User Support Co-ordinator - EHEC) put forward a different interpretation. In terms of the 'costings' involved with an outsourced technician Maria believes that, for this outsourcing exercise, *"there were no actual savings"*. Explaining she reports that it was necessary to provide Peter with *"an office along with computing equipment and a Blackberry"* to enable him to work effectively. However, part of EHEC's rationale behind the decision to outsource this position was the prospect of cost savings in terms that all training costs would be met by TechEd and not by EHEC. Formerly EHEC has met all costs of providing relevant training to in-house staff which is necessary due to the context sensitive nature of user support. EHEC argue that the €35,000 paid to TechEd only provides a "***Junior*** User Support Technician" with minimal experience or relevant education and thus further training is required. However, Maria reports that "*no training has been provided to the Outsourced Technician* (Peter)*, other than the initial induction training*". If a member of staff were to be employed internally for the same cost, it would be possible to recruit the equivalent of a 'Senior Technician' who would have a more strategic role and most likely would have superior education, in-house knowledge and experience. Michael suggests that for an equivalent outgoing cost it would have been possible to employ a technician with both the required and the desired skills identified by end-users.

However, analysis of the cost implications runs deeper than the fee that is paid to TechEd or alternatively, the prospective salary that could be paid to an internal employee. Maria reported that the same logistical costs are encountered regardless of whether the technician is internal or external, i.e. the cost of interviewing an external candidate is the same as for an internal candidate. This is supported by Michael who showed convincingly that the cost implications of the recruitment process of outsourcing are the same as potentially employing the candidate internally. Potential cost savings were perceived as removing the cost of financing of any ongoing training for the technician. Interestingly, as no training has actually been provided, this cannot be considered as a real cost saving. At this point the question can be posed, is the level of convenience provided by having outsourced the activity a large enough benefit to outweigh the consequences that have occurred? We surmise that the cost implications of utilizing outsourced staff instead of in-house staff are largely related to the fee paid by the EHEC to the *Supplier*, as the logistical and recruitment costs are unchanged and therefore in this outsourced technician activity, there is little evidence produced to substantiate any savings, irrespective of the skills level of an outsourced technician.

6 Conclusions

We have evidenced that the affects of outsourcing technical positions from an end-users perspective are significant. Smith (1776) raised the matter of the division of labour which occurs as a result of outsourcing and this study supports his views. Albeit from a relatively small case study, we found that an outsourced technician would have less of a strategic job role but a greater procedural responsibility than an internally employed equivalent familiar with the milieu. This is indicative of Smith's theory of division of labour. To delve deeper, it can be suggested that the division of labour has to a certain extent led to an element of isolation that has consequently contributed to the demotivation of the Outsourced Technician. The division of labour is not isolated to the Customer (EHEC) but is indeed also related to the Supplier (TechEd) as the empirical data indicates.

This study supports the framework put forward by Gingrich (2003). We surmise that a large number of the outsourced technical activities primarily require both organizational and communicational knowledge. However, the reported lack of interaction between the outsourced technician and the customer organization [HARF/EHEC] was reflected by the lack of organizational knowledge of the outsourced provision.

Benefits or disadvantages identified are context specific to the different roles/groups (KPMG (1995) involved within the case study setting (*End-users, Customer, Supplier, Outsourced Technician*). The *Customer* indicates that the benefits experienced by outsourcing (lack of a legal obligation, extended working day) were the basis for the outsourcing decision in the first place. The *Business Manager* stated that another incentive was the provision of ongoing training for the outsourced technician without cost to them. Interestingly however, the outsourced technician (Peter) did not received any ongoing training and there was also no evidence of the extended working hours which were theorized to be available due to the outsourcing exercise. A further aspect of outsourcing that was seemingly overlooked from a 'in practice' stance is the importance of opportunities for career progression for outsourced technicians. Interviews with TechEd provided a contrasting but theoretical picture. However it is evident that in the opinion of Peter (outsourced technician) his experience illustrates a negative perspective. It is clearly evident in our case study setting that he feels he has no opportunities within either organization, that he feels disadvantaged and this has also attributed low levels of motivation.

In accordance with the views put forward by Bulchand-Gidumal and Melian Gonzalez (2009), our study has shown that the role of user support is not limited to providing technical support. We evidenced that both end-users and business managers support the view that there is a strong "desire" to receive support which contains "the personal touch", "care" and "patience" which they feel are not being provided. The perception is that this is more easily provided by internal individuals than by outsourced provision. For our case it is possible to suggest that the demotivation experienced by Peter is a significant contributory factor in the lack of these desired attributes. In order to be practically valuable, this study proposes two important considerations that an organization should address when determining whether to outsource its IS/IT user support:

- Have all stakeholders been consulted regarding their needs from the User Support provision in terms of both necessary and desired requirements?
- Do the cost implications of outsourcing the user support provision provide the level(s) of benefit expected?

These questions are not definitive, as circumstances will dictate the specifics of any situation, however they are best utilized by being considering as enriching the decision making process when considering the option of outsourcing user support.

Acknowledgments. We thank all the participants for their participation and contributions.

References

Beisse, F.: A Guide to Computer User Support for Help Desk and Support Specialists. Cengage Learning, Boston (2009)

Bergkvist, L., Fredriksson, O.: Outsourcing Terms-A Literature Review from an ISD Perspective. In: Proceedings of the 16th European Conference on Information Systems, Galway, Ireland, June 9-11 (2008)

Beulen, E., Ribbers, P., Roos, J.: Managing IT Outsourcing. Routledge, Oxon (2011)

Bulchand-Gidumal, J., Melian-Gonzalez, S.: Redesign of the IS/ICT help desk at a Spanish public university. Higher Education 60(2), 205–216 (2009)

Collis, J., Hussey, R.: Business Research: A Practical Guide for Undergraduate and Postgraduate Students. Palgrave Macmillan, London (2003)

DeMarrais, E.: The materialisation of culture. Database for Advances in Information Systems 35(4), 6–102 (2004)

Fink, D., Shoeib, A.: Action: The most critical phase in outsourcing information technology logistics. Information Management 16(5), 302–311 (2003)

Gingrich, G.: Managing IT in Government, Business and Communities. IRM Press, London (2003)

Gonzales, R., Gasci, J., Llopis, J.: Information Systems Offshore Outsourcing. Industrial Management and Data Systems 106(9), 1233–1248 (2006)

Hirschheim, R.A., Lacity, M.: The Myths and Realities of IT Insourcing. Communications of the ACM 43(2), 99–107 (2000)

Huberman, A., Miles, M.: Qualitative Data Analysis: An Expanded Sourcebook. Sage, California (1994)

Johnson, K.A., Abader, S., Brey, S., Stander, A.: Understanding the outsourcing decision in South Africa regards to ICT. South African Business Management 40(4), 37–49 (2008)

Majchrzak, A., Rice, R.E., Malhotra, A., King, N., Ba, S.: Technology Adaptation: The Case of a Computer-Supported Inter-Organizational Virtual Team. MIS Quarterly 24(4), 569–590 (2000)

Kern, T., Willcocks, L.: Exploring Relationships in IT Outsourcing: The Interaction Approach. European Journal of Information Systems 11(1), 3–19 (2002)

Kuh, G., Vesper, N.: Do computers enhance or detract from student learning? Research in Higher Education 42(1), 87–102 (2001)

Kussmaul, C., Jack, R., Sponsler, B.: Research: Outsourcing How Well Are You Managing Your Partners? CIO Insight 1(33), 75–85 (2003)

Lacity, M., Hirschheim, R.: Implementing Information Systems Outsourcing: Key Issues and Experiences of an Early Adopter. Journal of General Management 19(1), 17–31 (1993)

Lacity, M., Willcocks, L.: Global IT Outsourcing – In Search for Business Advantage. John Wiley and Sons, Chichester (2001)

Marcella, R., Middleton, I.: The role of the help desk in the strategic management of information systems. OCLC Systems and Services 12(4), 4–19 (1996)

Mullin, R.: Managing the Outsourced Enterprise. Journal of Business Strategy 17(4), 28–32 (1996)

Olías de Lima, B.: La nueva gestión pública. Prentice-Hall, España (2001)

Patton, M.Q.: Qualitative Evaluation and Research Methods. Sage, UK (1990)

Salganik, M., Heckathorn, D.: Sampling and Estimation in Hidden Populations using Respondent-Driven Sampling. Sociological Methodology 34, 193–239 (2004)

Saunders, M., Lewis, P., Thornhill, A.: Research Methods for Business Students. Pearson Education Limited, Essex (2003)

Stake, R.: The Art of Case Study Research. Sage, London (1995)

Yin, R.: Case Study Research: Design and Methods. Sage, CA (2009)

Understanding Educational Administrators' Subjective Norms on Their Use Intention toward On-Line Learning

Tsang-Kai Chang[1], Hsi-fang Huang[2], and Shu-Mei Chang[3]

[1] Department of Education, Taipei Municipal University of Education, Taiwan
charlie1126@gmail.com
[2] Shuang-Yuang Elementary School, Taipei, Taiwan
hsifanghuang@gmail.com
[3] Ying-Qiao Elementary School, Taipei, Taiwan
Maychang720@yahoo.com.tw

Abstract. With the rapid growth of the Internet, it is much easier to access the web-based technology. The web-based technology has also dramatically influenced our life. Moreover, many institutions, including the government of Taiwan, required that their employees are capable of using technological tools to fulfill their job requirements. Public organizations in Taiwan are now widely utilizing web-based learning techniques to improve the quality of human capital and boost the productivities of public employees; the situation is the same in education field. Many previous studies showed that using the web-based technology efficiently enhances learners' performances, attitudes and motivation toward on-line learning. Therefore, with the trend of using web-based technology on learning and teaching, numerous education/training institutes and companies have dedicated great efforts and large amount of money to advance on-line learning programs for users. However, while many studies mentioned about the learners', teachers' and employees' acceptance toward on-line learning, few studies have reported the point of view from educational administrators, the crucial group of people who make the educational decisions. Therefore, the purposes of this study are using the theory of planned behavior and theory of reasoned action as background models to investigate the effect of the participants' subjective norms on their use intention toward on-line learning. The participants in this study were 176 educational administrators in Department of Education, Taipei City Government. A survey questionnaire was administered to understand their subjective norms, including "superior influence," "peer influence," and "regulations." The results demonstrate that peer influence has the most significant effects on participants' use intention, followed by superior influence. However, regulations have no significant effect on their use intention. Specifically, from the results of correlation analysis, there is a positive relationship between peer influence and their use intention toward on-line learning, however, a negative relationship has shown between superior influence and use intention toward on-line learning. In other words, it seems that the participants are influenced by their

L. Uden et al. (Eds.): 7th International Conference on KMO, AISC 172, pp. 267–273.
springerlink.com

peers on using on-line learning, thus expressing higher use intention. Contrary to the peer influence, the participants are less influenced or get discouraged by their superiors on using on-line learning.

Keywords: educational administrator, subjective norms, use intention, on-line learning.

1 Introduction

The use of web-based technology through the Internet has huge impact on nearly every aspect of life, including business operation, education, communication, entertainment, social activity, shopping, and so on (Cheung & Huang, 2005). In order to benefit from the use of web-based technology, the governments all over the world have instituted policies intended to increase the qualities of users' learning, teaching and working. Moreover, in many educational systems, the integration of technology has been recognized as one of the crucial drivers to the improvement of teaching and learning, thus prompting government to launch major initiatives and investment to build web-based technology in schools (Pelgrum, 2001).

In addition, many studies (e.g., Liu, Chen, Sun, Wible & Kuo, 2009; Teo, 2009) have proved that using technology acceptance model (TAM) (Davis, 1989) is an efficient way of learners' use intention toward on-line learning through "perceived ease," "perceived usefulness," and "compatibility." However, users' attitudes are not the only predictor of users' use intention toward on-line learning; users' behavioral intention will also be affected by other factors, such as the opinions of important individuals and subjective norms (Fishbein & Ajzen, 1975). According to Martin Fishbein and Icek Ajzen's (1975) "Theory of Reasoned Action" (TRA) and Ajzen's (1985) "Theory of Planned Behavior" (TPB), subjective norm is an individual's perception of social normative pressures, or relevant others' beliefs that he or she should or should not perform such behavior (Ajzen, 1991).

The TPB has been used successfully to understand a wide variety of human behavior, such as weight loss behavior (Schifter & Ajzen, 1985), and smoking cessation (Godin, Valois, Lepage, & Desharnais, 1992). Also, The TPB has also been used to explain teachers' intentions and behavior in the classroom. Thus, numerous studies (e.g. Lee, Cerreto & Lee, 2010; Lee, 2009; Teo, 2009) have adopted the Theory of Planned Behavior into the educational setting. The results showed that the model has the potential to help both students and teachers on their learning and teaching.

However, although the TPB is proved a valid model to predict people behavior in many different domains, few studies gave special attention to educational administrators' subjective norms on their use intention toward on-line learning. Subjective norms affect an individual's perception of social normative pressures, or relevant others' beliefs that he or she should or should not perform such behavior. Furthermore, it is recognized that educational administrators play an important role on making educational policies, there is no doubt that their subjective norms and use intention toward on-line learning are very critical on schools' technology-related policies. Hence, the purpose of this study is to investigate educational administrators' use intention toward on-line learning.

In sum, in order to forecast educational administrators' use intention toward on-line learning more efficiently, a research questionnaire was administered to explore the participants' subjective norms, including "superior influence," "peer influence," and "regulations." By gathering questionnaire responses from 176 educational administrators in Department of Education, Taipei City Government, this study addressed two research questions: (1) What are the relationships between the educational administrators' subjective norms and their use intention toward on-line learning? (2) What are the other factors that might affect educational administrators' subjective norms towards on-line learning?

2 Method

2.1 Sample

The participants of this study were randomly selected from the department of Education, Taipei City Government of Taiwan. The final sample included 176 educational administrators of which 61 (34.7%) were male and the remaining 115 (65.3%) were female. Among these participants, 37 (21%) were under 30 years old, 69 (39.2%) were 31-40 years old, 70 (39.8%) were over 41 years old.

2.2 Instruments

To assess the educational administrators' subjective norms and use intentions toward on-line learning, two instruments were implemented in this study.

The Subjective Norms toward On-line Learning Survey (SNOL) administered in this study was developed on the basis of Taylor & Todd (1995). The initial pool of items in the survey included a total of 12 items, which were presented by using a seven-point Likert mode (ranging from 1, "strongly disagree" to 7, "strongly agree"). Also, three scales were designed for SNOL. The details of the three scales are as follows:

1. Superiors influences scale: assessing perceptions of the extent to which educational administrators perceive that the impact of his/her superiors' opinions about the use of on-line learning.
2. Peers influences scale: assessing perceptions of the extent to which educational administrators perceive that the impact of his/her peers' opinions about the use of on-line learning.
3. Regulations scale: measuring the degree to which the regulations of the government request the educational administrators' learning and professional development.

The Use Intention toward On-line Learning Survey (UIOL) implemented in this study was adapted from Ajzen's (2002). Some items from his questionnaire were modified. Also, the 7 items were presented by employing a seven-point Likert scale (from 1, "strongly disagree" to 7, "strongly agree"). The survey measured the perceptions of the extent to which educational administrators' actual practice and willingness to use on-line learning.

2.3 Data Analysis

To fulfill the main purposes of this study, factor analysis, t-test, ANOVA and correlation were conducted as the statistical methods in this study. The factor analysis was utilized to reveal the scales of the instruments on the educational administrators' SNOL and UIOL. In addition, this study gathered educational administrators' information about their gender and age. By using t-tests, gender differences on educational administrators' SNOL and UIOL were analyzed. Also, the scales were examined via analysis of variance (ANOVA) to analyze age differences. Moreover, correlation analysis was utilized to examine the relationship between educational administrators' subjective norms and use intention toward on-line learning.

3 Results

3.1 Factor Analysis

The final version of the SNOL consisted of 12 questionnaire items with three scales. Through the factor analysis, the reliability coefficients for the scales respectively were 0.90 (Superiors influences, 4 items), 0.90 (peers influences, 4 items), and 0.91 (regulations, 4 items). The alpha value of the whole SNOL questionnaire was 0.91 and these scales explained 76.40% of variance totally. In addition, the aforementioned method was adopted to clarify the structure of use intention toward on-line learning (UIOL). The latest version of the UIOL consisted of 7 questionnaire items and the alpha value of the whole UIOL questionnaire was 0.93 and the factors explained 73.9% of variance totally. Therefore, these scales were deemed to be sufficiently reliable for assessing educational administrators' subjective norms and use intentions toward on-line learning.

3.2 Background Differences on SNOL & UIOL Scales

In this study, t-test and ANOVA tests were employed to examine the educational administrators' background differences, such as gender and age on the SNOL & UIOL scales. First, a series of t-tests were performed on the gender differences of educational administrators' mean scores for the SNOL and UIOL. The result of this study indicates that no significant differences are found on the scales of educational administrators' subjective norms and use intentions of participating on-line learning between two genders.

Moreover, in order to compare the possible differences derived from age, we categorized the participants into three major groups: less than between 30 years old, between 31-40 years old, and greater than 41 years old. The ANOVA tests, presented in Table 1, indicate that age differences play a statistically significant role in perceived usefulness of the SNOL scales (p <0.05) and UIOL scales (p <0.01). A series of Scheffe tests indicate that less than 30 years old educational administrators tend to express perceptions of superiors influences of subjective norms toward on-line learning than >41 years old educational administrators do. Moreover, the Scheffe tests further indicate that >41 years old educational administrators tend to express more positive use intentions toward on-line learning than 31-40 years old and <30 years old educational administrators. These comparisons indicate that younger educational administrators tend to follow their superiors' requests or behaviors than other senior educational administrators do.

Table 1 Educational administrators' subjective norms and use intentions toward on-line learning among different age groups

Age Group	(1) Less than 30 years old (mean, SD)	(2) 31-40 years old (mean, SD)	(3) greater than 41 years old (mean, SD)	F(ANOVA) Scheffe Test
Superiors influences	4.65(0.83)	4.30 (1.06)	4.03(0.99)	2.90*(1>3)
Peers influences	4.69(1.60)	4.48(1.04)	4.35(1.15)	1.23(n.s.)
Regulations	5.14(0.88)	4.71(1.08)	4.64(1.08)	1.77(n.s.)
Use intention	4.84(1.19)	4.90(1.03)	5.44(0.81)	4.09**(3>1; 3>2)

* $p < 0.05$ ** $p < 0.01$.

3.3 Correlation Attitudes and Use Intention toward On-Line Learning

The result of Pearson correlation indicates that the scales of the SNOL and UIOL are significantly positively correlated with each other (r>0.20, p <0.01), except that no statistical correlation is found on the superiors influences. These results in general support that educational administrators expressing stronger subjective norms would display more frequent use intention toward on-line learning. In particular, educational administrators' responses on the peers influences of subjective norms scales were relatively more highly correlated with use intention scale (r>0.32, p <0.01). This implies that educational administrators with stronger perceptions of peers influences may help themselves attain stronger intention of on-line learning. This may have been that these educational administrators through the support and encouragement from their peers in the adoption of on-line learning. However, the superiors' influences are not related to the educational administrators' use intentions toward on-line learning in this study. As far as the degree to which the superiors attributed to this influences, it is not significant in this study.

4 Discussion and Conclusion

This study aims to explore the relationship between educational administrators' subjective norms and use intentions toward on-line learning. To this end, two questionnaires were administered to assess educational administrators' subjective norms toward on-line learning (i.e., the SNOL) and use intention toward on-line learning (i.e., the UIOL). The results shows that the SNOL and UIOL implemented in this study are sufficiently reliable to assess educational administrators' subjective norms and use intention toward on-line learning.

Moreover, in this study, with the SNOL and UIOL, no gender differences are found in educational administrators' subjective norms and use intention toward on-line learning. It seems that both male and female educational administrators express similar levels of subjective norms and use intention toward on-line learning. Besides, the age differences are found significant on educational administrators' subjective norms and use intention toward on-line learning. The result indicates that younger educational administrators express stronger perceptions of superiors' influences of subjective norms toward on-line learning than senior educational administrators. On the other hand, the result also shows that senior educational administrators have stronger use intention toward on-line learning than younger educational administrators do. It may imply that educators should pay more attention to enhancing younger educational administrators' use intention toward on-line learning.

In this study, educational administrators' subjective norms toward on-line learning are positively correlated with their use intention toward on-line learning. Educational administrators with stronger subjective norms would display more positive willingness of frequent usage for on-line learning. This causal path is contradictory to Liker and Sindi's (1997) findings but implied by findings of Bansal and Taylor (2002), who reported that subjective norms played a vital role in use intentions. According to Venkatesh and Morris (2000), end-users have a tendency to comply with peers in the absence of direct information and experience with a new technology. Therefore, strategies to improve educational administrators' use intentions may have a measurably positive effect on their department. It seems that in a department culture where educational administrators perceive that their peers and superiors reinforce the value of on-line learning, positive use intentions toward on-line learning are more likely (Sivo, Pan & Hahs-Vaughn, 2007). The climate in which the perception of subjective norms is prominent and positive at the very beginning of the learning course may be most effective in sustaining positive use intentions across the government.

This is evidenced in the current study as subjective norm is a prominent factor in learners use intention toward web-based learning. And the results probably suggest that educators should try to create learning culture to improve educational administrators' learning behaviors in the web-based environments. It may be practicable for educators to enhance educational administrators' use intentions toward on-line learning by peers supports and regulations in the department. This study facilitates the understanding of educational administrators' subjective norms and use intention toward on-line learning. By using the SNOL and UIOL questionnaires, educators and researchers can assess and review educational administrators' subjective norms and use intentions toward on-line learning in a more effective way, with possibly higher validity.

References

Ajzen, I.: From intentions to action: A theory of planned behavior. In: Kuhl, J., Beckman (eds.) Action-Control: From Congnition to Behavior, pp. 11–39. Springer, Heidelberg (1985)

Ajzen, I.: The theory of planned behavior. Organizational Behavior and Human Decision Processes 50, 179–211 (1991)

Bansal, H.S., Taylor, S.F.: Investigating interactive effects in the theory of planned behavior in a service-provider switching context. Psychology & Marketing 19(5), 407–425 (2002)

Cheung, W., Huang, W.: Proposing a framework to assess Internet usage in university education: An empirical investigation from a student's perspective. British Journal of Educational Technology 36(2), 237–253 (2005)

Davis, F.: Perceived usefulness, perceived ease of use, and user acceptance of information technology. MIS Quarterly 13(3), 319–340 (1989)

Fishein, M., Ajzen, I.: Belief, attitude, intension and behavior: An introduction to theory and research. Addison-Wesley, Reading (1975)

Gondin, G., Valois, P., Lepage, L., Desharnais, R.: Predictors of smoking behavior: an application of Ajzen's theory of planned behavior. British Journal of Addiction 87(9), 1335–1343 (1992)

Lee, M.-C.: Explaining and predicting users' continuance intention toward e-learning: An extension of the expectation-confirmation model. Computers & Education 54, 506–516 (2009)

Lee, J., Cerreto, F.A., Lee, J.: Theory of Planned Behavior and Teachers' Decisions Regarding Use of Educational Technology. Educational Technology & Society 13(1), 152–164 (2010)

Liker, J.K., Sindi, A.A.: User acceptance of expert systems: a test of the theory of reasoned action. Journal of Engineering and Technology Management 14(2), 147–173 (1997)

Liu, I.-F., Chen, M.C., Sun, Y.S., Kuo, C.-H.: Extending the TAM model to explore the factors that affect Intention to Use an Online Learning Community. Computers & Education 54, 600–610 (2009)

Pelgurm, W.J.: Obstacles to the Integration of ICT in Education: Results from a Worldwide Educational Assessment. Computers & Education 37, 163–178 (2001)

Schifter, D.E., Ajzen, I.: Intention, perceived control, and weight loss: An application of the theory of planned behavior. Journal of Personality and Social Psychology 49(3), 843–851 (1985)

Sivo, S.A., Pan, C., Hahs-Vaughn, D.L.: Combined longitudinal effects of attitude and subjective norms on student outcomes in a web-enhanced course: A structural equation modeling approach. British Journal of Educational Technology 38(5), 861–875 (2007)

Teo, T.: The impact of subjective norm and facilitating conditions on pre-service teachers' attitude toward computer use: a structural equation modeling of an extended technology acceptance model. Journal Educational Computing Research 40(1), 89–109 (2009)

Venkatesh, V., Morris, M.G.: Why don't men ever stop to ask for directions? Gender, social influences, and their role in the technology acceptance and usage behavior. MIS Quarterly 24(1), 115–139 (2000)

An Investigation of Business and Management Cluster's Students' Motivation of Taking Technician Certification at Vocational High Schools in Central Taiwan

Chin-Wen Liao, Sho-Yen Lin, Hsuan-Lien Chen, and Chen-Jung Lai

Department of Industrial Education and Technology,
National Chunghua University of Education. No.2, Shi-da Road, Chunghua City,
500, Taiwan, R.O.C.
tcwliao@cc.ncue.edu.tw, linshoyen@gmail.com,
hsuan.lien@msa.hinet.net, tcivs74@yahoo.com.tw

Abstract. The study aims to investigate factors that motivate Business and Management Cluster's students to take secondary technician certification of their majors. With the literature review, a questionnaire survey was adopted. In total, 878 questionnaires were sent out and a valid response rate was 80.30%. After collecting the entire questionnaires, statistical software packages were used to analyze the collected data. The results showed that teacher's instructions and students' increasing sense of achievement and self-confidence by obtaining certification were the two main motivations students hold. Moreover, private school students' are much more motivated to obtain certifications than public school students. Based on the results, several suggestions are made as references for schools, teachers, parents and students.

Keywords: secondary technician certification, business and management cluster students at vocational high school, motivation.

1 Introduction

With the rapid change of industry and economy, the need to promote employee's professional technician and standard is crucial for companies and organizations. One of the criteria which companies select their personnel is certification, an indicator of their professional knowledge and practicum (Hsiao, 1990; Tsai, 2000). Due to the rapid change of industry and various demands on manpower, as a response to this, vocational high school students in Taiwan are encouraged to get professional certifications of their majors to meet with the competitiveness of their future workplace (Ministry of Education, 2000). It is claimed that getting related certifications of students' majors serve as a preparation for their future career (Wu, 2008; Dai, 2006). Getting the technician certification is not only a

L. Uden et al. (Eds.): 7th International Conference on KMO, AISC 172, pp. 275–281.
springerlink.com

representation of an individual ability, but a basis for job opportunities. It also has a positive effect on students' learning motivation and learning environment. Some schools even offer merits for students of they pass the technician certification test, such as giving extra points for finals or weaving practicum courses.

Several researches have indicated that motivation helps to guide and maintain a certain kind of behavior (Pintrich, Marx, & Boyle, 1993; Wen, 2006). The motivated behaviors tend to have a specific goal and can be used to examine why an individual undergo an act (Graham and Weiner, 1996). Deci (1975) categorized motivation into intrinsic motivation and extrinsic motivation. The former is driven by getting rewards or getting compliments from others while the latter is driven by an individual's pleasure and to take part in an activity without getting any rewards because it is out of personal interest or curiosity (Amabile et. al, 1994; Lahey, 2004). As Woolfolk (2007) maintained, intrinsic motivation is connected to a pursuit of overcoming challenges and extrinsic motivation is related to factors, such as getting good grades, avoiding punishment or pleasing teachers.

In addition to the above mentioned research, Marslow (1970) proposed that there is a hierarchy of people's needs, such as physiological needs, safety needs, love and belongingness needs, esteem needs, needs to know and understand, aesthetic needs and self-actualization needs. Among the seven needs, physiological needs, safety needs, love and belongingness needs belonged to lower-level needs while needs to know and understand, aesthetic needs and self-actualization needs belonged to higher-level needs. If the latter needs are fulfilled, a sense of happiness of fulfillment will be created.

Past studies have investigated the possible factors that motivated students on taking certification. It was found that learning motivation, learning attitude, learning methods, learning behavior and learning environment have positive correlations with students' taking certification (Hsieh 2003; Ma 2004) Hunag (2003) found that teachers at private schools have much to do with students' certification results because they tended to adjust their teaching to technician certification on their practicum classes so that students have more time on practice. Hsieh (2003) indicated that motivation played an important role on the effect of certification, such as career planning. However, gender differences and parents' educational levels seemed to have low correlations with students' motivation (Hsieh 2003). But if parents and teachers have higher expectations on certification, students may think taking certification worthwhile and are more willing to take it to meet their expectations. Furthermore, in her study, Wu (2008) maintained that most students pursued certification not by the influence of further studies but by the pursuit of self-esteem and fulfillment.

Based on the above literature review, the purpose of this research is to investigate the present situation of what motivates students at business and management clusters, including Department of Business and Management, Department of National Trade, Department of Accounting and Department of Data Processing, to take secondary technician certification, and to compare the influence of school types and background variables on motivation, and finally draw on suggestions as the references for vocational high schools, teachers, parents and students.

2 Study Design

Based on the literature review and investigation, a Student Motivation Questionnaire is established as the research framework with the following dimensions-self-esteem, parents' attitudes, teacher's expectations, administrative support, peer influence, social assessment and certification effectiveness. The samples are business and management cluster's students who take secondary technician certification from vocation high schools at central Taiwan. They are randomly sampled. After the first draft of the Student Motivation Questionnaire was established, it was reviewed by expert meeting by deleting inappropriate items and adding new items and was then distributed to third-grade students as pre-test. After collecting the data of pre-test, reliability and validity were examined and the final draft of questionnaire was set. To examine reliability and validity of the questionnaire, item analysis, factor analysis and Cronbach α were adopted. First of all, in item analysis, a critical ratio is used to determine the significance of each item. Besides, the make the consistency of the questionnaire, if Cronbach's α of any item is higher than the overall value, then that item would be deleted (Henson, 2001). Besides, Pearson's ratio is also adopted as a way to delete or retain the questionnaire items. After the examination, item 20 was deleted because it does not reach the significance level in critical ratio and Pearson's value.

A 5-point Likert Scale was adopted. Samples responded to questions according to their perceptions about their motivation on taking secondary technician certification. More points means higher motivation with strongly agree of 5 point, followed by agree of 4 points, neutral of 3 points and disagree of 2 points and strongly disagree of 1 point. Drawn on the literature review, the questionnaire is divided into three parts. The first part is about background information of the sample; the second part lists factors that motivate students to take certification, including self-esteem, parents' attitudes, teacher's expectations, administrative support, peer influence, social assessment and certification effectiveness. A total of 63 questions are listed. The third part of the questionnaire is a semi-opened questionnaire for samples to illustrate their opinions on this subject matter.

The independent variables include school background variables and personal information variable. School background variables contain school types, public school or private school and location of school as in the city, in the county, or in the township. Personal information variables include gender of male or female, departments with Department of Business and Management, Department of National Trade, Department of Accounting and Department of Data Processing, career planning with going for further studies or going into workplace, and parents' educational background with from junior high school graduates, senior high school gratuates or colleges graduates. The independent variables include self-esteem, parents' attitudes, teacher's expectations, administrative support, peer influence, social assessment and certification effectiveness.

Table 1 Items of factors that motivate students on taking secondary technician certification

Dimensions	Students' motivation on taking secondary technician certification	Items	Total number
1	Self-esteem	1-13	13
2	Parents' expectation	14-22	9
3	Teacher's expectation	23-30	8
4	Administrative support	31-37	7
5	Peer influence	38-46	9
6	Social assessment	47-53	7
7	Certification effectiveness	54-63	10
	Total		63

3 Data Analysis

A total of 878 questionnaires were sent out and 705 questionnaires were returned with a return rate of 80.30% after weaving 81 invalid ones. The average value of Cronbach's Alpha was 0.954. To undergo the statistical analysis, number of distribution, percentage, mean and standard deviation were adopted to examine the distribution of what motivate students on taking certification.

As illustrated in Table 2, the overall average point of Likert Scale is 3.99, ranging between 3.81 to 4.16, meaning that students are inclined to agree on the motivation items as defined in Chang (1990). The top motivation items in order are parents' expectation with a mean score of 4.16, followed by self-esteem of 4.15, and effectiveness of technician certification of 3.99, social assessment of 3.92, administrative support of 3.95, peer influence of 3.92, and parents' attitude of 3.81. It is clearly that students are more likely to be driven for taking technician certification out of parents' expectation and self-esteem while peer influence and parents' attitudes play a less significant role.

Table 2 Descriptive analysis of business and management students' motivation on taking secondary certification at central Taiwan

Factors	Average score of each level of factor	Number of items	Average score of each item	Standard deviation	Rank
Self-esteem	20.75	5	4.15	.49	2
Parents' expectation	22.85	6	3.81	.53	7
Teacher's expectation	24.93	6	4.16	.50	1
Administrative support	19.75	5	3.95	.51	5
Peer influence	23.49	6	3.92	.52	6
Social assessment	19.91	5	3.98	.53	4
Certification effectiveness	27.91	7	3.99	.51	3
Total	159.59	40	3.99	.37	
N=705					

Table 3 illustrates the analysis of school background variables of business and management cluster's students on their motivation of taking secondary technician certification. Overall, there was a significant difference on different categories of school types, t=-3.325, p<.01, while private school students' motivation, M=4.02, was relatively higher than public school ones, M=3.93. Students at private schools score higher in self-esteem, teacher's expectation and effectiveness of certification. Moreover, there is no significant difference in parents' attitudes t=-1.101, p>.05, peer influence (t=-1.816, p>.05), and social assessment (t=-0.728, p>.05). Therefore, it indicated that students at private schools are more motivated to take technician certification, and among the motivation items listed, they are more related to self-esteem and teacher's expectation since the mean score of the two items are above 4.

In terms of scale of city variable, it was found that there was no significant difference in self-esteem, parents' attitudes, peer influence, social assessment and effectiveness of technician certification with p<.05 respectively while in the item of teachers' expectation, students in township sector score higher, M>4.21, than those in city sector, M>4.11.

In terms of the effectiveness of certification, male students scored higher than female students. And regarding students of different departments, a significant difference was found in self-esteem, parents' attitudes, teacher's expectation and administrative support. Overall, students in Department of Data Processing score higher than Department of International Trade in teacher's expectation and administrative support. Students in Department of Data Processing got the highest score in parents' attitudes than Department of Business and Management.

Significantly, there was no significant difference in terms of the seven dimensions of motivation with different career plans. But in terms of parents' attitudes, students, whose parents graduated from colleges tended to have higher motivation than those whose parents graduated only from or below junior high school.

Table 3 Analysis of school background variables on the motivation of taking secondary technician certification of business and management cluster's students

Factors	School category	Scale of city
Self-esteem	(2)>(1)	—
Parents' expectation	—	—
Teacher's expectation	(2)>(1)	(3)>(1)
Administrative support	(2)>(1)	(2)>(1)
Peer influence	—	—
Social assessment	—	—
Certification effectiveness	(2)>(1)	—
total	(2)>(1)	—

N=705. School category:(1)Public school, (2) private school; Scale of city: (1) City, (2) County, (3)Township.

4 Conclusions and Suggestions

After statistical analysis on questionnaires, conclusions are made as follows.

4.1. It was found that the strongest motivations of business and management cluster's students when taking secondary certification are "teacher's guidance and emphasis" and "getting certification can increase one's fulfillment and confidence," followed by the effectiveness of technician certification, social assessment, administrative support, peer influence and parents' attitudes.
4.2. Students at private schools are more motivated to take secondary technician certification than those at public schools because the former ones tend to receive more guidance and training on certification from teachers.
4.3. There is no significant difference on students who intend to pursue further studies and those who do not or have not decided yet.

Based on the conclusions, suggestions are made as follows. First of all, it is essential to increase teacher's expectations and develop students' self-esteem and confidence by giving more care and support on them (Muller, 1999). Second, teachers should develop students' career planning concepts so that they are more aware of what the next step they should take either to continue their studies or stepping into workplace. Besides, develop students' good learning attitudes so that they can synthesize and organize the lessons they learn and develop their ability for problem-solving instead of doing mechanical drills with no integration ability (Vroom, 1963).

Moreover, parents should provide support and encouragement on children's learning situation automatically and create better interactions with their children to increase their willingness to take technician certification (Grolnick et. al, 1991). Finally, students should make efforts to get technician certification as a preparation for further studies or work because more and more enterprises select their personnel based on the technician certification an applicant holds. They should also be encouraged to take related certifications constantly to meet with the competitiveness of career needs and as a way to prove their own ability.

References

Amabile, T.M., Hennessey, B.A., Tighe, E.M.: The work preference inventory: Assessing intrinsic and extrinsic motivational orientation. Journal of Personality and Social Psychology 66, 950–967 (1994)

Dai, W.C.: The influence of technician certification on students' professional and practicum lessons. Department of Industrial Education and Technology, National Taiwan Normal University; Unpublished Thesis: Taipei (2006)

Deci, E.L.: Intrinsic Motivation. Plenum Press, London (1975)

Graham, S., Weiner, B.: Theories and Principles of Motivation. Macmillan, New York (1996)

Grolnick, W.S., Ryan, R.M., Deci, E.L.: Inner resources for school achievement: motional mediators of children's perceptions of their parents. Journal of Educational Psychology 83(4), 538–548 (1991)

Ground, N.E., Linn: Measurement and evaluation in teaching, 5th edn. Macmillan, N.Y. (1990)
Henson, R.K.: Understanding internal consistency reliability estimates: Aconceptual primer on coefficient alpha. Measurement and Evaluation in Counseling and Development 34, 177–189 (2001)
Hsieh, S.H.: Influence of cognition and learning types on students' accountingcertification. Department of Accounting, Chungyuan Christian University; Unpublished Thesis: Taoyuan (2003)
Hsiao, C.C.: Ways to promote and fulfill technician certification. Labor Committee of Executive Yuan, Taipei (1990)
Huang, H.C.: An investigation of the effect of technician certification on students of Electrical Engineering Departments. National Taiwan Normal University; Unpublished Thesis: Taiei (2003)
Lahey, B.B.: Psychology: an introduction. McGraw-Hill, NY (2004)
Maslow, A.H.: Motivation and Personality, 2nd edn. Harper & Row, New York (1970)
Ma, J.N.: The influence of technician certification on practicum lessons. Department of Industrial Education and Technology, National Changhua University of Education; Unpublished thesis, Changhua (2004)
Ministry of Education, On industrial and technology education. Taipei: Ministry of Education (2000)
Muller, C.: Investing in teaching and learning dynamics of the teacher-student relationship from each actor's perspective. Urban Education 34(3), 292–337 (1999)
Pintrich, P.R., Marx, R.W., Boyle, R.A.: Beyond cold conceptual change: The role of motivational beliefs and classroom contextual factors in the process of conceptual change. Review of Educational Research 63, 167–199 (1993)
Skinner, B.F.: Operant Behavior. American Psychologist 18, 503–515 (1963)
Tsai, T.C.: Dictionary of social works. Community Development Magazine, Taipei (2000)
Vroom, V.H.: Work and motivation. Wiley, New York (1964)
Wen, S.S.: Dictionary of Psychology. Sanmin Publisher, Taipei (2006)
Woolfolk, A.E.: Educational psychology, 10th edn. Allyn & Bacon, Boston (2007)
Wu, H.T.: The influence of technician certification on vocational students' career choice. Department of Industrial Education and Technology, Southern Taiwan University; Unpublished Thesis, Tainan (2008)

Investigation into a University Electronic Portfolio System Using Activity Theory

Wardah Zainal Abidin[1], Lorna Uden[2], and Rose Alinda Alias[3]

[1] Department of Informatics, Advanced Informatics School, Universiti Teknologi Malaysia, 54100, Kuala Lumpur, Malaysia
wardah@utm.my
[2] FCET, Staffordshire University, The Octagon, Beaconside, Stafford, ST 18 OAD. UK
l.uden@staffs.ac.uk
[3] Office of the Deputy Vice-Chancellor (Academics and Internationalisation), Universiti Teknologi Malaysia, 81380, Johor Bahru, Malaysia
alinda@utm.my

Abstract. The last few years have seen an enormous growth of interest in e-portfolios and the benefits they can bring to learners. While it is generally agreed that e-portfolios have great potential to engage students and promote deep learning, the research that has been conducted to date focuses very little on student perceptions of value of the e-portfolio for their learning. If students do not agree or wish to use the e-portfolio as an integral part of their educational experience, then the potential impact the e-portfolio have on learning will not be realised. This paper describes the development of an e portfolio system to promote reflective skills for engineering students in a university in Malaysia. The Activity Theory is used as a lens to explain the reasons for the failed adoption of the e portfolio system.

Keywords: Electronic Portfolio System, Activity Theory, Contradictions, Undergraduate Students.

1 Introduction

Universities today come under intense pressure to equip graduates with more than just the academic skills traditionally represented by a subject discipline and a class of degree. They have been urged to make more explicit efforts to develop the 'key', 'core', 'transferable' and/or 'generic' skills needed in many types of high-level employment[1]. Engineering and science disciplines graduates are expected to have appropriate work experience and evidence of commercial understanding when they start employment [2] which makes most universities (UK) included reflective learning skills in their curriculum to cater for this requirement.

The primary role of higher education (HE) is to educate students by enhancing their knowledge, skills, attitudes and abilities and to empower them as lifelong critical and reflective learners to equip them in their working life. This has led to

L. Uden et al. (Eds.): 7th International Conference on KMO, AISC 172, pp. 283–294.
springerlink.com

the widespread recommendation of using personal development planning [3] and [4], [5]. The primary objective of a PDP is to improve the capacity of individuals to understand how they learn, and to help them review, plan and take responsibility for their own learning [6]. This is particularly true for engineering students who need to be able to reflect on their achievements (or failures) and present these evidences to an employer and, later in their careers, to qualify for professional qualifications. These skills can be developed through the curriculum and specifically through the use of electronic portfolio (e-portfolio). Universities across the globe were slowly rising to the call of their important stakeholders (industry) to change the way they do their teaching and learning (T&L).

This paper seeks to find the answers to a dismal uptake of the e-portfolio by a university in Malaysia using Activity Theory. It begins with a brief overview of e portfolio and activity theory. This will be followed by the case study on the said e-portfolio system. Here, activity theory is used as a lens to explain the failures. The paper then proposed a new approach to the design of the e-portfolio system using co creation of value through the use of service dominant logic from service science.

2 Portfolio, Outcome-Based Education and e-Portfolio Systems

2.1 *Portfolios*

A portfolio is a collection of artefacts which are owned and gathered through the course of time by the portfolio owner for a specific purpose. It is a collection of evidence that is gathered together to show a person's learning journey over time and to demonstrate their abilities. Portfolios can be specific to a particular discipline, or very broadly encompass a person's lifelong learning. Many different kinds of evidence can be used in a portfolio such as samples of writing (finished and unfinished), photographs, research projects and reflective thinking. In fact, it is the reflections on the pieces of evidence, the reasons they were chosen and what the portfolio creator learned from them, are the key aspect to a portfolio [7–11].

In the UK, the portfolio is known as personal development planning (PDP) while it is called student portfolio in the US. The use of portfolios as a personal learning development tool was made mandatory for universities in the United Kingdom by the UK Quality Assurance Agency (QAA) since 2005 [12]. Portfolios are becoming increasingly popular as a tool for both learning and assessment in learning environments. The portfolio can have many purposes and a purpose determines how they are being populated by the portfolio owners. It can be a portfolio at the course level, at the programme level or the organisational level and it can either be for showcase, career or pre-service portfolios.

2.2 *E-Portfolio Systems*

In general the electronic portfolio or e-portfolio is the electronic version of the paper portfolios [13]. E-portfolio systems can be categorised together with other learning management systems (LMS) such as e-Learning. The e-portfolio is seen more as an individual's own learning space and signature. E-portfolio systems can be found on either open-systems, commercial platforms or home grown. They comprised of three (3) main components: the digital archive (repository of evidence), tools to support different processes, and different presentation portfolios developed for different purposes and audiences [14].

[15] listed down ten (10) 'value-added' features which an e-portfolio architecture and design has over paper-based portfolios. Among the features mentioned are: highly customisable; multiple structures/views; sharable – facilitating interaction with supervisors, peers and others; searchable; transferable data to support life-long learnin; downloading records in a variety of formats; and backup and reduced physical storage requirements. The most commonly used open-source e-portfolio system is the Mahara, a New Zealand university consortium project. [16] found that, in comparison with ELGG, Mahara was well received by the students in a survey – in terms of its attractiveness and functionality - in terms of its simplicity and function universality. Other than that they also concluded that the e-portfolio implementation for an institution is reasonable only when it has reached the 3rd level of maturity as suggested by [17].

[18] lists many benefits of e-portfolios including skill development, evidence of learning, feedback, reflection, psychological benefits, assessment, artefacts, maintenance, portability and sharing, access, audience, organisation, storage, cost and privacy. However many of these benefits are not dependent on an electronic medium although an e-portfolio may have advantages described above in providing a 'rich picture' of student learning and competencies, a broader range of pieces of evidence and more effective feedback and more efficient storage.

3 Activity Theory

Activity Theory originated in the former Soviet Union as a cultural-historical psychology by Vygotsky [19]. It incorporates the notions of intentionality, mediation, history, collaboration and development [20]. The unit of analysis is the entire activity. An activity consists of a subject and an object, mediated by a tool. A subject can be an individual or a group engaged in an activity. An activity is undertaken by a subject using tools to achieve an object (objective), thus transforming it into an outcome [21]. Tools can be physical, such as a hammer, or psychological such as language, culture or ways of thinking. An object can be a material thing, less tangible (a plan), or totally intangible (a common idea) as long as it can be shared by the activity participants [21]. Activity theory also includes collective activity, community, rules and division of labour that denote the situated social context within which collective activities are carried out. Community is made up of one or more people sharing the same object with the subject. Rules regulate actions and interactions with an activity. Division of labour informs how tasks are divided horizontally between community members. It also refers to any vertical division of power and status.

Activities always take place in a certain situation with a specific context [22]. He formulated activity context as a network of different parameters or elements that influence each other. Figure 1 shows [22] Engeström's (1987) model of an activity system.

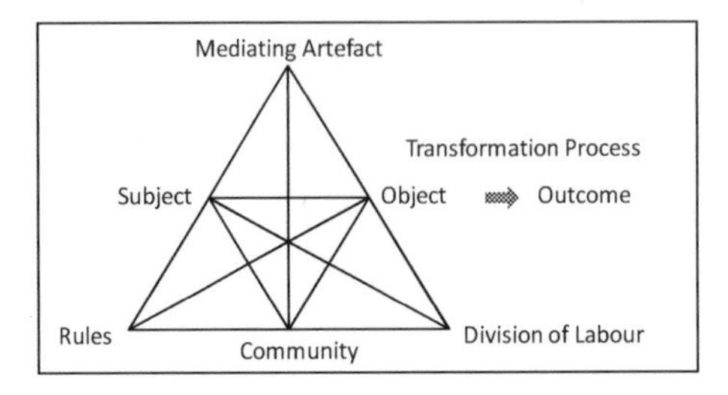

Fig. 1 Basic structure of an activity [22]

Because activities are not static, but more like nodes crossing hierarchies and networks, they are influenced by other activities and other changes in the environment. External influences change some elements of activities causing imbalances between them [21]. Contradictions are the terms given to misfits within elements, between them, between different activities or different developmental phases of the same activity. They manifest themselves as problems, ruptures, breakdowns, clashes etc. Activity theory sees contradictions not as problems but as sources of development. Activities are virtually always in the process of working through contradictions that subsequently facilitate change.

The concept of contradiction is important in activity theory. According to [22], any activity system has four levels of contradictions that must be attended to in analysis of a working situation. Level 1 is the primary contradiction. It is the contradiction found within a single node of an activity. This contradiction emerges from tension between use value and exchange value. Secondary contradictions are those that occur between the constituent nodes. For example, between the skills of the subject and the tool he/she is using, or between rules and tools. Tertiary contradiction arises between an existing activity and what is described as a more advanced form of that activity. This may be found when an activity is remodelled to take account of new motives or ways of working. Quaternary contradictions are contradictions between the central activity and the neighbouring activities, e.g. instrument producing, subject-producing and rule producing activities. The activity system for the university e portfolio went through a series of transformation as described in subsequent section.

4 E-Portfolio Case Study

The e-portfolio case study is taken from a public university in Malaysia which in 2005 had set its goal to become a world class engineering-focused university. One of the strategies is to ensure its graduates are highly employable and sought after both locally and globally. Employers demand graduates to be knowledgeable and socially skillful. These social skills are packaged by the university in the outcome-based education (OBE) philosophy and it is named as the graduate attributes (GA). Seven social skills of the GA are communication skills, critical thinking and problem solving skills, teamworking skills, life-long learning skills, entrepreneurial skills, leadership skills and ethics and professional skills.

For a typical engineering programme to be accredited by the professional bodies, the university need to show evidence of these skills are taught or acquired by the students. Hence in 2005 a task force was formed to look into the feasibility and development of the e-portfolio for undergraduates. A team comprising of academicians, administrators and system developers was set up to realise this project and in 2007 it was formally launched to the first cohort. The e-portfolio system was designed and built in-house by the task force. Coincidentally no other local universities were into e-portfolio so forming a consortium among universities was not an option for this taskforce.

At the same time too options for open source software on e-portfolio and journal articles on this matter were very limited. So the taskforce only had samples in the internet to base their designs and ideas. The system was built on the web platform using PHP language and the user interface (UI) was the typical web 1.0 menu with boxes to click and choose from and uses typical text editors for text writing. It is highly structured and students only need to fill in the available spaces with evidences which can comprised of images, videos, audio, pdf documents and text. There is not much choice for a student to show his individual creativity except for the given templates.

The end users of the e-portfolio were the university students with focus mainly on engineering students starting from first year till they graduate. Training was given to them during the first week of the semester and some faculties embed the e-portfolio system in their course work. However, despite the awareness campaign to students, academic advisors, deputy deans and academic administrators, the uptake of the system by the students after the end of the first semester was discouraging. The system went through several upgrading versions yet the usage among the undergraduate students was still very low. Unfortunate for the e-portfolio, the Facebook came into the picture around the same time. Several other contributory factors were identified as reasons for this failure and shall be dealt with in the next section.

5 Activity Theory as Lens to Analyse Failures

The researchers perceived that the earlier e-portfolio project failed partly due to lack on the part of the university management and the committee to view the e-portfolio system as a *service* to the students. The university management's initial objective and motive of introducing the e-portfolio was mainly solving a managerial issue and less of a learning strategy. The e-portfolio is seen as a mechanism for evidencing the acquirement of generic skills by the students. This evidence was critical in convincing the external accreditation bodies on the university's graduate level of achievement as stipulated in the programme objectives. This section attempts to analyse this failure using Activity Theory.

The first step is to translate the learning setting into the activity system by entitling its collective object of activity, different (groups of) actors who are involved in the learning environment, the way in which the labour has been divided among these actors, the mediating artefacts that are being used by the actors and the rules that apply between the actors involved. A typical activity system for this system can be drawn as in Figure 2. The components of activity systems are initially described from the perspective of one of the (group of) actors identified as the subject of the activity system.

What takes place in an activity system composed of object, actions and operations, is the context. Context is constituted through the enactment of an activity involving person (subject) and artefacts. Context is therefore the activity system and the activity system is connected to other activity systems.

In the design, it is important to understand how things get done in a context, and why. This is because different contexts impose different practices. To analyse context, we need to know the beliefs, assumptions, models and methods commonly held by the group members, how individuals refer to their experiences in other groups, what tools they found helpful in completing their problem etc. In addition, there are also external or community driven contexts.

The researchers divided the project activity into two (2) main phases and for each phase an activity system is investigated. Phase 1 is taken from the beginning of the project (2005), while Phase 2 is in mid-2007 when the beta version of the e-portfolio system was introduced to the first cohort of students. Table 1 summarises the findings of the two phases into the Activity Theory paradigm.

The components in the activity systems for Phase 1 and Phase 2 differ in their positions (roles) due to the different motivations found at each phase. But it is interesting to note that the students were never in the picture in Phase 1 and this may contribute to the lack of interest of the students in using the system.

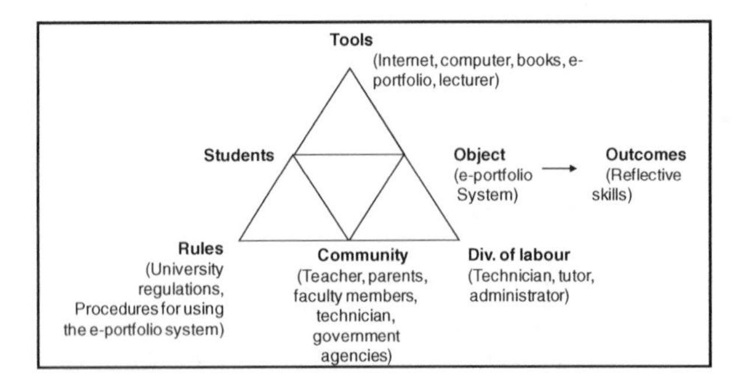

Fig. 2 The typical activity system for the e-portfolio system

Table 1 Summary of Activity components for Phase 1 and Phase 2

Elements of AT	Phase 1	Phase 2
subject [is the individual or group whose viewpoint is adopted (Engeström, 1987)]	e-portfolio taskforce	Students
object [the 'raw material' or 'problem space' at which the activity is directed at]	Univ e-portfolio system	Students e-Portfolios
outcomes [molded or transformed from object with the help of physical and symbolic, external and internal tools -- precedes and motivates activity. (Engeström, 1993)]	evidence of students acquirements of the GA	First university to use e-portfolio generic skills in students
tools [mediate the object of activity-- can be external, material or internal, symbolic; cause transformation of object into an outcome, (desired or unexpected); enable or constrain activity]	project plans; e-portfolio design architecture; computer technology specifications	Univ e-portfolio system University internet
community [participants of an activity system, who share the same object]	Lecturers	ePortfolio Taskforce
division of labour [division of tasks and roles among members of the community and the divisions of power and status]	Academic advisors ICT system team Administrators	Academic advisors Dep Deans CTL CICT
rules [explicit and implicit norms that regulate actions and interactions within the system (Engeström, 1993; Kuutti, 1996)]	University rules OBE guide	University rules Research Univ criteria
	See Figure 3	See Figure 4

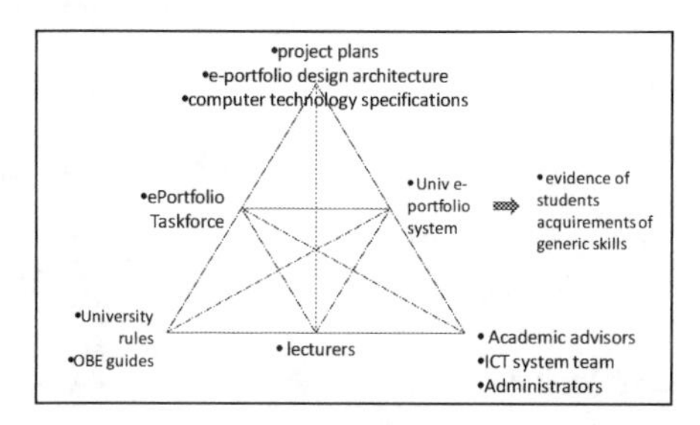

Fig. 3 Activity system in 2005- beginning of e-portfolio project

Activities are constantly evolving. To understand a phenomenon means to know how it is developed into its existing form [23]. This applies to all the elements of an activity. The current relationship between subject and object includes a condensation of the historical development of that relationship [21]. Historical analysis allows existing and emerging organisational structures to be examined as the result of their evolutionary development, sometimes intentional and sometimes not. This means that we must also describe and analyse the development and tensions within the activity system [24]. Some of the feedbacks taken from interviews with the stakeholders are listed in Table 2. Closer analysis on the components of the activity systems (Fig 2, 3 and 4) shows contradictions.

Because of the interests and needs of different stakeholders, contradictions are unavoidable. To identify what causes the e-portfolio system to fail, the internal contradictions as the driving forces behind disturbances, innovations and change of activity system are investigated. Inner contradictions of the activity systems

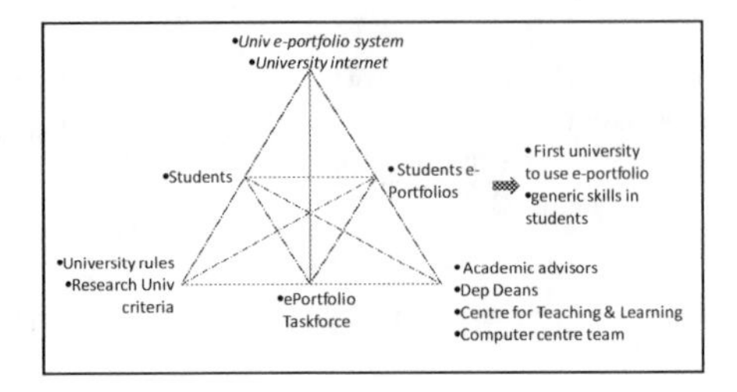

Fig. 4 Activity system in 2007 – e-portfolio rolled out to students

Table 2 Feedbacks from four (4) stakeholders of the e-portfolio system

Perspectives from:		
Students	Academic Advisors	e-Portfolio Taskforce
skills and knowledge needed to use e-portfolio awareness of system and its importance Appropriate functionality Reliability of ICT infrastructure Availability of other systems to complement the e-portfolio User training and support system usefulness, ease of use and its user friendliness weak wireless connection, other internet connections could not access Univ portals Don't know what/ how important it is to them Never think of using it Not familiar with it Don't really know the address (url) Don't have information on how to use it	Skills and knowledge needed to use e-portfolio Appropriate functionality system usefulness, ease of use and its user friendliness User training and support Staff time allocation	Skills and knowledge needed to develop and deliver e-portfolio faculty-wide e-portfolio strategy institution-wide strategy Funding priority for the e-portfolio development
Deputy Deans (Academic)		
Awareness of system and its importance		

shall be analysed as the source of disruption, innovation, change and development of that system. By identifying the tensions and interactions between the elements of an activity system, it is possible to reconstruct the system in its concrete diversity and richness, and therefore explain and foresee its development [25]. Several types of contradictions are identified as the causes of failure for the e-portfolio system (Fig 5) while descriptions of these contradictions are shown in Table 3.

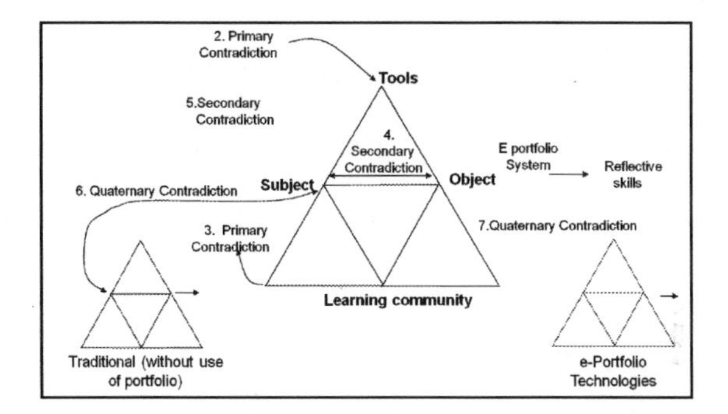

Fig. 5 Contradictions in the e-portfolio project

Table 3 Potential contradictions in the e-portfolio project

Primary Contradictions [Primary inner contradiction (double nature) within each constituent component of the central activity] [University of Helsinki, Center for Activity Theory and Developmental Work Research website]	Object node: there is tension as to the nature of the e-portfolio that would be suitable for the students, and the university at large. Tool node: there is the issue of using student-own software or the home-grown e-portfolio system Rule mode: there is the question of formalizing the assessment of the e portfolio into the student transcript or otherwise
Secondary contradictions [between the constituents of the central activity]	Tension between the different stakeholders and the motive of the system. Although all stakeholders share the same object of the need for the e portfolio system, different stakeholder has different motive. The lecturer wanted students to acquire reflective skills whereas for management, it is simply a means to satisfy the government outcome-based education. Tension between the tools and the subjects. Students are expected to use the e portfolio system with limited necessary training given to them mainly in the use of reflective skills.
Quaternary contradictions [between the central activity and its neighbour activities]	Fundamental contradiction between the use of e-portfolio system for learning and traditional classroom teaching. Tension between the availability of technology and the use of e-portfolio learning environment.

The contradictions identified above resulted from the different stakeholders having different motives and without the co creation of value. A better approach to the design of the e portfolio system would be to take the emerging research in service science through the co creation of value. Subsequent section proposes the new approach for our development of the new e portfolio system.

6 Proposed Approach for Designing an e-Portfolio System

The service sector is the dominant economic activity in developed countries today. According to [26], a service is a change in the condition of a person or a good belonging to some economic entity, brought about as the result of the activity of some other economic entity, with the approval of the first person or economic entity. Service science research seeks to find out how to design, build, operate, use, sustain and dispose of service systems for the benefit of multiple stakeholders.

Service system is defined as "a value – co-production configuration of people, technology, other internal and external service systems, and shared information (such as language, processes, metrics, prices, policies and laws)" [27]. Examples of service systems include people, organisations, corporations, etc. A key condition is that service systems interact to co-create value. Value is co-created in that results depend on both transportation contributed by the delivery service and objects and locations contributed by the clients [28].

An emerging new concept has been gaining popularity among service management concerning the role of the firm [29] called the service dominant logic in marketing to replace the traditional good-dominant logic. The new service dominant (S-D) logic is concerned with value-in-use. Value is always co-created between producers and consumers. Thus value is co-created through the combined efforts of firms, customers, employees, stakeholders and other related entities in any given exchange, but is always determined by the beneficiary (e.g. customer).

7 Conclusion

Students are encouraged to think beyond simply attaining a grade in a course to thinking about how to clearly articulate what they have learned as well as to identify areas for improvement or further learning. They have to reflect on their own learning and by doing so, will be able to better integrate their various learning experiences. The process of developing and implementing a successful e-Portfolio project—one that works and is adopted by users involves many challenges. Higher education will need to tackle those challenges and it is important to identify those ePortfolio system attributes that will lead to success. Learning from past experiences with other electronic educational tools and by studying difficulties already encountered in those earlier e-portfolio projects can help us to understand and address potential future problems. By using activity theory as a lens, it was possible for us to highlight why the e portfolio system failed. This allows us to look for new ways of designing e portfolio system that takes into account the failures using co-creation of value through service science.

Acknowledgments. The research team would like to express their sincere thanks to the Government of Malaysia via FRGS (Project No: 78654), Ministry of Science and Technology Innovation (MOSTI) for the financial support of the research work as well as the university via the RUG (Project No: PY/2011/00799).

References

1. The Council For Industry and Higher Education, Helping Students Towards Success at Work - An Intent being Fulfilled (1998)
2. Mason, G., Williams, G., Cranmer, S.: Employability skills initiatives in higher education: what effects do they have on graduate labour market outcomes? Education Economics 17(1), 1–30 (2009)
3. N.C. of I. into Gran Bretaña, H.E., Dearing, R., Barnett, R.: Higher education in the learning society. Her Majesty's Stationery Office London (1997)
4. Burgess, R.: Measuring and recording student achievement. Universities UK/SCOP (2004), `http://www.heacademy.ac.uk/learningandteaching/DecemberNewsletter.pdf` (abgerufen am 3.6.2006)
5. East, R.: A progress report on progress files. Active Learning in Higher Education 6(2), 160–171 (2005)
6. Halstead, A., Sutherland, S.: ePortfolio: A means of enhancing employability and the professional development of engineers. In: International Conference on Innovation, Good Practice and Research in Engineering Education (EE 2006), Liverpool (2006)
7. Loughran, J., Corrigan, D.: Teaching portfolios: A strategy for developing learning and teaching in preservice education. Teaching and Teacher Education 11(6), 565–577 (1995)
8. Wade, R.C., Yarbrough, D.B.: Portfolios: A tool for reflective thinking in teacher education? Teaching and Teacher Education 12(1), 63–79 (1996)
9. Klenowski, V., Askew, S., Carnell, E.: Portfolios for learning, assessment and professional development in higher education. Assessment & Evaluation in Higher Education 31(3), 267–286 (2006)
10. Abrami, P., Barrett, H.: Directions for research and development on electronic portfolios. Canadian Journal of Learning and Technology/La Revue Canadienne de l'Apprentissage et de la Technologie 31(3) (2005)
11. Smith, K., Tillema, H.: Clarifying different types of portfolio use. Assessment & Evaluation in Higher Education 28(6), 625–648 (2003)
12. R.A. CRA: PDP and e-Portfolio UK Newsletter (2008)
13. Barrett, H., Knezek, D.: E-Portfolios: Issues in Assessment. Accountability and Preservice Teacher Preparation (2003)
14. Barrett, H.C.: Differentiating electronic portfolios and online assessment management systems. In: Proceedings of the 2004 Annual Conference of the Society for Information Technology in Teacher Education (2004)
15. Cotterill, S., McDonald, T., Drummond, P., Hammond, G.: Design, implementation and evaluation of a'generic'ePortfolio: the University of Newcastle upon Tyne experience. CAL-laborate (2005)

16. Balaban, I., Bubas, G.: Educational potentials of ePortfolio systems: Student evaluations of Mahara and Elgg. In: 2010 32nd International Conference on Information Technology Interfaces (ITI), pp. 329–336 (2010)
17. Love, D., McKean, G., Gathercoal, P.: Portfolios to Webfolios and beyond: Levels of Maturation. Educause Quarterly 27(2), 14 (2004)
18. Butler, P.: A review of the literature on portfolios and electronic portfolios. Techreport, pp. 1–23 (2006)
19. Vygotsky, L.: Mind in society (1978)
20. Nardi, B.A.: Context and consciousness: activity theory and human-computer interaction. The MIT Press (1996)
21. Kuutti, K.: Activity theory as a potential framework for human-computer interaction research. In: Context and Consciousness: Activity Theory and Human-Computer Interaction, pp. 17–44 (1996)
22. Engeström, Y.: Learning by expanding. An activity-theoretical approach to developmental research (1987)
23. Kaptelinin, V.: Activity theory: implications for human-computer interaction. In: Context and Consciousness: Activity Theory and Human-Computer Interaction, pp. 103–116 (1996)
24. Boer, N.I., van Baalen, P.J., Kumar, K.: An activity theory approach for studying the situatedness of knowledge sharing. In: Proceedings of the 35th Annual Hawaii International Conference on System Sciences, HICSS 2002, pp. 1483–1492 (2002)
25. Engeström, Y., et al.: Activity theory and individual and social transformation. Perspectives on Activity Theory, 19–38 (1999)
26. Chesbrough, H., Spohrer, J.: A research manifesto for services science. Communications of the ACM 49(7), 35–40 (2006)
27. Spohrer, J., Maglio, P.P., Bailey, J., Gruhl, D.: Steps toward a science of service systems. Computer 40(1), 71–77 (2007)
28. Spohrer, J., Anderson, L., Pass, N., Ager, T.: Service science and service dominant logic. Otago Forum 2, 4–18 (2008)
29. Vargo, S.L., Lusch, R.F.: Evolving to a new dominant logic for marketing. Journal of Marketing, 1–17 (2004)

Sequence Compulsive Incremental Updating of Knowledge in Learning Management Systems

Syed Zakir Ali[1], P. Nagabhushan[2], R. Pradeep Kumar[3], and Nisar Hundewale[1]

[1] College of Computers and Information Technology, Taif University, Saudi Arabia
[2] Department of Studies in Computer Science, University of Mysore, Mysore, India
[3] Amphisoft Technologies Private Limited, Coimbatore, India
zakirsab@gmail.com, pnagabhushan@hotmail.com,
pradeep@amphisoft.com, nisar@computer.org

Abstract. Growing popularity of Learning Management Systems (LMS) coupled with setting up of variety of rubrics to evaluate methods of Learning, Teaching and Assessment Strategies (LTAS) by various accreditation boards has compelled many establishments/universities to run all their courses through one or the other forms of LMS. This has paved way to gather large amount of data on a day to day basis in an incremental way, making LMS data suitable for incremental learning through data mining techniques. The data mining technique which is employed in this research is *clustering*. This paper focuses on challenges involved in the instantaneous knowledge extraction from such an environment where streams of heterogeneous log records are generated every moment. In obtaining the overall knowledge from such LMS data, we have proposed a novel idea in which instead of reprocessing the entire data from the beginning, we processed only the recent chunk of data (incremental part) and append the obtained knowledge to the knowledge extracted from previous chunk(s). Obtained results when compared with teachers handling the modules/subjects match exactly with the expected results.

Keywords: Education and Training, Data Mining, Knowledge Management, Incremental learning, Regression distance measure, Multi-Source Temporal data.

1 Introduction

With the deployment of Learning Management Systems (LMS) such as Moodle [1], Blackboard [2], the traditional learning mechanism has moved into online learning, which is being considered as one of the latest innovations in the field of Education and Training. A typical web based LMS in its simplest form provides a platform for online learning where educators can post their learning materials, assignments, tests etc and monitor progress of their students who in turn have to log in to learn from the posted material. Various data mining techniques can be applied to learn from the generated data of the LMS [3-5] and extract useful information instantaneously.

L. Uden et al. (Eds.): 7th International Conference on KMO, AISC 172, pp. 295–306.
springerlink.com

An LMS can keep record of all the student activities through their log files which pave way to gather large amount of data on a day to day basis in an *incremental* way. The amount of data which is gathered incrementally is proportional to the number of students registered for each course, number of times each student logs in for each of the registered courses and number of courses that are handled by an LMS. We can find streams of log records getting generated every moment, which makes the real time knowledge extraction almost impossible. This is because of the usage of past data at each step of knowledge mining, which considerably slows down the system performance. As a counter measure to improve the system performance, the log records that were generated before a certain date can be ignored/discarded on a regular basis. But this limits the knowledge mining capacity of the LMS up to a specific date.

Using the data of a particular day for the extraction of knowledge of that day is acceptable without any arguments. However, while generating the weekly report, reprocessing the data of all the days of that week is not at all acceptable for two reasons - (i) *What has happened to the generated knowledge of every day of that week?* (ii) *How much effort was applied in getting the daily reports?* After spending resources (processor time and memory) had the generated knowledge been saved, it could have been utilized effectively. But one may argue that by storing the knowledge, memory requirement may go up and hence system performance will degrade further. Our argument is that *when the generated knowledge is available, what is the necessity of reprocessing the data?* However, if the same data might be required for some other objectives of knowledge mining, we can think of neglecting it once the knowledge for all the objectives is obtained. Moreover, *knowledge mining objectives are pre-defined and there is meager or no chance of objectives getting added at a later stage.*

Had the knowledge of every day been retained, getting knowledge of the entire week would have been a simple summation of knowledge of all the days of that week, which would even have encouraged the divide and conquer based parallelism in obtaining the overall knowledge. Further, *once the knowledge of a week is available what is the necessity of knowledge of all the days of that week? In a similar way, knowledge of a month can be a summation of knowledge of four weeks, knowledge of a year could be the summation of knowledge of twelve months of that year.* Once knowledge of higher scale is available, knowledge of the lower scale could be destroyed which keeps the memory requirement to store the knowledge under control. The above idea can be applied in a real time environment as *the data to be processed is completely under control and the time to work with knowledge parameters will be essentially lesser than working with data.*

Hence in this research, it is proposed to update the generated knowledge with the next set of incoming chunk/data without reprocessing the past data in arriving at the overall knowledge. The process of deriving knowledge in phased manner without re-indenting the past data or with minimal re-indenting in unavoidable cases is termed as *Incremental Learning* [6]. The method of data mining that is applied for incremental learning is *clustering*. Knowledge of resulting clusters is retained in histogram based regression lines. Section 2 provides reasons for selection of clustering as method of incremental learning and selection of histogram based regression line as a knowledge representative. Further, section 2 also highlights the challenges involved in the incremental update of knowledge and presents a brief literature review. Sections 3 present mathematical formulations for incremental update of histograms and regression lines. Section 4 furnishes the details of experimental set up. Section 5 gives experimental results. Section 6 provides discussion and conclusion.

2 Literature Review and Challenges Involved in Incremental Update of Knowledge

When the intention is to extract knowledge from large volumes of data, undoubtedly the term data mining pops up, since data mining is the field which can handle very large volumes of data and can derive useful knowledge from it [7-11]. To incorporate incremental learning concepts in data mining, Michalski [6, 12] has observed that provisions for inserting *background knowledge* and knowledge about *goal* of data mining into the database should be made. Generated knowledge packets if visualized can be of great help in understanding the problems at hand [12]. To address the research direction that aims at incorporating above mentioned tasks into data mining, Michalski [12] has coined a phrase called Knowledge Mining or Incremental Knowledge Mining and has suggested the following model for the knowledge mining process.

DATA + PRIOR_KNOWLEDGE + GOAL → NEW_KNOWLEDGE

Where, GOAL is the encoding of the knowledge needs of the user(s), and NEW_KNOWLEDGE is the knowledge satisfying the GOAL. Such knowledge can be in the form of *decision rules, association rules, decision trees, conceptual or similarity based clusters, regression, equations, statistical summaries, visualizations, natural language summaries*, etc.

Since learning in an unsupervised scenario is more demanding [13-15], we would like to employ clustering as a mechanism to demonstrate incremental learning. Various attempts to extract knowledge incrementally using clustering have been made [16-20]. However none of the above works concentrated on the *incremental agglomeration* of knowledge.

The important challenge in design of incremental clustering algorithms is *making clustering result insensitive to the order of the presentation of the data samples* [21]. Since Density Based Spatial Clustering of Applications of Noise (DBSCAN) [22] algorithm is insensitive to the order of presentation of data samples, we would like to employ it for the purpose of incremental learning. *Nagabhushan et al* [24-26] lists excellent motivating factors for the usage of DBSCAN algorithm for incremental learning.

A deeper look at the type of collected data reveals the different types/features which makes the knowledge mining process more complex. Since histograms have the generic ability to characterize most of the data types [27-28], it has received significant attention as *summarization/representative* object. Since it has been shown that computational complexity can be drastically reduced by transforming a histogram to regression line [29-32], in this research a set of histogram based regression lines is considered as a representative object of a group of samples (cluster). Usage of histogram based regression lines as a representative object (to represent the knowledge of clusters) is established [23-26].

Identifying proper parameters to represent the knowledge which supports incremental update is quite a challenging task. A detailed analysis which is available in [23-26], it is proposed to have the combination of *number of elements, mean (μ), standard deviation (σ) and regression line (L)* as a simple knowledge packet (KP) to describe each feature of the cluster. Without researching further, we would like to import the idea of KP as it is to describe the clusters obtained by LMS data. For comparison between the KPs/clusters, distances between the corresponding elements of the *KP*s have to be computed. For μ and σ, the standard Euclidian distance can be used and for histogram based regression lines (*histo-regression* lines) the *Histo-Regression-Distance* [33] is used.

In machine learning community, updating of knowledge or knowledge improvement via incremental learning is achieved in three forms [6] viz., zero, partial and full instance memory learning; depending on the volume of data from which the earlier knowledge was generated is re-processed or is available for re-processing. We would like to have experiments with zero instance memory learning as it is considered as most optimal.

Let us say KP_i and KP_j are the nearest two knowledge packets that are considered for merging. If n_i represents number of elements in KP_i and σ_i represents the standard deviation of KP_i then the density of KP_i is given by $\rho_i = (n_i / \sigma_i)$ [34]. Similarly the density of KP_j is given by $\rho_j = (n_j / \sigma_j)$.

Merging of the nearest KPs could cause uneven distribution of knowledge requiring the splitting of KP(s) which requires access to the raw data. Since the idea is not carrying the data along, there should be some mechanism at the time of merging itself which can avoid splitting of the KPs at a later stage. In [26], the following criterion has been applied for merging. If the density of the proposed merge is greater than or equal to the minimum of the existing densities of the *KP*s under consideration, then the *KP*s are merged. Otherwise, the process of merging is terminated. We would like to utilize the same concept for merging of the *KP*s of the LMS data.

Rapid growth in the amount of data necessitates the study of scalability of the data mining / Incremental mining techniques and is being referred to as *Scalable Data Mining (SDM)* [35]. In the following section, we provide scalable equations for incremental update of histogram and regression line variables/features.

3 Scalable Equations for Some of the Univariate Statistics of Different Types of Features

3.1 Scalable Equations of Histogram Features for Incremental Learning

Histogram arithmetic (addition, subtraction, multiplication and division of histograms) and regression line arithmetic has been dealt in detail [36]. In continuation of it, we propose here the formulations for univariate statistics of histogram and regression lines in incremental mode.

3.1.1 Count of Histogram Data

Total count of k subsets of a Histogram *H*, each with B number of bins is scalable and is given by:

$$N = \sum_{i=1}^{k} (\pm n)_i \tag{1}$$

wheren_i is the number of elements in i^{th} histogram and is given by $n_i = \sum_{j=1}^{y} B_j^i$

where j is the number of bins.

3.1.2 Scalable Mean of Each Bin of the Histogram

Let the number of subsets be M and the number of elements in each subset be n_s ; let $H(i)$ indicates the ith bin of histogram and $\overline{H(i)}$ indicates the mean of ith bins of the available histograms; then the scalable mean of the ith bin of a histogram H of subsets 1 to M is given by:

$$\overline{H_{1..M}(i)} = \left[\frac{\overline{H_{1..(M-1)}(i)} * \sum_{s=1}^{M-1} n_s + \sum_{k=1}^{n_s} H_M^k(i)}{\sum_{s=1}^{M-1} n_s + n_M} \right] \tag{2}$$

Where *k* is the number of elements in the subset *M*.

3.1.3 Scalable Standard Deviation (SD) of Each Bin of the Histogram

$$SD\left[H_{1..M}(i)\right] = \sqrt{\frac{1}{N+n_m}\left(N\left(\left(SD\left[H_{1..(M-1)}(i)\right]\right)^2 + \left(\overline{H_{1..(M-1)}(i)}\right)^2\right) + \sum_{k=1}^{n_s}\left(H_M^k(i)\right)^2\right) - \left(\overline{H_{1..M}(i)}\right)^2} \tag{3}$$

where M is the number of subsets; n_m is the number of elements of the subset M;.

3.2 Scalable Equations for Regression Line Features for Incremental Learning

A regression line can be represented by its slope and intercept. However, it can also be represented by the co-ordinates $(x_{\min}, y_{\min})$ and $(x_{\max}, y_{\max})$. This representation is convenient in deriving the statistical formulations for SDM. Using the formulations given by Sayad et al [35], we can derive formulations for regression line features as:

3.2.1 Count of Regression Lines

The total count of k subsets of a Regression line L is scalable and remains same as equation 1 above in which n_i is the number of elements in i^{th} regression line.

3.2.2 Scalable Mean of Regression Lines

Let the line L_1 be represented by $\left[x^1_{\min}, x^1_{\max}, y^1_{\min}, y^1_{\max}\right]$ and the line L_2 be represented by $\left[x^2_{\min}, x^2_{\max}, y^2_{\min}, y^2_{\max}\right]$. If we fix the co-ordinates of y axis between 0 and 1, we can represent the lines L_1 and L_2 as: $L_1 = \left[x^1_{\min}, x^1_{\max}, 0, 1\right]$ and $L_2 = \left[x^2_{\min}, x^2_{\max}, 0, 1\right]$.

The mean of given two regression lines can be obtained as:

$$\left[\frac{L_1 + L_2}{2}\right] = \left[\frac{x^1_{\min} + x^2_{\min}}{2}, \frac{x^1_{\max} + x^2_{\max}}{2}, \frac{0+0}{2}, \frac{1+1}{2}\right] = \left[\frac{x^1_{\min} + x^2_{\min}}{2}, \frac{x^1_{\max} + x^2_{\max}}{2}, 0, 1\right] \quad (4)$$

Proof and the mathematical formulations for the regression line have been dealt in detail in [36] while discussing the similarity between a histogram and a regression line. We can obtain the scalable mean for regression line features using the formulations of the scalable mean for histogram features (eq. 2) as:

$$\overline{L_{1..M}(x_{\min})} = \left[\frac{\overline{L_{1..(M-1)}(x_{\min})} * \sum_{s=1}^{M-1} n_s + \sum_{k=1}^{n_s} L^k_M(x_{\min})}{\sum_{s=1}^{M-1} n_s + n_M}\right] \quad (5)$$

where M is the number of subsets; n_M is the number of elements of the subset M;. n_s is the number of elements in each subset; $L(x_{\min})$ indicates the $x_{\min}$ of the regression line of the subset M and $\overline{L(x_{\min})}$ indicates the mean of $x_{\min}$ of the regression line of subset M ; $\overline{L_{1..M}(x_{\min})}$ is mean of the $x_{\min}$ of a regression line L of subsets 1 to M.

Similarly, we can obtain the scalable mean of $x_{\max}$.

3.2.3 Scalable Standard Deviation (SD) for Regression Line Features

The expression for the scalable standard deviation of histogram feature (eq. 3) can be modified to fit the regression line feature as:

$$SD\left[L_{1..M}\left(x_{\min}\right)\right]=\sqrt{\frac{1}{N+n_m}\left(N\left(\begin{array}{l}\left(SD\left[L_{1..(M-1)}\left(x_{\min}\right)\right]\right)^2\\+\left(\overline{L_{1..(M-1)}\left(x_{\min}\right)}\right)^2\end{array}\right)+\sum_{k=1}^{n_s}\left(L_M^k\left(x_{\min}\right)\right)^2\right)-\left(\overline{L_{1..M}\left(x_{\min}\right)}\right)^2} \quad (6)$$

Similarly we can obtain $SD\left[L_{1..M}\left(x_{\max}\right)\right]$.

4 The Proposed Approach and Experimental Setup

In this research, for the sake of experimentation, the quiz data pooled by an LMS called *EkLuv-Ya* [Ekl09] is used. For the sake of simplicity, we have considered the marks scored by students while attempting to answer quiz questions on-line for different subjects of their study. The quiz data under consideration is the marks scored in different quizzes by the 52 students of sixth semester of Bachelor of Engineering from *Aditya Institute of Technology,* Coimbatore (Tamil Nadu), India [38]. The subjects of study in the order of appearance are: *Computer Graphics, Mobile Computing, Numerical Methods, Object Oriented System Design* and *Open Source Software*. Maximum marks for each quiz is 10. Marks of students who could not take up the quiz in a particular subject have been marked as '-'. Each quiz is conducted at different time instants over time. Therefore it is temporal data. Each subject represents a source of data. Therefore it is multi-source data. Overall it is a typical example of *multi-source* temporal data.

As data of each quiz of a particular subject is available over a period of time, learning from each quiz of a particular subject is essentially an application of *Sequence Compulsive Incremental Learning (SCIL).* The objective of knowledge mining is multi-folded. The first objective is to find natural groups/clusters of students based on the marks scored by them in each of the quizzes of a particular subject at a time and to visualize the change in groups when the data of subsequent quizzes become available. The second objective is to use the knowledge of groups/clusters to get the cumulative knowledge in SCIL mode. Other alternatives could also become possible depending on the need.

A simplified pictorial representation of the process of assimilation of knowledge of one subject is depicted in Fig. 1. The same mechanism is applied on the overall cumulative knowledge of each subject in obtaining the final knowledge. It is to be noted that each outlier of the DBSCAN algorithm [22] is considered as a cluster with one element.

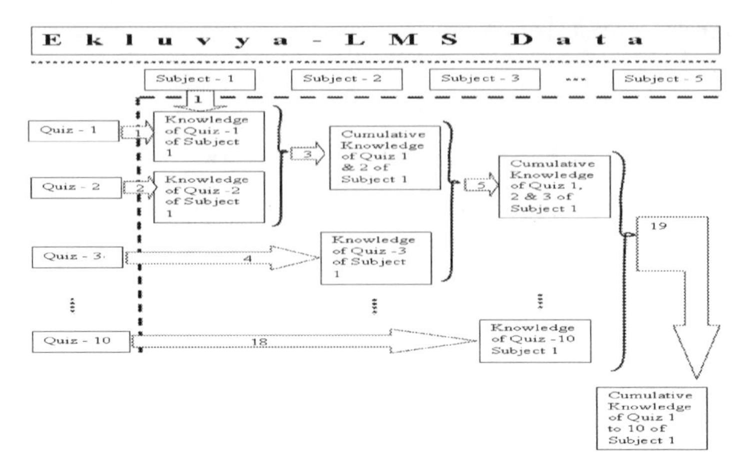

Fig. 1 A Simplified Pictorial Representation of a Process of Quiz wise Incremental update of Knowledge of a particular subject

5 Results of the Experimentation

The bar graph showing the number of students verses the mean of each cluster of the first subject is as shown in Fig.2. A similar bar graph of the cumulative knowledge is as shown in Fig.3.

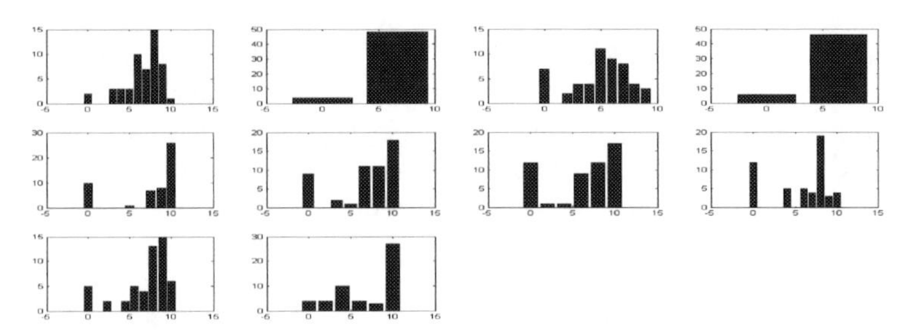

Fig. 2 Bar Graphs of Number of students verses the mean of each cluster of quizzes 1 to 10 of Subject-1

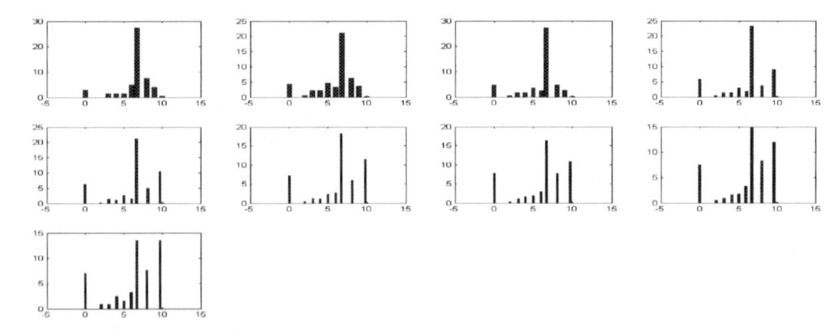

Fig. 3 Bar Graphs of Number of students verses the mean of each cluster of the cumulative knowledge of quizzes 1 to 10 of Subject-1

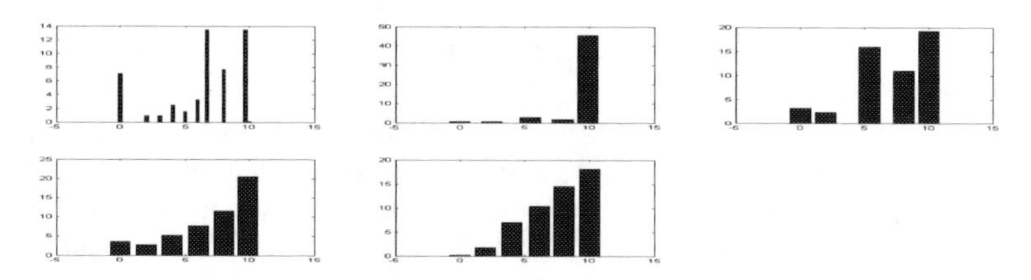

Fig. 4 Bar Graphs of Number of students verses the mean of each cluster of the final cumulative knowledge of quizzes 1 to 10 of each of the five Subjects

In Fig.2, the four bar graphs of first row belong to knowledge of quizzes 1 to 4, second row fit-in as the knowledge of quizzes 5 to 8 and the last row blend in for quizzes 9 and 10. In Fig.3, the first bar graph of the first row shows the cumulative knowledge of quizzes 1 and 2, the second bar graph depicts the cumulative knowledge of quizzes 1 to 3, the third bar graph portray the cumulative knowledge of the quizzes 1 to 4 and so on. The bar graphs of the knowledge of remaining four subjects and bar graphs of the cumulative knowledge can be obtained in a similar way.

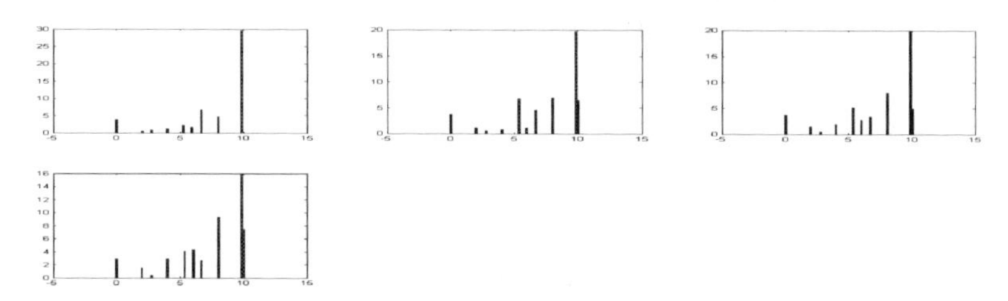

Fig. 5 Bar Graphs of Number of students verses the mean of each cluster of the final cumulative knowledge of subjects 1 to 2, Subjects 1 to 3, Subjects 1 to 4 and Subjects 1 to 5

6 Discussion and Conclusion

The purpose of above experimentation is multi folded. At the first instance, it helps the teacher who is handling the subject to visualize the groups in the entire class at the end of each quiz and also to visualize the formation of cumulative groups at the end of subsequent quizzes in real time. Visualization of the first few cumulative groups will help the teacher in obtaining the trend in that subject and take corrective measures to improve it rather than realizing the mistake at the end. However, this experimentation will not help the teacher in finding out the students of a particular group and counsel them as the information about serial number of students is not retained as knowledge.

At the second instance, by looking at the distribution of classes in the visualized results, the administrators such as the Head of the Department/Institution can direct the teacher to take corrective measures in increasing / decreasing the complexity of questions while setting up the question papers. The administrators can also get an idea whether a teacher is over liberal or is pessimistic by relative comparison of the visualized results of different subjects.

At the third instance, the incremental assimilation of final knowledge of each subject by the SCIL model will help the administrators in getting the estimate of the final overall knowledge in advance without getting into the overall knowledge of all the subjects as the final overall knowledge depends on the same population.

To conclude, in this research, we have identified reasons to update the knowledge incrementally while learning from LMS data. Challenges involved in incremental update of knowledge using clustering have been addressed with the help of DBSCAN algorithm and histogram based regression lines. Mathematical formulations for incremental update of histogram and regression line features/variables have been derived, which aid in extraction of knowledge from heterogeneous data. We are exploring further to support our proposed theory when different options of sequencing are possible with Sequence Optional Incremental Learning (SOIL).The objective of incremental knowledge mining was to find the natural groups from the available data and update the obtained knowledge of groups with the next set of incoming data by incremental agglomeration of knowledge. Results obtained were compared with the module leaders handling the subjects and found to be very much accurate.

References

1. `http://moodle.org/` (last accessed December 10, 2011)
2. `http://www.blackboard.com/` (last accessed December 10, 2011)
3. Romero, C., Ventura, S., Garcia, E.: Data Mining in Course Management Systems: Moodle Case Study and Tutorial. Computers and Education. Article in Press (2007)
4. Clari, P.M., Arevalillo-Herraez, M., CeveronLleo, V.: Data Analysis as a Tool for Optimizing Learning Management Systems. In: Ninth IEEE International Conference on Advanced Learning Technologies (2009)
5. Miranda, E.: Data Mining as a technique to analyze the learning styles of students in using the Learning Management System. In: SNATI 2011, June 17-18 (2011)
6. Maloof, A.M., Michalski, R.S.: Incremental Learning with partial instance memory. Artificial Intelligence 154, 95–126 (2004)
7. Piatestsky, S.U., Smyth, M.A., Uthurusamy, R.: Advances in Knowledge Discovery and Data mining. A text Book by AAAI/MIT Press (1996)
8. Han, Kamber: Data Mining – Concepts and Techniques, 2nd edn. Elsevier (2006)
9. Adhikari, A., Rao, P.R.: Synthesizing heavy association rules from different real data sources. Pattern Recognition Letters 29, 59–71 (2008)
10. Masseglia, F., Poncelet, P., Teisseire, M.: Incremental Mining of Sequential Patterns in Large Databases. In: Proceedings of the Tenth ACM SIGKDD International Conference on Knowledge Discovery and Data Mining (2004)
11. Seeman, W.D., Michaski, R.S.: The CLUSTER3 System for goal-oriented conceptual clustering: method and preliminary results. WIT Transactions on Information and Communication Technologies 37 (2006)
12. Michalski, R.S.: Knowledge Mining: A Proposed New Direction. In: Invited talk at the Sanken Symposium on Data Mining and Semantic Web, March 10-11. Osaka university, Japan (2003)

13. Dubes, R.C., Jain, A.K.: Algorithms for Clustering data. Prentice Hall, Englewood Cliffs (1998)
14. Jain, A.K., Murthy, M.N., Flynn, P.J.: Data Clustering: A Review. ACM Computing Surveys 31(3) (September 1999)
15. Li, Y., Dong, M., Hua, J.: Localized feature selection for clustering. Pattern Recognition Letters 29, 10–18 (2008)
16. Eric, C., Akanksha, B., Tig, C., Anhai, D., Jeffrey, N.: A Relational Approach to Incrementally Extracting and Querying Structure in Unstructured Data. In: Proceedings of the 33rd International Conference on VLDB, pp. 1045–1056 (2007)
17. Fazli, C., Edward, A.F., Cory, D.S., Robert, K.F.: Incremental Clustering for Very Large Document Databases: Initial MARIAN Experience. An International Journal of Information Sciences—Informatics and Computer Science 84(1-2) (1995)
18. Hammouda, K.M., Kamel, M.S.: Incremental Document Clustering Using Cluster Similarity Histograms. In: IEEE/WIC International Conference on Web Intelligence, Halifax, Canada, pp. 597–601 (October 2003)
19. Ester, M., Kriegel, H.-P., Sander, J., Wimmer, M., Xu, X.: Incremental Clustering for Mining in a Data Warehousing Environment. In: Proceedings of the 24th VLDB Conference, New York (1998)
20. Murthy, M.N., Sridhar, V.: A Knowledge-based Clustering algorithm. Pattern Recognition Letters 12, 511–517 (1991)
21. Chen, C.-Y., Hwang, S.-C., Oyang, Y.-J.: A Statistics-Based Approach to Control the Quality of Subclusters in Incremental Gravitational Clustering. Pattern Recognition 38(12), 2256–2269 (2005)
22. Ester, M., Kriegel, H.-P., Sander, J., Xu, X.: A Density-Based Algorithm for Discovering Clusters in Large Spatial Databases with Noise. In: Proceedings of 2nd International Conference on Knowledge Discovery and Data Mining (KDD 1996), pp. 226–231. AAAI Press (1996)
23. Zakir Ali, S., Nagabhushan, P., Pradeep Kumar, R.: Regression based Incremental Learning through Cluster Analysis of Temporal data. In: The 5th International Conference on Data Mining, DMIN 2009, Las Vegas, Nevada, USA, July 13-16, pp. 375–381 (2009)
24. Nagabhushan, P., Zakir Ali, S., Pradeep Kumar, R.: A New Cluster-Histo-Regression Analysis for Incremental Learning from Temporal Data Chunks. International Journal of Machine Intelligence 02(01), 53–73 (2010)
25. Nagabhushan, P., Zakir Ali, S., Pradeep Kumar, R.: Intelligent Methods of Fusing the Knowledge During Incremental Learning via Clustering in A Distributed Environment. International Journal of Advanced Networking and Applications 02(02), 581–596 (2010)
26. Nagabhushan, P., Zakir Ali, S., Pradeep Kumar, R.: Incremental Updating of Knowledge Through Histo-Regression Lines using DBSCAN. In: The Bilateral Russian-Indian Scientific Workshop on Emerging Applications of Computer Vision, EACV 2011, Moscow, Russia, November 1-5 (2011)
27. Nagabhushan, P., Gowda, K.C., Diday, E.: Dimensionality Reduction of Symbolic Data. Pattern Recognition Letters 16, 219–223 (1995)
28. Diday, E.: An Introduction to Symbolic Data and Analysis and the SODAS software. Journal of Symbolic Data Analysis (July 2002)
29. Rangarajan, L.: Some New Approaches for Image Pattern Analysis Through Dimensionality Reduction. PhD Thesis of the University of Mysore, India (September 2004)

30. Sanjay Pande, M.B.: An Algorithmic model for exploratory analysis of trace elements in cognition and recognition of neurological disorders; PhD thesis of the University of Mysore, India (2004)
31. Pradeep Kumar, R., Nagabhushan, P.: An Approach Based on Regression Line Features for Low Complexity Content Based Image Retrieval. In: International Conference on Computing: Theory and Applications (ICCTA 2007), pp. 600–604 (2007)
32. Gowda, S.D., Nagabhushan, P.: Equivalence between two document images through iterative histo analysis of the conventional entropy values. International Journal of System, Cybernetics and Informatics, 23–33 (October 2008)
33. Pradeep Kumar, R., Nagabhushan, P.: An Approach Based on Regression Line Features for Low Complexity Content Based Image Retrieval. In: International Conference on Computing: Theory and Applications (ICCTA 2007), pp. 600–604 (2007)
34. Kadampur, M.A., Somayajulu, D.V.L.N., Dhiraj, S.S.S., Satyam, S.G.P.: Privacy Preserving Clustering by Cluster Bulging for Information Sustenance. In: 4th International Conference on Information and Automation for Sustainability, ICIAFS 2008, pp. 240–246 (2008)
35. Sayad, S., Balke, S.T., Sayad, S.: Scalable Data Mining. University of Toronto, `http://chem-eng.utoronto.ca/~datamining/SDM/` (last accessed December 10, 2011)
36. Pradeep Kumar, R.: Wavelets for Knowledge Mining in Multi Dimensional Generic Databases. PhD Thesis of the University of Mysore (2006)
37. `http://ekluvya.adithyatech.edu.in/aitlms/` (last accessed December 10, 2011)
38. `http://adithyatech.edu.in:8080/website/` (last accessed December 10, 2011)

A Service Quality Framework for Higher Education from the Perspective of Service Dominant Logic

Najwa Zulkefli and Lorna Uden

FCET
Staffordshire University,
The Octagon, Beaconside,
Stafford, AT 18 OAD

Abstract. Despite progress made through research and debate, there is still no universal consensus on how best to manage quality within Higher Education Institution (HEI). The key reason is that quality is a complex and multi-faceted construct, particularly in HEI. Quality of service offered cannot be directly observed before consumption. This makes comparing universities extremely difficult, or practically impossible. The aim of this study is to develop a conceptual framework for measuring service quality in HEI by applying the emerging concept of service dominant logic. This paper provides a new paradigm for the measurement of service quality in Higher education based on the emerging discipline of service science. It describes a framework that we have proposed that can be used to measure the quality of service in HEI. The framework is developed through the co-creation of value between students and the staffs.

1 Introduction

Service quality in higher education has been the subject of considerable interest and debate by both practitioners and researchers in recent years. For universities to remain competitive, it is important to actively monitor the quality of the services they offer and to commit to continuous improvements for students and the revenue they generate. The issue of service excellence is becoming particularly important because it has a major influence on students' purchasing intentions and, as a result, constitutes an essential competitive factor. Recent report by the Higher Education Policy Institute (Bekhradnia et al., 2006) confirms that non-EU overseas respondents were considerably less satisfied than others with the value for money received on their course. Against this background, it is clearly important for universities to understand what students' value in their university experience, including those from overseas.

There is a great need for universities to develop valid and consistent measures of students' perceptions of ser-vice quality in order to gain a better understanding of the quality issues that impact on students' experiences (O'Neill and Palmer,

L. Uden et al. (Eds.): 7th International Conference on KMO, AISC 172, pp. 307–317.
springerlink.com

2004; Joseph et al., 2005). The challenges are: to determine how students evaluate university's service quality, to identify the dimensions that students use as reference points for their evaluations, and to establish the relative importance of each of those dimensions.

Most of the existing service quality measurement tools (i.e. SERVQUAL, SERVPERF, TQM, and Balanced Scorecard) were developed during good-centered economy. Methodologies developed at that time to measure and assessing service quality are considered to be less appropriate to be applied to current service economy that is more customer-centric and market-driven. Although both provider and customer are aware of the implication of good service quality may provide positive impact for both parties but still number of problems encountered. First, both parties could not reach mutual agreement or understanding of what makes good service quality. Second, it is very difficult to determine the impact of service quality to customer satisfaction as well as to supplier's performance. Thirdly, the difficulty to identify which dimensions of service quality are seem to be important and suitable for all sector of service. These tools neglected the crucial concept of current service-based economy, the co-creation of value. For that reason, this study intends to develop an alternative tool that measures customer satisfaction in HEI by applying the main concept of SDL, which is value co-creation.

In order to design a framework that can be used to measure the quality of HE service, we need to define the variables that consumers use in their evaluations of service quality. This paper provides a new paradigm for the measurement of service quality in Higher education based on the emerging discipline of service science. It de-scribes a framework that we have proposed that can be used to measure the quality of service in Higher Education Institution (HEI). The framework is developed through the co-creation of value between students and the staffs. To do this, we used a case study to show how the co-creation of value can be achieved. , The study aims to investigate relationship between service quality dimensions from existing measurement tools and students satisfactions as well as integrate the critical factor that believe to be important in the current concept of service quality measurement. This paper begins with a review of the importance of service quality for HEI. Subsequent section reviews service dominant logic and why it is needed to revise the former service quality models. The next section describes a case study showing how the framework was developed. Subsequent section discusses the evaluation of the framework. The paper concludes with suggestions for further research.

2 Service Quality in Higher Education Institutions

The nature of competition in education differs from the standard market solution. Scholars and governmental agencies have come up with numerous methods and indicators for estimating the "quality" of universities, it is important to bear in mind the purpose and possible personal biases of report-writers. One could simply rely on official accreditations, which correspond to international standards, and leave the issue of quality-aside. Education is concerned with experience good" rather than "searching for good". Quality of the service offered cannot be directly observed before consumption. This makes comparing universities extremely difficult, or practically impossible.

(Joseph and Joseph, 1997) conducted study in New Zealand business students' perceptions of service quality in education and identified 7 important determinants of service quality: (1) Physical aspects of university (i.e. accommodation facilities, campus layout, etc.); (2) Cost/time; (3) academic issues such as standard and degree reputation, excellent academic staffs; (4) programme issues related to structure and content of the programmes, availability of options, and flexible entry requirements; (5) career opportunities as in producing employable graduates and provide information and guidance on career opportunities; (6) location; (7) others (i.e. peers/family recommendation, word-of-mouth). Empirical research conducted by (Hill et al., 2003) found four most influence factors in provision of quality in education: (1) Quality of the lecturers in terms of flexibility in teaching delivery, having proficient knowledge in their area and competencies that they possessed as well as maintain relationship with students and being very helpful; (2) students engagement with learning; (3) social/emotional system. Student found support that they need from HEI and wanted to be surrounded by positive atmosphere during their learning experience; (4) Resources of IT and library.

Many other researches apply SERVQUAL in their study to measure service quality in HEI. (Cuthbert, 1996) testing modified version of SERVQUAL to explains the need for valid and reliable instrument for course man-ager to evaluate their product/service with student body as customer. In his findings, tangibility dimension scored highest followed by assurance, reliability, responsive and empathy. However, Cuthbert believes tangibility is not a major contributor towards student satisfaction regardless their highest score. This supports by (O'Neill and Palmer, 2004) suggested that although tangibility scored highest but it has been ranked as the least importance by students compare to process and empathy. (Perisan and McDaniel, 1997) describes assurance and reliability as the most important dimension.

(Becket and Brooke, 2008) conducted a study to present a review of current quality management practices within HEIs. They see Total Quality Management (TQM) encompass quality perspectives in both external and internal stakeholders in an integrated manner. They also tested other models that equalised to TQM. These include (1) EFQM excellence model; (2) Balanced scorecard; (3) Malcolm Baldrigde award; (4) ISO 9000 series; (5) Business process re-engineering; (6) SERVQUAL. All these models are relevant as they incorporate students' perspectives. However, all of these models have limitations. Most of these applications have drawbacks due to convoluted bureaucratic structures and some self-directed academics feel intermediated with the utilities of this model that relies heavily on team-based approach (Srikanthan and Dalrymple, 2002).

However, the main criticism of these models is that they are based on the traditional service concept. In the last few decades there has been work going on by researchers trying to differentiate between goods and services. According to (Vargo and Lusch, 2004), traditional good-dominant logic is inadequate to understand current markets, economic exchange and marketing. They argue that traditional good-dominant logic focuses on tangible resources and embedded value and transactions. Instead of this, the authors argue that the service dominant logic proposed by them should be used for economic exchange and marketing. The new perspective of service-dominant logic focuses on intangible resources, the co-creation of value and relationships.

3 Service Dominant Logic

Traditional economics is based on the dominant logic of exchange of goods that are usually manufactured out-put. This dominant logic focuses on tangible resources, embedded value and transactions. A new paradigm known as service dominant logic (SDL) has revised logic focused on intangible resources, the co-creation of value and relationship (Vargo and Lusch 2004). According to (Vargo and Lusch, 2004), marketing s increasingly shifting its dominant logic away from the exchange of tangible goods (manufactured things) towards the exchange of intangibles, specialised skills and knowledge processes (doing things for and with). This approach points toward a more comprehensive and inclusive logic, one that integrates goods with services. (Vargo and Lusch, 2004) define service as the application of specialised competences (knowledge and skills) through deeds, processes and performances for the benefit of another entity or the entity itself. This also includes tangible output (goods) in the process of service provision. (Spohrer et al., 2007) define services informally as collections of resources that can create value with other service systems through shared information.

The new notion of service, the service dominant logic, views service as being about resources, exchange and human action (Sporher et al 2008). Service is now seen as the process of doing something for and with another party. Value creation is a collaborative process and is always co-created. (Spohrer et al., 2008) argue that the purpose of economic exchange in service is service provision for (and in conjunction with) another party in order to obtain reciprocal service – that is, service is exchanged for service. Although goods are involved in the process, they are appliances for service provision – they are conveyors of competencies. According to these authors, services provided directly or through a good are the knowledge and skills (competencies) of the providers and beneficiaries that represent the essential source of value creation, not goods that are sometimes used to convey them. Because of this, (Spohrer et al., 2008) argue that service involves at least two entities, one applying competence and another integrating the applied competence with other resources and determining benefits (value co-creation). These interacting entities are service systems.

In SDL, the distinction between the producer and consumer disappears and all participants contribute to the creation of value for themselves and for others. All economic and social actors are resource integrators (Vargo and Lusch, 2008). The ten foundational premises (FPs) of Service Dominant Logic as suggested by (Vargo and Lusch, 2008); Service is fundamental basis of exchange (FP1); Indirect exchange masks the fundamental basis of ex-change (FP2); Goods are distribution mechanism for service provision (FP3); Operant Resources are the fundamental source of competitive advantage (FP4); All economies are service economies (FP5); Customer is always a co-creator of value (FP6); The enterprise cannot deliver value, but only offer value propositions (FP7); A service-centered view is inherently customer oriented and relational (FP8); All economic and social sector are re-source integrator (FP9); Value is always uniquely and phenomenological determined by the beneficiary (FP10).

In SDL, value is co-created between provider and consumers through the utilisation of resources and application of competence (Vargo and Lusch, 2004). Firm apply their knowledge and skills to a particular good to make it valuable while customers apply their knowledge and skills in the use of it and integrate their resources (money) to provide their part of service. Thus, value co-created by having this mutual beneficial relationship (Vargo et al. n.d.). It is our belief that the premises of SDL can provide us guidelines for the development of a conceptual framework for service quality in HEI.

4 Co Creation of Value in Service

User satisfaction is a widely recognized indicator of any organization's quality of service and performance. A given service reaches its level of excellence when it meets the needs of customers. In order to measure service quality in HEI, It is important to understand what students' value in their university experience. (Grönroos, 2000) argues that value for customers is created throughout the relationship by the customer, partly in interaction between the customer and service provider. The focus is not on product, but on the customer's value creating processes where value emerges for customers and is perceived by them. We should support value creation rather than simply distributing ready-made value to customers.

The conventional way of business for a firm is to organise its activities around creating goods and service and unilaterally create value for the customer through its marketed offerings. The role of the customer is that of a consumer simply passively consuming what the firm has to offer. However, in today's service dominant market, he has moved away from conventional value chains towards empowered customer interactions with a firm's offering and with other customers. These interactions centred on the individual rather than the firm. These interactions are where the customer experience is, and what the company must understand to design the technology platforms and information infrastructure to support compelling experience environments. We believe that to develop an effective service quality for HEI, it is important that we adopt SDL concepts of co creation of value. Our study aims to develop a framework based on SDL premises that can be used for service quality for HEI.

In SDL, service is about resources, exchange and human action (Spohrer et al., 2008). Service is now seen as the process of doing something for and with another party. Value creation is a collaborative process and is always co-created. Service involves at least two entities, one applying competence and another integrating the applied competence with other resources and determining benefits (value co-creation).

According to (Ramaswamy and Gouillant, 2008), there are three principles for co-creating value:

- Individuals' experiences are central to conceiving value propositions. It is important to look beyond product and service offerings to see how customers experience value propositions— and connect the two.
- It is also vital to go bèyond processes to interactions. That is, we must focus on the interactions be-tween the individual customer and company, and on generating outcomes of value to individuals. These authors proposed a model known as DART.
- We must involve and engage the customer. Companies must involve and engage with individuals, start-ing with customer-facing employees and expand to customers and other stakeholders.

In HEI context, university should view students as operant resources that assist in producing and creating value as well as exchanging their competence.

5 Research Methodology

This section briefly reviewed the development of the framework using a case study at Staffordshire University.

Our research is limited to the study of a university, Staffordshire University in UK. The focus was on non-EU postgraduate students with at least 2 years' experience in HEI. Descriptive methodology was used for the re-search. It involves five consecutive steps:

1. Primary Data Collection
2. Secondary data collection
3. Survey development and data collection
4. Data analysis
5. Evaluation

5.1 Primary Data Collection

The main purpose of this step was to find out what constitutes service quality at university. Direct observational methods were used to learn about students and staffs behaviours and identify their needs. Data were collected from observing students' teaching-learning in lecturing halls and classrooms from the first week of registration (start of term) until Christmas of 2010. Observation also included how students interacted with non teaching staffs in their activities. (administrative and technical supports for students). This provides us the context of our research. Informal interviews were conducted with small set of respondents (N=10) to get information that can be used as research problems. Interviews were carried out with 10 international students and 5 employees from library, information desks, international student office and financial office. The interviews were semi-structured and open-ended. This allow the interviewee to speak freely and more flexible. Critical Incident Techniques (CIT) Flanagan (1954) was employed to collect the information from students.

5.2 Secondary Data Collection

During this step, secondary research was conducted. Literature reviews were carried out on service quality. The aim was to analyse methodologies that have been used by researchers to measure service quality. Critique of the literature on service quality was done to determine what the necessary factors, dimensions and attributes that are needed to carry out service quality measurement for HEI. Results obtained from the research would be used in the proposed framework. From the research conducted, we have identified 11 determinants to be tested in HEI environment: Reliability, Responsiveness, Competence, Credibility, Security, Accessibility, Communication, Understanding, Courtesy, Offering and Empathy. Items representing these 11 determinants were later generated to develop 46 items to measure providers' performance and to identify the importance of factors underlying service quality.

5.3 Survey Development and Data Collection

Findings from critical incident techniques conducted during primary data collection were used to describe and defined service quality management at the university. The analyses also portray experiences of key players (service provider and users). Results from CIT were later transformed into survey questions. Taken findings from the primary data collection and combined with analysis from literature review, a set of 44 items from 11 determinants were developed. These items are used to construct two sets of questionnaires to be distributed to students and to staffs. The items were designed to evaluate and assess service quality provided by academic staffs, technical support staffs, facilities, library facilities and library support staffs, study skills staffs, front desk, international student officers, and student counsellors. Two surveys were developed: one for students and one for staffs.

5.4 Data Analysis

Data collected from both groups of respondents (student and staff) were analysed to examine the dimensionality of 44 service quality items using factor rotation analysis; orthogonal rotation analysis and oblique rotation analysis. The factor analysis was performed to seek underlying unobserved variables that affecting the other observable variables. It aims to reducing the large set of variables into more manageable dimensions by looking at group of closely related items. The analyses were carried out using SPSS (Statistical Package for Social Science).

Before conducting factor analysis, Cronbach's alphas were computed to purify the scale. Each dimension was computed separately. Later, the variables were analysed based on the inter-correlation between variables. Variables that did not correlate was excluded in the test. Pearson correlation and Spearman correlation were used to explore the strength between variables (Field, 2005; Pallant, 2007). To verify that the data set was suitable for factor analysis, Kaiser-Meyer-Olkin (KMO) measure of sampling adequacy and Bartlett's Test of Sphericity were used. Orthogonal analysis (using Varimax in SPSS) was run o check if the factors were meant to be inde-pendent. The emergent dimensions were later renamed and examined using simple regression analysis. Pair T-Test was performed to measure gaps between perceived importance and perceived performance. Finally gaps between perceived importance and perceived performance for the dimensions were tested by comparing mean ratings of both importance and performance variables and then determine the gap by plotting the graph.

5.5 Evaluations and Validation

To evaluate and validate the emerged dimensions, two sets of questionnaires for students and staffs were sent out to 25 students and 15 staffs. For students' questionnaires, 33 items under 8 factors was included and for staffs' questionnaires, 27 items under 7 factors are included. The questionnaires were distributed to 25 non-EU postgraduates' students and 15 staffs. The result of this evaluation test was used in designing the conceptual framework.

5.5.1 Framework for Service Quality in HE

It is important for us to determine 1) what are the roles of consumers in value co-creation process? 2) How does value co-creation process work? In attempt to answer these questions, we start by looking at the student requirements when they purchase the service. The value co-creation activities in this university based on (Prahalad's, 2004) five value co-creation activities. Figure 1 show preliminary model that can be used in assessing service quality in HEI by adapting the concept of Service Dominant Logic (SDL).

SDL emphasizes the process of value co-creation between suppliers and consumers. The model framework depicts how the determinants from staffs and students perspectives through value co-creation activities could lead to loyalty behaviour, increment level of satisfaction and positive overall service quality. The service provided by university does not only based on teaching and learning services but also providing other ancillary services. These services are consumed through the stated 5 value co-creation activities that suggested by (Prahalad, 2004). The framework suggests that university should take critical attention on the determinants of both member of staffs and students perspectives in assessing and managing their service quality. Determinants suggested from Student perspectives are: Offering, Reliability, Courtesy, Empathy, Understanding, Security, Competency, and Responsiveness. While determinants on member of staff perspectives are: Offering, Courtesy, Responsiveness, Reliability, Communication, Empathy, and Accessibility.

Customer (in this context, student) play their role as value creator during consumption of service and interaction with suppliers. While supplier play their active role as value facilitator in providing customer with the foundation of value creation as well as value creator during service consumption (Gronroos, 2008).

These roles were adapted from (Gronroos's, 2008) 'value-fulfilment' model. Both staffs and students engage in five value co-creation activities at the HEI (Prahalad, 2004) – customer engagement, self-service, customer experience, problem solving and co-designing. These framework suggest guidance in utilising all five activities of value co-creation for HEI and provide understanding of supplier and customer role respectively in order to achieve the ultimate goal of customer loyalty, positive service quality and customer satisfaction (representing in oval shape).

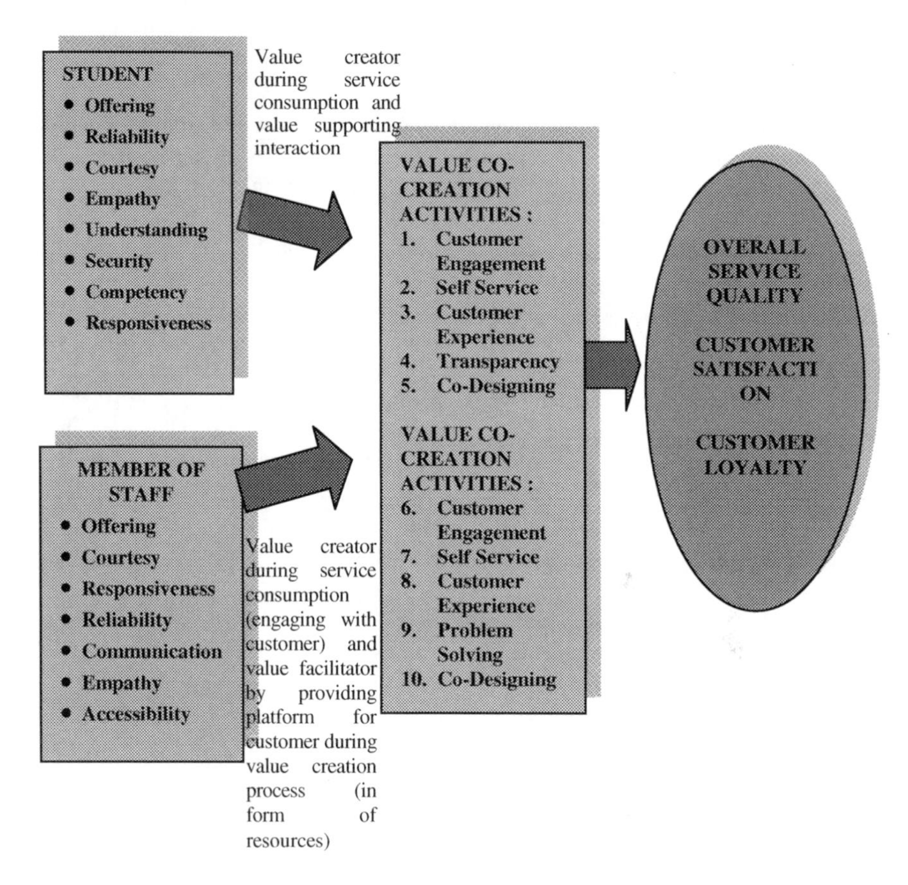

Fig. 1 Service Quality Framework

6 Conclusion

This paper outlines empirical study involving international students from non-EU countries taking a range of postgraduate's programmes in university. The aim of this study is to investigate what are the importance dimensions that contribute to the development of a service quality framework in the HEI context. The dimensions investigated were taken from both student and staff perspectives.

From the empirical study, we found that Offering dimension was ranked the top from both staff and students. This is due to the fact that students look at what university could offer in terms of accreditation, study experience, socialising experience, as well as facilities provided. Thus, it is very important for university management to understand student requirements from the moment they making decisions to which university they would be interested in enrolling.

We conclude that, from the student perspectives, service quality at the university consists of 8 dimensions, Offering, Reliability, Courtesy, Empathy, Understanding, Security, Competency, and responsiveness. On the contrary, from the staff perspectives service quality dimensions consist of: Offering, Courtesy, Responsiveness, Reliability, Communication, Empathy and Accessibility. All these dimensions for service quality scale shows high reliability and validity as well as interrelated to each other. It therefore indicates that all the dimensions should be accomplished to obtain or assure good service quality. Our empirical findings show that service quality is antecedent to customer satisfaction. The study also provides information on gaps between performance and importance. University should consider closing the gaps that exists on the Courtesy, Credibility, Competency, and Responsiveness dimensions from student point of view. While on staff's perspectives, gaps should be closed in reliability and empathy dimensions.

Students play active role as both value creator during value generating process and value-supporting interactions. While member of staffs play their active roles as value facilitator by providing customer a platform for value creation process (in form of resources), they also play active role as value creator during direct engagement with customer during value generating process. This empirical findings support that customer is always a co-creator of value".

References

Balanced Scorecard Report 10(4), 9 (viewed on February 12, 2010), https://ssl.instantaccess.com/bscol/articles/docs/5.pdf

Becket, N., Brooke, M.: Quality Management Practice in Higher Education – What Quality Are We Actually Enhancing? Journal of Hospitality, Leisure, Sport & Tourism Education 7(1), 40–54 (2008)

Bekhradnia, B., Whitnall, C., Sastry, T.: The Academic Experience of Students in English Universities. Higher Education Policy Institute, London (2006)

Parasuraman, A., Zeithaml, V.A., Berry, L.L.: A Conceptual Model of Service Quality and its implications for future research. Journal of Marketing 49, 41–50 (1985)

Chell, E.: Critical incident technique. In: Symon, G., Cassell, C. (eds.) Qualitative Methods and Analysis in Organisational Research: A Practical Guide. Sage Publications, London (1998)

Churchill, G.A.: A Paradigm for Developing Better Measures of Marketing Constructs. Journal of Keting Research 16, 64–73 (1979)

Cuthbert, P.F.: Managing Service Quality in HE: is SERVQUAL the answer? Part 1. Managing Service Quality 6(2), 11–16 (1996)

Grönroos, C.: Service management and marketing. John Wiley & Sons, West Sussex (2000)

Hill, Y., Lomas, L., McGregor, J.: Students Perceptions of Quality in Higher Education. Quality Assurance in Education 11(1), 15–20 (2003)

Joseph, M., Joseph, B.: Service Quality in Education: a Student Perspective. Quality Assurance in Education 5(1), 15–21 (1997)

Joseph, M., Yakhou, M., Stone, G.: An educational institution's quest for service quality: customers' perspective. Quality Assurance in Education 13(1), 66–82 (2005)

Kotler, P., Fox, K.: Strategic Marketing for Educational Organizations. Prentice-Hall, Englewood Cliffs (1995)

O'Neill, M., Palmer, A.: Importance-performance analysis: a useful tool for directing continuous quality improvement in higher education. Quality Assurance in Education 12(1), 39–52 (2004)

Parasuraman, A., Zeithaml, V., Berry, L.: A Conceptual Model of Service Quality and Its Implications for Future Research. Journal of Marketing, 41–50 (Fall 1985)

Parasuraman, A., Zeithaml, V.A., Berry, L.L.: SERVQUAL: A multiple item scale for measuring consumer perceptions of service quality. Journal of Retailing 64(1), 14–40 (1988)

Perisau, S.E., McDaniel, J.R.: Assessing service quality in school of business. International Journal of Quality and Reliability Management 14(3), 204–218 (1996)

Prahalad, C.K., Ramaswamy, V.: The future of competition: Co-creating unique value with customers. Harvard Business School Publishing, Boston (2004a)

Ramaswamy, V., Gouillart, F.: Co-creating strategy with experience co-creation (2008)

Spohrer, J., et al.: Steps Toward a Science of Service Systems, pp. 71–77. IEEE Computer Society (2007)

Sporher, J., Vargo, S.L., Caswell, N., Maglia, P.P.: The service system is the basic abstraction of Service Science. In: Proceedings of the 41st International Hawaii Conference on System Sciences (2008)

Srikanthan, G., Dalrymple, J.: Developing a Holistic Model for Quality in Higher Education. Quality in Higher Education 8(3), 216–224 (2002)

Stauss, B.: Using the critical incident technique in measuring and managing service quality. In: Scheuing, E., William, F.C. (eds.) The Service Quality Handbook, pp. 408–427. American Management Association, New York (1993)

Vargo, S.L., Lusch, R.F.: Evolving To a New Dominant Logic for Marketing. Journal of Marketing (68), 1–17 (2004)

Vargo, S.L., Lusch, R.F.: Service-Dominant Logic: Continuing the evolution. Journal of the Academic Marketing Science 36, 1–10 (2008a)

Vargo, S.L., Lusch, R.F.: A Service Logic for Service Science. In: Hefley, B., Murphy, W. (eds.) Service Science, Management and Engineering: Education for the 21st Century, pp. 82–87. Springer (2008b)

Vargo, S.L., Lusch, R.F.: The Service Dominant Mindset. In: Hefley, B., Murphy, W. (eds.) Service Science, Management and Engineering: Education for the 21st Measurement of Service Quality on Higher Education Institution by Applying Service Dominant Logic Concept Page 166 century, pp.88-94 (2008c)

The Use of Web 2.0 Technology for Business Process

Aura Beatriz Alvarado Gaona[1], Luis Joyanes Aguilar[2], and Olga Najar Sanchez[3]

[1] Ministerio de Educación Nacional, Bogotá, Colombia
[2] Universidad Pontificia Salamanca, Campus Madrid, España
[3] Universidad Pedagógica y Tecnología de Colombia, Tunja, Boyacá, Colombia
{aura_beatriz_a,olnasa}@hotmail.com, joyanes@gmail.com

Abstract. This article presents the results of one of the first phases of the doctoral thesis proposal, which includes the study of technologies that are part of the paradigm of Web 2.0, to propose a model of software development with BPM approach 2.0. The researcher [1] points out that it would help analysts to understand business requirements and apply them directly in the process by using existing IT systems. The research involves areas such as software engineering, specifically all that has to do with the business process management and Web 2.0 technologies. These paradigms promote the creation of business management applications focused on the modeling, execution, administration and monitoring of business processes and a new generation of communication services which are based on the social web.

Keywords: Business Process, BPM, Web 2.0 Technologies, Web 2.0 Applications.

1 Introduction

According to studies carried out by the Ministry of Information and Communication Technologies in Colombia (ICT Ministry), the number of people who use the Internet and have access to social networks in Colombia has significantly increased, although the incorporation of ICT in the field of small and medium-sized exporting enterprises (SMEs), has not been significant. Companies, especially those classified as SMEs, need to implement mechanisms that allow them to be competitive, reduce operation costs, improve product and / or services quality, to make the processes flexible and foster an environment of collaboration among all employees, suppliers or customers. To achieve these goals, it is necessary an adequate business processes management.

Another factor that affects the organizations to achieve their goals efficiently is related to employees and information systems which should go in the same direction, being the business processes the ones that facilitate this collaboration.

L. Uden et al. (Eds.): 7th International Conference on KMO, AISC 172, pp. 319–331.
springerlink.com

Business processes are essential to understanding how an organization operates regarding the design and implementation of flexible information systems. These systems provide the basis to adapt the existing functionality to business requirements.

According to Gartner, [2], "the next evolution will be mainly self-adjusted processes based on the brightness of patterns in the user preferences, consumer demand, prediction capabilities, trend, competitive analysis and social connections." Regarding the use of processes models in organizations, the author states that business executives and professionals, who influence the process models, will have more direct control on their areas of operations, and thus, they will achieve competitive status.

BPM focuses on the Administration of business processes. It is understood as a methodology that guides efforts to optimization of business processes, seeking to improve their efficiency and effectiveness through systematic procedures. These processes must be continuously modeled, automated, integrated, monitored and optimized [3]. The official page of Club-BPM [4] presents the results of a study on this technology implementation in Spanish companies, and the researchers conclude that "BPM plays a key role for companies to face the current economic crisis by generating a complete control of processes, visibility of the company for a better decision-making, and strategic orientation for the achievement of short and long term objectives."

Although the concept Web 2.0 was articulated in 2004, people are starting to use it now, not as a series of tools but as underlying concepts of listening, interaction and network connection The Web 2.0 technologies and their main applications have been adopted by people in an individual way for their daily tasks, but most of the organizations don't know how to apply them and their corresponding benefits.. Large companies are incorporating these tools, both externally and internally, to support the processes of communication, Management of knowledge as well as collaboration platforms.

In 2006, Andrew McAfee, [5], a professor at Harvard Business School, published the article "Enterprise 2.0: The Dawn of Emergent Collaboration" in the Sloan Management Review, which set a new paradigm: Enterprise 2.0. The term Enterprise 2.0 is simply the application of many Web 2.0 ideas and web 2.0 technologies to the company. According to the definition of McAfee, Enterprise 2.0 is the implementation of the attributes and characteristics of Web 2.0 in business. It represents a new way of working in corporations to overcome the limits of those communication tools of the previous model by means of the use of new technologies and business practices.

This new paradigm represents a huge competitive advantage to those companies using social software, which allows collaboration and simplifies the exchange of information between employees' networks and networks of suppliers, customers and other stakeholders. As stated by McAfee, the tools of Web 2.0 are producing a radical change for corporations, as they channel the collective intelligence and impact on business innovation, productivity and efficiency [6].

Experts in BPM, [7] have introduced a new paradigm in BPM architecture, it is called BPM 2.0. This new business model helps analysts to understand business requirements and apply them in the processes of existing IT systems. In addition, BPM 2.0 allows the execution of the process of Web 2.0, providing a new way to communicate inside and outside the company. Therefore, they allow the business models of any company to respond quickly to strategic changes. BPM 2.0 has taken some of the features of Web 2.0 technologies to make BPM more powerful and relevant to the process of business analysts. The vision of BPM 2.0 is to create agile and dynamic processes that can act as a basis for greater flexibility and innovation. Thus, its main purpose is to take advantage of BPM as a catalyst for the alignment between business architecture and IT services, and continuously adapt them to changes more quickly,

With the use of Web 2.0 technologies, BPM 2.0 users' levels of participation in the creation, modification and management of a business process has increased. Each user has access to the business content to make adjustments or modifications. This makes the content administration could be more decentralized. For example, in BPM 2.0, a business user can control the process of replacing the content on the dashboard of an RSS of a different process to control a particular condition in the business process. This higher level of content exchange and users' participation can create a different culture in the processes management within the organization, which may take time for its adoption.

As it was previously mentioned, communication tools of Web 2.0 such as blogs, wikis and RSS feeds facilitate collaboration and information sharing and they are used in a company to help employees participate actively in the collaboration, communication and knowledge management. Likewise they allow easy access, exchange and visualization of information and data by means of Internet, intranets, portals, e-mail, digital magazines, readers / aggregators and RSS enterprise content management [8].

This article is aimed at reviewing these Web 2.0 tools that can be included in the Modern Business (Enterprise 2.0, according to McAfee), According to what has been proposed by [9] web 2.0 tools can generate:

- Processes of interactivity when a user can add and share information with others.
- Processes of combinatorial content interaction, when they make possible the interrelation of content from different databases.
- Processes of interaction of interface, whether esthetic preferences or functions where the user can place the content in different places on the screen or decide what contents should appear, or those of generative preference, when the information system decides what data should be presented.

Section 2 summarizes what a Business Process is. Section 3 describes the technologies of Web 2.0 that have been used the most in the business field and Section 4 presents some applications of Web 2.0 that can be adapted to the enterprise level.

2 Business Processes

A business process is a structured set of activities designed to produce a particular output or achieve a goal. This implies a strong emphasis on how work is performed within an organization, in contrast with the product approach in which the emphasis is on what is produced. Therefore, the process is a specific sequence of work activities trough time and space, with a beginning, an end, and clearly defined inputs and outputs: a structure for action. Processes describe how work is done in the company and they are characterized by being observable, measurable, improved and repetitive [10].

A Business Process can be represented (Figure 1) as an intermediate layer and a major axis of communication between roles and systems, relating the lower layers which are Legacy Systems (CRM, ERP, Host, etc.) with the first layer which represents the roles involved in the Business Process (employees, suppliers, partners, public, customers, etc.).

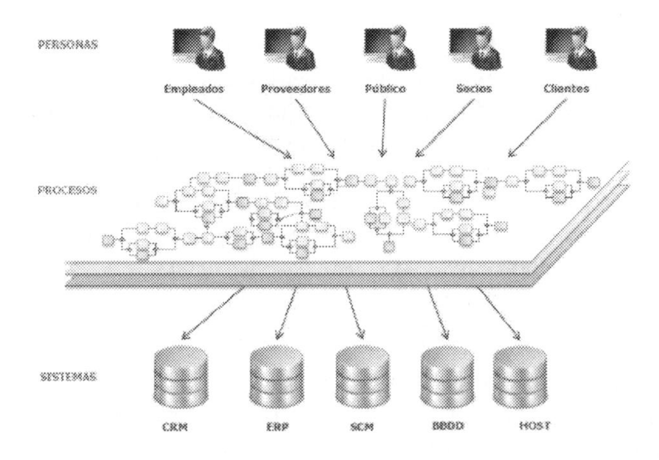

Fig. 1 Business Process as a coordination axis. Taken from [11].

Focus on processes requires the utilization of conventions and standard notations so they can be understood by the director of operations and IT director. With this methodology we can say that software development is the process of constructing, maintaining and improving the business processes [12]. BPM can directly incorporate information systems and existing assets, coordinating their use in the process layer accessible to everyone in the organization, by means of B2B systems and tools to integrate and reuse any of the existing applications [13].

2.1 Stages of Process Management with BPM Technology

According to [12], companies that implement BPM technology for business processes management are based on the definition of a life cycle, which is manifested in the development of certain stages:

- Design. It involves modeling, manipulating and redesigning processes to train and then to inform the organization about potential discoveries or suggested improvements. It integrates activities, rules, participants and their interactions. Its features are: composition, decomposition, combination restructuration and transformation.
- Deployment. Socialization of knowledge with all participants, including the concepts, applications and other business processes.
- Interaction. It uses processes that allow all staff to fully interact with business processes. This includes the administration between the interface, manual work (workflow) and automation. In this administration the work falls on the allocution, tasks administration and the way the data are integrated.
- Monitoring and control. Integrate processes with the process management system on which people are operating. This includes the necessary tasks to maintain optimum development of processes, from both perspectives: technical perspective and resources use perspective.
- Optimization. It combines the design and analysis processes to provide feedback on the execution of the processes regarding the current situation.
- Analysis. Controls the presentation of the process to provide the metric, analysis and business intelligence needed to manage the best practices and strategies, and discover innovative opportunities.
- Execution. Ensures that the new process can be developed by all participants (people, information systems, organizations and other processes). It is responsible for process management system.

Many features of the BPM technology are combined to meet the BPM lifecycle, driven directly by organizational goals. This fusion of technologies in a transparent integrated design environment (IDE) provides the level of abstraction necessary for both the specialist in technology and the specialist in business to use the same language.

3 Web 2.0 Technologies

In a research study [14], identifies a number of technologies considered 2.0, because they tend to be present on the websites 2.0:

Ajax

The combination of XML and JavaScript, enable the creation of web application executable on the client, greatly reducing data traffic and server workload with greater interactivity [15]. Examples of applications where this language is used are Google Maps, GMail or Flickr.

The Mashups

They are also known as web hybrid applications. Their function is the transparent integration of data from other web services, which have created APIs to develop new applications that access the free data. The most cited example is

HousingMaps.com ms, which combines information from a real estate database with Google Maps

The folksonomies

In some applications 2.0, users become indexers of information. They relate information with keywords or marks (tags) that they freely choose. This user's freedom to label the information has been given the name of "folksonomies" [9]. Thus, the priority use of "folksonomies" lies in how contents are sorted and localized.

The Social networking

It is one of the key elements in the development of Web 2.0, especially as regards the use of collective intelligence. It refers to the set of applications that enable communication between users through the web [16].

This category of social software includes the tools and services that enable sharing information and digital objects, such as videos of YouTube or Flickr photographs. The three components that have a decisive influence in the Web 2.0 are blogs, forums and wikis.

The Collective Intelligence

Collective Intelligence means the ability of a group to solve problems that each individual, will not be able to solve or even understand. Activities involving collective intelligence can be divided into three groups [9]:

- The production of Contents
- The optimization of resources
- The control exerted over content and individuals

RSS

It is an XML format for syndicating Web content; it allows people to describe news, blog entries, etc. Web-based RSS readers and aggregators are used to receive and read the feeds as they are published. It was popularized by blogs and is currently implemented in websites of institutions, in most magazines, portals, databases and library catalogs [17]. According to [9] an RSS is a document that contains metadata related to a particular website. RSS files are structured in items, with the title, abstract and a link to the information that describe and eventually can contain other data such as date of publication of the document and the author's name, among others.

4 Applications Web 2.0

Since its origins, the Web has had a major impact and through the time we have tracked the rising adoption of Web 2.0 technologies, as well as the ways organizations are using them as a means of communication, because of the velocity with which information can be exchanged, and because it breaks all geographic boundaries, enabling organizations to expand their businesses. This has meant that every day more web applications are implemented in government agencies, educational institutions and business management.

Not only is the development of web 2.0 tools which is increasing, but also their use in different companies. There are many examples of companies in Second Life that are deriving measurable business benefits from their investments in the Web with the use of blogs, forums or wikis for the coordination of work teams, social networking, social bookmarking to share common information that enable them to form contact networks.

In his doctoral thesis, [18] presents the concept of application Web 2.0, from two perspectives:

- *A collaborative aspect in which the end user is the central axis of the Web application.* In Web 2.0 sites, the user is in charge of creating the site's content, as well as determining the quality of its contents, and establishing a categorization. This change, has led an active role in the exponential growth of content on the Web in the form of: opinions, news, blogs and personal experiences, definitions of concepts in wikis, videos and photos on pages of multimedia content. Simultaneously, the so-called Social Web has emerged, establishing an analogy with the traditional Web connecting different users. Thus, virtually social networks have been created to connect people who have similar interests (hobbies, work relationship, place they study or work, etc.). The participation of the user is essential to provide a precise interaction with the application which has become an essential requirement [18].
- *An advanced technological aspect. It facilitates the interaction of the end user with the Web application.* The interfaces of the most popular Web 2.0 sites have a high level of usability; they have technologies that allow users to implement multimedia interfaces, user-generated content and more complex interactions. According to [19] cited by [18], the use of these technologies, has led to the RIA applications that reside on a Web server and presentation layer in the web client. It is stated by[20], also cited in [18] that the most prominent RIA technologies are: (1) AJAX to get data from the server on demand to avoid full charge of a page, (2) JavaScript frameworks, for adding business logic and (3) RIA development environments, to implement advanced interfaces which include animations, multimedia and complex interactions.

Given these realities, it is very important to mention that [18] concluded that "The compliance, to a greater or lesser extent, of both aspects is what gives the status of Web 2.0 to an application accessible from the Internet." Thus, Google Maps is cataloged as a Web 2.0 as a result of advanced AJAX-based interface and JavaScript, although the basic content from the application and the maps are not created by users. On the other hand, Web 2.0 site Twitter, which is a tool for the micro blogging would not be in the domain of the RIA. However, the easy interaction with users makes it a clear example of Web 2.0.

A Web application 2.0, as a software product with specific characteristics, can be developed by using the principles of Software Engineering, although traditional methods are not adapted to meet the requirements of these specific types of applications, which should include the creation of mashups, the exploitation of social networking and the incorporation of multimedia content or the definition of folksonomies.

The technologies to develop Web 2.0 applications described by [8], include some components, such as server software, messenger protocols, syndication of content and several client applications and browsers based on standards with ' plug-in and extensions. The author emphasizes the need to use techniques such as, wiki software and weblog software designed to create and maintain blogs, mashup tools, RSS, RIA technologies and folksonomies tools for the creation of personalized web sites with constantly updated information, social systems, bookmarking and Social Networks services. The same author states some uses of these tools:

- Wikis, blogs, podcasts and intranet knowledge base to facilitate professionals' education.
- The use of blogs and RSS to broadcast corporate news and information.
- Wikis to develop shared documentation.
- Podcasts for reviewing and disseminating recordings of training sessions, product demonstrations, support for internal services and education.
- Social networks, content 'tags' and 'social bookmarking' to share knowledge and resources of interest.

According to [21] organizations not only use new social technologies to regularly communicate with customers, employees and vendors, but also to do it with other target audience (employees of a company, suppliers and shareholders, among others) The effectiveness of such internal channels has been demonstrated by examples such as IBM, which has an extensive network of internal blogs to manage its employees knowledge.

Based on what [22] points out, there are some applications of web 2.0 technologies in business, which are summarized in table 1.

Table 1 Business Applications of Web 2.0 tools. Adapted from [22].

WEB 2.0 TOOLS	BUSINESS APPLICATION	EXAMPLE
RSS Readers	**Increase in the employees' knowledge** Readers allow you to read in a centralized way news or information to keep employees informed about competitors, market, tendencies. **Dissemination of information of common interest** It is possible to share information - in a centralized way with customers or employees by adding, for example, a customizable clip that shows the most recent shared items in the side column of the company's web site.	Google Reader

Table 1 (*continued*)

WEB 2.0 TOOLS	BUSINESS APPLICATION	EXAMPLE
	External communication The use of a RSS generator gives the firm the opportunity to spread their message via internet, at the same time that others disseminate it (through blogs) easily. **Analysis & Marketing** Statistics of web pages in general do not do follow-up of users who have added the "feed" of the page, they simply make follow-up of visits to the page. This can produce a difference between the target audiences of the current page as it may be that many do not access directly the page, because they can read the information through the feed.	FeedBurner
Geolocators	**Locating companies** It is possible to obtain the location of a particular company in a map. It is very useful for the displacement of the commercial department at little-known areas. **Location of their own company** It is possible to enter the location of our own company, so that other users of Google Maps can locate it on a map easily.	Google Maps
Specialized Search Engines 2.0	**Time saving search of flights or hotels** Search engines display all the information quickly, in a very organized and centralized way. **Cost Reduction of flights or hotels:** Trabber allows employees to find flights and hotels at the best price for the company, because it searches the best 34 websites of flights simultaneously, compares them and book the flight without extra charges	Trabber
Video Sharing	**Transmission of information on the company for advertising purposes.** More businesses are using You tube today to promote their goods and services because it disseminates videos that display information about the company, its products, services or anything else that might be of interest to the actual or potential clients.	YouTube
	Dissemination of private information It offers information about company meetings, congresses or any event confidential and of interest to employees. **Dissemination of long term visual content** Extensive dissemination of videos with advertising content or any other type of information that might be useful to customers.	Net2u_ Web 2.0 tools
Photosharing	**Advertising** Business are taking advantage of social networking websites to market their business online, for example flickr allows companies to show off photos and videos of their products and put them in front of many potential customers.	Flickr

Table 1 (*continued*)

WEB 2.0 TOOLS	BUSINESS APPLICATION	EXAMPLE
	Dissemination of internal information to the company Companies can upload massive amounts of pictures of employees and events and label them with tags so they're easy to find **Dissemination of high quality photos** This application allows companies to show off high quality photos with advertising content or any other type of information that might be useful to customers.	Net2u_ Web 2.0 tools (photo tool)
Social networking websites	**Communication with employees** It is advisable to create a "page for the company" and present information about it. The administrative staff can have the curricula vitae of all employees on line. **Recruiting employees** The company can access information in more than 10 million users and hire the best employees based on their profiles. They can also read the comments that other users make on the candidate. **Management of business contacts** The company directors can access to the contacts of their contacts to open new business opportunities. "social networking"	Linkedin
	Review of documents The company can find the best quality and widest selection of documents used to start, grow, and manage small business. For example Docstoc provides a platform to upload, share and even sell documents, and maintains an ever-growing repository of both free and purchasable legal, business, financial, technical, and educational documents that can be easily searched, previewed and downloaded in different languages.	docstoc
	Marketing The company can upload professional, financial and legal documents for the business community and at the same time increase the possibility that potential customers share and see them. **External Communication** It is possible to know the opinion of customers through the comments on the presentations that have been uploaded.	slideshare
Blogs creators	**Advertising** It is possible to create a company blog to publish news and useful information for customers. This Web site allows companies to share their business with the world from the comfort of their own offices. **Internal communication** It is possible to restrict access to only the employees, so that you can use the blog as a means to publish information of interest and receive comments.	Blogger

Table 1 (*continued*)

WEB 2.0 TOOLS	BUSINESS APPLICATION	EXAMPLE
Agendas organizers	**Agenda Organization** It is possible to organize shared agendas in a group of people, so that everyone is informed of the meetings and tasks for the same period of time. Companies let employees, customers, and friends see their calendar, and view schedules that others have shared with them.	Google Calendar
Instant messaging	**Internal communication** This tool offers an instantaneous transmission of text-based messages from sender to receiver. Communication is much more immediate and dynamic that the communication carried out through email, it is cheaper and less intrusive than the telephone transmission.	MSN Web Messenger

In short, ICT can not only make the service management more effective, but also improve communication with citizens, providers and staff, facilitating the creation of business management networks [23].

5 Conclusions

Business Process Management BPM is a recent approach that has derived some research fields, among which are mentioned, the effort to adapt the standards of quality in Software Processes for measuring business effectiveness and efficiency while striving for innovation, flexibility, and integration with technology. BPM attempts to improve processes continuously and represents an interoperability project in Information Systems and a metrics proposal for conceptual models of business processes.

New social networks are established around organizations; employees with ideas and desire to develop them can work smarter if they are in touch with one another, maybe when the employee has a question, he or she can post it to the group and find useful information from very knowledgeable people. Social networking is a more interactive and collaborative way of emphasizing peers' social interaction and collective intelligence, because the company can use it as an approach to build better user groups among co-workers, foreigners, suppliers and customers.

References

1. Roychowdhury, P., Dasgupta, D.: Take advantage of Web 2.0 for next-generation BPM 2.0, en DeveloperWorks (2008) Disponible en, http://www.ibm.com/developerworks/webservices/library/ws-web2bpm2/index.html
2. Gartner Inc. Key Issues for Business Process Management, Marzo (2009) Disponible en, http://www.gartner.com/DisplayDocument?id=910814&ref='g_fromdoc'

3. Díaz, F.: Gestión de procesos de negocio BPM (Business Process Management), TICs y crecimiento empresarial Qué es BPM y cómo se articula con el crecimiento empresarial En Revista Universidad & Empresa, Bogotá (Colombia) 7(15), 151–176 (2008) Disponible en, http://revistas.urosario.edu.co/index.php/empresa/article/viewArti-cle/1061
4. Club-BPM. Tercer Estudio sobre la Implantación de las Tecnologías BPM en España "El estado del BPM y las tendencias en España" (Julio de 2010) Disponible en, http://www.club-bpm.com/Documentos/2010TercerEstudioImplantacionBPM_ESP.doc
5. McCafee, A.: Enterprise 2.0: The Dawn of Emergent Collaboration. En MIT Sloan Management Review, Hardware: MIT 47(3) (Spring 2006)
6. Jerez, E.: Sostenibilidad 2.0: Capitulo 14.- Bienvenidos a la empresa 2.0 (2009) Disponible en, http://www.elviajedeodiseo.com/bienvenidosa_la_empresa_2.0.pdf
7. Roychowdhury, P., Dasgupta, D.: Take advantage of Web 2.0 for next-generation BPM 2.0, en DeveloperWorks (2008) Disponible en, http://www.ibm.com/devel-operworks/webservices/library/ws-web2bpm2/index.html
8. Costa, R., Sallan, J., Fernández, V.: Herramientas de Comunicación Web 2.0 en la Dirección de Proyectos. In: 3rd International Conference on Industrial Engineering and Industrial Management. XIII Congreso de Ingeniería de Organización Barcelona-Terrassa (2009) Disponible en, http://upcommons.upc.edu/e-prints/bitstream/2117/6370/1/Costa.pdf
9. Rives, X.: La Web 2.0: El valor de los Metadatos y de la Inteligencia Colectiva. Fundación Telefónica. Revista Telos (73) (Octubre-Diciembre 2007), http://sociedadinformacion.fundacion.telefonica.com/telos/articuloperspectiva.asp@idarticulo=2&rev=73.htm
10. Jimenez, C.: Indicadores de Alineamiento entre Procesos de Negocios y Sistemas Informáticos. Tesis de Magíster, Universidad de Concepción (2002)
11. Noy, P., Pérez, Y.: La actualidad de la Gestión de Procesos de Negocio: Business Process Management (BPM). En Revista Estudiantil Nacional de Ingeniería y Arquitectura 1(3) (2010) Disponible en, http://renia.cujae.edu.cu/index.php/revistacientifica
12. Smith, H., Fingar, P.: Business Process Management: The Third Wave. Meghan-Kiffer Press, Tampa (2003)
13. Garimella, K., Lees, M., Bruce, W.: Introducción a BPM para Dummies. Wiley Publishing, Inc., Estados Unidos (2008)
14. Margaix, A.: Conceptos de web 2.0 y biblioteca 2.0: origen, definiciones y retos para las bibliotecas actuales. El profesional de la información 16(2), 95–106 (2007), http://www.oei.es/noticias/spip.php?article593
15. Lerner, M.: At the forge: Google Maps. Linux Journal (146) (2006), http://www.linuxjournal.com/article/8932
16. Tepper, M.: The Rise of Social Software. netWorker 7(3), 19–23 (2003), http://portal.acm.org/citation.cfm?doid=940830.940831
17. Wusteman, J.: RSS: The Latest Feed. Library Hi Tech. 22(4), 404–413 (2004) Disponible en, http://www.ucd.ie/wusteman/lht/wusteman-rss.html
18. Valverde, G.F.: OOWS 2.0: Un Método de ingeniería Web Dirigido por Modelos para la Producción de Aplicaciones Web 2.0. Tesis Doctoral, Departamento de Sistemas Informáticos y Computación, Universidad Politécnica de Valencia, España (2010)

19. Duhl, J.: Rich Internet Applicattions - IDC Report (2003)
20. Noda, T., Helwig, S.: Rich Internet Applications - Technical Comparison and Case Studies of AJAX, Flash, and Java based RIA (2005), http://www.uwebc.org/opinionpapers/archives/docs/RIA.pdf
21. DcorporateCom. Web 2.0 y Empresa. Manual de Aplicación en Entornos Colaborativos: Web 2.0 y la comunicación corporativa (Capitulo 10) (2008) Disponible en, http://www.tecnologiapyme.com/ebusiness/manual-web-20-y-empresa
22. ANEI. Web 2.0 y Empresa. Manual de Aplicación en entornos Colaborativos. Asociación Nacional de Empresas de Internet (2008) Disponible en, http://www.a-nei.org/noticias/web-2.0.-manual-de-aplicacion-en-entornos-corporativos.html
23. Graells i Costa, J.: La wikiAdministración. Revista if...La Revista de Innovación (57) (Octubre 2007) Disponible en, http://www.infonomia.com/if/articulo.php?id=180&if=57

Applying Social Networks Analysis Methods to Discover Key Users in an Interest-Oriented Virtual Community

Bo-Jen Chen and I-Hsien Ting

Department of Information Management
National University of Kaohsiung, Taiwan
seraph0331@hotmail.com, iting@nuk.edu.tw

Abstract. In recent years, with the growth of Internet technology and virtual community, the users in virtual community not only play as the information receiver but also very important role to provide information. However, information overload has becoming a very serious problem and how to find information efficiently is also an important issue. In this research, we believe that users in a virtual community may affect each other, especially those with high influence. Therefore, we propose an architecture to discover the key users in a virtual community. By applying the architecture, it would be a very efficient and low cost approach. In the architecture, social networks analysis and visualization technique will be the main methods to discover the key users. In this paper, we also present an experiment to demonstrate the proposed method and the analysis results.

1 Introduction

With the rapid growth of Internet and communication techniques, the interaction between people in Internet has a breakthrough development. Furthermore, with the introduction of Web 2.0 concept, the so-called "online social networking website" has also becoming a very hot and popular application for users to interact and share[1].

The information that provided by the users who have high influence on online social networking websites may affect the behavior of other users. For example, more and more buyers usually search comments or ratings from other buyers before they make decision to purchase a product, especially the new products. Recently, it is very convenient for users to find information on the web. However, how to filter out necessary, useful and suitable information from the web is a rising problem under this background. Therefore, in this paper, we intend to find a method to assist users to acquire necessary information to support the decision making process under the situation of information explosion.

According to the study of this research, the information resources on the web including Blogs, community websites and rating websites, etc. Among these

L. Uden et al. (Eds.): 7th International Conference on KMO, AISC 172, pp. 333–344.
springerlink.com

websites, virtual community websites provide huge amount of information on the web. Armstrong and Hagel (1997) mentioned it is necessary for a company to establish a virtual community in order to increase the time that users spent on the website or to increase the loyalty of their customers [2]. Since there are large amount of information in virtual community, how to extract useful and valuable information from virtual community is considered as a hard problem. In this research, we assume that there are differences between users' influence and information quality. Some users' comment maybe fair and accurate than others. The information from these users is very helpful for users to make decision. Thus, if the key users can be discovered from the virtual community, it would be valuable for the virtual community or company.

In the research area of social networks analysis, there are many methods can be used to analyze the structure and relationship in the virtual community. Scott (2002) has organized many important methods of social networks analysis [3] and therefore these methods will be considered to discover key users in a virtual community. Thus, the title of the paper has been set as "Applying Social Networks Analysis Methods to Discover Key Users in an Interests-Oriented Virtual Community". The research objectives of the paper are 1) To understand the characteristics of key users in a virtual community 2) To analyze the relationship between the comments of key users and product sales 3) To observe the influence of users' comments.

The structure of this paper is as follows. Section 1 describes the background and motivation of this paper; some related literatures are discussed in section 2 including virtual community, social network researches based on movie and social networks analysis methodologies. Section 3 deals with our research methodology about how to discover key users from a virtual community. Section 4 presents and discusses an experiment. Section 5 concludes the paper and provides a brief outline of the future works.

2 Literature Review

In this section, related literatures will be reviewed including virtual community; social network researches based on movie and social networks analysis methodologies.

2.1 Virtual Community

In 1993, Rheingold made a definition about the term of virtual community. He mentioned that virtual community is social aggregations which aggregate enough Internet users. These users involve in discussion for a period of time and pay sufficient human feeling. [4]. Alder and Christopher consider virtual community as a group of Internet users who have similar interests. The users exchange comments, communicate and exchange information is a public space, such as WWW. The users in a virtual community share the culture, valuable and regulation with each other in the platform [5]. In this paper, we define virtual community based on the definition of the two literatures above. Virtual community is a social relationship network which consists of a group of Internet users with similar interests to exchange different comments and to share information in a public Internet space.

In 1997, Armstrong and Hagel classified virtual community into four different classes according to the behavior of virtual community users, which are *Transaction-Oriented virtual community*, *Interest-Oriented virtual community*, *Relationship-Oriented virtual community* and *Fantasy-Oriented virtual community* [2]. However, in recent years, more and more virtual communities are developed as multi-purposes virtual community, such as Yahoo! and Facebook.

In this paper, we will focus on interest-oriented virtual community and the main reason is that it developed based on common interests. There are no complicate social relationship and transaction in interest-oriented virtual community and therefore it is easier for us to understand the behavior of users.

2.2 Social Network Researches Based on Movie

In this paper, we select a very popular interest-movie as the target of this research and try to study how to discover key users in this kind of virtual community. Recently, there are many researches about movie based social networks and communities. Ahmed et al. (2007) have tried to use data from a very famous movie-based social networking website-IMDb to extract a visualized social network and also developed an analysis method to discover the relationship between a particular event (e.g. 911) and movie[6]. Some researchers also focus on how to use the content-based data in the website to improve the performance of a recommendation system [7].

Debnath et al. (2008) proposed a research to study the important factors that may affect the discussion in movie based virtual community [8]. In their research, the data in IMDb have been used as the main source to discover valuable factors. In this paper, we also select IMDb as the target and according the factors that proposed by Debnath et al. to understand the social relationships as well as to identify and discover the key users.

2.3 Social Networks Analysis Methodologies

The history of SNA (social networks analysis) related researches is long and Lewin has started researches on social networks analysis since 1925. The research about SNA intends to illustrate the relationships in a network by the mean of visualization. Scott proposed a so-called “Sociogram” in 2002 as showed in figure 1 and 2. Figure 1 is a simple directed sociogram and figure 2 is a simple undirected sociogram [9].

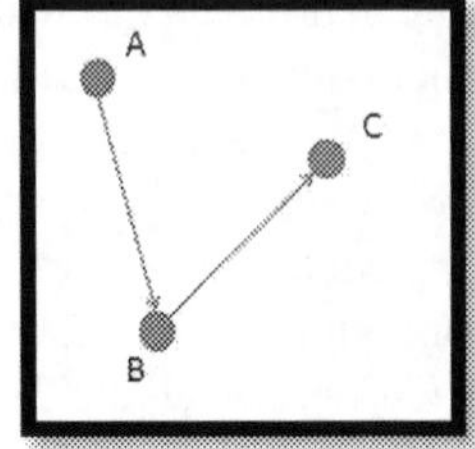

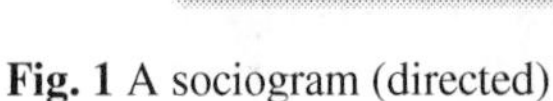

Fig. 1 A sociogram (directed)

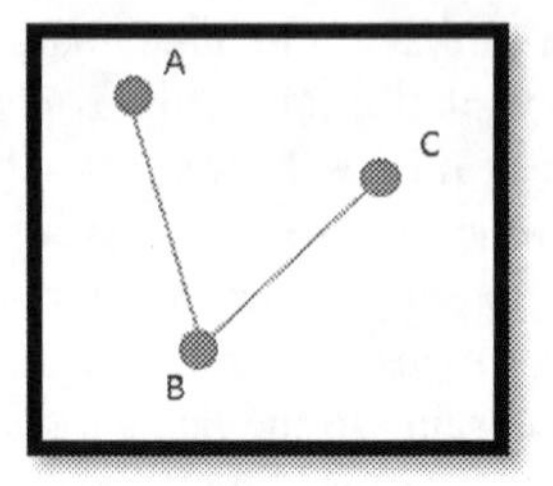

Fig. 2 A sociogram (undirected)

In a sociogram, 'Node' is the basic unit in a network. For example, a node may denote a person, role or user in a network, such as A, B and C in figure 1 and 2. Edge is a line to connect any two nodes in a network and it denotes the relationship between two nodes. For instance, figure 1 is a directed sociogram and the line that connect node A and B means that user A has the outward relationship with B, but B hasn't. User B has outward relationship with C but C hasn't. Figure 2 is an undirected sociogram and it only show the relationship between A, B and C. However, we can't find the direction of relationship in this figure.

In 1967, Milgram proposed a very famous 'Six Degrees Separation' theory and the concept of 'Small World' which means the distance between any two nodes (even they don't know each other) in a small world is less than six connections [10]. Watts also proved that many real social network cases do really match the theory of small world [11]. He also proposed some measurements to present the characteristics of a small world network, such as *Clustering Coefficient.* In a small world network, it always contains very high clustering coefficient which can be measured by equation 1. In this equation, E(x) denotes the number of edges in a network and E_{max} is the maximum number of edges in the network. The concept of clustering coefficient can be presented as figure 3.

$$\text{Clustering Coefficient} = E(x)/E_{max} \tag{1}$$

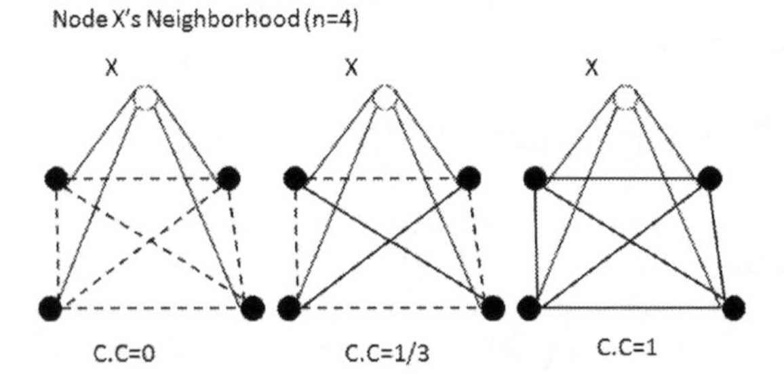

Fig. 3 The concept of clustering coefficient in a sociogram

Another character of small world network is low degree of separation which means the distance between any two nodes in the network is short. In another words, we can say the diameter in a network is short. The degree of separation is designed to measure the average distance in a network. Watts also mentioned that the graph of a small world network is between random graph and regular graph. The categories and characteristics of graphs are shown in table 1.

Table 1 The categories of sociograms (clustering coefficient-degree of separation)

	Low clustering coefficient	High clustering coefficient
Low degree of separation	Random Graphs	Small World Graphs
High degree of separation	N/A	Regular Graphs

Density is another famous measurement and has been used in most researches about SNA. It is usually used to measure the clustering situation of an organization or network. Figure 4 shows the concept about the density measurement.

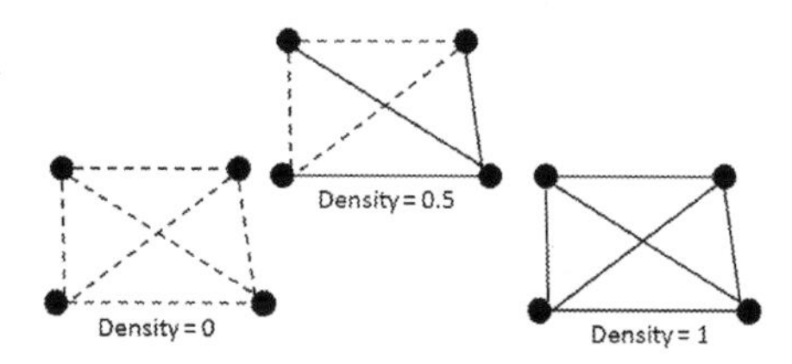

Fig. 4 The concept of the density measurement

For example, Lewis et al. proposed a methodology to combine the concept of density and regression model to study the connectivity of the famous social networking website- Facebook. In the study, it shows the average density in Facebook is 22.4 % [12]. Kumar et al. also observed the changing of density and nodes in different time (weekly) in Flickr and Yahoo! 360. They found that in a virtual community, the users usually show the enthusiasm when they just join the community. The users usually invite many of their friends to join the community together and therefore the density and degree of the community will be increased with the changing of the time. Therefore, the concept of time difference in a social network will be utilized in this paper to show the changing of density, the changing of degree and the network structure, which will also be used to discover the key users in virtual community.

3 Research Methodology

The research architecture of this paper is shown in figure 5 which can be separated into three different phases. The three phases are *Data Processing*, *User Processing* and *Verifying and Adjusting*. The detail of these three phases will be discussed in the following section. The main concept of this research architecture is to collect and process the data. Then, we will try to build the social model of the website and applying the methods of social networks analysis to discover the so-called key users. The experiment results will then be discussed to verify the relationship between the key users and the sales of the box office.

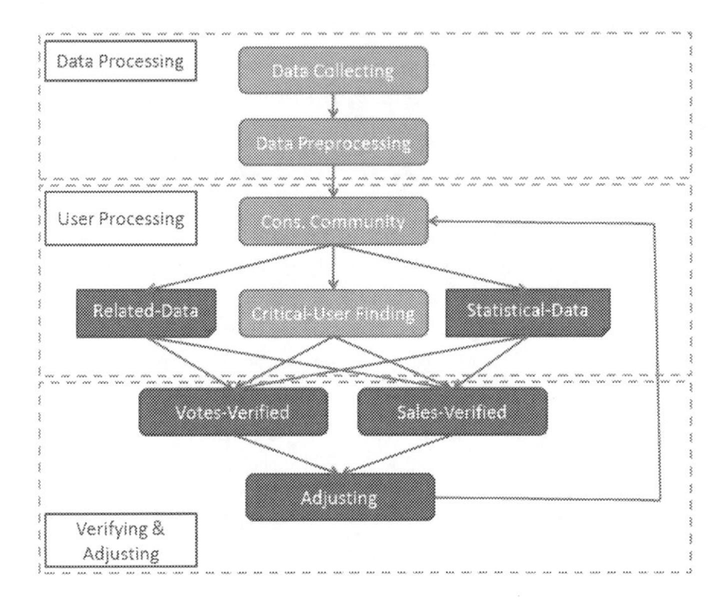

Fig. 5 The research architecture of this paper

3.1 Data Processing

Data Processing is the first phase of the research architecture, and there are two steps include data collection and data pre-processing.

3.1.1 Data Collection

In this research, we collect the data from the largest movie discussion website-IMDb, and the duration of data collection is 17 months (from 2010/1 to 2011/5). IMDb records the detail information of each movie and allows users to post their comments and judgments. Users can also vote for each movie. Figure 6 is a sample page for a movie in IMDb.

In this paper, we collected two different types of data in IMDb. The two types of data are movies' background information and users' data. Movies' background data include the category of the movie, actors, directors, box office, and duration. The users' data include rating, review and rating for other users' comments.

Fig. 6 A sample movie page in IMDb (data source: IMDb, http://www.imdb.com/)

3.1.2 Data Preprocessing

In the first step, the data were collected by using a crawler, which is a program designed by us to collect data automatically from IMDb. However, the collected data are full of noises. Therefore, we need to preprocess the data before it can be used for analysis. The collected raw data will be preprocessed to transform into analyze-able information. Basically, data will be pre-processed into two difference types of information including film node and user node. Table 2 shows film-node data and table 3 shows user-node data. In table 2 and 3, the sample data are from a movie "The Girl with Dragon Tattoo".

Table 2 Film-Node data after data preprocessing

Title	The Girl with Dragon Tattoo
Director(s)	David Fincher
Writer(s)	Steven Zaillian, Stieg Larsson
Star(s)	Daniel Craig, Rooney Mara, Christopher Plummer
Date	2011/12/20
Rating	8.3 (17250)

Table 3 User-Node data after data preprocessing

User ID	Movie-(year)	Date	Rating	Useful
Carflo_ur2509775	Alice in Wonderland -(2010)	2010/8/18	7/10	1/2
	Million Dollar Baby (2004)	2010/10/2	10/10	1/1
	Aliens (1986)	2004/12/8	10/10	224/264
	The Shawshank Redemption (1994)	2003/11/26	10/10	1290/1552

3.2 *User Processing*

3.2.1 Community Modeling

In order to discover the key users in a virtual community, we need to model the community first. In the step of community modeling, the categories of movies will be treated as nodes in the community. If the two movies both reviewed by a user, then we will create an edge between the two nodes. Figure 7 shows the concept of community modeling. In figure 7, M_1 and M_2 denote two movies and the link between M_1 and M_2 means user's co-reviews.

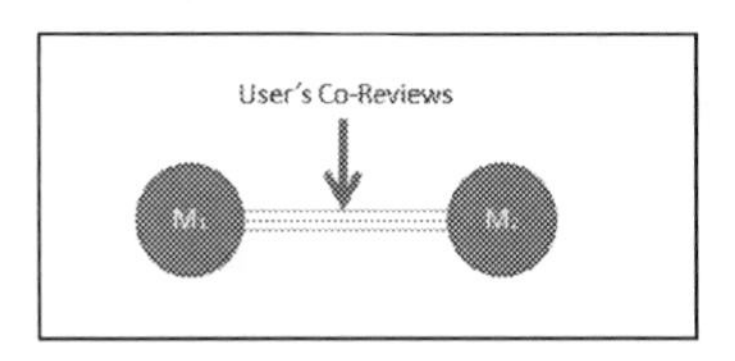

Fig. 7 The concept of community modeling

The processing of community modeling is in order to understand which movies are highly focused in a movie category. We will then use the concept of network density in SNA to categorize those highly focused movies. A measurement T_m will be used as a threshold to decide which is a highly focused movie. The collection of the highly focused movies has been denoted as Ch.

3.2.2 Key Users Discovery

After community modeling, we can now try to discover key users from the highly focused movies (Ch). In order to discover the key users, we have to identify the meaning of the term 'key users'. Normally, we consider the characteristics of key users should match the following points: 1) the users express very enthusiastic to post reviews. 2) the posts of key users are usually rated as useful reviews. 3) the ratings of key users are very fair. The process and methodology about how to discover the key users has been shown in figure 8.

In figure 8, the key users will then be discovered by test their attendance and useful reviews. If the attendance is higher than 50% then the next stage of user's review test will start to the percentage of average useful reviews. If the percentage of average useful review is higher than 50 % then this user will be recognize as a key user in this community. The percentage in this paper has been set to 50%, however, it can be different to difference researches.

In most cases, the box office of a movie may be affected by the actors and directors at the beginning (the first month). However, the comments and reviews from other audiences are the critical factor for a success movie, especially the reviews and comments from key users. From this study, we think the life cycle of a movie's box office should present as the movie's comment accumulation life

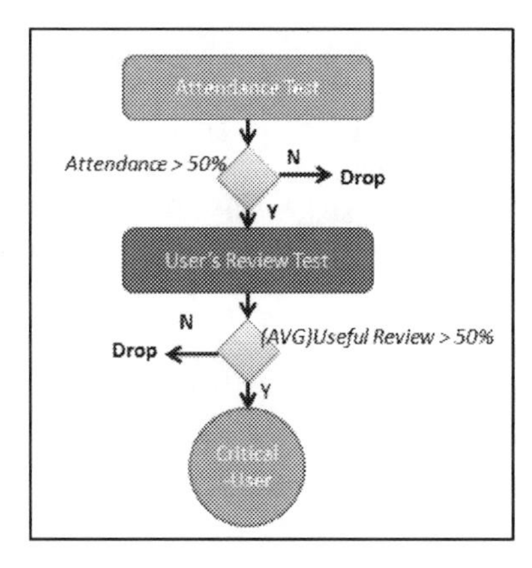

Fig. 8 The methodology to discover key users

cycle in figure 9. In figure 9, there are three different phases for a movie's life cycle, which are accumulated stage, growth stage and decline stage. The x-axis in figure 9 is the duration of the movie and the y-axis is the number of reviews.
The accumulated stage is considered as the stage that key users may affect the box office of a movie. Therefore, we can easily from the comments of key users to predict the box office. If a movie can't receive enough positive comments and reviews, then it usually can't achieve idea box office. In another word, we can observe the comments and reviews from the key users to roughly predict the box office of a movie. The movie theater can also use this information to make decision about the in theaters duration of a movie.

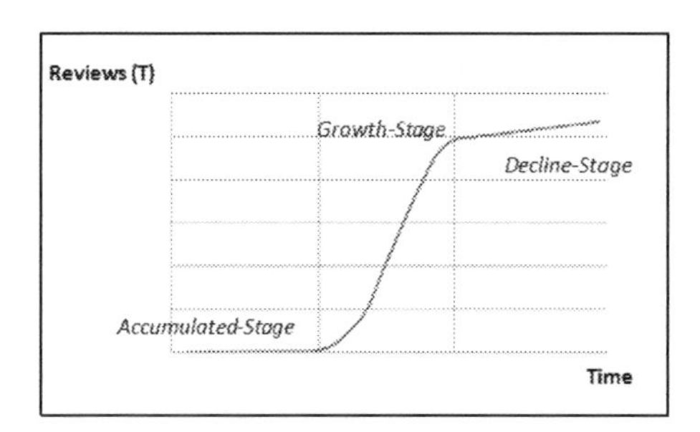

Fig. 9 The graph of comments accumulation for a movie

4 Experiment Results

In order to test the methodology that proposed in this paper, we designed an experiment. Due to some tasks in the experiment need automatic process, we therefore also developed some programs to perform the tasks. We use the client-server architecture for the experiment. In server side, the main task is to collect data from

IMDb website automatically and to quantify the data. In client side, the main task is to organize and visualize the data by using UciNET.

From IMDb, we use Perl and Java language to design programs to collect data automatically. The duration of data collection is from January 2010 to May 2011, which is 17 months in total. The statistical information of the collected data is presented in table 4. Table 5 shows the top 10 movie categories and number of movies for each category.

Table 4 The statistical information of the collected data

Number of Movies	402
Number of Users	17388
Number of Reviews	243074
Number of Genres	22

Table 5 The top 10 movie categories

Category	Number of movies
Drama	213
Comedy	150
Action	78
Romance	75
Thriller	69
Crime	57
Adventure	54
Fantasy	34
Mystery	34
Documentary	33

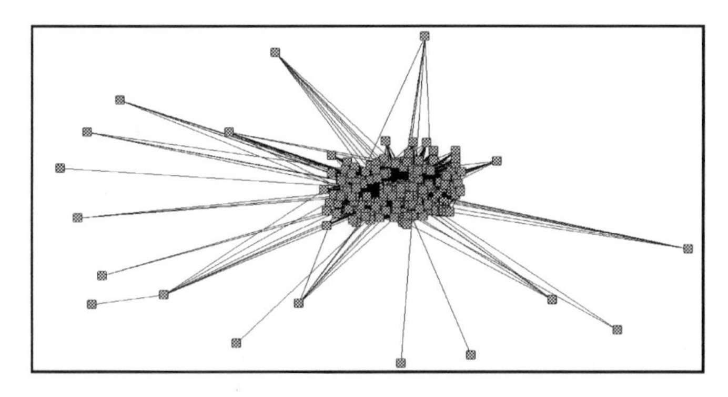

Fig. 10 The co-review network from the collected data

In order to discover the key users in a virtual community, we have to discover the highly focused movies. The method that discussed in section 3.2.1 will be used to study the co-review behavior. Figure 10 shows the experiment results.

In figure 10, there are 339 nodes and 38000 edges, which is an undirected sociogram. From this figure, it shows that the focused movies really exist in IMDb. Figure 11 is the network of the discovered key users. However, there are degree differences between these key users. The users located in center have higher degree and which means the key users are more important. Thus, the comments and reviews from these users are very important to understand the rating for a particular movie and to predict the box office accordingly. In this figure, there are four different degrees and labeled with different colors.

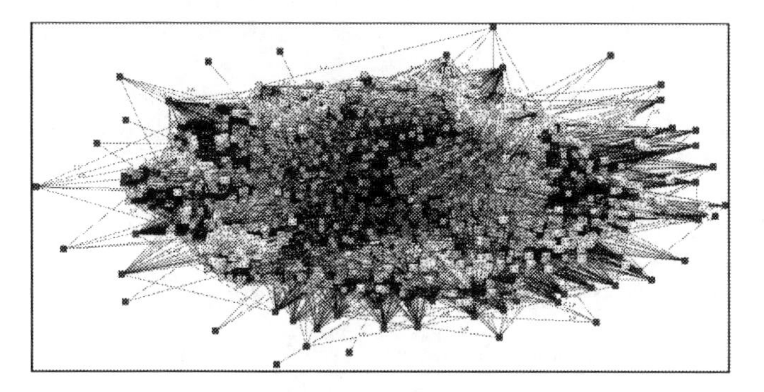

Fig. 11 The discovered key users and the network

5 Conclusion and Future Research

In this paper, we proposed the architecture about how to discover key users from a virtual community and used a movie database as case study (IMDb). The hypothesis of this paper is that highly focused products or movies in a virtual community should also reflect in their sales. Therefore, we firstly discussed about how to find the highly focused movies from IMDb and then to identify the term of key users and its meaning in social networks and virtual community. The methodology is proposed in this paper as well as the experiment has been performed. In the experiment, it shows that highly focused movies really exist in IMDb and can be discovered from the data in the website. Furthermore, it also shows the key users for these highly focused movies can be discovered accordingly and we also categorized these key users into four different degrees.

In the future, we suggest related researchers could focus on how to verify the results and to prove the key users. Furthermore, future researches can also use the proposed approach to discover key users from other different virtual communities and to compare the differences. Finally, it is also an interesting topic about how to utilize the discovered key users to study how to apply the results in real environment.

References

1. Adamic, L.A., Adar, E.: Friends and neighbors on the web. Social Networks 25, 211–230 (2003)
2. Armstrong, A.G., Hagel III, J.: Net Gain: Expanding Markets Through Virtual Communities. Harvard Business School Press, MA (1997)
3. Scott, J.: Social Network Analysis: A Hand Book. SAGE Publication (2002)
4. Rheingold, H.: The Virtual Community: Homesteading on the Electronic Frontier. Addison-Wesley, MA (1993)
5. Adler, R.P., Christopher, A.J.: Internet Community Primer Overview and Business Opportunity, http://www.digiplaces.com
6. Ahmed, A., Batagelj, V., Fu, X., Hong, S.-H., Merrick, D., Mrvar, A.: Visualisation and Analysis of the Internet Movie Database. In: Asia-Pacific Symposium on Visualisation, pp. 17–24 (2007)
7. Jakob, N., Weber, S.H., Muller, M.C., Gurevych, I.: Beyond the Stars: Exploiting Free-Text User Reviews to Improve the Accuracy of Movie Recommendations. In: TSA 2009, Hong Kong, China (2009)
8. Debnath, S., Ganguly, N., Mitra, N.: Feature Weighting in Content Based Recommendation System Using Social Network Analysis. In: Proceedings of the 17th International Conference on World Wide Web (2008)
9. Scott, J.: Social Network Analysis: Critical Concepts in Sociology. Routledge, New York (2002)
10. Milgram, S.: The Small-World Problem. Psychology Today, 61–67 (1967)
11. Watts, D.J.: Collective Dynamics of "Small World" Networks. Nature 393, 440–442 (1998)
12. Lewis, K., Kaufman, J., Gonzalez, M., Wimmer, A., Christakis, N.: Tastes, Ties, and Time: A New Social Network Dataset Using Facebook.com. Social Networks (2008)

The Framework of Web 3.0-Based Enterprise Knowledge Management System

Hongbo Lai, Yushun Fan, Le Xin, and Hui Liang

Department of Automation, Tsinghua University, Beijing, P.R. China
{laihb05,xin-109}@mails.tsinghua.edu.cn,
fanyus@tsinghua.edu.cn, danaha@163.com

Abstract. Internet has been running into the era of Web 3.0 equipped with the development of semantic technology. Therefore, how to manage the enterprise knowledge that suits Web 3.0 becomes an urgent problem. In this paper, we first summarize the evolution of Internet and describe the feature of Web 3.0. After analyzing the practice of linked data applied in Web 3.0, we propose a new framework of enterprise knowledge management system combining the theory of linked data and the close loop of knowledge management. Moreover, we analyze each layer's function in the framework and discuss the implementation of the knowledge publish layer in detail. As a remarkable result, the proposed enterprise knowledge management system can extend the knowledge sharing to a new level, and reduce the cost of enterprise collaboration caused by heterogeneous data as well.

1 Introduction

With the rapid development of Internet, knowledge has been growing exponentially these years. According to the American Society of Training and Documentation (ASTD), the amount of knowledge was doubling approximately every 18 months by 2004 with the advent of Web [1]. Since knowledge is a key factor for the enterprise to achieve success in the highly competitive commercial environment, how to design the Knowledge Management (KM) model adapted to the internet evolution trend will be an important topic for enterprises.

Knowledge can be categorized as "Know-what", "Know-why", "Know-how" and "Know-who" by OECD [2]. The former two kinds are codified knowledge, or explicit knowledge in other words, and the latter two kinds are tacit knowledge implying more difficult to be codified and measured. The explicit knowledge and the tacit knowledge can be transformed with each other. Takeuchi and Nonaka [3] believe that knowledge creation is a spiral process of interactions between explicit and tacit knowledge, and the interactions between the explicit and tacit knowledge lead to the creation of new knowledge. Considering most of the knowledge is tacit, the research on improving the knowledge management efficiency and making tacit knowledge explicit are still on-going. Ribino [4] proposed a KM system based on ontology, allowing of specializing the reasoning capabilities and

L. Uden et al. (Eds.): 7th International Conference on KMO, AISC 172, pp. 345–351.
springerlink.com

of providing ad hoc behaviors. Lavoué [5] studied the model of the interconnection of communities of practice, and implemented this model to develop a platform relying on a knowledge management tool and a social networking service. Li and Zhang [6] showed the behavior modes of multiple agents in KM system based on complexity theory, including the definition, structure, composition of different agents. De Boer and Vliet [7] shared the experience with using semantic wikis for architectural knowledge management in e-government and distributed software development.

In this paper, we introduce a framework of Web 3.0-based Enterprise KM system and the implementation method. This paper is structured in three main sections: the evolution of Web and the feature of Web 3.0 are described in 2, the linked data applied in Web 3.0 is explained in 3, the proposed framework of KM system is shown in 4 and finally the conclusion is covered in 5.

2 Web 3.0 Overview

2.1 Web Evolution

The historical origin of the Web can be traced back to 1993 with the release of the WWW to the public, and the evolution of the Web is segmented into three phases until now (see Table 1). The main feature of Web is one-way flow of information at the beginning, such as Yahoo. This kind of Web is called Web 1.0. Then, Web is evolved along two dimensions, comprising socialization dimension and semantic dimension. Web 2.0, emphasizing the user interaction and collaboration, is first proposed in 1999 and rises in popularity after 2004 with the emerging of Facebook.

Table 1 The comparison between Web 1.0, Web 2.0 and Web 3.0

	Web 1.0	Web 2.0	Web 3.0
Released time	1993	1999	2006
Main feature	Information publish	User interaction and collaboration	Knowledge connection, Semantic Web
Information flow direction	One-way	Two-way	Multi-way
Technology	Html	Ajax, Flash, Flex, XML	RDF [8], OWL [9]' SPARQL [10]
Application	Yahoo	Facebook	iGoogle

On the other hand, in order to make the internet data structured, semantic technology has been researched in the recent years, and Tim Berners-Lee [11] gives the definition of Web 3.0, indicating that users will get an overlay of scalable vector graphics - *everything rippling and folding and looking misty* - on Web 2.0 and access to a semantic Web integrated across a huge space of data in Web 3.0.

2.2 *The Features of Web 3.0*

Web 3.0 is known as "Web of Data", dedicating to make all the data, such as traditional webpage and document, structured and easy to be utilized by AI, to integrate all the data into one invisible database. The main features of Web 3.0 are listed as follows:

Intelligent: In web 3.0, all the information is categorized and stored in such a way that a computer can understand it as well as a human. It will tell computer the mean of data, so that the artificial technology on computer can utilize this information, to push the interested information or knowledge to the user automatically and precisely.

Integration: Web 3.0 can integrate the User Generated Content by Mash-up, to strengthen the characteristic of the content and simplify the information search. All the applications adapted to the uniform standard can be integrated on the Web 3.0 platform. Moreover, the application of aggregation technology, such as Tag and RSS, will play a greater role in the content aggregation platform construction in Web 3.0.

Personalization: Web 3.0 will process and analyze user's personal preference, to display the interested information in the home page. Therefore, every user can control and integrate the information on their personal portal freely, meaning a high degree of autonomy for the information, to satisfy the individual demand.

3 Linked Data Applied in Web 3.0

To realize the Web 3.0, the first and foremost step is making the data structured on Internet. Linked data is a recommended best practice for exposing, sharing, and connecting pieces of data, information, and knowledge on the Semantic Web using URIs and RDF. It can publish various open data sets as RDF on the Web, and provide a uniform platform for the application to utilize the data on the Web, indicating one possible implementation way of Web 3.0.

The linked data principles [12] are listed as follows:

- Use URIs as names for things.
- Use HTTP URIs, so that people can look up those names.
- When someone looks up in a URI, provide useful information, using the standards (RDF, SPARQL).
- Include links to other URIs, so that they can discover more things.

When publishing linked data on the web, all the data follows the format of RDF. In RDF data model [13], the description of resources is represented as a number of triples (see Fig. 1).

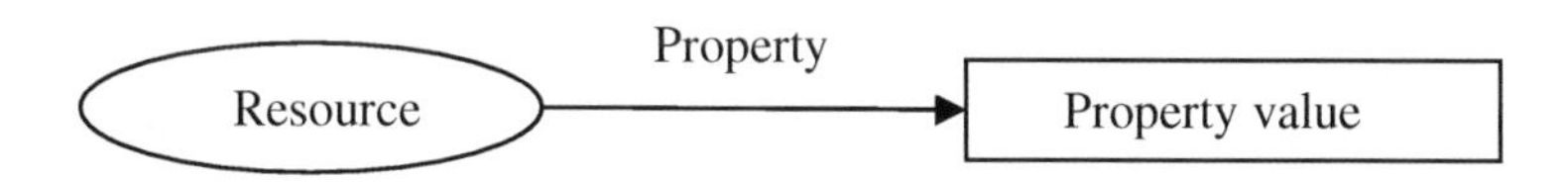

Fig. 1 The basic RDF model

The subject of a triple is the URI identifying the described resource, the object is simple literal value or the URI of another resource, and the predicate indicates the relation between the subject and the object in the way of URI. In this way, all the published resources can be linked with each other, to generate the web of data around the world.

In 2007, Chris Bizer and Richard Cyganiak submitted the application of Linking Open Data (LOD) to W3C SWEO, representing the start of linked data development. 295 data sets have been published and interlinked by the project as of Sep. 1st, 2011, consisting of over 31 billion RDF triples which are interlinked by around 504 million RDF links.

As for the enterprises, publishing the knowledge as linked data will not only realize the knowledge sharing to a great extent inside the company, but also utilize the outside knowledge that has been published conveniently. Moreover, the knowledge management based on linked data will reduce the cost of enterprise collaboration caused by heterogeneous data.

4 Framework of Web 3.0-Based KM System

Commonly, the enterprise KM process comprises the following key activities: knowledge storage, knowledge publish, knowledge sharing, knowledge valuation, knowledge acquisition and knowledge innovation, to form the closed loop in KM. The framework of Web 3.0-based KM system is shown as Fig. 2, combining the theory of linked data and the KM closed loop.

4.1 The Function of Every Layer

The different layers' functions of the framework are listed as follows:

Knowledge innovation layer: in the knowledge innovation layer, users are divided into two roles, comprising knowledge engineer and ordinary system user. Both of them are able to create new knowledge and save them in the knowledge database during the daily work. Moreover, knowledge engineer is responsible for maintaining and categorizing the knowledge in the database.

Knowledge acquisition layer: the main KM function modules are arranged in this layer, such as expert map, knowledge reference. Combing the modules freely, users can acquire the demanded knowledge if they have the access. All the modules in this layer get the data from the knowledge valuation layer in the way of SPARQL or RDF API.

Knowledge valuation layer: The Web knowledge data needs to be treated with suspicion due to the open nature of Web. In this layer, the origin knowledge data is accessed by HTTP, and then pretreated through data cleaning, vocabulary mapping and quality evaluation, to provide the knowledge acquisition layer with reliable knowledge data.

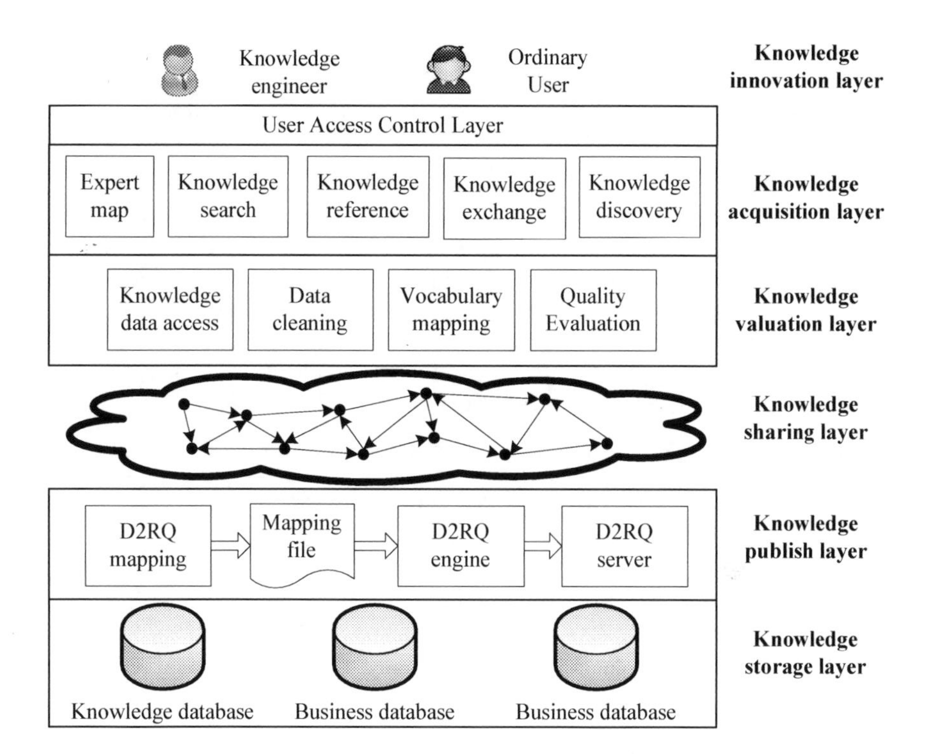

Fig. 2 The framework of enterprise KM system

Knowledge sharing layer: All the knowledge data of enterprise is published on the Web as RDF, to construct the Web of knowledge inside the enterprise. It can further link the knowledge data with other data published by other institutions, so that users can get the related data outside the enterprise, indicating the knowledge sharing to a great extent.

Knowledge publish layer: The knowledge stored in the relation database is published to the web of knowledge in this layer. This paper publishes the linked data of knowledge and business data related with the knowledge in the enterprise by D2R [14] (see Section 4.2).

Knowledge storage layer: Commonly, the enterprise knowledge is stored in the knowledge database. For every piece of knowledge, it not only connects with other pieces of knowledge, but also relates with the business data which is stored in the business database.

4.2 The Implementation Method of Knowledge Publish

According to the theory of Integrated Enterprise Modeling System Architecture [15], enterprise model is described in multiple views, including process view, function view, information view, organization view and resource view. Since all the knowledge in the enterprise is embedded in the normal activities, every piece of knowledge in the knowledge database will be related with the cells in one or more views. Therefore, the related business data has to be published unavoidably when publishing the enterprise knowledge.

In this paper, we publish the linked data of knowledge by D2R:

Step 1: The mapping file is generated through the executing script provided by D2R. All the tables storing the information of the views mentioned above will be mapped as the resource classes, and the tables storing the relations between the views will be mapped as the relations between the resource classes.

Step 2: D2R server calls the D2R engine automatically to publish the data in relation database based on the mapping file. The applications can access the published data in many ways, such as SPARQL, to acquire the useful information.

In this way, the knowledge and the related business information are published to form the web of knowledge (see Fig. 3). In addition, the ordinary user or the knowledge engineer can construct the relations between the enterprise knowledge and other data published by other institutions.

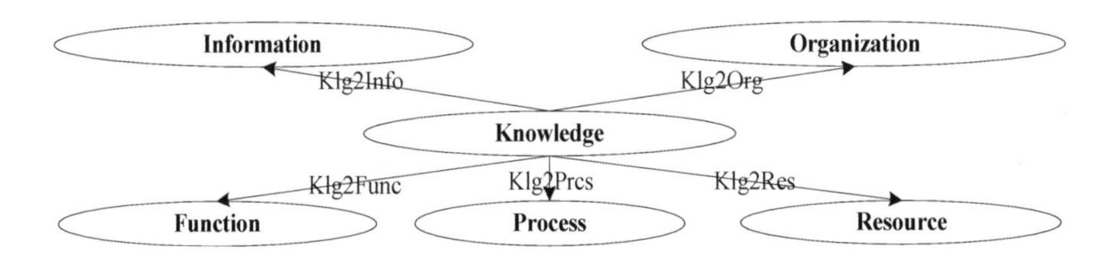

Fig. 3 The knowledge data concept model

5 Conclusion

Semantic and Intelligence is the future development trend of Internet, to improve the utilization efficiency of the information emerging every day. So, the theory of linked data, which is a possible way to realize the Web 3.0 application, has been proposed to make the data interlinked and easy to be understood by the computer. Applying the concept of Web 3.0 and the technology of linked data in enterprise KM can solve the problem of knowledge sharing, especially for the tacit knowledge, in order to acquire the interested knowledge as soon as possible.

In this paper, we design the framework of the Web 3.0-based enterprise KM system and the method to publish the knowledge by D2R. Our work presented in this paper is only a basic research for the enterprise KM system and the detail implementation method to realize all the functions of this system needs to be discussed in the future work.

Acknowledgments. This work is supported by the National Natural Science Foundation of China (No.61174169 and No. 61033005), and the National Key Technology R&D Program (No. 2010BAH56B01).

References

1. Gonzalez, C.: The Role of Blended Learning in the World of Technol-ogy. University of North Texas (2004), http://www.unt.edu/benchmarks/archives/2004/september04/eis.htm (accessed December 12, 2011)
2. OECD, The Knowledge-based Economy. OECD, Paris (1996)
3. Takeuchi, H., Nonaka, I.: Hitotsubashi on Knowledge Management. John Wiley & Sons, Tokyo (2004)
4. Ribino, P., Oliveri, A., Re, G.L., Gaglio, S.: A Knowledge Manage-ment System based on Ontologies. In: 2009 International Conference on New Trends in Information and Service Science, pp. 1025–1033 (2009)
5. Lavoué, É.: A Knowledge Management System and Social Networking Service to Connect Communities of Practice. In: Fred, A., Dietz, J.L.G., Liu, K., Filipe, J. (eds.) IC3K 2009. CCIS, vol. 128, pp. 310–322. Springer, Heidelberg (2011)
6. Li, Y.-X., Zhang, L.: The Behavior Modes of Multiple Agents in Knowledge Management System based on Complexity Theory. In: 2011 International Conference on Intelligence Science and Informa-tion Engineering, pp. 53–56 (2011)
7. de Boer, R.C., van Vliet, H.: Experiences with SemanticWikis for Ar-chitectural Knowledge Management. In: 9th Working IEEE/IFIP Conference on Software Architecture, pp. 32–41 (2011)
8. Manola, F., Miller, E.: RDF Primer. W3C (2004), http://www.w3.org/TR/2004/REC-rdf-primer-20040210/ (accessed December 25, 2011)
9. McGuinness, D.L., van Harmelen, F.: OWL Web Ontology Language Overview. W3C (2004), http://www.w3.org/TR/owl-features/ (accessed December 25, 2011)
10. Prud'hommeaux, E., Seaborne, A.: SPARQL Query Language for RDF. W3C (2008), http://www.w3.org/TR/rdf-sparql-query/ (accessed December 26, 2011)
11. Shannon, V.: A 'more revolutionary' Web. The New York Times (2006), http://www.nytimes.com/2006/05/23/technology/23iht-web.html (accessed December 20, 2011)
12. Heath, T., Bizer, C.: Linked Data: Evolving the Web into a Global Data Space. Morgan & Claypool (2011), http://linkeddatabook.com/editions/1.0/ (accessed October 11, 2011)
13. Klyne, G., Carroll, J.J.: Resource Description Framework (RDF): Concepts and Abstract Syntax. W3C (2004), http://www.w3.org/TR/rdf-concepts/ (accessed January 11, 2012)
14. Bizer, C., Cyganiak, R.: D2R Server Publishing Relational Databases on the Semantic Web. Freie University, Berlin (2010), http://www4.wiwiss.fu-berlin.de/bizer/d2r-server/ (accessed January 15, 2012)
15. Fan, Y., Wu, C.: Research of System Architecture and Implementation Method for Integrated Enterprise Modeling. Control and Decision 15(4), 401–405 (2000)

Customer Knowledge Management in the Age of Social Networks

Dario Liberona[1], Manuel Ruiz[2], and Darcy Fuenzalida[3]

[1] Universidad de Santiago de Chile (Usach.cl)
[2] Universidad de Lleida, Spain (Udl.es)
[3] Universidad Santa María (Usm.cl)

Abstract. Knowledge is increasingly being recognized as a vital organizational resource that provides competitive advantage (Davenport, T. & Prusak, L. 1998, Edvinsson, L. & Malone, M. 1997., Stewart, T. 1997, Argote, L. and P. Ingram 2000, Choo 1998). Managing knowledge assets can be a challenge for organization, specially their external Knowledge like their customer base information. The use of information technology (IT) in knowledge management (KM), is an essential consideration for any company wishing to take advantage of modern technologies to manage their knowledge assets. IT tools have been recently created to listen and learn from customers in the public and private social networks. This paper presents research about KM IT tools and social monitoring tools, a survey and interviews have been conducted to identify the technologies that are currently used to manage customer knowledge in Chilean companies and how valuable is this Knowledge for companies. A questionnaire was distributed among Chilean organizations in order to obtain general data about the level of use and the benefits of using such tools. This approach was supplemented by interviews to reveal richer data about the nature of IT for Customer KM (CKM), in organizations. The research revealed that conventional technologies, such as the telephone, are used more frequently, than more radical IT tools, such as social CRM and social listening tools, but great importance and attention to this new tools is taking place. In Chilean organizations, the potential benefits of IT for Customer KM, are not fully exploited and many have expressed a need for greater implementation of social analysis tools and Customer Knowledge Management practices.

Keywords: Knowledge Management, Customer Knowledge Management, Social CRM, Social Metrics, Enterprise Content Management (ECM).

1 Introduction

There is currently a compelling debate about the changing nature of business environments and the sources of competitiveness in advanced economies. It is asserted that knowledge is fast overtaking capital and labor as the key economic

L. Uden et al. (Eds.): 7th International Conference on KMO, AISC 172, pp. 353–364.
springerlink.com

resource in advanced economies (Edvinsson, 2000). The intangible assets in an organization are widely celebrated as vital elements in improving competitiveness (Egbu, 2000; Edvinsson, 2000). This has compelled academics and practitioners to discuss the way in which knowledge assets are managed.

Although relatively new on the scene, social media is spreading fast after the expansion of Facebook in 2006 and other social media platforms and tools. By some estimates, social networking now accounts for 22 percent of all time spent online in the US (NielsenWire 2012). Nearly one in 10 Internet visits ends up at a social network; Fecebook has more than 800 million users and they expend on average 15 hours and 33 minutes a month on the site. (Search Engine Journal)

It didn´t take long for the KM practitioners to realize that there is a lot of information and knowledge outside the firm, in the customers base of a company, this is the so called fourth stage of KM were attention to external knowledge is emphasized (Srikantaiah K. ,2008).

Since many conversations are taking place on these social networks, there is a lot of information being shared and created on this networks, this Social knowledge could build or erode your brand, affect companies reputation, create or destroy leads, and affect sales and revenue of firms. Outspoken and respected individuals can influence masses of friends and followers to change their opinions about you. They can trash you or defend you.

Common questions that arises for practitioners are, how do you Measure and Manage the Impact of Social Media? how do you leverage this collective knowledge ?, and for researches the question is: is there a new trend in the field name Customer Knowledge Management ?

Knowing that KM means much more than IT (Malhotra, Y., 2004, Avram, G 2006), but there is evidence that the IT relevance in KM is increasing (Egbu 2001, Srikantaiah K. ,2008), and IT has being taking a more relevant role in the use and implementation of KM programs. Nowadays several millions of messages are being share in Twitter, blogs or other social networks and the amount of information to be process is massive, this requiring new tools to process this information, some that even require supercomputing technology. Customer Knowledge Management (CKM) has become crucial and fundamental to the collection, gathering and analysis of ever increasing amounts of customer information. Even that there are various very important KM enablers (Figure 1), the Technology one, plays a crucial role on the gathering, sorting and analysis of this customer data information.

Empowered customers are disrupting every industry; and competitive barriers like manufacturing strength, distribution power, and information mastery don´t seem to be enough to cope with this new customers. In this so call age of the customer (Martin R, 2010, Bernoff, 2011), to have customer knowledge and engagement with them, could become a sustainable competitive advantage.

This new tools provide a lot of real-time customer intelligence information to improve the customer experience and customer service, to provide sales channels that deliver customer feedback; and there is a lot of useful content and information that the marketing departments could use to learn more from customer, being the most interested in creating Customer Knowledge Management.

The purpose of this paper is to explore the role of IT tools for Customer KM. Chilean organizations have been slow to acknowledge the benefits of social customer analysis tools in managing Customer knowledge. This paper begins by exploring the potential of IT social analysis tools in managing knowledge, drawing from multi-disciplinary literature and previous research. The empirical evidence will be analyzed, identifying the existing technologies used in some Chilean companies and how effective these are in managing Customer knowledge. Finally, some discussion about the future usefulness of social tools to manage customer knowledge in the Chilean industry will be considered.

2 Importance of KM It Tools

KM should be understood how the processes by which knowledge is created, acquired, communicated, shared, applied and effectively utilized and managed, in order to meet existing and emerging needs related to Customers, to identify and exploit existing and acquired knowledge assets (Egbu 2001).

There are various KM models, that attempt to describe the 'enablers' of knowledge management. Five 'enablers' – Technology, Leadership, Culture, Measurement, and Process – are mentioned enablers in most of the models. (Figure 1.)

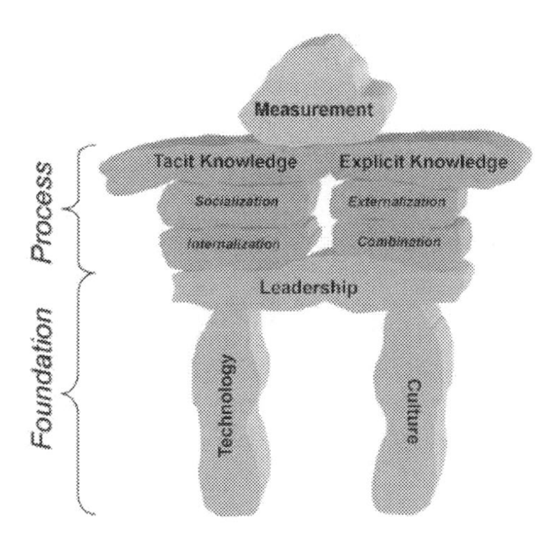

Fig. 1 Inukshuk: Knowledge Management Model (John Girard)

Technology is one of the most common KM models enablers, if customers want things faster, better, cheaper, and with a higher degree of service; technology makes it possible for them to get what they want. We have concentrated our efforts on reviewing the Technology enabler since is the one that has evolve faster and more dramatically than the others, and has created the possibility of social networks and social analysis.

Consumers have been embracing the power of social technology — US consumers make 500 billion impressions on one another about products and services every year.1 And 5.3 billion people — 76% of the world's population — are connected to each other and information through mobile devices.

The result of all of this people power is disruption. A tweeter in Pakistan scooped the president of the US in announcing the attack on Osama bin Laden's compound. Corporations can't easily defend themselves — technology-driven shifts in consumer preferences killed Circuit City and Tower Records and bankrupted Borders, Kodak, Palm and Blockbuster. All succumbed because their strengths — their valued business models, distribution, and supplier relationships — weren't sufficient to keep them competitive in the face of rapidly shifting customer expectations (Forrester 2011).

2.1 KM Tools Types

KM related IT tools have been developed for a while, and are deeply related to WEB 2.0 technologies and recently Social Networks technologies.

Before Social Networks, KM it Tools were related to the basic process of KM models, this is Acquisition of Information, Sharing and using it, this fundamental KM tools were, groupware applications (intensive use of smart mailing systems), document management (Content Management Systems), intranets, connected to workflow, search engines and information retrieval tools, this mostly develop during 2000 to 2005.

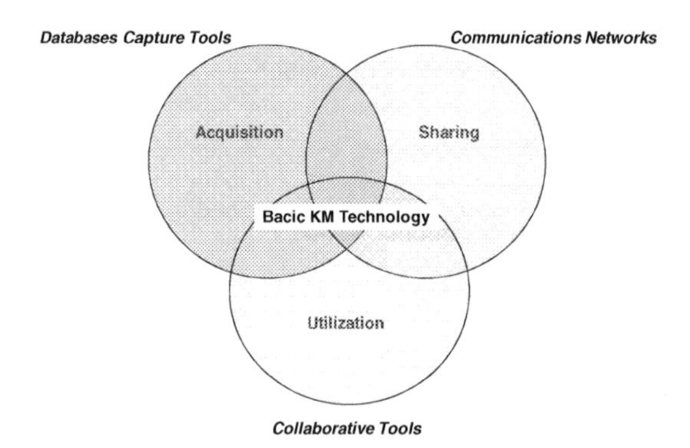

Fig. 2 Knowledge use process, - The three fundamental processes of knowledge management according to Amrit

KM technology has evolve over the past years integrating the information Management (Acquisition), the Information and communication (Sharing), the organizational learning (Utilization) and adding another dimension, the Human Resource Management (Culture and Talent), this new dimension brings important aspect such as Change Management and Performance Management to the KM program at organizations.

During 2006 until 2011, collaborations suites and Content Management systems evolved into ECM platforms allowing to have more efficient KM tools.

Enterprise Content Management or ECM is a term to describe vendors that amassed the ability to offer not just one type of content management but (usually through the acquisition of other CMS vendors) also able to provide multiple types of solution to manage different types of content.

Some of the ECM platforms available are: Share Point, Jive Engage Platform, Lotus Notes collaboration suite, Tellingent, SalesForce.Com, OpenText, Attlasian, Saba, Success Factors, Drupal, Social Text

This platforms are and umbrella of corporate functions like document management, web content management, search, collaboration, records management, digital asset management (DAM), work-flow management, capture and scanning. ECM is primarily aimed at managing the life-cycle of information from initial publication or creation all the way through archival and eventually disposal.

The benefits to an organization include improved efficiency, better control, and reduced costs. For example, many banks have converted to storing copies of old checks within ECM systems versus the older method of keeping physical checks in massive paper warehouses. Under the old system a customer request for a copy of a check might take weeks, as the bank employees had to contact the warehouse to have someone locate the right box, file and check, pull the check, make a copy and then mail it to the bank who would eventually mail it to the customer. With an ECM system in place, the bank employee simply searches the system for the customer's account number and the number of the requested check. When the image of the check appears on screen, they are able to immediately mail it to the customer—usually while the customer is still on the phone.

3 The Fourth Stage of KM

Koening and Srikantaiha (2004) recognized three stages of KM in terms of evolution. The first stage of KM is driven by IT, in particular large organization couldn´t cope with all the information they were gathering, the wheel was being reinvented inside different areas of the organization. They concentrated on Internet-Intranet and Best practices or lessons learned. The second stage of KM, added the complexity of human and cultural dimensions. During this stage Communities of practice were widely implemented, and the knowledge creation and retrieving was accompanied by Knowledge Sharing.

The third stage was a call to action and better use of relevant information, they need to find the right information right away, Content Management, During this stage Competitive Intelligence and Project Management were important concerns, and ECM platforms gain maturity.

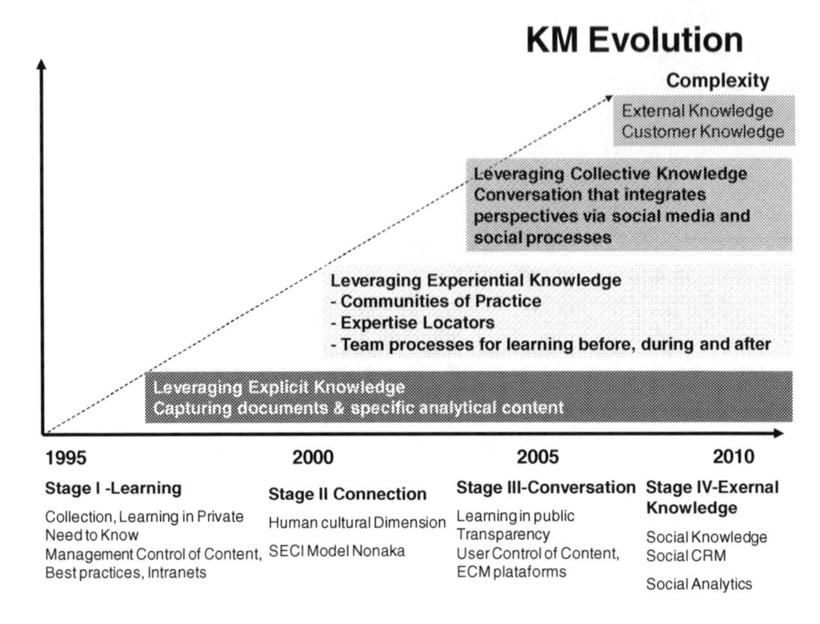

Fig. 3 Knowledge Management stages. (Srikantaiha, 2008).

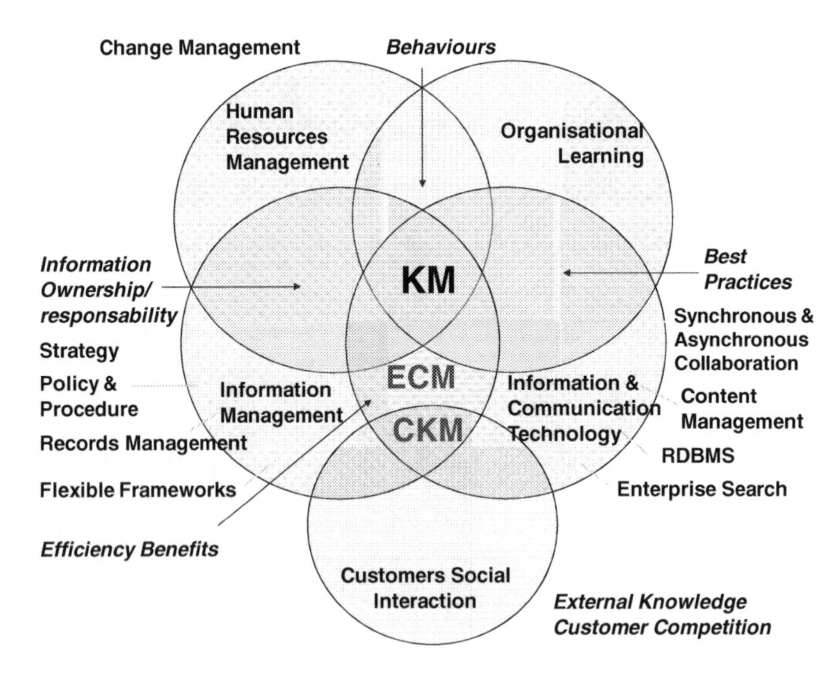

Fig. 4 Customer Knowledge Management CKM tools space.(External Knowledge). Enterprise Content Management in a Knowledge Management Context (Cawthorne 2010).

The fourth stage of KM (Koening 2008) is associated with KM maturity, and is related to the importance of information and knowledge external to the organization (Figure 3.). Even though that external knowledge is not new to KM, its relevance and the tools that are available now, are relatively new.

The result is a greatly increased emphasis upon external information, and specially regarding customer information or customer tacit knowledge that has been able to emerge thanks to a strong trend in Social Networks adoption, with and important role in the age of the customer.

Companies now have access to the information, beliefs, thoughts and opinions of their customers or potential customers in a way that has never been experience before.

The KM fourth stage is incorporating Customer Knowledge Management (CKM) and new powerful IT tools to manage the information and translate it into Knowledge (Figure 4.).

3.1 The Social Trend

Social Networks are here to stay, Facebook has more than 800 million users, 86% of Internet users under 30 years old, use Social networks, 53% of employers research potential job candidates on Social Networks in United States (Career Builder, 2009), Wikipedia had a 21% in articles publications on 2011, for 2012 is estimated that 43% of companies in United States will be using Social Media (source Search Engine Journal).

A 2009 IBM study of 250 chief marketing officers revealed that organizations are shifting significant amounts of money away from traditional advertising and into public relations, particularly mobile and online channels.

This new Social Media Metrics tools are helping companies to:

- Marketing – building awareness, generating leads and sales.
- Communications – engaging with people in support of civic or safety mission.
- Perception–improving the organization's relationships, reputation or positioing.
- Service – improving the service giving to customer, quickly finding out points of friction.

CKM (Customer Knowledge Management) is trying to understand your audiences and what motivates them. This knowledge determines which channels to focus on, what tone of voice to adopt, and what types of responses, offers and online content to provide.

This relatively new analytical tools have various metrics Click-troughs, unique visitors, repeat visitors, number of friends, followers, comments, repeat comments, tweets and retweets. They help organizations to conduct research, and use a number of metrics that will vary by goal, audience and vertical market. The tools mine some actionable insights out of structured data (hits, unique visitors, frequency of visits, etc.) and unstructured data (free-form text in comments, tweets, blog entries, articles and comments).

3.2 Exploit the Volumes of Information Found in Social Media

There is valuable insight contained in all the user-generated content on social media sites

These analysis tools fall into three broad categories in general:

- Descriptive statistics that clarify activity and trends, such as how many followers you have, how many reviews were generated on Facebook, or which channels are being used most often.
- Social network analysis that follows the links between friends, fans and followers to
- Identify connections of influence as well as the biggest sources of influence.
- Text analytics examines the content in online conversations to identify themes, sentiments and connections that would not be revealed by casual surveillance.

An analytics tool should be able to show you trends in positive, negative and neutral activity at least, and enable you to drill down and see the actual comments. By seeing changes over time, companies can correlate social media trends with the circumstances that triggered those trends, such as: traditional marketing activities, organizational or product changes, world events and market conditions.

3.3 Social Metrics and Listening Tools

This tools allows companies to translate the millions of online consumer conversations taking place each day into real-time insight and enterprise response to manage risk, engage with customers and drive innovation. Most of this tools started their development after 2007 when social networks started to gain consumer preference.

Some of the Social Metrics tools more commonly use are ListenLogic, Radian 6, Collective Intellect, Lithium, Sysomos and Attensity360.

There are many benefits of using this tools, Public Relations, find out Crisis events or problems, Search Engine Optimization, Corporate Marketing use and Brand Building, Industry Competitive Insight, Customer Service improvement, Identify New Markets, Sales Lead Generation with lower cost.

4 Objectives and Results

The objectives of this study was to unveil the importance that Chilean companies give to Social Networks and CKM and the levels of related KM IT tools that they are familiar with and use.

What are the technological KM tools more commonly use in companies to support KM, what are the types of tools being in place and their plans.

Through personal interviews to Marketing Directors of seven companies a more insightful research was conducted aiming to find out the relevance of the use of social networks, and the success of their current KM initiatives.

4.1 The Survey

A survey was conducted among Managers and TI officials at a large numbers of companies in Chile, the scope was, to review how much they knew of KM, the survey had 17 questions, and was answer by 98 companies, of diverse industries and sizes.

Table 1 Results of % use and knowledge of ECM tools

ECM Plataform	Vendor	Year Founded	% of Use	% Have some Knowledge
Share Point	Microsoft	2001	7%	37%
Jive Engage	Jive Software	2001	0%	5%
Lotus Live collaboration suite	IBM	2005	3%	16%
Tellingent	Telligent Systems	2004	0%	1%
SalesForce.Com	Salesforce.com	1999	1%	26%
OpenText	Open Text Corporation	1991	0%	2%
Lifecycle		2004	0%	0%
Attlasian	Attlasian Software	2002	0%	2%
Saba	Saba software	1997	2%	15%
Success Factors		2002	0%	0%
Drupal	GNU General Public	2001	9%	40%
Social Text		2002	0%	1%

In Tables 2 there are the results of using ECM platforms and Social Metrics Tools, and the percentage of familiarity with the applications.

Table 2 Percentage of Social Analytics Tools use

Social Analytics Tools CRM	Vendor	Year Founded	% of Use	% Have some Knowledge
ListenLogic:	Listen Logic	2007	0%	8%
Radian 6	SalesForce.com	2006	3%	14%
Lithium	Lithium Technologies	2001	0%	1%
Sysomos	Sysomos Inc.	2007	1%	4%
Attensity360	Attensity	2000	0%	2%
Alterian SM2	Techrigy Inc	2009	0%	0%
Crimson Hexagon	Crimson Hexagon	2007	0%	0%
Spiral16	Spiral 16	2007	0%	1%
Webtrends	Webtrends	1993	3%	8%
Spredfast	SpredFast	2008	0%	0%
NM Incite	Nielsen McKinsey company	2009	0%	0%
Converseon	Converseon	2001	0%	0%
Dna13	CNW Group Co.	2005	0%	0%
Attentio	Attentio	2006	0%	0%
Visible Technologies	Visible Technologies	2006	0%	0%
Cymfony	Kantar Media company	2008	0%	0%
Buzzcapture	Buzzcapture	2006	0%	0%
BuzzLogic	Twelvefold Media	2004	0%	0%
Meltwater Buzz	Meltwater Group	2009	0%	0%
Brandwatch:	Brandwatch	2007	0%	0%

5 Summary and Conclusion

The ability to leverage from your customer base knowledge has become increasingly important for companies. The analysis reveal that Customer KM tools are being rapidly develop to get a better knowledge and process increasingly big amounts of information, on a real time fashion that has never being done before. The use of this new tools could have and important implication for Customer KM practice, they contribute to process immediately, events that are affecting the company or the brand, the results supports a number of recent industry trends, and are at the essence of KM, Customer knowledge is created, acquired, communicated, shared, applied and effectively utilized on a real time bases.

With the emerge of new Customer KM systems, and services that provide supercomputing abilities, organizations will be able to learn more and transform this new customer learning into actions to provide better services and products. Chilean companies are very aware of the benefits of implementing CKM programs and utilizing this new tools, but they have not taken action into the use of this programs and have not deferred funds for 2012 to engage in related projects.

Given the short history of this new systems there is no cases of success so far, and there is not the professional expertise to better use this new customer knowledge, being this the main barrier to incorporate this tools in Chile. The trend in the United States is to not buy software but to hire services, companies do not favor to make capital investments that are going to require an expensive engineering team to operate them, services are expenses that they can tune when and how much they need in the opinion of the interviewed experts.

In summary, information technology can support effective organizational learning by providing persistent and well-indexed tools for collaborative knowledge management and social and knowledge network analysis. However, tools are not enough: an organization needs to have some kind of systematic practice that will use the tools appropriately to monitor performance, anticipate and attend to feedback and outcome measures, design avenues for change, and then take action. There is a need to adequate training and culture.

Chilean companies seems to be basically in the first and second stages of KM (Srikantaiah K. 2008), the bigger companies are implementing ECM systems where SharePoint and Saba are the most commonly use platforms. In terms of CKM and social metrics, most of the companies have not really incorporated the use of this tools with the exception of Radian 6 that has been tried out by some marketing departments and ListenLogic that is known to some bigger companies. Chilean companies are far from reaching KM maturity in their practices.

Customer Knowledge Management is not properly addressed in the age of Social Networks in Chilean organizations, they recognized the importance of incorporation CKM and Social Tools to their procedures, practices and culture, and executives realize the potential benefits of IT for Customer KM, but they are not fully exploited and many have expressed a need for greater implementation of social analysis tools and Customer Knowledge Management practices, and declare insufficient training and education of staff.

Customer knowledge Management will be very relevant for companies in the present decade, the companies that could implement innovative and smart ways of mastering and using their customers external information could develop a valuable strategic advantage.

References

Tiwana, A.: The Knowledge Management Toolkit. Prentice Hall (2002)

Argote, L., Ingram, P.: Knowledge transfer: A basis for competitive advantage in firms 82(1), 150–169 (2000)

Avram, G.: At the crossroads of knowledge management and social software. The Electronic Journal of Knowledge Management 4(1), 1–10 (2006), http://www.ejkm.de (retrieved April 17, 2010)

Choo, C.W.: Information management for the intelligent organization: the art of scanning the environment. Information Today Inc., Medford (1998)

Career Builder, 50% of Employers Use Social Networking Sites to Research Job Candidates (2009), http://www.sociableblog.com/2010/01/17/employers-use-social-networking-sites/ (retrieved November 16, 2011)

Cawthorne, J.: Enterprise Search Summit (2010), http://www.ecm-stuff.blogspot.com (retrieved December 10, 2011)

Davenport, T., Prusak, L.: Working knowledge: how organizations manage what they know. Harvard Business School Press, Boston (1998)

Edvinsson, L., Malone, M.: Intellectual capital: realizing your company's true value by finding its hidden brainpower. Harper Business, New York (1997)

Egbu, C., Gaskell, C.: The role of organizational culture and motivation in the effective utilization of information technology for team working in construction. In: Proceedings of the 17th Annual Conference of the Association of Researchers in Construction Management, ARCOM, September 5-7. University of Salford, UK (2001)

Forrester Research, Competitive Strategy in the Age of the Customer (2011), http://blogs.vanderbilt.edu/marketing/wp-content/uploads/2011/08/2011-6-6-Forrester-Competitive-Strategy-In-The-Age-Of-The-Customer.pdf (retrieved January 10, 2012)

Girard, J.: The Inukshuk: A Canadian Knowledge Management Model. KMPRO Journal (2005), http://johngirard.net/johngirard/The%20Inukshuk%20KM%20Pro%20Vol%202%20No%201.pdf

Bernoff, J.: Competitive Strategy in the Age of the Customer. CMO & Marketing Leadership Professionals, Forrester (June 10, 2011)

Koenig, M., Srikantaiah, K.: Knowledge Management Lessons Learned: What Works and What Doesn't. Information Today, Inc., Medford (2004)

Koening: Knowledge Management in Practice. Information today, Inc., Medford (2008)

Malhotra, Y.: Why knowledge management systems fail. In: Koenig, M.E.D., Srikantaiah, T.K. (eds.) Enablers and Constraints of Knowledge Management in Human Enterprises (2004)

Martin, R.: The Age of Customer Capitalism. Harvard Business Review (January 2010), http://hbr.org/2010/01/the-age-of-customer-capitalism/ar/1 (retrieved January 07, 2012)

Miron, M., Cecil, J., Bradicich, K., Hall, G.: The myths and realities of competitive advantage. Datamation 34(19), 71–82 (1988)

Nielsen Wire (January 30, 2012), http://blog.nielsen.com/nielsenwire/global/social-media-accounts-for-22-percent-of-time-online/

Paine, K., Chaves, M.: Social Media Metrics, Listening, Understanding and Predicting the Impacts of Social Media on Your Business. In: Insights from a May 2010 Workshop on Social Media Metrics (2010)

Paine, K.: Social Media Metrics. SAS The Power to Know (2011), http://www.sas.com/resources/whitepaper/wp_19861.pdf (retrieved 11-19-2011)

Search Engine Journal. Social Media Info graphic (August 30, 2011), http://www.searchenginejournal.com/the-growth-of-social-media-an-infographic/32788/

Srikantaiah, K., Koenig, M. (eds.): Three stages of knowledge management. Knowledge Management; Lessons Learned, pp. 3–9. City of Learned Information Inc, Singapore (2004)

Stewart, T.: Intellectual capital: the new wealth of organizations. Doubleday, New York (1997)

Sveiby, K.: The new organizational wealth: managing and measuring knowledge-based assets. Berrett Koehler, San Francisco (1997)

Tanriverdi, H.: Information technology relatedness, knowledge management capability, and performance of multibusiness firms 29(2), 311–334 (2005)

Can a Wiki Be Used as a Knowledge Service Platform?

Fu-ren Lin[1], Cong-ren Wang[2], and Hui-yi Huang[1]

[1] Institute of Service Science, National Tsing Hu University
[2] Institute of Technology Management, National Tsing Hua University,
101 Sec. 2 Kuangfu Road, Hsinchu City, 300, Taiwan
frlin@mx.nthu.edu.tw,
{jnwang0703,huanghuiyimiranda}@gmail.com

Abstract. Many knowledge services have been developed as a matching platform for knowledge demanders and providers. However, most of these knowledge services have a common drawback that they cannot provide a list of experts corresponding to the knowledge demanders' need. Knowledge demanders have to post their questions in a public area and then wait patiently until corresponding knowledge providers appear. In order to facilitate knowledge demanders to acquire knowledge, this study proposes a knowledge service system based on Wikipedia to actively inform potential knowledge providers on behalf of knowledge demanders. This study also developed a knowledge activity map system used for the knowledge service system to identify Wikipedians' knowledge domains. The experimental evaluation results show that the knowledge service system is acceptable by leader users on Wikipedia, in which their domain knowledge can be identified and represented on their knowledge activity maps.

Keywords: Knowledge service, Wikipedia, Knowledge activity map, HAC+P.

1 Introduction

With the advance of information technology, especially the emergence of Web 2.0, the Internet has become a popular knowledge aggregation and distribution platform. In order to facilitate knowledge seekers to acquire knowledge via the Internet, knowledge services which emphasize on the knowledge search, provision, answering, solution seeking, matching, and pricing are getting popular although the sustainability varies in different initiatives. There have been many knowledge service initiatives, such as Answers.com, WikiAnswers, InnoCentive, Google Answers, and Yahoo! Answers, etc. These websites utilized the concept of Web 2.0 to build their platforms on which knowledge seekers can ask questions or seek problem solutions from knowledge providers.

A knowledge service should actively refer knowledge seekers to a list of candidate knowledge providers who are capable of solving the problems. To realize this idea, we selected Wikipedia as the experimental platform. Wikipedia (http://en.wikipedia.org) is a well known website where people can contribute and

L. Uden et al. (Eds.): 7th International Conference on KMO, AISC 172, pp. 365–376.
springerlink.com

share their knowledge. A person who contributes contents to Wikipedia is called a Wikipedian. Through editing the web pages on Wikipedia, Wikipedians can easily contribute their knowledge to others. In addition, someone else may modify the content of the topics if they have other ways to describe them. After each modification, the older version of the article is saved as history which allows people to trace version changes. In Wikipedia, each topic has a discussion page to allow people to discuss the related issues online. Besides, it also retains a Wikipedian's editing history.

We have three reasons to use Wikipedia as the experimental platform. First, the contents of Wikipedia are highly qualified via peer review. Second, individual editing history is kept and can be easily analyzed. Finally and the most importantly, a platform like Wikipedia was mainly designed for knowledge providers, instead for knowledge demanders. For example, the contents of Wikipedia mainly depend on what knowledge providers want to share instead of what knowledge demanders need. Therefore, it is necessary to develop a knowledge service system to facilitate knowledge demanders to actively acquire knowledge.

In this study, we propose a knowledge service system based on Wikipedia to help knowledge demanders to identify domain experts when they need advanced knowledge about certain domains. Moreover, this system provides the matching function to identify potential knowledge providers according to knowledge seekers' questions. Through the proposed service system, we expect that knowledge demanders would participate in the knowledge sharing process on Wikipedia and have quality interaction with knowledge providers. In addition, it will help the knowledge providers to contribute the knowledge needed by knowledge seekers.

As stored documents are accumulated, it is necessary to represent the evolution of knowledge [2]. We further propose a knowledge activity map system responsible for building the knowledge activity maps for individual users by clustering personal editing records and edited articles. Moreover, it can update the map incrementally without rebuilding a new one. The results of experimental evaluation from this study extend the boundary of the literatures reviewed by [1], which reviewed knowledge management systems merely in organizational settings. Thus, based on their review on organizational knowledge management, we can position the uniqueness and contribution of this study. The evolution of Wikipedia from one type of knowledge repository to a networked domain expert location system or hybrid of them can be realized if the proposed knowledge service system enabled by the built knowledge activity map is accepted by Wikipedians. That is the major research question of this study.

2 Related Work

This study proposes a knowledge service system based on Wikipedia, and develops a knowledge activity map system serving as the kernel of the proposed service system.

2.1 *Knowledge Services and Wikipedia*

Knowledge services facilitate knowledge seekers to obtain the designated knowledge objects. These knowledge objects need to be discovered, retrieved, evaluated, and selected [9]. Wikipedia is a well known website where people can contribute and share their knowledge. It is a multilingual, web-based, and content-free encyclopedia project (Wikipedia, http://en.wikipedia.org) which was founded by Jimmy Wales and Larry Sanger in 2001. Wikipedia's articles provide links to guide the user to related pages with additional information. Wikipedia is written collaboratively by volunteers around the world. Anybody can edit an article using a wiki markup language that offers a simplified alternative to HTML [5]. Although anyone can participate in editing articles, the Wikipedia community takes issues of content quality very seriously. The Wikipedia community reviews and selects the articles which are well-organized and well-edited in the Wikipedia standard as featured articles. Feature articles are used as a means of setting a quality standard against general articles.

Since its creation in 2001, Wikipedia has grown rapidly into one of the largest reference websites, and has been given much attention by researchers. For example, Denoyer and Gallinari [5] built a Wikipedia XML corpus to support many XML information retrieval/ machine learning activities, such as ad-hoc retrieval, categorization, clustering and structure mapping. Lih [16] analyzed the crucial technologies and community policies that had enabled Wikipedia to prosper. For example, Wikipedia's articles which have been cited in the news media as a source of information for the public. Krötzsch et al. [13] proposed the concept of a typed link to describe the relationship between two linked articles on Wikipedia. As categories are now used for classifying articles, typed links would then be used for classifying links. Gabrilovich and Markovitch [8] applied machine learning techniques to building a text categorization system based on Wikipedia.

There are also many studies about the nature of Wiki. Voss [23] provided a detailed statistical analysis of Wikipedia including articles, authors, links, edits, etc., which discovered the characteristics of its service. Denning et al. [4] described several issues in articles on Wikipedia, such as accuracy, motives, uncertain expertise, volatility, coverage and sources. Stvilia et al. [21] discussed information quality of articles on Wikipedia. From reviewing these prior studies, we found that the information quality on Wikipedia and the authors' motivations were already discussed. As noted above, Wikipedia has identified featured articles by Wikipedians as a means to set a quality standard against general articles. Thus, the authors and editors of featured articles could be a good source of active knowledge contributors.

2.2 *Knowledge Map*

A knowledge map is a visual display of captured information and relationships, which enables efficient communication and knowledge sharing at multiple levels of detail [22]. Also, it gives a useful blueprint for implementing a knowledge management system with captured knowledge [12].

Knowledge maps have been used for knowledge sharing and acquisition in organizations. NakSun system is a web bulletin board system enhanced with the knowledge map to facilitate collaborative learning [19]. O'Donnell et al. [20] summarized the assistance of knowledge maps for active learning. Knowledge maps were also used by problem solving systems in several domains [11, 14].

Eppler [6] classified knowledge maps into five types: knowledge source map, knowledge asset map, knowledge structure map, knowledge application map, and knowledge development map. The knowledge activity map used in this study can be classified as a knowledge asset map visually expressing a Wikipedian's knowledge assets and dynamic knowledge structure along a timeline.

2.3 *Knowledge Activity Map Creation*

The purpose of the knowledge activity map is to depict a Wikipedian's knowledge structure along a timeline. This study adopts clustering techniques to retrieve individual edited articles to represent their knowledge structures. In order to facilitate knowledge seekers to trace knowledge categories with the timeline, we adopted a clustering algorithm called HAC+P [3] to create knowledge activity maps. The HAC+P algorithm represents the clusters by a multi-way-tree hierarchy. In order to make the clusters human-readable, we further adopted the normalized paragraph dispersion [15] to extract the theme of each cluster in terms of informative terms.

Hierarchical Clustering Algorithm: HAC+P

The HAC+P algorithm consists of two parts: the HAC-based clustering algorithm [18] to build a binary-tree cluster hierarchy, and the min-max partitioning algorithm [3] to transform the original hierarchy into a multi-way-tree hierarchy. HAC is an agglomerative clustering algorithm which clusters data via combining two clusters in each hierarchy (bottom-up). The main idea of the min-max partition algorithm is to find a particular level that minimizes the inter-similarity among the clusters produced at the level and maximizes the intra-similarity of all those clusters.

There are two criteria used to determine the best cut level. The first one is cluster set quality which is used to measure the intra-similarity and the inter-similarity between all-pair clusters. The equation of cluster set quality is defined as $Q(C)=\frac{1}{|C|}\sum_{C_i\in C}\frac{sim_A(C_i,\overline{C_i})}{sim_A(C_i,C_i)}$, where C is a set of clusters, $\overline{C_i}=\bigcup_{k\neq i}C_k$ is the complement of Ci, $sim_A(C_i,C_i)$ is the intra-similarity within a cluster Ci, and $sim_A(C_i,\overline{C_i})$ is the inter-similarity between Ci and $\overline{C_i}$. Note that a smaller value of the $Q(C)$ represents a better quality of the given set of clusters. The second criterion is cluster-number preference, denoted as

Nclus, which is used to guarantee that the generated clusters are neither too few nor too many. Because the preference of cluster number is varied by people, it is hard to use a predefined constant as Nclus. A simplified gamma distribution function to determine Nclus based on the number of the given set of clusters [3]. The suitable cut level is chosen as the level l with a minimum value of Q(LC(l))/N(LC(l)), where LC(l) is the set of clusters produced after cutting the binary-tree hierarchy at level l.

Normalized Paragraph Dispersion (NPD)

The paragraph dispersion, denoted as $PD_{i,d}$, is used to measure the degree of a word i appearing across different paragraphs in a document d [7]. In the equation, the smaller value of $PD_{i,d}$ denotes that the term appears more uniformly in every paragraph. In contrast, if the term mainly appears in a portion of paragraphs, its $PD_{i,d}$ value is larger. A concept of $PD_{i,d}$ to extract the informative words in a document [15]. They revised the original dispersion formula by adding a negation sign. It lets the informative terms have a larger PD value. Furthermore, to normalize the paragraph dispersion value to be small and around 0, it employs the logarithm value and adds 0.01 to avoid the occurrence of $log(0)$. The revised formula called normalized paragraph dispersion (NPD) is defined as

$$NPD_{i,d} = -LOG\left(\sqrt{\frac{\sum_{j=1}^{M}(tf_{i,j,d} - \mu tf_{i,d})^2}{M} + 0.01}\right), \mu tf_{i,d} = \frac{\sum_{j=1}^{M} tf_{i,j,d}}{M}$$

, where $tf_{i,j,d}$ denotes the term frequency of word i in paragraph j of document d, $\mu tf_{i,d}$ denotes the average term frequency of word i in document d, and M denotes the total number of paragraphs in document d.

3 The Architecture of a Knowledge Service System on Wikipedia

This study adopted the guideline of design science [10] to create and evaluate the artifact, a knowledge service system, based on the contents and users of Wikipedia. The system was built and operated in an environment where Wikipedia users can access the service via an internet connection. The knowledge base was theories on knowledge sharing and was implemented by text mining techniques. The obtained system was evaluated using an experimental design.

In order to facilitate knowledge seekers to acquire advanced knowledge from Wikipedians, we propose the architecture of a knowledge service system based on Wikipedia as shown in Figure 1. The knowledge activity map system performs expert identification and recommendation for knowledge seekers. Knowledge activity map system is the kernel of the knowledge service system. It constructs the knowledge activity maps of Wikipedians by analyzing their edited articles on Wikipedia. A Wikipedian's knowledge activity map demonstrates his/her actions on editing wiki pages in certain knowledge domains. Expert identification function identifies experts of a certain knowledge domain from the population of Wikipedians. The main evidence used to identify domain experts is individual knowledge activity maps. Expert matching function performs the recommendation operation for knowledge seekers to ask a domain expert for advanced knowledge which may not yet be publicized on Wikipedia. This matching function compares the question asked by a user and the knowledge activity maps of candidate Wikipedians.

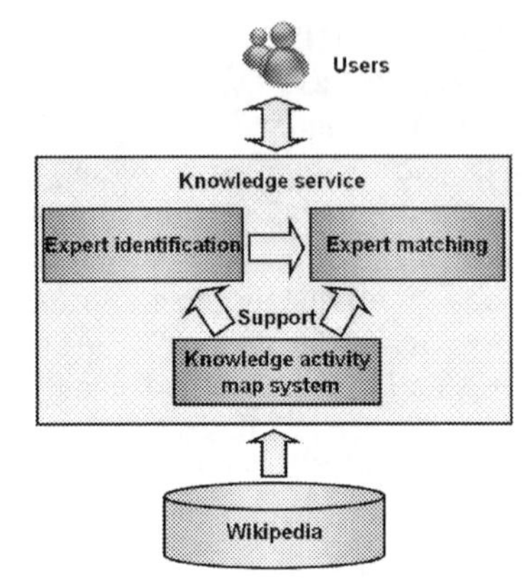

Fig. 1 The architecture of a knowledge service system based on Wikipedia

4 The Architecture of the Knowledge Activity Map System

The architecture of the proposed system is depicted in Figure 2. The knowledge activity map generator is responsible for constructing the knowledge activity map. The knowledge activity map viewer provides a visualized view of the knowledge activity map for users.

4.1 Knowledge Activity Map Generator

Knowledge activity map generator is composed of data collection, data translation, clustering, and theme labeling as shown in Figure 2. First of all, we retrieved the wiki pages edited by candidate Wikipedians to create individual knowledge activity maps. We then retrieved the topics from these wiki pages, and then accessed the corresponding articles on Wikipedia. Note that we filtered out two kinds of trivial pages: the pages irrelevant to the owner's knowledge, *e.g.*, discussions and images, and the edited records marked as the minor edition.

Before constructing the knowledge activity map, the system transforms the original data into the Vector Space Model (VSM) as the input data to the clustering analysis. In order to present a Wikipedian's knowledge activities on a timeline, we counted one day as a unit. For each day, we selected the topics and the terms with hyperlinks from the articles edited in the day as its features. After this step, the original data were transformed into a $m \times n$ matrix, where n denotes the number of columns as the total number of distinct features in the data collection, and m denotes the number of rows as the total number of days. Each day is represented as a vector of features. The weight of each element of a vector is the term frequency of the corresponding feature appearing for that day.

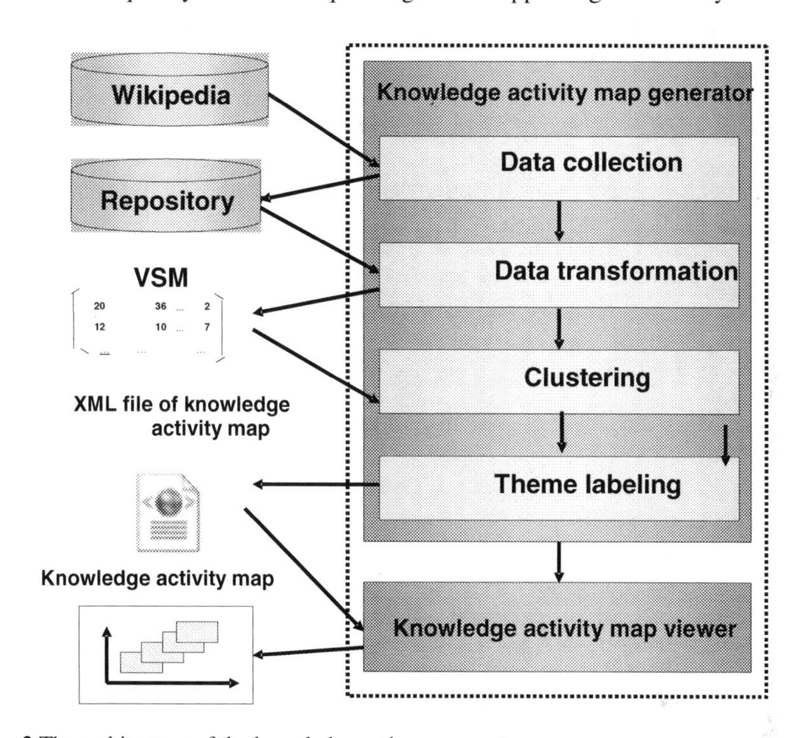

Fig. 2 The architecture of the knowledge active map system

We observed that the articles a Wikipedian edited in neighboring days may contain similar knowledge categories. We used the concept of n-gram to firstly connect those adjoining days with similar knowledge categories into a cluster. We compared knowledge features of two separated days with cosine similarity. If the knowledge categories of two days are similar, we connect them as a cluster.

After the previous n-gram clustering, days with similar knowledge categories are grouped into a period of days. The next step is to further cluster periods with similar patterns. We used the HAC+P algorithm to cluster these data and determine the most suitable number of clusters. We set the parameters in the min-max partitioning algorithm with the same values as [3].

After the previous process, a Wikipedian's knowledge structure will be revealed by the clusters of personal edited articles. Every cluster represents a pattern of knowledge structure. We then extracted the theme of each cluster with the most representative terms to specify its main concepts.

To retrieve representative terms we used normalized paragraph dispersion mentioned to extract the most informative words for each cluster. We treated a cluster of documents as an article, where the edited documents on a day in the cluster were treated as a paragraph of the article in the original NPD. That is, the term more equally distributed among paragraphs will have a high probability to represent the theme of the cluster. In this study, we selected the top 5 terms to specify a cluster's theme. Finally, we output the result as a XML file as inputs to the knowledge activity map viewer.

4.2 Knowledge Activity Map Viewer

The knowledge activity viewer is responsible for visualizing the results generated by the knowledge activity generator. We constructed the knowledge activity map viewer using an open source java-based project named JFreeChart (http://www.jfree.org/jfreechart/). A knowledge activity map is a bar chart, in which the x-axis shows the timeline and the y-axis shows the number of feature terms representing the quantity of knowledge. The color of a bar represents the category of a certain cluster during a period of time.

5 Evaluation

5.1 Subject Sampling

We set two criteria to select subjects for an experimental evaluation. First, a Wikipedian selected as a subject must have edited wiki pages in the recent one month to make sure that his/her knowledge activity map presents his/her up-to-date knowledge categories. Second, the majority of wiki pages edited by the Wikipedian are domain knowledge related contents. A user's wiki pages with the discussion of topics, the profiles of users, the information of figures, and the minor editing records are treated as domain independent contents.

We directly chose subjects from the candidate users whose articles have been selected as featured articles. The candidate users are more likely to fit our aforementioned criteria. We sent the invitation message to 70 Wikipedians, and received confirmative messages from 20. We invited these 20 Wikipedians as our experimental subjects. We retrieved 500 editing records on each subject's history page.

5.2 Evaluation Design

The experimental evaluation consists of two tasks. One is to evaluate the feasibility of the proposed knowledge service system based on Wikipedia. The other is to evaluate the fitness between a user's knowledge categories and the generated knowledge activity map.

To investigate users' perception on the feasibility of the knowledge service system based on Wikipedia, we designed a questionnaire to ask users' attitude toward the service denoted as Task 1. The questionnaire was used to evaluate a user's ability, willingness, and confidence on answering questions which we assigned to him/her. The questionnaire also contains items in assessing a user's willingness to spend efforts on searching answers to exchange for compensation.

The questions for a specific user were composed from two sources. First, three of six questions are derived from the articles the user edited in three different time periods, i.e., recent, middle, and long periods of time, respectively. We expected to evaluate whether time is the influential factor on the user. Second, we randomly picked three questions irrelevant to the user's knowledge domain also in different time periods. We mixed these six questions up to test whether users preferred to answer the questions which they were familiar with. Moreover, we set two items, the prices of a can of Coke Cola and a move ticket, as the utility benchmarks to evaluate the relations among the willingness, the effort, and the compensation for a user to contribute the answer.

We evaluated the fitness of user perceived knowledge categories denoted as *Task 2* to evaluate if the knowledge categories on the knowledge activity map were consistent with the perception of the corresponding user. We firstly constructed a knowledge activity map for each user according to the articles s/he edited. Then, we randomly selected 20%, at most 30, of the articles which the user edited as testing articles. Finally, we asked each user to assign testing articles to corresponding categories on the map according to the informative terms tagged with the categories. Users were allowed to assign testing articles to multiple categories if they wanted.

5.3 Evaluation Results

Note that we obtained 10 and 12 subjects out of the 20 subjects to finish the evaluation for Tasks 1 and 2, respectively. The result of Task 1 is shown in Table 1. We used a t-test to verify the users' confidence level in answering questions which they were familiar with versus those they weren't. The resulting p-value is 0.026, which indicates the confidence of users in answering two kinds of questions is significantly different. The result indicates that knowledge seekers facilitated by the knowledge activity map have more chance to get the correct answers by passing their questions to those knowledge providers who are familiar with the questions.

Table 1 The result of a *t*-test of subjects' confidence level

Questions	User #1	User #2	User #3	User #4	User #5	User #6	User #7	User #8	User #9	User #10
Q2	NULL	8	4	0	10	10	10	10	8	6
Q4	NULL	10	8	1	10	10	8	6	8	0
Q6	NULL	10	7	7	10	10	10	10	8	8
Q1	8	10	2	10	10	10	10	10	9	8
Q3	10	8	3	0	0	9	NULL	6	3	0
Q5	NULL	9	2	5	2	10	7	7	9	2

Note: Q2, Q4, and Q6 are domain relevant questions; Q1, Q3, and Q5 are domain irrelevant questions; NULL means that the subject did not answer the question; p-value =0.026; α= 0.1; range = 1~10.

In evaluating the users' willingness to answer the questions, we have two interesting findings. First, most of the subjects were willing to answer the questions no matter if they were familiar with the subject matter or not. Second, seven subjects set the compensation of each question as "free," regardless of the difficulty of the questions. It might be possible that subjects who were willing to accept our evaluation invitation were those who were enthusiastic to help others on Wikipedia.

Table 2 The result of Task 2

*User #**	*Hits*	*# of testing articles*	*Accuracy*
1	15	30	0.50
2	13	30	0.43
3	10	30	0.33
4	10	30	0.33
5	15	30	0.50
6	6	20	0.30
7	7	15	0.47
8	14	30	0.47
9	8	18	0.44
10	11	30	0.37
11	4	12	0.33
Average	10	25	0.41

* The result from user#12 is an outliers and was dropped.

In Task 2, we adopted an accuracy formula proposed by [17] to measure the accuracy of the result. The accuracy (AC) is defined as $\frac{\sum_{i=1}^{N} \delta(\alpha_i, l_i)}{N}$, where N denotes the total number of tested articles in the evaluation. For each tested article i, α_i is the cluster categorized by users, li is the cluster categorized by the proposed algorithm, and δ(x,y) denoted as the delta function is one if x and y are categorized into the same cluster, and is zero, otherwise.

The result of Task 2 is shown in Table 2. The average accuracy was 0.41 after discarding the result from user #12 only achieved 0.1 accuracy. This result indicates that the proposed knowledge activity map system can capture about 41% of a user's perceived knowledge categories.

6 Conclusion and Future Work

This study aims to develop a knowledge service system facilitated by a knowledge activity map based on the contents and users of Wikipedia. From the experimental evaluation results, we found that some Wikipedians are willing to answer questions regardless of their familiarity with the subject matter. However, the Wikipedians are more confident in answering the questions belonging to their own knowledge domains. This result indicates that knowledge seekers have more chance to find correct solutions to their queries if facilitated by the matching function to identify knowledge providers with relevant domain knowledge. Therefore, the knowledge service based on Wikipedia is feasible and necessary.

The major contribution of this study is to demonstrate the uniqueness and usefulness of a knowledge service system based on a knowledge repository, such as Wikipedia. The knowledge activity map system built and evaluated using design science approach is a feasible facilitator to identify domain experts willing to serve. The evolution of wiki knowledge repositories to include a networked domain expert location system can be realized by the proposed knowledge service system enabled by the knowledge activity map.

References

1. Alavi, M., Leidner, D.E.: Review: Knowledge Management and Knowledge Management Systems: Conceptual Foundations and Research Issues. MIS Quarterly 25, 107–136 (2001)
2. Barthelme, F., Ermine, J.L., Rosenthal-Sabroux, C.: An Architecture for Knowledge Evolution in Organisations. European Journal of Operational Research 109, 414–427 (1998)
3. Chuang, S.L., Chien, L.F.: Taxonomy Generation for Text Segments: A Practical Web-based Approach. ACM Transactions on Information Systems (TOIS) 23, 363–396 (2005)
4. Denning, P., Horning, J., Parnas, D., Weinstein, L.: Wikipedia Risks. Communications of the ACM 48, 152–152 (2005)
5. Denoyer, L., Gallinari, P.: The Wikipedia XML Corpus. ACM SIGIR Forum 40, 64–69 (2006)
6. Eppler, M.J.: Making Knowledge Visible Through Intranet Knowledge Maps: Concepts, Elements, Cases. In: Proceedings of the 34th Annual Hawaii International Conference on System Sciences, HICSS-34, p. 4030 (2001)
7. Fumiyo, F., Yoshimi, S.: Event Tracking Based on Domain Dependency. In: Proceedings of the 23rd Annual International ACM SIGIR Conference on Research and Development in Information Retrieval, pp. 57–64 (2000)
8. Gabrilovich, E., Markovitch, S.: Overcoming the Brittleness Bottleneck Using Wikipedia: Enhancing Text Categorization with Encyclopedic Knowledge. In: Proceedings of the 21st National Conference on Artificial Intelligence, pp. 1301–1306 (2006)
9. Gregoris, M., Kostas, K., Panos, G.: Knowledge Services on the Semantic Web. Communications of the ACM 50, 53–58 (2007)
10. Hevner, A.R., March, S.T., Park, J., Ram, S.: Design Science in Information Systems Research. MIS Quarterly 28, 75–105 (2004)
11. Hsu, C.K., Kuo, R., Chang, M., Heh, J.S.: Implementing a Problem Solving System for Physics based on Knowledge Map and Four Steps Problem Solving Strategies. In: IEEE 2nd International Conference on Advanced Learning Technologies, ICALT 2002, Kazan, Russia (2002)
12. Kim, K., Suh, E., Hwang, H.: Building the Knowledge Map: an Industrial Case Study. Journal of Knowledge Management 7, 34–45 (2003)

13. Krötzsch, M., Vrandecic, D., Völkel, M.: Wikipedia and the Semantic Web-The Missing Links. In: Proceedings of Wikimania, Frankfurt, Germany (2005)
14. Kuo, K., Chang, M., Dong, D.X., Yang, K.Y., Heh, J.S.: Applying Knowledge Map to Intelligent Agents in Problem Solving Systems. In: World Conference on Educational Multimedia, Hypermedia and Telecommunications, pp. 24–29 (2002)
15. Kuo, J., Chen, H.: Multidocument Summary Generation: Using Informative and Event Words. ACM Transactions on Asian Language Information Processing (TALIP) 7, 1–23 (2008)
16. Lih, A.: Wikipedia as Participatory Journalism: Reliable Sources? Metrics for evaluating collaborative media as a news resource. In: Proceedings of the 5th International Symposium on Online Journalism (2004)
17. Liu, X., Gong, Y., Xu, W., Zhu, S.: Document Clustering with Cluster Refinement and Model Selection Capabilities. In: Proceedings of the 25th Annual International ACM SIGIR Conference on Research and Development in Information Retrieval, pp. 191–198 (2002)
18. Mirkin, B.: Mathematical Classification and Clustering. Kluwer Academic Press (1996)
19. Nagai, M., Shiraki, K., Kato, H., Akahori, K.: The Effectiveness of a Web Bulletin Board Enhanced with a Knowledge Map. In: Proceedings of the International Conference on Computers in Education, p. 268 (2002)
20. O'Donnell, A.M., Dansereau, D.F., Hall, R.H.: Knowledge Maps as Scaffolds for Cognitive Processing. Educational Psychology Review 14, 71–86 (2002)
21. Stvilia, B., Twidale, M.B., Gasser, L., Smith, L.: Information Quality in a Community-based Encyclopedia Knowledge Management: Nurturing Culture, Innovation, and Technology. In: 2005 International Conference on Knowledge Management, pp. 101–113 (2005)
22. Vail, E.F.: Knowledge Mapping: Getting Started with Knowledge Management. Information Systems Management 16, 16–23 (1999)
23. Voss, J.: Measuring Wikipedia. In: Proceedings of the 10th International Conference of the International Society for Scientometrics and Informetrics, pp. 221–231 (2005)

Understanding and Modeling Usage Decline in Social Networking Services

Christian Sillaber, Joanna Chimiak-Opoka, and Ruth Breu

University of Innsbruck, Austria
{christian.sillaber,joanna.opoka,ruth.breu}@uibk.ac.at

Abstract. In this paper, we propose a new research model that describes why users leave social network services. Multiple models and theories were proposed in the past that describe why users join in on innovation, become active users of new IT solutions and adopt technology. But no concise model has been established that explicitly focuses on why users withdraw from social network services – or web based technology in general. The goal of this paper is to develop, based on established theories of information systems research, a research model that provides a predictive set of rules for analyzing user withdrawal from social network services.

1 Introduction

Since their introduction, many social network services have attracted millions of users, many of whom have integrated these sites into their daily practice. As with each new technology being introduced, the growth and adoption by the users follow distinct patterns. The fast rise and deep fall of social network giants like Myspace and Digg has demonstrated that not only an accurate prediction method for their growth but also the users' withdrawal from such systems is of high commercial and academic interest. Several models and theories were developed to describe why and how users join in on innovation. However due to the specific characteristics of today's social network services the predictive capabilities of existing models are either limited or cannot be directly applied to social network services. Therefore, to enable further research in this area, we present and propose a research model that is directly applicable to social network services and capable of describing user withdrawal from such services.

Structure: A literature overview is given in Section 2. Section 3 introduces social network services and defines the semantics of *leaving* a social network service in

L. Uden et al. (Eds.): 7th International Conference on KMO, AISC 172, pp. 377–388.
springerlink.com

the context of this paper. In Section 4 we assess withdrawal from social network services and present the research model we propose. Section 5 discusses practical design implication for social network services based on the proposed research model that help mitigating user withdrawal. Section 6 provides a short summary.

2 Literature Overview

Several theories in the field of information systems research can be found that model how users come to use new technology. This section presents the models and concepts that build the foundation for our research model. Due to limited space we can only describe each model briefly and refer the reader to the given literature for more information on them.

The **Fit-Viability Model** [33] was proposed for evaluating organizational adoption of Internet initiatives. The model separates between *fit* (the extent to which new applications are consistent with core competences, value and culture of the organization) and *viability* (value added potential of the new application) to propose general operative strategies. The framework provided by the **Social Cognitive Theory** allows for the understanding, prediction and change of human behavior. Human behavior is identified as an interaction of the environment, general behavior and personal factors [26].

In the **Structuration Theory** the concepts of structure (*domination, legitimation*) and social systems are introduced to describe how the actions of individual agents are related to the structural features of a society [13]. The model proposed by the **Adaptive Structuration Theory** adapts the general Structuration Theory to the context of technology. It describes the interaction of groups and organizations with information technology [8].

The **Theory of Reasoned Action** enables the prediction of an individual's *behavioral intention*, *attitude* and *behavior* [31]. The **Theory of Planned Behavior** describes *attitude* and *behavior* as the governing concepts behind an individual's action. It is an extension of the Theory of Reasoned Action [3].

The **Technology Acceptance Model** extends the more general Theory of Reasoned Action to the field of Information Systems. It introduces the concepts of *perceived usefulness* and *perceived ease of use* to predict the *actual system use* of an individual [7]. As its name suggest, the **Technology-Organization-Environment Framework** links the elements *technology*, *organization* and *environment* to predict technological decisions [34].

The **Unified Theory of Acceptance and Use of Technology** unifies several ideas behind previously presented theories to create a model that allows the prediction of *use behavior* based on four key constructs: *performance expectancy*, *effort expectancy*, *social influence* and *facilitating conditions* [36].

To varying degree these models can not only be used to model *successful* establishment of new technology but can also be used to model.

3 Social Networks

Social networks in general are associations of people drawn together by family, work or hobby [28]. Social network services are online services that aim at reproducing classic social networks in a virtual environment [4]. As their real world counterparts, they also focus on building and maintaining social relations among people who share common interests, activities and knowledge. Although multiple flavors of social network services exist, almost all offer virtual representations of each user, means of communication and interaction over a multitude of services [15].

3.1 Classification of Social Network Services

Beneath rudimentary functions that are shared among all social network services, each focuses on different user groups and therefore provides a multitude of different functions. We have identified four main types of social network services.

- **General purpose social networks:** They provide means of interchange and communication among registered users without any governing topic. For example Facebook[1], Orkut[2] and Myspace[3].
- **Categorized social networks:** Social network services that focus on a special group of like minded people. Examples are StudiVZ[4] for Students, or, LinkedIn[5] for professional contacts.
- **Functional social network services:** Their social networking capabilities are an environment built around a core functionality like image sharing (Flickr[6]), Video sharing (Youtube[7]), etc.
- **Recommendation based systems:** Although belonging to the category of functional social network services, the degree to which social involvement governs recommendation based systems justifies the creation of this category. Examples are Digg[8], Reddit[9] and Delicious[10].

3.2 Two Examples: Myspace and Digg

In this section we use two social network services to spotlight reasons for users leaving: Myspace and Digg.

[1] http://www.facebook.com
[2] http://www.orkut.com
[3] http://www.myspace.com
[4] http://www.studivz.net
[5] http://www.linkedin.com
[6] http://www.flickr.com
[7] http://www.youtube.com
[8] http://www.digg.com
[9] http://www.redd.it
[10] http://del.icio.us

Myspace: A (originally) general purpose social network service that reached its peak in popularity back in June 2006[11] and was overtaken by its main competitor Facebook two years later.

Digg: A social network service focused on social news, where users can submit articles and other users can vote on them. After an unsuccessful redesign of Digg in August 2010, thousands of users left Digg for its competitor, Reddit[12] (cf. Figure 1).

These two social network services will be used in the following sections to exemplify on reasons why users leave.

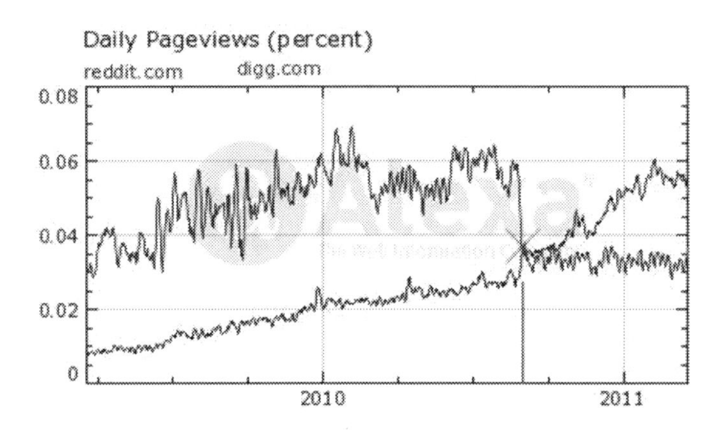

Fig. 1 Daily Page views of Reddit (Blue, growing) and Digg (Red) in comparison (in % of global Page views). Courtesy to Alexa.com.

3.3 *Defining User Withdrawal*

The first question to ask before examining reasons users have to leave social network services is *why do people use* social network services in the first place?

As can easily be imagined by the sheer amount of social network services currently available on the Internet, reasons for use vary heavily [10]. Therefore a valid answer to the posed question depends a lot on the social network services in question, its user groups and the multitude of factors governing the individual's decision in becoming an active member.

Multiple studies can be found that highlight the intentions of use for multiple user groups and social networks such as Facebook [10, 37, 29, 21], online social games [17], and several others [22, 18, 25, 24]. Based on the findings in these studies the general statement can be made that most people use social network services to keep in touch with people they know from real life, communicate, share information and create new relationships.

[11] `http://mashable.com/2006/07/11/myspace-americas-number-one/`

[12] `http://www.guardian.co.uk/technology/blog/2010/jun/03/digg-dead-falling-visitors(Blog article, accessed: 30.12.2011)`

The general states of user activity can be divided into four main categories: Others (non-members), Guests, Passive Members, Active Members and the additional group of Internal Members including administrators, general staff and developers as proposed in [32] and depicted in Figure 2.

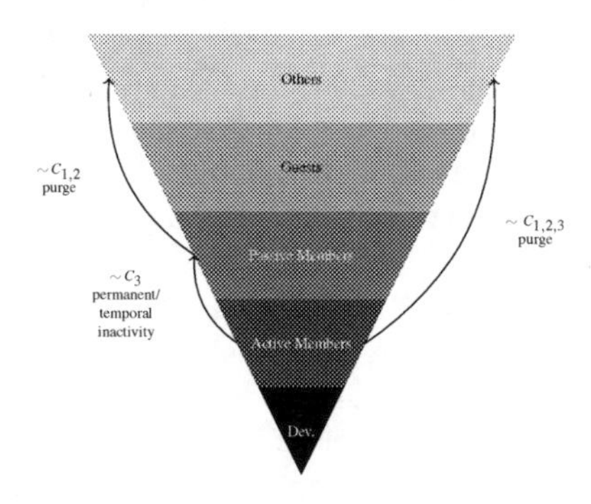

Fig. 2 User withdrawal from social network services

This paper focuses on user withdrawal, defined according to the transitions $\sim C_3$, $\sim C_{1,2}$ and $\sim C_{1,2,3}$. The transitions represent three categories of user withdrawal:

1. **Purge:** Leaving the social network service, deleting all previously submitted content, abandoning all social bonds previously formed via the social network service and deleting the account: $\sim C_{1,2,3}$ and $\sim C_{1,2}$.
2. **Permanent inactivity:** Leaving the social network service forever without burning all bridges and leaving at least the profile and previously submitted content available: $\sim C_3$.
3. **Temporal inactivity:** Leaving the social network service for a unspecified period of time while being available for contact in the social network service ($\sim C_3$, but with the possibility to reverse). The return of the user can be triggered by internal events (curiosity, needs, etc) or external triggers such as notification about activities in the social network service in relation to the user (another user posted a message, tagged the user on an image, etc).

For two reasons classical theories of *substitution of goods and services* [11] can not be applied to the notion of users leaving a social network service. The fundamental theorem of substitution states that the rate of consumption falls as the price of goods rises. Since there is **no pecuniary incentive** to totally abandon a social network service. Profiles can remain defunct - even if the social network service is no longer used. Second, due to the **low cost of usage** (not accounting for time required), social network services are often used in parallel, without completely substituting each other in the classical sense.

4 The Proposed Research Model to Understand User Withdrawal

In this section we combine and adapt selected theories and concepts (cf. Section 2) from the domain of user behavior to a unified research model that describes which factors influence users in their decision to leave social network services.

By using the well established Technology Acceptance Model and by utilizing its extensions, this section discusses an application to social network services and the investigation of user withdrawal. The Technology Acceptance Model is a theory that describes how users accept and use a new technology. This model suggests that when users are confronted with a new technology, multiple factors influence their decision about how and when they will use it and therefore it allows to postulate reasons for leaving. Although the original Technology Acceptance Model has been widely criticized in literature for its lack of falsifiability, limited explanatory and predictive power and little practical value [6], its findings are helpful and with the support of its extension, the Unified Theory of Acceptance and Use of Technology [36], a solid framework for our research model exists.

The two main factors, that we built our research model on, from the Technology Acceptance Model are:

Perceived usefulness which was originally defined in [1] as "the degree to which a person believes that using a particular system would enhance his or her job performance". And **Perceived ease-of-use** which was defined as "the degree to which a person believes that using a particular system would be free from effort" [1]. The notion of perceived ease-of-use can be transferred to the context of social network services without any further modification. However the notion of perceived usefulness must be adapted to the context of social network services due to the fact the term *job* as used in the original version stems from a business oriented environment. As already outlined in Section 3, tasks executed by users with the help of social network services are predominately the exchange of personal information, communication, sharing information on common interests and activities and taking part in an online social life [20, 23].

Perceived ease-of-use is determined by the four anchors of computer self efficacy, perception of external control, computer anxiety and playfulness and two adjusting factors of enjoyment and objective usability [35]. In order to appeal to users, avoid frustration and establish a behavioral usage intention for the social network services it is therefore necessary to consider these influences to the perceived ease-of-use and usefulness in our research model.

Furthermore, to assess the internal motivation of users, [16] recommends to evaluate their individual **expectancy** to **effort** (how much effort the user believes to need) and **performance** (how well the user thinks he is going to perform) and are therefore included in the model.

We came to the conclusion that another important factor to consider in our model is the loyalty of already existing users (cf. [9, 14]). **Customer loyalty** is viewed as the strength of the relationship between an individual's relative attitude toward an entity (in this case a social network) and repeated patronage behavior [9]. Based on work originating from classic customer loyalty modeling, Rui, Lih-Bin and Kanliang propose in [14] a model that describes customer loyalty for social networking sites using five influences that are adapted to the social networking environment:

- **Utilitarian value:** the value of useful functions of the social network services.
- **Hedonistic value:** the amount of pleasure given to the user from the social network services.
- **Satisfaction:** The contentment resulting from using the social network services.
- **Objective knowledge:** The user's knowledge that can not be contested or refuted in the context of the social network services.
- **Subjective knowledge:** True beliefs held by the user.

The authors propose that the *utilitarian value* of a social network services most commonly derives from the ability to interconnect with other users through sharing pictures, commenting on posts, pictures, profiles and group capabilities. The *hedonistic value* on the other hand is described as a result of spontaneous responses which are highly subjective and personal. We have observed that social network services appeal to this by providing challenging games and interesting applications that allow the users to fulfill their social needs.

The previously presented concepts build the foundation of our research model as depicted in Figure 3: Direct dependencies between the determinants are depicted by arrows. Negative changes in the six identified direct determinants of intention directly correspond to a negative behavioral intention thus decreasing usage behavior. The model contains four moderating factors (experience, voluntariness, gender and age) that were suggested by [36, 6] and are directly applicable to the context of social network services. They have varying effect on the model's direct determinants (see [36] for further details) for intention of use. The same goes for facilitating conditions that directly influence use behavior. Since they heavily depend on the context, we concluded that they must be adapted to each social network service individually

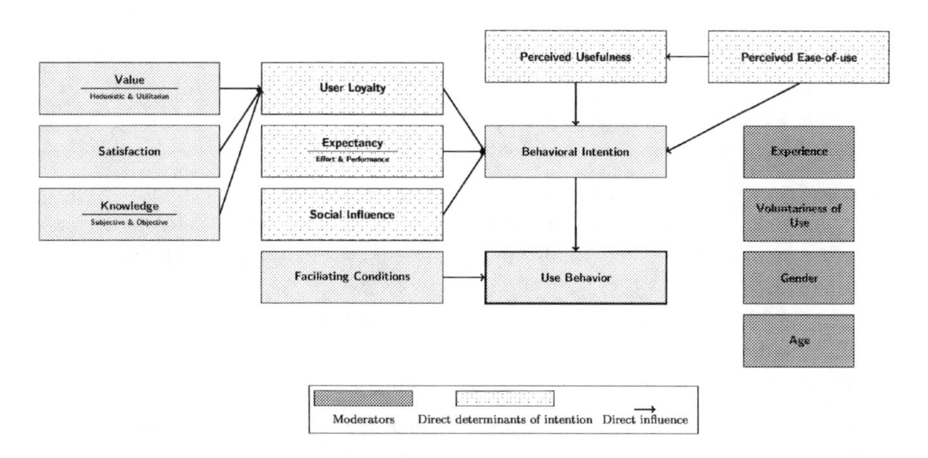

Fig. 3 Suggested Model to describe reasons for users to leave social network services

and may include regulatory factors, monetary incentives (social network services of the employer), etc.

The first moderator, *experience*, states that the use of the platform heavily depends on the experience level of the user (e.g. [19, 30]). The better the overall experience of the platform is, the more it will be used. Furthermore, we deemed the general properties of *gender* (according to [12]) and *age* (according to [27]) as equally important moderators. Especially in the context of organizational social networking services it is important to take the *voluntariness of use into account*. Users who are forced, for instance by company rules, to use specific systems are more reluctant than users who discover those systems, their values and functionality on their own terms (see also [2]).

To account for additional influences that depend on the organizational context and can neither be attributed to the user himself nor the platform, such as company policies, environmental or organizational factors, we decided to introduce the generic *faciliating conditions*.

5 Applying the Findings - Design Principles to Mitigate User Withdrawal

Due to the diverse nature of social networks in general, it is almost impossible to use the proposed research model as depicted in Figure 3 to describe withdrawal from each and every social network available without further adaption. Based on the findings of the previous sections, we present in this section a general approach to mitigate user withdrawal from general purpose social network services which can also, to a varying degree, be applied to other kinds of social network services [5].

5.1 Deriving Four Rules to Decrease User Withdrawal

Since previously discussed models focus on usage reasons, it can be conversely be concluded that missing incentives for use bring users to leave social network services. Therefore four rules can be established that foster user activity and conversely the lack thereof increases user abandonment:

User engagement: By giving users something to do, for instance through promoting interesting or relevant content from their peers, the impression of activity and perceived usefulness is given. Real-time features such as regularly changing content (e.g. status updates of peers) keep the user engaged and interested. This primarily satisfies the determinants of *user loyalty* and *perceived usefulness* from the proposed model. It is obvious that engaged and interested user perceive the social network services to be more useful and are more likely to remain loyal to it.

User expression: Through profile pages and the promotion of individuality throughout the social network service, user satisfaction can be fostered. Again, a high level of user expression within a social network service satisfies the determinants of *user* *loyalty*, *social influence* and *perceived usefulness* from the proposed model. If the social network services does not provide expressive means of communication among it's user base, the perceived usefulness by an individual is much lower than compared to other social network services. This also decreases the social influence to use this social network services and increases the social pressure to migrate to another social network services.

Establishing communication channels: The usage of manually created as well as automatically created groups tightens the bond between like minded users inside a social network service. By providing open programming interfaces, third party providers can provide additional applications inside the social network services ecosystem thus increasing the overall usefulness. Higher utilization can be achieved through additional communication channels such as SMS, mobile versions of the social network services and email. The availability of many communication channels that the user can use as he desires (of course he should also be able to choose to not use) improves the *perceived usefulness*, *user loyalty*, increased *value* and *satisfaction* as well as *social influence* (many communication channels mean many potential communication partners). This also increases *perceived ease-of-use* since the user can use the means of communication that he is most familiar with.

Keep everything easy: Avoiding clutter and removing unnecessary visual distraction mitigates user frustration and the ability to create custom filters increases the utilitarian value. A simple user interface that emphasizes on relevant information and guarantees ease-of-use is important. By using Avatars for users and simple means of finding relevant other users the overall user experience can be optimized. This primarily increases individual's *perceived ease-of-use* and indirectly *user loyalty* by increasing the user's *satisfaction* and *value*.

6 Conclusion

In the field of information systems research several theories have been developed that are capable of describing what drives users to technology and how new software products are adapted. With the ever growing popularity of social networking sites and currently ongoing re-consolidations on the social networking market in connection with the linked downfall of formerly major players the question of why users withdraw from such sites has become of great interest.

To answer this question, a research model was presented in this paper that, based on established theories from the field of information systems research, can outline why users leave social network services. Starting from a market analysis for social network services, established theories from the field of IS research were analyzed, adapted and combined to the resulting research model. The verification of the proposed model against existing social network services is the subject of an upcoming evaluation.

The proposed research model has several important implications for research on social networking services. This model is grounded in information systems theory and, as such, shares the advantage of systems models: it mirrors reality. Here, the object of interest is the usage decline in social networking services that it is understood within the context of its environment.

Acknowledgements. The research herein is partially conducted within the competence network Softnet Austria II (`www.soft-net.at`) and funded by the Austrian Federal Ministry of Economy, Family and Youth (bmwfj), the province of Styria, the Steirische Wirtschaftsförderungsgesellschaft mbH (SFG) and the city of Vienna in terms of the center for innovation and technology (ZIT).

References

1. Adams, D., Nelson, R., Todd, P.: Perceived usefulness, ease of use, and usage of information technology: a replication. MIS Q. 16(2), 227–247 (1992)
2. Agarwal, R., Prasad, J.: The role of innovation characteristics and perceived voluntariness in the acceptance of information technologies. Decision Sciences 28(3), 557–582 (1997)
3. Ajzen, I.: The theory of planned behavior. Organizational Behavior and Human Decision Processes 50(2), 179–211 (1991), doi:10.1016/0749-5978(91)90020-T
4. Boyd, D.M., Ellison, N.B.: Social Network Sites: Definition, History, and Scholarship. Journal of Computer-Mediated Communication 13(1), 210–230 (2008)
5. Chapman, C.: Social Network Design - Examples and best practices (2010)
6. Chuttur, M.: Overview of the Technology Acceptance Model: Origins, Developments and Future Directions. Technology 9(2009) (2009)
7. Colvin, C.A., Goh, A.: Validation of the technology acceptance model for police. Journal of Criminal Justice 33(1), 89–95 (2005), doi:10.1016/j.jcrimjus.2004.10.009
8. DeSanctis, G., Poole, M.S.S.: Capturing the complexity in advanced technology use: Adaptive structuration theory. Organization Science 5, 121–147 (1994)
9. Dick, A.S., Basu, K.: Customer Loyalty: Toward an Integrated Conceptual Framework. Journal of the Academy of Marketing Science 22(2), 99–113 (1994), doi:10.1177/0092070394222001
10. Ellison, N.B., Steinfield, C., Lampe, C.: The Benefits of Facebook "Friends:" Social Capital and College Students' Use of Online Social Network Sites. Journal of Computer-Mediated Communication 12(4), 1143–1168 (2007), doi:10.1111/j.1083-6101.2007.00367.x
11. Epstein, L.G., Zin, S.E.: Substitution, Risk Aversion, and the Temporal Behavior of Consumption and Asset Returns: An Empirical Analysis. The Journal of Political Economy 99(2), 263–286 (1991)
12. Gefen, D., Straub, D.W.: Gender differences in the perception and use of e-mail: An extension to the technology acceptance model. MIS Quarterly 21(4), 389–400 (1997)
13. Giddens, A.: The constitution of society: outline of the theory of structuration. University of California Press (1984)
14. Gu, R., Oh, L.-B., Wang, K.: Determinants of Customer Loyalty for Social Networking Sites. In: Sharman, R., Rao, H.R., Raghu, T.S. (eds.) WEB 2009. LNBIP, vol. 52, pp. 206–212. Springer, Heidelberg (2010)

15. Hargittai, E.: Whose Space? Differences Among Users and Non-Users of Social Network Sites. Journal of Computer-Mediated Communication 13(1), 276–297 (2008), doi:10.1111/j.1083-6101.2007.00396.x
16. Heckhausen, H.: Achievement motivation and its constructs: A cognitive model. Motivation and Emotion 1(4), 283–329 (1977)
17. Hsu, C.: Why do people play on-line games? An extended TAM with social influences and flow experience. Information & Management 41(7), 853–868 (2004)
18. Joinson, A.N.: Looking at, looking up or keeping up with people? Motives and use of facebook. In: Proceeding of the Twenty-Sixth Annual SIGCHI Conference on Human Factors in Computing Systems, pp. 1027–1036. ACM (2008)
19. Jonassen, D.H., Peck, K.L., Wilson, B.G.: Learning with technology: A constructivist perspective, vol. 16. Prentice-Hall (1999)
20. Kumar, R., Novak, J., Tomkins, A.: Structure and evolution of online social networks. In: Link Mining: Models, Algorithms, and Applications, pp. 337–357 (2010)
21. Lampe, C., Ellison, N.B., Steinfield, C.: Changes in use and perception of facebook. In: Proceedings of the ACM 2008 Conference on Computer Supported Cooperative Work, CSCW 2008, p. 721 (2008), doi:10.1145/1460563.1460675
22. Lerman, K.: User Participation in Social Media: Digg Study. In: 2007 IEEE/WIC/ACM International Conferences on Web Intelligence and Intelligent Agent Technology - Workshops, pp. 255–258 (2007), doi:10.1109/WI-IATW.2007.68
23. Livingstone, S.: Taking risky opportunities in youthful content creation: teenagers' use of social networking sites for intimacy, privacy and self-expression. New Media & Society 10(3), 393–411 (2008), doi:10.1177/1461444808089415
24. Maia, M., Almeida, J., Almeida, V.: Identifying user behavior in online social networks. In: Proceedings of the 1st Workshop on Social Network Systems, pp. 1–6. ACM (2008)
25. Mankoff, J., Matthews, D., Fussell, S., Johnson, M.: Leveraging Social Networks To Motivate Individuals to Reduce their Ecological Footprints. In: 2007 40th Annual Hawaii International Conference on System Sciences, HICSS 2007, p. 87 (2007), doi:10.1109/HICSS.2007.325
26. Miller, N.E., Dollard, J.: Social Learning and Imitation, vol. 39. Yale University Press (1941), doi:10.1037/h0050759
27. Morris, M.G., Venkatesh, V.: Age differences in technology adoption decisions: Implications for a changing work force. Personnel Psychology 53(2), 375–403 (2000)
28. Nooy, W.D.: Social Network Analysis, Graph Theoretical Approaches. Network, 1–23 (1994)
29. Raacke, J., Bonds-Raacke, J.: MySpace and Facebook: applying the uses and gratifications theory to exploring friend-networking sites. Cyberpsychology & behavior: The Impact of the Internet, Multimedia and Virtual Reality on Behavior and Society 11(2), 169–174 (2008), doi:10.1089/cpb.2007.0056
30. Schmitz, J., Fulk, J.: Organizational colleagues, media richness, and electronic mail: A test of the social influence model of technology use. Communication Research 18(4), 487–523 (1991)
31. Sheppard, B.H., Hartwick, J., Warshaw, P.R.: The Theory of Reasoned Action: A Meta-Analysis of Past Research with Recommendations for Modifications and Future Research. The Journal of Consumer Research 15(3), 325–343 (1988)
32. Sillaber, C., Chimiak-Opoka, J.: Social Requirements Engineering Overview. Computer, 1–13 (2011)
33. Tjan, A.K.: Finally, a way to put your Internet portfolio in order. Harvard Business Review 79(2), 76–85 (2001)

34. Tornatzky, L.G., Eveland, J.D., Boylan, M.G., Hetzner, W.A., Johnson, E.C., Roitman, D., Schneider, J.: The process of technological innovation: Reviewing the literature (1983)
35. Venkatesh, V.: Determinants of perceived ease of use: Integrating control, intrinsic motivation, and emotion into the technology acceptance model. Information Systems Research 11(4), 342–365 (2000)
36. Venkatesh, V., Morris, M.G., Davis, G.B.: A Unified Theory of Acceptance and Use of Technology. MIS Quarterly (2003)
37. Walther, J.B., Van Der Heide, B., Kim, S.Y., Westerman, D., Tong, S.T.: The Role of Friends Appearance and Behavior on Evaluations of Individuals on Facebook: Are We Known by the Company We Keep? Human Communication Research 34(1), 28–49 (2008), doi:10.1111/j.1468-2958.2007.00312.x

Connecting Customer Relationship Management Systems to Social Networks

Hanno Zwikstra, Frederik Hogenboom, Damir Vandic, and Flavius Frasincar

Erasmus University Rotterdam, P.O. Box 1738, NL-3000 DR Rotterdam, The Netherlands
265948jz@student.eur.nl,
{fhogenboom,vandic,frasincar}@ese.eur.nl

Abstract. As the popularity and the commercial potential of social networks such as LinkedIn and Facebook increase, we present a framework that aims to reuse social networks data within a customer relationship management (CRM) application. The framework has been implemented in *LinkedInFinder* that pulls data from LinkedIn into the Microsoft Dynamics CRM system. Our proof-of-concept implementation demonstrates the use of the proposed framework, based on a use case to find second-degree connections within one's network that work at a specific company of interest. A survey amongst target users suggests that the application is useful and adequately designed for the intended use.

1 Introduction

With the advent of Web 2.0 [6], there has been a growing importance regarding the social aspects of the Web. Even though social networks have been existing as long as there have been societies, digital networks – such as Twitter, Facebook, and LinkedIn – experienced a substantial growth over the last decade. Due to their promising commercial potential, there has been put an increasing amount of effort and research into social networks on the Web. Web 2.0 social networks are defined as sets of social entities (e.g., people, organizations, etc.) connected by a set of social relationships (e.g., friendship, co-working, information exchange, etc.) [4].

The rich potential of social networks comes from two aspects. First, social networks can be utilized to push information to a target group, e.g., company advertisements, blogging, tweeting at Twitter, etc. Second, pulling information from social networks into a customer relationship management (CRM) system is also possible. An example is monitoring social network sites for content (conversations, blogs, tweets on Twitter, etc.) in which a company or brand is mentioned, providing

L. Uden et al. (Eds.): 7th International Conference on KMO, AISC 172, pp. 389–400.
springerlink.com

potentially valuable information and means to interact with customers. Microsoft already provides an add-on for its CRM system that imports tweets from Twitter in which a company is mentioned, into the CRM application. Oracle, the world's leader in CRM, advocates Social CRM actively by offering Oracle Social CRM Applications.

Due to the increasing popularity of social networks as well as their commercial potential, in this paper we present a framework that aims to reuse LinkedIn data within a CRM application. We evaluate the framework by means of a proof of concept implementation. In our development, we aim for simplicity, yet we also allow for future extendability. The remainder of this paper is organized as follows. Related work is described in Sect. 2. We then discuss our framework and its implementation in Sects. 3 and 4. The implementation is evaluated in Sect. 5. Finally, we conclude and provide directions for future work in Sect. 6.

2 Related Work

With the rise of Web 2.0, many new features emerged that enriched the Web environment, e.g., blogs, wikis, (social) tagging in folksonomies, and mashups that combine data from one or more other sources to create a new application, usually using an Application Programming Interface (API). One of the most noticeable features of Web 2.0 are social networks, like LinkedIn and Twitter. Social networks – also known under the common denominator as Social Networking Sites (SNS) – are a specific type of Social Media. Their content is generated and maintained by its visitors, without central coordination. Typically, in Social Media Web sites people form relationships with other users and interact with them [2]. Figure 1 depicts the number of unique visitors for LinkedIn and Twitter in the Netherlands, illustrating the immense growth in terms of visitors in the past 18 months. Interestingly, Twitter is still less popular than LinkedIn in the Netherlands, while in the rest of the world it is the other way around.

In the early 1990's the term CRM began to emerge [1]. Nowadays, many CRM providers exist, which offer Software as a Service solutions, e.g., Salesforce, Microsoft Dynamics CRM online, Oracle CRM On Demand, and SageCRM. The difference between CRM and traditional marketing is in the focus on customer

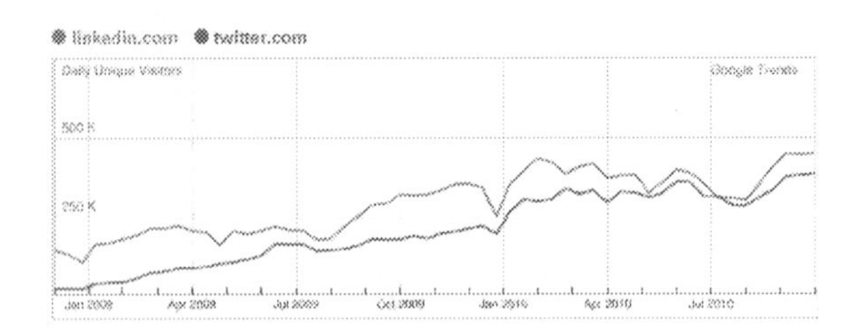

Fig. 1 Trend lines of unique visitors of LinkedIn and Twitter

relationships. Traditional marketing's main focus is on acquiring new customers, while CRM focuses on developing long term relationships with existing customers. According to Payne and Frow [7], three views CRM exist, ranging from a narrow and tactical view where CRM is a specific technical solution or tool, to a broad and strategic, or even philosophical view where CRM is a holistic approach for managing customer relationships.

Driven by changes in customers, starting in 2007 and more rapidly in 2008, CRM began to transform into what is now known as CRM 2.0 or Social CRM (sCRM) [3]. Both terms hint to Web 2.0 which is also called the Social Web. Today's customers are no longer passive customers, but have become social customers, who are active on the Internet, writing blogs, using Twitter, having discussions and informing themselves on social networking sites etc. and thus are well-informed about companies and products and services that they offer. CRM 2.0 is about joining the conversation [5] and extending CRM out of the business offices into the Social Media on the Internet, implying a change in strategy from a focus on customer transactions to a focus on both customer interactions.

3 Framework

This section discusses a general approach to social network data pulling called *SocialCRMConnector*. The *SocialCRMConnector*, which is a mashup between an arbitrary social network and a CRM system, can be used for pulling data from social networks using available APIs to build sCRM applications that use this data. Data can be pulled from a social network by sending a request to the API, which returns a response with the requested data. Our framework is targeted at retrieving profile data of social network members, i.e., personal information about social network members that is available by using the API of the social network. Although such profile data is also available on the social network sites themselves, there are three reasons why a company might want to retrieve the data by using an API. The first is simplicity. An application can contain specific business rules that are executed without input that is required from the user. This makes it easier to retrieve the desired data. The second is control. Data that is retrieved from an API can be processed in other applications or (temporarily) stored in internal information systems for further processing. Third, APIs enable applications to keep information up-to-date by allowing real-time access to data.

3.1 Entities and Relationships

At its core, a social network consists of *Users*, i.e., the people who have an account at the social network, and inter-user *Connections* define the relationships between users. Users have several attributes, i.e., a unique ID, nickname and real name, birthday, and likely a number of other attributes. With these entities it is already possible to create a network, though the possibilities of such a network are of course very limited at this point. Two *Users* with a *Connection* between them form a very small network. This simple model can be extended to better reflect real world online social networks. We have examined several online social networks, with a focus on Facebook and LinkedIn. To visualize the model, we have created an entity-relationship diagram for the model as an example, which is shown in Fig. 2. We focus only on the entities that may be useful sCRM applications; in practice social networks have a more complex structure, with additional entities such as *Event* or *School* and support for multiple media types such as videos, photos, and music.

Compared to the simple model that consists only of *Users* and *Connections*, we have introduced some new entities that are common for social networks. The figure illustrates that the *User* is the central entity in a social network. The first entity that we added is *Message*. Being able to send messages to other members of a social network is a very important feature of social networks. Without the ability to send messages, it would be impossible to interact with other users. Messages can be sent from a user to another user, but also to other entities, for example a user can comment on a photo that is posted by someone else, or post a comment to a discussion in a *Group*. For companies who want to monitor social networks for comments on their brand, the *Message* entity is an important one.

Instead of an entity that may have been called "*MediaItem*" to reflect all types of media items, we have added an entity called *Photo*, because photos are supported by all sites, even if it is only to add a picture to the user profile. We assume that the only useful media item to pull from a social network in relation to sCRM would be a profile picture. Comments that may have been added to profile pictures by other users are not really relevant in a business application, nevertheless for completeness reasons we decided to keep the link between *Message* and *Photo*.

Another feature of many social network sites is the concept of *Groups*. Groups are a convenient way for users to connect with other users who share a similar interest. Groups can have many different forms, for example companies, schools, brands, persons, and virtually anything else. Groups are basically a collection of users, and are in many ways similar to individual users, but the concept and purpose is too

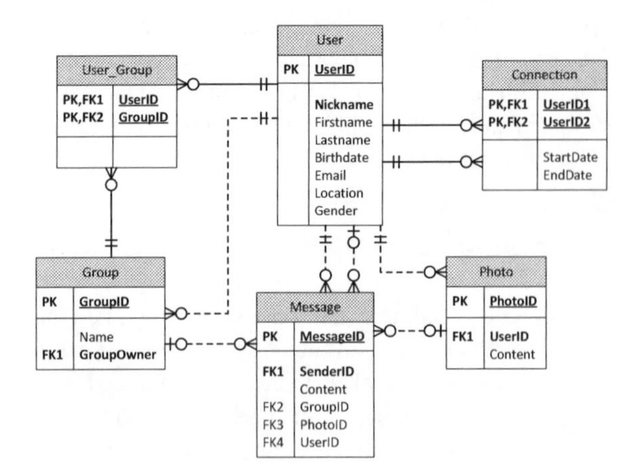

Fig. 2 Entity-relationship diagram of the social network model

distinct to treat them as one entity. The connections in groups are not direct connections (users do not become "friends" when joining the same group), but it makes it easier to meet new people. There is a many-to-many relationship between *Users* and *Groups*. There is also a direct relationship between *User* and *Group*, because a group is created by a user who then becomes the group owner. This relationship might also be displayed as a many-to-many relationship if the group can be owned by multiple users. Groups make it much easier for companies to find the consumers that they are interested in and the companies can participate in relevant groups to join the conversation with the consumers.

CRM databases are relational databases and usually consist of many tables, such as companies, persons, opportunities, products, orders, quotes, support calls, appointments, etc. Like the social network structure, our focus is on those entities that might be useful in the *SocialCRMConnector* framework. The CRM structure is shown in Fig. 3. Since we target social applications, the *Person* entity is an important entity in our CRM model, just like *User* is in a social network. However, *Companies* are the central entity in a CRM system. A company has a specific type that indicates the relationship between that company and the super company in which the current company is listed. Specific company types can be customers, competitors, suppliers, partners, government, etc. A third entity called *Lead* is also added. Leads, or prospects, are defined as potential customers that do not yet have a relation with the company. Finally, an *Opportunity* entity is used to record a potential sale. An opportunity is given an estimated value and expected probability that the sale will be made. Linked to an opportunity must be one or more *Products* that specify what is going to be sold.

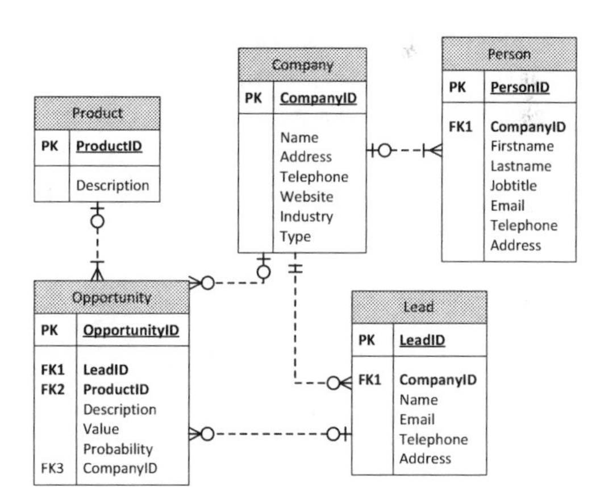

Fig. 3 CRM entity-relationship diagram (simplified)

3.2 Framework Architecture

After defining a model that covers the part of social networks that is relevant for our research, as well as a model that covers the structure of a CRM system, we can define the framework architecture. The framework consists of the following steps, which are also depicted in Fig. 4:

1. Generate query for retrieving profile information from a social network;
2. Retrieve data by sending the query to the API;
3. Process the data so it can be used in an application;
4. Present information from the application to the user;
5. Store data in the CRM system (optional, depending on user's choice).

The content of a query needed for retrieving the necessary data from a social network depends on a number of input parameters, for example the current user that uses the application, entity attributes that must be retrieved (dictated by business rules), and optional other parameters to refine the query such as a specific company or person to search for. Because the APIs flatten the data before returning it, queries to retrieve data are relatively simple. Flattening data is the opposite of normalizing data in the database. For example, a normalized *User* in a database may store a user's country with a foreign key to a record in a *Country*. The flattened data will return the data for a specific user with a country attribute that contains the country name. The pseudo SQL query for retrieving user data is:

```
SELECT attribute1, attribute2, ..., attributeN
FROM User, ...
WHERE condition1, condition2, ..., conditionN;
```

Depending on the type of application, information can be inserted into the CRM system. Some applications might only display additional information without the need to store anything in the CRM system. Caution is required when data is going to be saved, because while it is technically possible to store data in the CRM system once it has been retrieved from a social network, not just any data can be stored without considering issues concerning privacy or obsoleteness. In contrast to

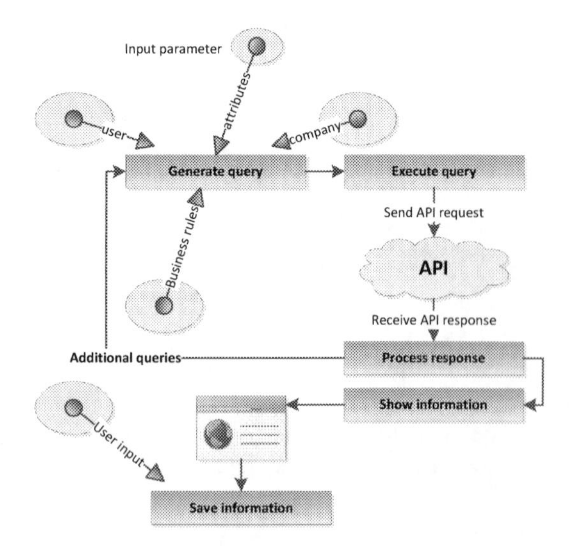

Fig. 4 Steps within the *SocialCRMConnector* framework steps

contents of social networks, the IDs of records never change, and hence only insert queries (i.e., no update queries) are required. An insert query in pseudo SQL is:

```
UPDATE Person
SET Person.socialnetworkID = 'ID'
WHERE condition1, condition2, ..., conditionN;
```

4 Implementation

Now that we have defined the general framework, this section introduces the implementation of the *SocialCRMConnector*. *LinkedInFinder*, which integrates Microsoft Dynamics CRM with the LinkedIn social network, supports a use case where CRM users aim to find people that have a job at specific companies that are not yet listed within the CRM and with whom the company that uses the CRM would like to get into contact. These employees, which are supposedly the connection to a company of interest, are connected through first-degree connections with the CRM users. Connections of a second-degree or higher do not really make sense as it makes communication difficult. For demo and evaluation purposes, we have developed a stand-alone version of the *LinkedInFinder* application, which is available at `http://linkedin.hantheman.tk`. The main flow of the application is as follows.

After presenting the user a login page hosted by LinkedIn in order to authorize *LinkedInFinder*, first, a list of LinkedIn members that work at a certain company is retrieved with a call to the LinkedIn Search API. The list is then filtered so that only second degree connections are displayed to the user. Second degree connections have a distance of 2 to the user, which is measured as the number of steps from the user to the LinkedIn member. This means the distance to the own profile is 0, the user's connections have a distance of 1 and their connections have a distance of 2. Second, after selecting a name from the list by the user, more detailed information is displayed, including how the user is connected to this person. This connection is presented as a list of one or more of the user's first-degree connections. This list of connections actually is a list of mutual connections between the user and the person that is viewed.

4.1 Application Back-End

The *LinkedInFinder* application is integrated with Microsoft Dynamics CRM, which is an ASP.NET Web application that uses .NET Web services to communicate with the CRM database. For easy integration, our application is therefore also an ASP.NET Web application. It is built in Visual Studio 2010, using the (ASP).NET 3.5 Framework and C# as programming language. For connecting and interacting with the LinkedIn API, we employ the LinkedIn Developer Toolkit, which is an open source library for using the LinkedIn API in .NET applications. This library provides .NET support for the LinkedIn API, by providing .NET wrapper methods for most LinkedIn API methods and an implementation of the OAuth (Open

Authentication) protocol that is used by many social network sites to authenticate requests to their APIs. It should be noted that the LinkedIn API has a number of technical restrictions that limit the possibilities of building applications that use the API. More complex applications that require more API calls are expected to suffer more from these limitations than simple applications that only use a few simple API calls. First, LinkedIn restricts the use of its API by limiting the number of calls to the API methods, i.e., throttling. Second, the number of results returned for a people-search query is limited by the account-level of the LinkedIn member.

The entity-relationship diagram underlying our implementation as depicted in Fig. 5 is a modified version of the entity-relationship diagram discussed in Sect. 3 (Fig. 2), reflecting the LinkedIn structure. In LinkedIn, media items are not as important as they are in other social networks like Facebook. Only photo items are available for storing a profile picture of the members. A user profile in LinkedIn has a large number of attributes and links to other entities. We have added the attributes and entities that are relevant for our application. For privacy reasons, LinkedIn does not return exact addresses, but areas, and hence we have added a *location* attribute to the model. The email address of LinkedIn members cannot be retrieved by any API calls, so we have removed it from the model. Phone numbers are optional and stored in a separate entity *PhoneNumber* that stores the actual phone number and type of phone number (e.g., "mobile"). Subsequently, we include a *URL* attribute, that stores the public profile URL. Furthermore, the *industry* attribute indicates in which industry the person is active; this may give some additional background information about the person's job. Then, another important entity in our application is *Company*. The relationship between *Users* and *Companies* is a many-to-many relationship, i.e., users can work at multiple companies. In our application we search for LinkedIn members that work at a specific company. The last entity that we need in our model is *Position*. A position describes the actual job that the LinkedIn member

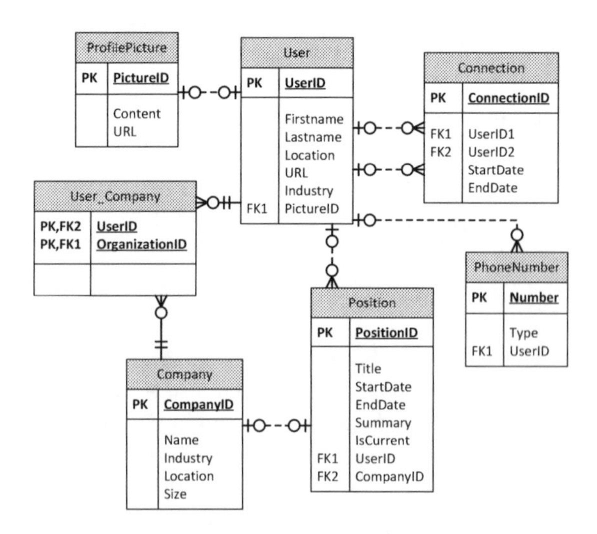

Fig. 5 Entity-relationship diagram of LinkedIn

has at a company. The *iscurrent* attribute indicates whether the position is a current - or past position. In the LinkedInFinder application we only use current positions.

In our implementation, the application is started from a company record in the Microsoft Dynamics CRM system, which is opened in a Web form. The company name of the record is sent to the application as input parameter. The most important items of the implementation are the API requests that are sent to LinkedIn. The application uses a Search API request and a Profile API request. The API request has three variable parameters, which are the `company-name`, `start` and `count`. The `company-name` specifies the company name to search for. The `start` and `count` parameters specify the indexes for a subset of data. For example `start=0`, `count=10` retrieves the first 10 results. The other parameters will remain the same for each request. A search for the first ten people that currently work at "Erasmus" would yield the following request:

```
http://api.linkedin.com/v1/people-search:(people:(id,distance,
first-name,last-name))?company-name=Erasmus&current-company=1&
sort=relevance&start=0&count=10
```

This request is divided into several parts. The first part, i.e., `http://api.linkedin.com/v1/`, is the base URL of the LinkedIn API. The second part is the API function that is called, i.e., `people-search:`. The third part contains field selectors and utilizes a JSON-like syntax, i.e., `(people:(id,distance,first-name,last- name))`. The rest of the query, i.e., `?company-name=Erasmus¤t-company =1&sort=relevance&start=0&count=10`, comprises an optional query string with the three variable parameters. In this case we specify that the company name is Erasmus, it must be the current company, search results are ordered by relevance (other options are number of connections and distance), and the number of results that is returned per request is limited to ten.

The aforementioned request returns ten profiles (assuming at least ten people work at Erasmus) with the `id`, `distance`, `first-name`, and `last-name` fields. The distance field tells the distance between the user that executes the query and the person that is returned by the API. People with a distance of 1 are first-degree connections, and people with a distance of 2 are second-degree connections. We use the distance field to filter the search results to contain only second-degree connections (we cannot specify this in the search query).

The Profile API request has only one variable parameter, which is the profile id. The request is as follows:

```
http://api.linkedin.com/v1/people/id=nnl7Qkt7Kb:(last-name,
first-name,num-connections,num-connections-capped,phone-num
bers,three-current-positions,picture-url,location:(name),re
lation-to-viewer:(distance,num-related-connections,related-
connections),positions:(title,company:(name)))
```

The response for this request contains the relevant profile details and is parsed and displayed to the user who can then decide to use this information or to discard it and view another profile.

4.2 Application Front-End

The front-end of the application has three main windows. First, when opening an account record within the CRM tool, there is a button that will open the *LinkedIn-Finder* application shown in Fig. 6. Upon first use, the user will be redirected to a LinkedIn page to authorize the application using OAuth. Subsequently, the application redirects to a secure page on the LinkedIn Web site, and after entering the LinkedIn account credentials, the user is redirected back to the application. Second, there is a Web page (`Search.aspx`) displaying the results of the search query as a list of hyperlinks. The names of the people that are returned by the search query are displayed as the hyperlink text. Third, another Web page (`Details.aspx`) is opened when a hyperlink is clicked. This page, as depicted in Fig. 7, displays the detailed information for that person, which requires a Profile API call.

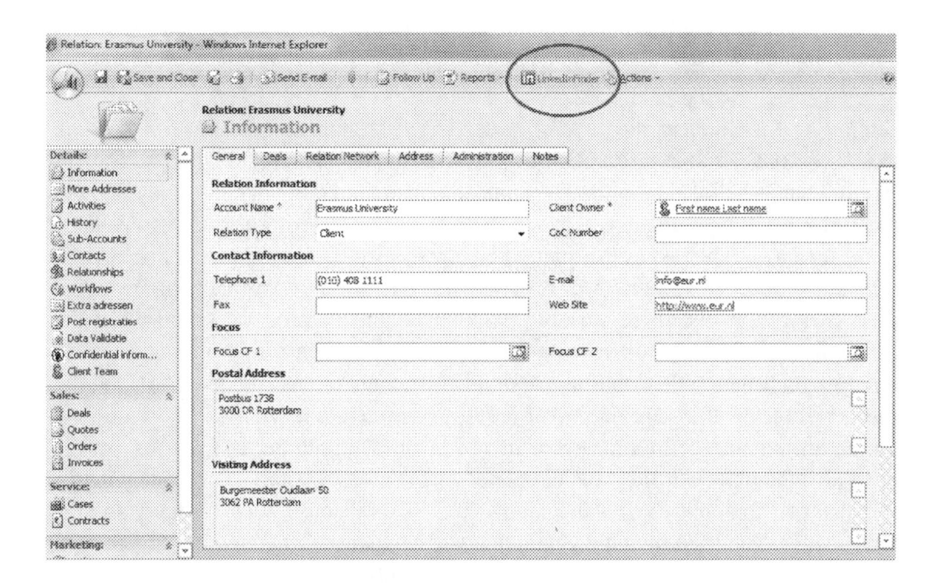

Fig. 6 Account form in Microsoft Dynamics CRM

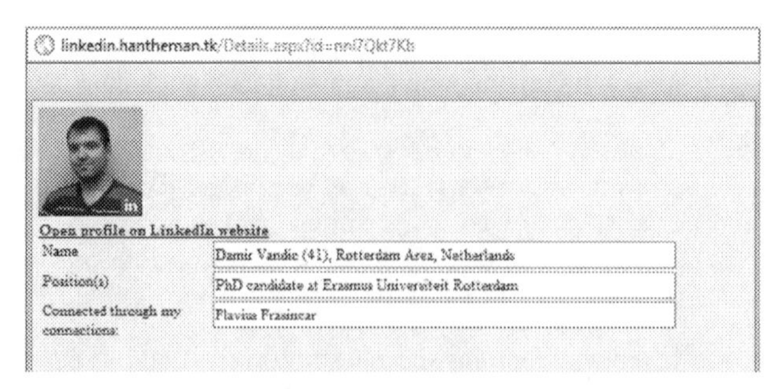

Fig. 7 Person details in the *LinkedInFinder* application (`Details.aspx`)

5 Evaluation

In terms of evaluation, we first validate our implementation by comparing search results of *LinkedInFinder* with the LinkedIn Web site search engine (using advanced search). For several companies, the search results are identical in all cases. Interestingly, when using the standard search function on the LinkedIn Web site, search results are different in terms of display order, pointing to a different ranking algorithm.

In order to evaluate the *LinkedInFinder* tool qualitatively, we conduct a survey in which participants are asked for their opinion about the following statements in a questionnaire:

Statement 1: I find the application useful.
Statement 2: The application shows enough information to be useful.
Statement 3: The application offers enough functionality to be useful.
Statement 4: I would use the application for my job.

A five-point Likert scale is used for the available answers (ranging from 1 [totally disagree] to 5 [totally agree]), which provides an ordinal scale. The results of the survey (amongst 17 participants that are CRM users and members of the authors' LinkedIn network) are depicted in Fig. 8. Furthermore, we define two hypotheses to evaluate the *LinkedInFinder* application, i.e.:

Hypothesis 1: The *LinkedInFinder* application is useful.
Hypothesis 2: The *LinkedInFinder* application design is adequate.

To assess hypothesis 1, we use statements 1 and 4. To assess hypothesis 2 we use statements 2 and 3. When employing the χ^2-test to determine whether the attitude towards the *LinkedInFinder* is neutral, we obtain significant outcomes to reject this null hypothesis (with p-values lower than 0.01) for all statements, which means we can accept both hypotheses.

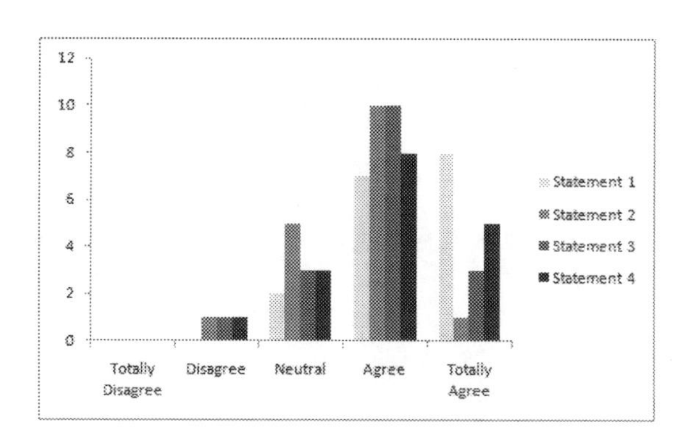

Fig. 8 *LinkedInFinder* evaluation results

6 Conclusions

In this paper, we presented the *SocialCRMConnector* framework that aims to feed CRM applications with Web 2.0 social networks data. The framework has been implemented as a tool called *LinkedInFinder* that pulls data from LinkedIn into the Microsoft Dynamics CRM system. Our implementation validates the proposed framework by means of a use case to find second-degree connections within one's network that work at a specific company of interest. Results from our evaluation based on a user survey indicate that the application is useful and adequately designed for the intended use. As future work, we envision implementations of our framework using other social networks (possibly for other purposes as well) and CRM systems. Also, one could employ data pulled from social networks to other applications, such as personalization tools.

References

1. Buttle, F.: Customer Relationship Management: Concepts and Technologies, 2nd edn. Butterworth Heinemann (2009)
2. Gilbert, E., Karahalios, K.: Predicting Tie Strength With Social Media. In: 27th International Conference on Human Factors in Computing Systems, CHI 2009, pp. 210–220. ACM (2009)
3. Greenberg, P.: The impact of CRM 2.0 on Customer Insight. Journal of Business & Industrial Marketing 25(6), 410–419 (2010)
4. Laura Garton, C.H., Wellman, B.: Studying Online Social Networks. Journal of Computer-Mediated Communications 3(1) (1997)
5. Leary, B.: The Tweet Is Mightier than the Sword. CRM Magazine 13(1), 48 (2009)
6. O'Reilly, T.: What is Web 2.0: Design Patterns and Business Models for the Next Generation of Software. International Journal of Digital Economics 65(1), 17–37 (2007)
7. Payne, A., Frow, P.: A Strategic Framework for Customer Relationship Management. Journal of Marketing 69(4), 167–176 (2005)

Advances in Intelligent and Soft Computing: Potential Application of Service Science in Engineering

Norsuzailina Mohamed Sutan, Ibrahim Yakub, Siti Nor Ain Musa, and Asrani Lit

Faculty of Engineering, Universiti Malaysia Sarawak, 94300, Kota Samarahan, Sarawak, Malaysia
{msnorsuzailina,yibrahim,msnain,lasrani}@feng.unimas.my

Abstract. This paper discusses the potential of emerging service science with engineering applications. First the definition and classification of service for engineering discipline are detailed and elaborated. Based on that, this paper focuses on the potential application of service science in the Construction Industry namely Building Information Modeling (BIM). Elaborate discussion on the service value of BIM leads to suggestion for further research in specific areas namely the interaction between experts from the world's major BIM player in order to improve the implementation of BIM.

Keywords: Service, engineering, Building Information Modeling (BIM).

1 Introduction

Since engineering is the application of multi-types of skills and knowledge, its business is highly related to service, an important field in the development of almost all industries ranges from agriculture to manufacturing and construction. Indeed, IBM has cross-related these two fields (engineering and service) into a field wide enough to aid the quality of the industrial growth, called Service Science, Management and Engineering or simply SSME (Xie 2011).

Engineering field has undoubtedly emerged into a higher phase from day to day alongside the amplification of information and communication technology because of the growth of competitive businesses. The survival pave for the field relies on the quality of services provided that differentiates their performance and reliability. This is where service science plays its role as the integrated disciplines of management science, social science, decision making, operational research, computer science and jurisprudence (Huang, Lin, and Peng 2010).

Generally, service science is a study of service design, quality and innovation which requires the combination of fundamental sciences and engineering with management field (Macbeth and Opacua 2010). The service science analyzes and understands the business process, constructs a systematic measurement of the

L. Uden et al. (Eds.): 7th International Conference on KMO, AISC 172, pp. 401–407.
springerlink.com

quality and manages the complex hybrid of skills and knowledge (Liu, Xu and Ma 2010). Hence, the implementation of this approach in engineering industry is important for increasing productivity and creating greater value for all stakeholders.

Though many researches have been done lately on service science, none ever specifies the application of services science in engineering in detail. As an initiation to the application of service science in engineering, this paper aims in proposing a proper definition of service as a tool in widening of the field of service study in engineering. Based on the definition, all engineering industries are classified as specific as possible to distinguish between service related and non-related service activities. Finally, a case study is be used to display the example of the application.

2 Definition of Service for Engineering

Till date, many definitions have been given to service but few are exceptional because of their generality and reliability. For instance, IBM has defined service as interactions between provider and client which create and capture value for both parties (IBM, 2011). However, there are many engineering activities that involve no direct interaction between provider and client such as wastewater treatment project. The client for this system is not only the related industry, but also the community who enjoy clean river which in many cases have no idea of the service provider company.

Besides, Sampson's model of Unified Services Theory has separated manufacturing industry from services because co-production between client and provider is needed in services (Grandison and Thomas 2008). Nevertheless, manufacturing industry relies highly on the market environment that consists mainly of their client which brings them back into the equation. Hence, there is a need for a general definition that could include more activities in the study of service science and/or a reliable definition that could specify some activities that can be used for scientific study. Fortunately, Lau et al. (2011) have provided a definition that is both general and reliable which is;

"A service is a process by which the provider fulfils a mission for a client so that value is created for each of the two stakeholders." (Lau, Wang and Chuang 2011)

This definition consists of four elements which have been thoroughly defined by the authors. For the benefit of engineering study, this paper qualifies the elements further as follows;

Provider is a qualified individual or structured organization that possesses the ability to accomplish a mission. For example, a researcher with Mechanical Engineering qualification is a provider because of the ability to carry out an engineering research. However, other personnel with non-engineering background can also become a provider for engineering services in condition the client knows the profile, skill and knowledge of the provider which qualify its reliability to engineering discipline.

Mission comprises of a single task or several tasks that is planned, managed and carried out by provider to the client whom expects the outcome of the mission by certain requirements. The nature of the mission is engineering which includes the application of science, mathematics, economics, art, design, social and practical knowledge.

Client is the acceptor of the mission, regardless of the realization of the service provider and/or the involvement in the service process. Client can be an individual or a group of people or an organization that is affected by the mission carried out by the provider. For example, the government knows and involves in the road construction projects but the industries only know of the provider without giving any direct input to the provider while the citizen may or may not know and/or involve in the road design and construction.

Value is the tangible or intangible effect of the mission on both the provider and client. Tangible value includes the deliverables prepared by the provider for the client such as blueprint and proposal report while the example of intangible values includes the advice and suggestion given by the provider to the client. Besides, the service must also create value for the provider such as payment and experience that can be used to continue the business. Here, payment is a type of direct value for the provider and experience is created indirectly.

3 Proposed Classification of Services in Engineering

Many of the engineering applications are services in nature. This paper intended to classify those engineering applications based on the definition in Section 2. According to Lau et al., 2011, direct recipients and designated action to achieve the target are the keys to successfully implement the service activities. Lau et al. have identified that services can be categorized into client, non-client, client with ownership and client with no ownership. Based on similar classification, services in engineering are proposed to be classified into three categories; i) service in which the client with engineering ownership is having direct relation with provider, ii) service in which client has engineering ownership but does not realize or aware of provider, and iii) services provided to non-engineering client but aware of provider. This classification is a useful tool for further development in service science in engineering field. Table 1 shows the classes of services that are related to engineering field.

In Class A, the provider has engineering background and has direct relationship with the client in which the client knows and chooses the provider. However in Class B, the client knows its provider but is not necessarily needed to be involved in the process of accomplishing the mission that is to generate electricity. For Class C, the provider must obtain input from the client but the client may not know their existence. This is a back-stage service where engineering provides

Table 1 Classification of Services in Engineering

Provider's Skill & Knowledge	Client		Class	Example
	Relation with provider	Direct input for mission		
Engineering background	Realize of provider	Involve in direct input	A	Provider consults the client in construction
		Not involve in direct input	B	Citizen use electricity generated by provider
	Not realize of provider	Involve in direct input	C	Biomedical equipments that require biological data from human are manufactured by provider and used by patients
		Not involve in direct input	D	Passengers use airplane constructed by provider
Non-Engineering background	Realize of provider	Involve in direct input	E	Architect draws conceptual design for infrastructure that is built by provider
		Not involve in direct input	F	Wastewater treatment operator uses available activated carbon in the market which has been developed by chemist

technology for hospital use to cure patients. Furthermore in Class D that is also a back-stage service, the provider does not require input from the client to accomplish the manufacturing of aeroplane nor requires the client to know the profile of the provider. This is a type of service because the passengers do not buy the plane but only use it. Classes E and F include the non-engineering provider that is required to accomplish engineering mission where the clients for these classes are usually engineering individual or organization. Hence, they must realize the profile of the provider to determine the reliability of the provider's skill and knowledge so it can be applied in engineering field.

4 Case Study: The Service Value of Building Information Modeling (BIM)

Definition of Building Information Modeling (BIM)

Building Information Modeling (BIM) was introduced nearly ten years ago for the purpose of distinguishing the information rich architectural 3D modeling from the traditional 2D drawing. Once merely an industry catchphrase regarding the future of design, it is quickly becoming the central, guiding principle of the design process in the Construction Industry (Hammond 2007). It is a building design and documentation methodology that relies on an internally consistent, coordinated, and computable digital representation of the building and is used for design decision making, and accurate construction document production, planning, and performance predictions. In other words, it is a digital representation of the physical and functional characteristics of a facility. It is also a shared knowledge resource for information about a facility, forming a reliable basis for decisions during its life-cycle; defined as existing from earliest conception to demolition (Deke Smith 2007).

A basic premise of BIM is a collaboration of different stakeholders at different phases of the life cycle of a facility to insert, extract, update or modify information in the BIM to support and reflect the roles of stakeholder. It is about intelligent design that let the stakeholder explore a project's physical and functional, by faster and stronger visualization and more cost effective management of buildings (Golzarpoor 2010). With BIM, engineers are able to keep information co-ordinated, up-to-date, and accessible in an integrated digital environment, which increases profitability, reduces risks, and eliminates inefficiencies in the building design, construction, and management of any project. The model is a shared

Table 2 Benefits of the services provided by BIM (Jin)

Stakeholder	Benefits
Owner	Improved design quality, better performing building (systems coordination, engineering analysis), Fewer change orders (building systems clash detection), Schedule optimization (construction schedule simulation), Schedule compression (digital assisted fabrication, offsite fab), Efficient handover (data exchange for operations/maintenance), Risk reduction (more transparency)
Contractor	Provides a good estimation during bidding and procurement ,improves coordination in construction sequencing ,Effective marketing presentation of construction approaches ,helps in identifying possible conflicts that may arise during building construction ,Risk reduction (better understanding of complexity), Building systems coordination (clash detection, layout) ,Shop drawing reduction (model to fabrication), Digital fabrication (steel, HVAC ducts, piping), Construction schedule optimization (visual schedule simulation), Cost estimating (quantity takeoffs)
Engineer	Design prototype (space arrangements, assemblies, materials), Building systems coordination (space reservation, clash detection), Analysis (space, lighting, energy, structural), Drawing production quality (flexible, exploits automation, better coordinated), Design exploration/interrogation (data rich visual models),Engineering accuracy (measurement, context)
Team	Efficient processes (forces process re-engineering), Improved collaboration (concurrent design, process to rally around) ,Better decisions (better coordinated information),Accelerated understanding (graphics + data), Efficiencies from reuse of data (enter once use many),Improved data quality (less data re-entry, less human error)

digital representation founded on open standards for interoperability. It may be a database made up of a set of interrelated files and not just one entity. The concept of Building Information Modeling is to build a building virtually, prior to building it physically, in order to work out problems, and simulate and analyze potential impacts (Azhar 2008).

In order for it to be successful, Information Modeling requires comprehensive breadth and depth, immersive interaction, simulated performances and trusted deliverables. All of these requirements are services to stakeholders hence service science principles are required. BIM applies and gives services to the entire municipal, utility and building life cycle. The benefits of the services provided by BIM to owner, contractor, engineer and team are listed in Table 2.

5 Conclusion and Suggestion for Future Research Direction on the Service Value of BIM

This paper has given a proposed definition of services for application in engineering together with the classification of the industry in services. The definition is important for the commencement of service science in engineering which could bring the industry to higher level of service performance. Based on the definition, a broad classification of services application in engineering was proposed. The classification basically can cover all activities of interest in engineering field. The discussion on the application of BIM has shown the needs of interaction between experts from the world's major BIM player that will help to understand how the operational and planning of BIM. Since active interaction between local planners and player of issues is the key to successfully provide better service in implementing BIM, a clearer and specific definition is needed to serve as initial platform. The BIM example indicates that service science can as well be adopted in engineering field especially those related to manufacturing and structure fields. Based on the proposed definition, further in-depth analysis is to be conducted to study the potential of emerging service science with engineering field. Additionally, the proposed classification can serves as base of emerging service science with service engineering.

Acknowledgments. The authors would like to express their sincere gratitude to Research and Innovation Management Centre (RIMC), Universiti Malaysia Sarawak (UNIMAS) for the financial support and Prof. Lorna Uden for her generous idea.

References

1. Azhar, S., Hein, M., Sketo, B.: Building Information Modeling: Benefits, Risks And Challenges. In: The 44th Asc National Conference Alabama, USA (2008)
2. Deke Smith, A.: An Introduction to Building Information Modeling (BIM). Journal of Building Information Modeling (2007)

3. Golzarpoor, H.: Application of Bim in Sustainability Analysis. Faculty of Civil Engineering. Johor, Universiti Teknologi Malaysia. Master of Science (Construction Management): 135 (2010)
4. Grandison, T., Thomas, J.O.: Asset reuse and Service Science: The Delicate Balance. In: PICMET 2008 Proceedings, pp. 27–31 (2008)
5. Hammond, S.: Getting to BIM. StructureMag, USA (2007)
6. Huang, P., Li., P., Peng, H.: Towards a Virtual Market Paradigm for Service Science, pp. 15–19. IEEE (2010)
7. IBM. Services Science, Management and Engineering: Services definition (2011), `http://researchweb.watson.ibm.com/ssme/services.html` (retrieved January 25, 2012)
8. Lau, T., Wang, H.-C., Chuang, C.-C.: A Definition of Service as Base for Developing Service Science, pp. 49–53. IEEE Computer Society (2011)
9. Liu, Y., Xu, X., Ma, R.: Service Science Knowledge System Bottom-up Constructed Closely with Service Industry. In: International Conference on Service Sciences, pp. 272–276 (2010)
10. Macbeth, D.K., Opacua, A.I.: Review of Services Science and Possible Application in Rail Maintenace. European Management Journal, 1–13 (2010)
11. Xie, X.: Service Quality Measurement for Customer Perception Based on Services Science, Management and Engineering. System Engineering Orocedia 1, 337–343 (2011)

Entropy- and Ontology-Based E-Services Proposing Approach

Luka Pavlič, Marjan Heričko, and Vili Podgorelec

University of Maribor, Faculty of Electrical Engineering and Computer Science,
Institute of Informatics, Smetanova 17, 2000 Maribor, Slovenia
{luka.pavlic,marjan.hericko,vili.podgorelec}@uni-mb.si

Abstract. E-services have significantly changed the way of doing business in recent years. We can, however, observe poor use of these services. There is a large gap between supply and actual e-services usage. This is why we started a project to provide an environment that will encourage the use of e-services. In the paper we propose an idea of intelligent e-services platform. In addition to established possibilities of searching (e.g. keyword searching, manual classified knowledge browsing), we also propose our own original approach. The ontology and algorithm for proposing appropriate e-services are described in the paper. We use expert knowledge in form of question-answer pairs. It is used by the algorithm to dynamically guide a dialog with user. Intelligently selected sequence of questions is used to suggest the e-service that could help user at a given situation. Ontologies and semantic web technologies are used heavily therefore.

1 Introduction

E-services (IT-supported services, e-services are typically available via a computer network) market has grown considerably. In the case of the Republic of Slovenia, this is also clearly stated in European Commission report for digital economy i2010 [4]. Although the report is limited to e-government services, it also gives a good insight into the whole area of e-services. E-services selection is satisfactory to both businesses consumers and individuals. However, we can observe poor use of these services: e-services are used in less than three quarters of businesses consumers and only little over quarter of individual users [4].

Obviously, there is a large gap between supply and actual e-services usage. This is why we started the project "Ontology-based E-Services Adoption Improvement". Its aim is to provide an environment that will encourage the use of e-services. One of the most challenging features, knowledge management area is trying to address, is searching for knowledge. The searching approach used the most includes browsing categorized knowledge assets (with support of taxonomies) and full text searching [7]. We believe, today's modern technologies and knowledge management approaches have capabilities and opportunities for supporting more intelligent knowledge searching. In this paper, however, we are

L. Uden et al. (Eds.): 7th International Conference on KMO, AISC 172, pp. 409–420.
springerlink.com

not addressing knowledge searching in general. We focus on searching e-services in well defined situation. We established a platform, where searching becomes proposing. With the platform we enabled technology, where a problem solution can be proposed by the "searching" engine. The platform enables searching for knowledge asset with possibility to have explanation of the solution. From the user's point of view it is used regardless of the knowledge representation structure or the location in the network.

This platform is used in order to reduce the gap between supply and use of e-services - we are proposing an appropriate method of formal presentation of knowledge about e-services and applications on this basis. We believe that formal notation is necessary to improve the use of e-services significantly. At the same time, we will ensure that the new notation will not demand any additional activities of e-services users or providers. To enable our methodology, we have built a prototype platform. It allows users to use advanced components, such as intelligent proposing component to discover appropriate e-service or combination of them. System will also allow verification, if there is any alternative e-service or a collection of e-services to be selected. In this paper the main focus is put to ideas and formulas behind intelligent component for e-services proposing. The algorithm is presented in details.

This paper is organized as follows. Section 2 gives a brief overview on related work. In section 3 we advocate our decision to base formal e-services description on ontologies. Underlying ontology is also presented. Section 4 highlights ontology, used for proposing e-services. In section 5 we present mathematical formulas and algorithm for guiding the dialog. In the section 6, we give a brief introduction to the platform, which implements our e-services representation and algorithm. Finally conclusions are given.

2 Related Work

The gap between the high level of offering and low level of e-services usage has been detected and tried to be addressed by many authors. The problem is not local, since it is observed throughout the European Union and in other IT developed countries.

Wang et al. [10] is proposing the use of specific vocabulary, based on ontologies, which allows intelligent search and automatic integration of services on the Internet. The author is limited only to online services in the technical sense ("pure" Web services with software interface and without user interface).

Certain authors have already partially addressed the search of the e-services. Shan-Liang et al. [7] are exposing that just indexing keywords cannot serve for searching the appropriate e-services. Therefore they propose the statistical ontological matching of e-services. This could have been a basis of intelligent e-services inquiries. The study is interesting because the authors are not limited to Web services, but to e-services as a whole. However, they request a specific e-Service interface (OWL-S).

There are also more tries of using ontologies as a helping tool. For example article [11], which proposed the ontological description of Web services. For each domain it suggests its own ontology, which provides a basis for Web agent's search for appropriate e-services. Thus, accumulated knowledge can be used to facilitate the search of e-services. Author addresses only the collection of data about services in certain domains, and does not address the search of e-services.

An similar approach of searching for knowledge assets can be found in paper "An Algorithm Selection Approach for Simulation Systems" [1]. It proposes expert system that can be used by developers, choosing algorithm, based on performance requirements. The idea of intelligent selection mechanism is introduced. It integrates knowledge from many sources and exposes them as a development environment plug-in. We also integrate knowledge, but we upgrade the approach with additional expert knowledge and whole different approach on using it.

The paper "PYTHIA-II: A Knowledge/Database System for Managing Performance Data and Recommending Scientific Software" [3] discusses the problem o finding appropriate software for scientists. They enable automatic knowledge discovery as well as the system, used for searching for appropriate software. User can describe requirements, based on that the system proposes software. Approach on integrating data as well as selecting right solution is different than ours, it is more straightforward. We integrate knowledge, based on ontologies and leave many possibilities for using it – the approach used here can be treated as only one possibility.

As we are mentioning later, the entropy from information theory [8] is used in our platform. Other authors have used it also, for example in paper "Question Answering Using Maximum Entropy Components" [5] authors use the knowledge from local encyclopedia to find answer for users question. They use entropy as a measure for answer relevance.

The paper "An expert system for suggesting design patterns: a methodology and a prototype" [6] describe a methodology for constructing an expert system, that suggests design patterns to solve a design problem. The focus is on the collection and analysis of knowledge on patterns in order to formulate questions, threshold values and rules to be used by an expert shell. What we propose is an ontology-based e-services expert knowledge description and an open platform for building supporting tools.

3 Ontology-Based E-Services Representation

In philosophy, ontologies allow inter-linking and formal description of concepts in any area of human involvement. In computer science, the term ontology has much narrower meaning. With the help of ontologies, we can do advanced classification of individual information elements. We can also interconnect them and create an arbitrary metadata system [2].

The term ontology is mainly used in the field of knowledge management - specifically for classification of individual information objects. There are also simpler and less capable classification methods (e.g. controlled vocabulary, taxonomy, and dictionary). Ontologies not only allow hierarchical and network links between information objects, but also allow specification of axioms, rules and other restrictions for specific information object. One of the most important capabilities of ontologies is the ability of interconnecting objects with arbitrary relations. They allow ontology to formally describe knowledge.

The main advantages of using ontologies are as follows:

- Ontology based formal descriptions are easily processed by a computer and are thus suitable for automated processing.
- Transformation of mathematical notation to a user friendly output is easy to achieve.
- Semantic Web technologies, as a technological foundation for ontologies, are standardized and the adaptation rate is increasing rapidly.
- Knowledge sharing from a technical viewpoint as well as from a conceptual viewpoint is basically achieved automatically.
- Ontology based data is distributed by default.
- There is a wide range of tools that are supporting technologies from the Semantic Web stack.

Ontology as a conceptual foundation in the project has been chosen upon following reasons:

- We want to achieve greater degree of connectivity with existing formal notations and other solutions.
- We want to establish extensible and standard design for intelligent solutions.
- We want to establish a simple mechanism for capturing expert knowledge.
- We want to establish a simple data transformation mechanism.
- Notation and the platform should support automatic exchange of knowledge.
- They allow relative simple machine processing of knowledge.
- We want to support existing formal notations for representing knowledge about e-services.

Semantic web techonologies, which are supervised by the W3C consortium [9], are represented as a stack of standards which allow use of ontologies from a technical viewpoint. One of the core components is RDF (Resource Description Framework), which is based on the markup language XML (eXtensible Markup Language). Concepts in RDF are denoted by URIs (Uniform Resource Identifier - another W3C standard, which is designed for globally unique naming). RDF allows the construction of a data model for resources and the relationships between them. Unified naming on URI is not enough for use of knowledge in intelligent agents. Ontologies fill this gap. There are several languages available for constructing ontologies. OWL (Web Ontology Language) is most widely used one and is also recommended by W3C. The whole stack of semantic technologies is available on the Web pages of W3C consortium [9].

The core of our ontology is shown in Figure 1. We have classified services to be simple services, not supported by ICT at all, or e-services. We also manage services, which are composed with several services. Ontology also covers special e-services, supported by web applications or exposed with web services. Ontology enables capturing several service providers. They can also be classified using taxonomies and folksonomies. Please note that Figure 1 gives only basic insight on some classes and their relations in larger ontology.

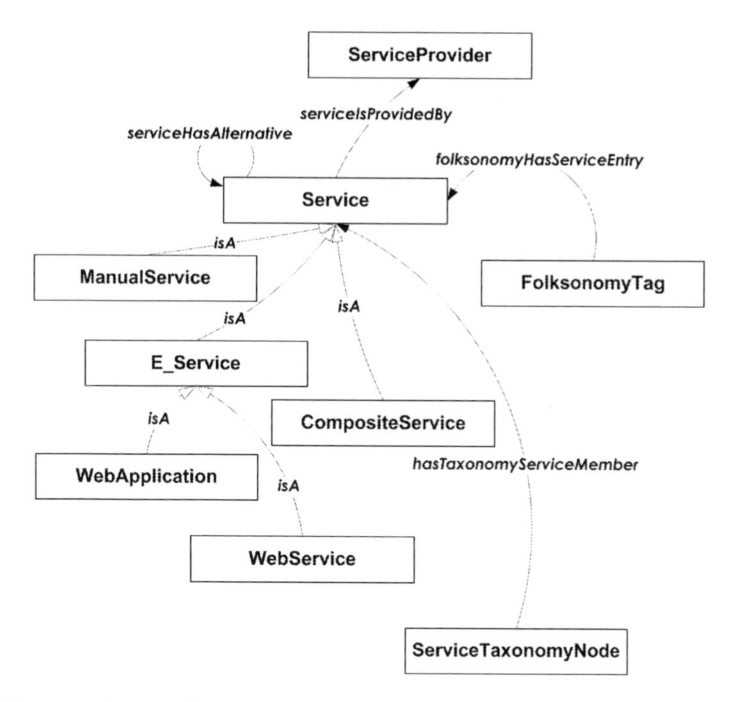

Fig. 1 The core of our ontology

4 E-Services Proposing Ontology

One of the key components, supported by the ontology is "e-service proposing component". Knowledge that is required by proposing component is consistent with ontology, presented in platform (see Figure 2). We introduce concept of Solution. Solution class is represented with Service class in Figure 1. Since we include possibility of related solutions at platform level, proposing component is able to propose not only potentially interesting solutions, but also solutions that can be related to proposed one.

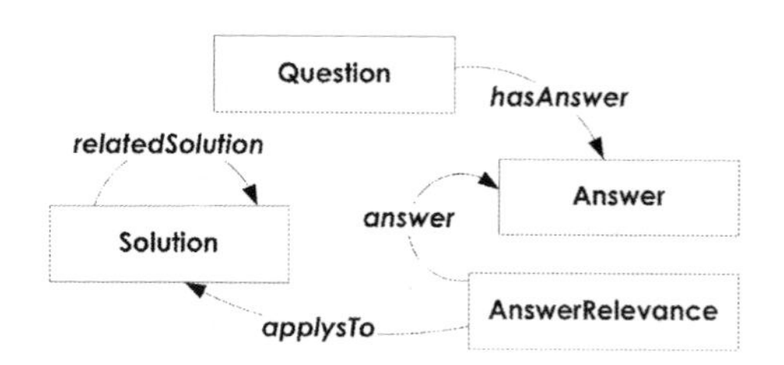

Fig. 2 Proposing ontology structure

Furthermore, the user knowledge aspect is also supported by the presented ontology in quite straightforward way. Users can provide experiences (knowledge) in question-answer pairs, which enables them to capture their implicit knowledge. Not only users can give experiences to tell which service is used in a particular situation ("Question" class), they can also specify more possible services to a situation ("Answer") with specified probability ("AnswerRelevance"). This value ranges from 0% to 100% and tells the user how likely it is that their particular candidate ("Solution") is used when the answer to a given question is confirmed as positive. Answers and possible e-services can easily be updated or added to questions at any time with the aid of a rich user-friendly web interface.

In order to connect several hints (question-answer pairs) into a dialogue, an oft-used technique in expert systems is to connect them into a decision tree, which can be used for guiding dialogues in pursuing a final solution. We believe that such an approach is a) quite expensive, since you need a lot of work from your domain experts b) inappropriate in the long-term, since updating complex decision trees with new questions is not an easy task and c) impractical, since experts would not be particularly motivated to connect questions to form a decision tree, whereas it is not as taxing to just provide a hint or two.

5 E-Services Proposing Algorithm

In this chapter we are presenting our algorithm, which enables us to connect previously mentioned hints into guided dialogues. The main objective of our algorithm is to dynamically connect separate hints into a guided dialogue as an alternative to constructing fixed decision trees. There could, however, be several techniques on how to select an appropriate service, based on answers given to the posed questions. One could simply ask all possible questions and use simple math to select an e-service that was given the highest relevance through such a dialogue. A second technique would be to ask random questions for as long as some kind of measure could not be found that a particular e-service had significantly greater value than others. We decided to guide a dialogue more cleverly: after the user answers a question, we would choose the next most promising question. We wanted to ask as few questions as possible. In an ideal scenario, we would ask as few questions as would be asked if we would have a fixed decision tree. The main idea of the algorithm is therefore to select such a question at every step, which would maximally increase the captured knowledge about current design issue.

Our algorithm has three possible termination states: a) one e-service is chosen based on the predefined threshold, b) the user terminates the algorithm manually, or c) there are no more questions to be asked.

Let us present a set of e-services with vector $P = [p_1, p_2, p_3, \ldots, p_n]$, where every service is presented with $p_i\,(1<=i<=n)$. The number of all possible services is n.

In the algorithm itself and as an output, the vector C is used. Its structure is as follows: $C=[c(p_1), c(p_2), c(p_3), \ldots, c(p_n)]$. Let us call vector C a "certainty vector", where $c(p_i)$ is the current certainty for service p_i. The vector values $c(p_i)$ are within the range [0,1]. The number 0 means that it is certainly not likely that the service is the final solution, 1 means it is certainly likely that the service is the final solution, while with 0.5 we have no information about the e-service whatsoever. It is obvious that the initial value of vector C is $c(p_i)=0.5$ for all i. It is the goal of the algorithm to employ questions until vector C is transformed as much as possible.

When vector C is evaluated with a value that is deemed good enough (at the moment it is the experimental determined threshold), the algorithm is finished. If this is not the case, the next most promising question is selected.

When the answer is given, vector $A=[\ a(p_1),\ a(p_2),\ a(p_3),\ \ldots,\ a(p_n)]$ is constructed. The values $a(p_i)$ are the values of relevance for a particular service in the given answer. For e-services that are not mentioned by answer, we put value 0.5 (which means that we have no information on the e-services). When changing vector C (current certainty vector) with regard to vector A (the current answer), several guiding factors should be followed, which are captured by Equation 1:

$$c_{new}(p_i) = \frac{c(p_i) + a(p_i)}{2};\ a(p_i) \neq 0.5$$

$$c_{new}(p_i) = c(p_i);\ a(p_i) = 0.5$$

Eq. 1 Combining certainty vector with an answer vector

This simple method meets all the needed factors and is in addition also quite resistant to contradictory answering: if user gives answer in favor of a particular service and later an answer that is not in favor of the same e-service, we are again close to 0.5 (the "do not know"-zone).

The evaluation of the current dialogue state (vector C) is used to a) determine if there is one e-service that can be significantly chosen as a final solution and b) select the next question to be asked. In the first case, we calculate the evaluation value of vector C, and if the value is lower (a lower value is better) than the predefined threshold, the dialogue can stop. When choosing the next question, we calculate the values of vector C with all possible answer vectors A. The algorithm selects those questions that definitely and maximally lower the value of vector C in every case.

We designed the algorithm in such a way that the evaluation method is separated from the algorithm itself. This is how we were also able to compare several competing formulas. The first idea was to use the formula to calculate information entropy from vector C, as defined by Shannon [3]. The process of the dialogue would then basically be the process of lowering the entropy value: when the user starts the dialogue with the platform, the entropy value is maximal, since we have no information regarding the proposing of the e-service. Theoretically, if the user answers all the questions held by the platform, the entropy would reduce to the minimum possible level and it would be quite simple to propose an e-service. Entropy formula was a good starting point, but it showed some weaknesses in our case: dynamic threshold value and a lack of capturing e-services relations. This is why we developed our own formula, which works on our domain even better than the formula for entropy. But all original ideas about entropy and lowering entropy are preserved. The relevancy of vector C is calculated as:

$$Rel(C) = 1 - (c_{max} - \bar{c}) \cdot D$$

$$D = \frac{4 \cdot (n-1) - 3 \cdot count(c_{05}) - count(c_{max})}{4 \cdot (n-1)}$$

Eq. 2 Calculating the relevance value of vector C

c_{max} is the maximum value of vector C

$\bar{c}$ is the average value of vector C, calculated as $\bar{c} = \frac{\sum_{i=1}^{n} c(p_i)}{n}$

and D is the factor that captures the current »progress« of vector C: $count(c_{05})$ means the number of e-service certainties, that share the value of 0.5 and $count(c_{max})$ means the number of e-service certainties, sharing the maximum value in vector C.

The algorithm as a whole looks like follows. The function "guide_dialog" is the primary one (see figure 3). It is used to select questions until termination by a) the user b) the solution is selected or c) there are no more questions. Since the user observes the progress of vector C all the time, it is easy to see when a particular e-service is starting to gain. The user is also given the opportunity to skip questions that he or she does not understand or does not know the answer. The threshold value of 0.45 for *Rel(C)* was determined in experiments and was shown to work well. This value can, however, be easily changed in order to investigate its ideal value.

The function "select_question" (not presented in details) receives a list of all unasked questions and the current state of vector C. The output is a question to be asked, if possible. It is selected according to formulas described before: selected question would lower the value *Rel(C)* evenif the worse answer is given. This approach is well known from game-playing theory.

```
create a set of all questions QUE
create a set of all e-services P
guide_dialog(QUE, P)
        THRESHOLD=0.45
        create vector C, c_i=0.5, 1<=i<=n, n=number of e-services
        while QUE not empty
                Q=select_question(QUE,C)
                if (Q==empty) print C, terminate
                remove Q from QUE
                collect answer ANS
                if (ANS != »do not know«)
                        create vector V_ANS from answer ANS
                        C=combine(C,V_ANS)
                        print C
                        if (user select to finish) terminate
                        if (rel(C)< THRESHOLD) terminate
                else continue
        do
end
```

Fig. 3 The e-services proposing algorithm

6 Open Semantic Services Platform

The prototype (Open Semantic Services Platform) enables hosting of intelligent services, which will empower use of e-services. The platform will also be available through application programming interfaces (API), in this manner the prototype itself will be an e-service available for reuse in third party applications. The prototype is implemented using open-source and freely available software. The base of the platform is semantic web technologies.

The platform itself among others allow knowledge integration from distributed data sources (catalogues of e-services, World Wide Web, etc.) and management of the gathered knowledge. Additionally, the platform also:

- enable providers and users of e-services annotation of services with additional knowledge,
- integrate data from the World Wide Web, and from additional data sources,
- transform data encoded in RDF language to a user-friendly format,
- manage index of words, which were used in the platform and additional data sources in order to enable searching based on keywords,
- provide all the data in raw RDF format, which will in turn enable newly built components to easily reuse information,
- manage a collection of real-world examples.

The prototype implementation contains 3 services, which is running on the presented platform (see figure 4):

- full text search (data is not indexed only from local sources but also from external sources, e.g. World Wide Web),
- proposition of e-services or collection of e-services by using questions and answers approach,
- search for alternative e-services or alternative collections of services.

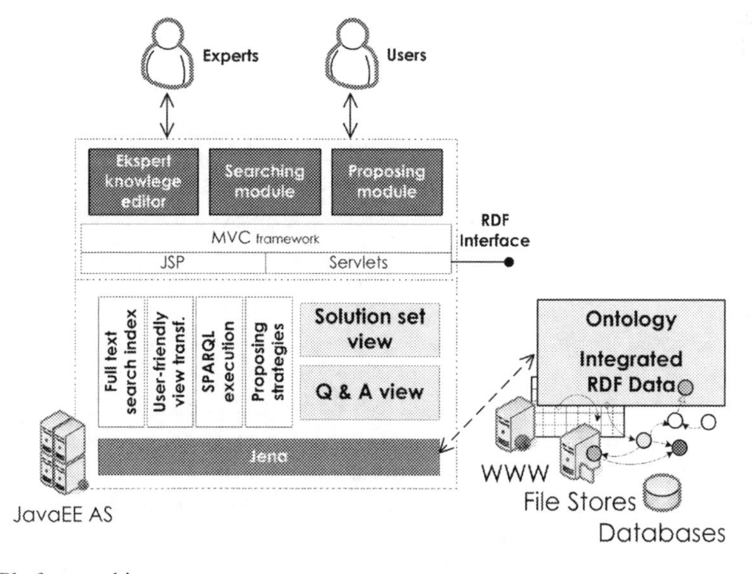

Fig. 4 Platform architecture

Proposing module implements the algorithm, discussed in section 5. In order to do the experimenting, there is also possibility to implement another strategy of asking questions. One could be asking random questions until measure of success meets certain threshold. Another could also build fixed decision tree based on question-answer pairs and decisions available. This strategy is actually available in the platform to support comparison of human-designed question sequence with calculated ones.

All knowledge, held or integrated by the platform, is also available through RDF interface. This enables possible future integrations with similar systems in quite straightforward way.

At the moment, we are testing core functionalities of the platform. Users have entered 549 e-services so far at this phase. They have also classified them in the means of different providers, interaction types, economy activities, domains, types etc. They were actively involved in defining taxonomy also. Taxonomy itself is not closed and final but can, on the other hand, be altered at any time.

Based on that 549 e-services, we can get quite good first insight in the state of e-services in the Republic of Slovenia. The data might, however, be interesting for international audience also: majority of collected e-services is provided by private enterprises; services are mainly oriented in business-to-customer manner; users discovered mainly services form the area of vehicle trade; finance and insurance economy activities; test users mainly discovered e-services from the domains of traffic, education and culture (if we ignore unclassified services).

Entering expert knowledge in terms of question-answer pairs is a work in progress. Based on preliminary tests we are confident in the method of using that knowledge in presented algorithm.

7 Conclusions

It is our goal, that we will provide an e-services management methodology at the end of the project. It will in addition to e-services repository management also allow advanced, easy to use mechanisms for search and proposal of e-services. The project outcomes will allow user collaboration and linking of e-services. At the moment, platform itself is almost complete, so we can really focus on research on its benefits.

Firstly, the methodology approaches and the platform will be demonstrated in real life. We will address e-services that are targeted at student population.

The work on the project will be verified using established scientific research methods. In the final stages of the project, we will conduct a controlled experiment and a large number of surveys on the test population. In this manner, we will determine whether and to what degree our approach helps promoting the use of e-services. At this time, out platform will go public.

In this paper we presented a gap between supply and actual e-services usage. This is why we started a project to provide an environment that will encourage the use of e-services. This paper showed the origins of our project and its current position. The main focus was put to intelligent component for proposing e-services and its underlying ontology and algorithm. We also discussed the decision of using semantic web technologies and their potential to improve using of e-services.

In this paper we presented our own original approach on finding solutions. It is based on minimal expert involvement and intense involvement of modern, intelligent technologies.

At the moment, the project can serve with some original contributions. The most important are:

- the ontology-based method for describing e-services and knowledge about them,
- the holistic methodology, which cover capturing, management and using knowledge about e-services,
- the semantic web-based prototype platform for hosting intelligent components based on knowledge about e-services,
- collected knowledge about ready-to-use e-services,
- operable intelligent component with novel approach for e-services proposing.

At the moment of testing we can summarize that even at this stage, the outcomes of the project are promising.

References

1. Ewald, R., Himmelspach, J., Uhrmacher, A.M.: An Algorithm Selection Approach for Simulation Systems. In: 22nd Workshop on Principles of Advanced and Distributed Simulation (2008)
2. Gruber, T.R.: Towards Principles for the Design of Ontologies used for Knowledge Sharing. In: Guarino, N., Poli, R. (eds.) International Workshop on Formal Ontology, Padova, Italy (1993)
3. Houstis, E.N., Catlin, A.C., Rice, J.R., Verykios, V.S., Ramakrishnan, N., Houstis, E.C.: PYTHIA-II: a knowledge/database system for managing performance data and recommending scientific software. ACM Transactions on Mathematical Software (TOMS) 26(2), 227–253 (2000)
4. i2010 Report | EUROPA, `http://ec.europa.eu/information_society/eeurope/i2010` (accessed March 20, 2012)
5. Ittycheriah, A., Franz, M., Zhu, W., Ratnaparkhi, A., Mammone, R.J.: Question answering using maximum entropy components. In: Second Meeting of the North American Chapter of the Association For Computational Linguistics on Language Technologies, Pittsburgh, Pennsylvania, June 1-7. North American Chapter Of The Association For Computational Linguistics, pp. 1–7. Association for Computational Linguistics, Morristown (2001), `http://dx.doi.org/10.3115/1073336.1073341`
6. Kung, D.C., Bhambhani, H., Shah, R., Pancholi, R.: An expert system for suggesting design patterns: a methodology and a prototype. In: Software Engineering With Computational Intelligence (2003)

7. Shan-Liang, et al.: A Hybrid Web Service Selection Approach Based on Singular Vector Decomposition, Services - I. In: 2009 World Conference, pp. 724–731 (2009)
8. Shannon, C.E.: A mathematical theory of communication. Bell System Technical Journal 27 (1948)
9. W3C Semantic Web, `http://www.w3.org/2001/sw` (accessed March 20, 2012)
10. Wang, X.: Representation and Discovery of Intelligent EServices. In: Intelligence. SCI, vol. 37, pp. 233–252 (2007)
11. Xiaogang, J.: Research on Web Service Discovery Based on Domain Ontology. Computer Science and Information Technology, pp. 65–68 (2009)

Why Projects Fail, from the Perspective of Service Science

Ronald Stanley[1] and Lorna Uden[2]

[1] Asia e University, School of ICT, No 4 Main Block, Jalan Sultan Sulaiman, 50000 Kuala Lumpur, Malaysia
ronald.stanley@aeu.edu.my
[2] L.Uden@staffs.ac.uk

Abstract. This article examines Project Failure in the IT Sector, specifically regarding projects being undertaken in organisations where project tasks clash for resources with concurrent regular ongoing "Business as Usual" IT operations. Projects also clash due to the project altering the organisational "status quo" in many other areas of the firm which are also causes of Project Failure which are examined. A background theoretical review of current Service Science thinking is undertaken, statement of current issues is made and then suggestions for areas of further research to attempt to address the issues from a Service Science perspective are discussed.

Keywords: Service Science, Service Dominant Logic, Project Failure, SDL, S-D, Project Management.

1 Introduction

It is no secret in the computer industry that information systems projects are more likely to fail than not. Information system breakdowns, financial market failures, accidents, natural disasters and planning failures, are common subjects. Bignall and Fortune (1984) called these failures 'systems failures. Systems failures are recognised as occurring from a complex interaction of technical and human factors set in a social situation rather than as the result of the failure of one particular component; human or technical (Stanforth 2010).

Although there is much work being done in describing the failures of many IT projects, there are few theoretical underpinnings given to the failures. Many have consequently chosen to avoid addressing the issue of failure directly and have couched their discussion in terms of what is needed to achieve success. It is the authors' belief that projects failed because there is a lack of understanding about value creation. This paper examines the failures of IT projects from the perspectives of service science. It is our belief that by understanding the co-creation of value from service science, it is possible to resolve many of the known pitfalls in project management. This paper proposes a paradigm that can be used to

L. Uden et al. (Eds.): 7th International Conference on KMO, AISC 172, pp. 421–429.
springerlink.com

help IT projects to avoid failures; it begins with a brief review of IT project failures, followed by description of the personal experiences of the author in project implementations. Afterwards reviews the emerging field of service science and service dominant logic, concluding with suggestions for further research.

2 Project Failures

IT projects have a bad reputation for going over budget and schedule, not realising expectations and for providing poor return on investment (Jaques, 2004; McManus and Wood-Harper, 2008). A study of 50 government projects over 20 years found that those involving technical innovation and systems development historically ran over budget by up to 200% and over the original contract duration by 54% (Bronte-Stewart 2009). The cost of project failure across the European Union was €142 billion in 2004 (McManus and Wood-Harper, 2008).

The Office of Government Commerce (OGC), part of the Efficiency and Reform Group within the Cabinet Office, in 2005 identified eight common mechanisms which lead to project failure:

- Lack of clear links between the project and the organization's key strategic priorities (including agreed measures of success)
- Lack of clear senior management and ministerial ownership and leadership
- Lack of effective engagement with stakeholders
- Lack of skills and a proven approach to project management and risk management
- Too little attention to breaking development and implementation into manageable steps
- Evaluation of proposals driven by initial price rather than long-term value for money (especially securing delivery of business benefits)
- Lack of understanding of, and contact with, the supply industry at senior levels in the organization
- Lack of effective project team integration between clients, the supplier team and the supply chain.

Many articles have been written about the extent and causes of IS/IT project failure and numerous studies have discussed a range of recognised risk factors including those concerning project leadership and management, organisational culture and structure, commitment and patterns of belief, user involvement and training, developer expertise, technology planning, scope and objectives setting, estimation and choice/use of methodology, (Block, 1983; Boehm, 1991; Willcocks and Margetts, 1994; Sumner, 2000; Cusing, 2002; Hashmi, 2009).

IS/IT projects are unique in that they are conceptual in nature and require the intense collaboration of several different groups of stakeholders including IS staff, users and management. IS projects are usually undertaken in teams and therefore subject to the vagaries of group dynamics, interactions, coordination and communication. The diverse backgrounds and training of the various team members and other people associated with the project make the ability to communicate and coordinate the activities of the group extremely important. (Ewusi-Mensah 1997).

Roger Elvin, (Senior 2003) argued that the management complexity arises from the necessity to deal simultaneously with several tensions:

- Innovation versus risk
- Learning versus control
- The need for organisational change to deliver business benefit versus stakeholder resistance to change
- Multiple stakeholder perceptions of the purpose of the project
- The need to deliver value to the organisation versus managing effectively to satisfy time, quality and cost objectives
- Managing detail and the "big picture".

From the literature reviews above, it can be seen that there are many reasons given for project failures. The author has experienced many failures personally. Subsequent sections of the paper briefly review some of the causes of project personally encountered by one of the authors.

3 Personal Experiences in Multiple Organisations

How many Information Technology Project fail? The truth is that the failed percentage is quite high, but an even higher percentage are delivered with much less in terms of useful outputs and results than was originally envisaged, often in a much longer timeframe than originally expected and at the same time at much higher incurred costs, also with ongoing long term issues ("Technical Debt" to use an Agile Software Development term).

The first author of this paper has experienced of many project failures during his 15 years of working with different organizations. Before we proceed with analyzing the causes of failures through the lens of service science, it is suffice to introduce the important emerging subject of service science. This will enable us to propose a new framework for the co creation of value for project management.

Many failed IT projects have been 'explained away" in the past in many organisations on the basis of poor Project Management and over the past decade there has been a concerted push by the IT industry as a whole to address this issue, to the point where this is no longer a completely valid excuse. Most IT Project Managers have undertaken the now almost mandatory PMI courses to the point where having a PMI certification is no longer a differentiator in the market place in terms of jobs skills, but the poor performance of IT projects continues unabated. Although the addressing of poor project management skills has managed to alleviate one pressing issue for the industry, the industry is yet to come to terms with issues surrounding the fundamental issue of poorly conceived projects in the first instance, which are then thrust onto project managers with the expectation that the Project Manager is some type of miracle worker who will either bring the project in successfully, in which case they will become a "Company Hero" or on the other side of the coin, they can be used as a scape goat to take the blame.

4 Service Science

The service sectors now represent 80% of the developed economies. Currently there is a lack of understanding of the science underlying the design and operation of service systems. In response to this, IBM has advocated a new discipline known as Service Science, or Service Science Management and Engineering (SSME). Research in service science seeks to find out how to design, build, operate, use, sustain and dispose of service systems for the benefit of multiple stakeholders such as customers, shareholders, employees, partners and society (IBM 2007). The aim of service science includes addressing issues such as to what extent organizations can be restructured, how to manage service innovation, and others.

There are many different definitions given to services. Services are deeds, processes and performance (Zeithaml et al 2006). Vargo & Lusch (2006) define service as the application of specialised competences (knowledge and skills) through deeds, processes and performances for the benefit of another entity or the entity itself (p.4).

A "Service System" is defined as "a value co-production configuration of people, technology, other internal and external service systems, and shared information (such as language, processes, metrics, prices, policies and laws.)" (Spohrer et al 2007). Examples of service systems include people, organisations, corporations, cities, families etc. A key condition is that service systems interact to co-create value.

4.1 Service Dominant Logic

Central to service science is the concept of service dominant logic. In the last few decades there has been work going on by researchers trying to differentiate between goods and services. An emerging new concept has been gaining popularity among service management concerning the role of the firm. Vargo and Lusch (2004) argue for evolving a service dominant logic in marketing to replace the goods-dominant logic of the traditional view.

The new service dominant (S-D) logic is concerned with value-in-use. Value is always co-created between producers and consumers. Thus value is co-created through the combined efforts of firms, customers, employees, government agencies, stakeholders and other entities related to any given exchange, but is always determined by the beneficiary (e.g. customer).

In S-D logic, the notion of value co-creation suggests that there is no value until an offering is used. Experience and perception are essential to value determination (Vargo & Lusch 2006, p.44). Offerings must be integrated with other market facing (from other firms) and non-market facing (e.g. personal/private and public) resources for value to be created. Value creation typically requires resources beyond a two-party system, often involving a firm, its customers, suppliers, employees, stakeholders, and other network partners (Lusch & Vargo 2006).

According to these authors, services provided directly or through a good are the knowledge and skills (competencies) of the providers and beneficiaries that represent the essential source of value creation, not goods that are sometimes used to convey them. Because of this, Sporher and others (2008) argue that service involves at least two entities, one applying competence and another integrating the applied competence with other resources and determining benefits (value co-creation). These interacting entities are service systems. Therefore, a service system is defined as a dynamic value co-creation configuration of resources, including people, organisations, shared information (language, laws, measures, methods) and technology, all connected internally to other service systems by value propositions. These authors further explained that each service system engages in three main activities:

- Proposing a value co-creation interaction to another service system (proposal).
- Agreeing to a proposal (agreement) and
- Realising the proposal (realisation).

4.2 Co-creation of Value

Understanding what customers value within a particular offering, creating value for them and then managing it over time have long been recognized as essential elements of firms' business strategy (Drucker, 1985). Porter (1985) argued that a firm's competitive advantage stems from its ability to create value for its customers that exceeds the firm's cost of creating it. The dominant logic of business is now being challenged. Today consumers want to interact with firms to co-create their own experiences. Interaction is the basis for value creation. Companies no longer act autonomously to design products and services. Customers want to interact with firms to co-create their own experiences. Today we have the new logic of business (Prahalad & Ramaswamy 2004).

Coates (2009) argues that the aim of co-creation is to enhance organisational knowledge processes by involving the customer in the creation of meaning and value. Co-creation also transforms the consumer into an active partner for the creation of future value.

From the general management perspective, Prahalad et al., (2004) have developed a holistic generative framework describing the fundamental building blocks of value co-creation practices, including Dialog, Access, Risk management and Transparency (thus, DART framework):

- The open Dialog between the multiple actors within the value network encourages knowledge sharing and mutual understanding. It provides an opportunity for customers to interject their view of value into the value creation process and helps companies understand the emotional, social, and cultural contexts of end user experiences.
- Access challenges the notions of openness and ownership. Providing customer access to resources, information, tools, assets and processes at multiple points across the value network provides companies with innovative ideas about new products and services, new business opportunities and new potential markets. As customers become co-creators of value, they become more vulnerable to risk

- Risk and demand more information about the potential risks associated with the design, manufacturing, delivery and consumption of particular products and services. Proactive risk communication and management offers companies with new opportunities for competitive differentiation.
- Transparency builds trust between both institutions and individuals. It enables a creative dialogue in which trust emerges (Tanev et al 2011).

The value co-creation process not only occurs within a provider and customer dyadic relationship but also involves several participants as dynamic operant resources in a many-to-many perspective (Gummesson, 2008; Mele, Colurcio and Russo Spena, 2009). We need to start to think about it from the customer's perspective and to work out how to co-create more value together with customers in project management. In order to do this, we need knowledge as knowledge is the Source of Business Value. "Knowledge includes all the valuable concepts and vital know-how that shape a business to be wanted and needed by customers". It is our belief that co-creation of value from SDL can be used to help overcome the problems associated with project management. A framework for co-creation of value is proposed.

5 Causes of Failures in Project Management through the Lens of Service Science

Through the lens of service science, there are several reasons why project fail:

Firstly, the various stakeholder group representatives often act divisively in terms of attempting to satisfy self interest in as much as they are attempting to achieve goals that have been set for their own group and these goals are focused towards individual identifiable achievement for the group rather than for the organisation as a whole. Each stakeholder has his or her value ignoring the needs of others, that is; there is a lack of co creation of value.

Secondly, most of the stakeholders were only interested in how much money or cost they can save rather than providing service. That is using their knowledge and skills to benefit the other partners. The value perceived here is that of value in exchange rather than value in use. Thirdly, the relational link between each of the stakeholders is that of translational rather than relationship from the service perspective. Fourthly, a firm cannot create or deliver value alone. It always requires co-creation with customers and it cannot be embedded in the manufacturing process.

How do we start to address the above problems that besiege project managers? It is our belief that a new approach is needed. A possible solution could be from the perspective of service science.

6 Framework for Co-creation of Value for Project Management

The proposed framework should be based on the co-creation of value of DART and other factors. To achieve competitive advantage it is not knowledge itself that is important, but rather the firms' capacity to apply this knowledge effectively in order to create new knowledge (Grant, 1996). Dynamic capabilities are important because knowledge flows from one capability to another, through the reconfiguration of organizational capabilities, leading to new knowledge that enables the firm to create superior customer value.

To co-create value for sustainability; firms need to take a more comprehensive view of the environment in which they must compete. This view not only includes buyers and suppliers but the local market for skilled workers, financial institutions, universities, legal system, and the domestic political situation. This requires the concept of dynamic capabilities. The dynamic capabilities emphasise the "soft assets" that management needs to orchestrate resources both inside and outside the firm. This includes the external linkages that have gained in importance, as the expansion of trade has led to greater specialization. It recognizes that to make the global system of vertical specialization and co-specialization work, there is an enhanced need for the business enterprise to develop and maintain asset alignment capabilities that enable collaborating firms to develop and deliver a joint "solution" to business problems that customers will value (Teece 2010).

We argue that in order to maintain competitiveness it is not enough for firms to be in possession of valuable resources and capabilities; they also require dynamic capabilities to develop and renovate their organizational resources and capabilities (Teece et al., 1997). The authors have developed a conceptual framework for co-creation of value for project management. The dynamic capabilities value co-creation framework should consist of the following capabilities:

- Customer knowledge capabilities
- Collaborative networks capabilities
- Organizational capabilities
- Market orientation capabilities
- Management of technology capabilities.

Each of these require dynamic capabilities to enable the co-creation of value using DART and other features. The authors are currently working on this framework.

7 Conclusion

IT project failure is a common thing. There are many tales of IT projects that have failed. IT projects have difficulties with completion on time or on budget or on scope. In fact many are cancelled before completion or ultimately not implemented. There are many different reasons for the failure of IT projects, the most common reasons are rooted in the project management process itself. Projects fail because they do not meet one or more of the following criteria for success: delivered on time, or on budget, or satisfactory to user requirements.

Central to the problems in project failure is that there are many different stakeholders involved in the project and each often has conflicting interests. It is our belief that successful project management requires the co-creation of value between the different stakeholders of the project. The emerging field of service science can provide the answer and the development of the framework started with the recognition that of the centrality of co-creation of value is the process. Co-creation can be viewed as set of processes and resources with which the company seeks to create value propositions. Processes include the procedures, tasks, mechanisms, activities and interactions which support the co-creation of value. It is our belief that the co-creation framework proposed will provide project management with a means of addressing many of the failures. Currently we are working on the framework and empirical studies will be conducted using real cases to demonstrate its effectiveness.

References

Bignall, V., Fortune, J.: Understanding Systems Failures. The Open University. Manchester University Press, Manchester (1984)

Block, R.: The Politics of Projects. Yourdon Press, Prentice-Hall, Englewood Cliff, New Jersey (1983)

Boehm, B.: Software Risk Management: Principles and Practices. IEEE Software 8(1), 32–41 (1991)

Bronte-Stewart, M.: Risk Estimation From Technology Project Failure. In: 4th European Conference on Management if Technology, Glasgow, Scotland, September 4-6, pp. 4–6 (2009)

Coates, N.: Co-Creation: New Pathways to Value: An Overview (2009), http://www.promise.corp.com/newpathways (accessed on September 12, 2010)

Cusing, K.: Why projects fail, IT project leadership report. The Cloverdale Organisation (2002)

Drucker, P.F.: Innovation and Entrepreneurship. Harper and Row, New York (1985)

Elvin, R.: Advice: There is no 'magic bullet' solutions (2003), http://www.computerweekly.com; http://www.computerweekly.com/feature/Directors-notes (accessed on October 21, 2006)

Ewusi-Mensah, K.: Critical Issues in Abandoned Information System Development Projects. Communications of the ACM 40(9) (1997)

Grant, R.M.: Toward a Knowledge-Based Theory of the Firm. Strategic Management Journal 17, 109–122 (1996)

Gummesson, E.: Extending the new dominant logic: from customer centricity to balanced centricity. Journal of the Academy of Marketing Science 36(1), 15–17 (2008)

Hashmi, M.: High IT Failure Rate: A Management Prospect. Blekinge Tekniska Hogskola Sektionen for Management (2009)

IBM, Service Science, Management and Engineering (2007), http://www.research.ibm.com/ssme (accessed on September 23, 2009)

Jaques, R.: UK wasting billions on IT projects (2004), http://vnunet.com

McManus, J., Wood-Harper, T.: A study in project failure. BCS, Articles, Project Management (2008)

Mele, C., Colurcio, M., Russo Spena, T.: The 2009 Naples Forum on Service: Service-Dominant Logic, Service Science and Network Theory, Naples, June 16-19 (2009)

NAO/OGC, Common Causes of Project Failure. National Audit Office and the Office of Government Commerce (2005)

Porter, M.E.: Competitive advantage: creating and sustaining superior performance. Free Press, New York (1985)

Prahalad, C.K., Ramaswamy, V.: Co-Creation Experiences: The Next Practice in Value Creation. Journal of Interactive Marketing 18, 5–14 (2004a)

Prahalad, C.K., Ramaswamy, V.: The Future of Competition: Co-Creating Unique Value with Customers. Harvard Business School Press (2004b)

Spohrer, J., Maglio, P.P., Bailey, J., Gruhl, D.: Steps Toward a Science of Service Systems. IEEE Computer 40(1), 71–77 (2007)

Stanforth, C.: iGovernment Working Paper Series. Analysing e-Government Project Failure: Comparing Factoral, Systems and Interpretive Approaches. Manchester Centre for Development Informatics, iGovernment Working Paper 20 (2010), `http://www.sed.manchester.ac.uk/idpm/research/publications/wp/igovernment/documents/iGovWkPpr20.pdf` (accessed on January 4, 2012)

Sumner, M.: Risk Factors in Enterprise-wide/ERP projects. Journal of Information Technology 11, 317–327 (2000)

Tanev, T., Bailetti, T., Allen, S., Milyakov, H., Durchev, P., Ruskov, P.: How do value co-creation activities relate to the perception of firm's innovativeness? Journal of Innovation Economics (7), 131–159 (2011), `http://www.cairn.info/revue-journal-of-innovation-economics-2011-1-page-131.htm`, doi:10.3917/jie.007.0131 (accessed on January 4, 2012)

Teece, D.J., Pisano, G., Shuen, A.: Dynamic Capabilities and Strategic Management. Strategic Management Journal 18(7), 509–533 (1997)

Vargo, S.L., Lusch, R.F.: Evolving to a New Dominant Logic for Marketing. Journal of Marketing 68(1), 1–17 (2004)

Vargo, S.L., Lusch, R.F.: Service-dominant logic: What it is, what it is not, what it might be. In: Lusch, R.F., Vargo, S.L. (eds.) The Service-Dominant Logic of Marketing: Dialog, Debate and Directions, pp. 43–56. M.F. Sharpe, Armonk (2006)

Zack, M.H.: Rethinking the knowledge-based organization. MIT Sloan Management Review 44(4), 67–71 (2003)

Zeithaml, V.A., Bitner, M.J., Gremler, D.D.: Service Marketing: Integrating customer focus across the firm, 4th edn. McGraw-Hill International, London (2006)

A System for Cyber Attack Detection Using Contextual Semantics

Ahmed AlEroud and George Karabatis

Department of Information Systems, University of Maryland, Baltimore County (UMBC)
1000 Hilltop Circle, Baltimore, MD 21250, USA
{Ahmed21,Georgek}@umbc.edu

Abstract. In this paper, we present a layered cyber-attack detection system with semantics and context capabilities. The described approach has been implemented in a prototype system which uses semantic information about related attacks to infer all possible suspicious network activities from connections between hosts. The relevant attacks generated by semantic techniques are forwarded to context filters that use attack context profiles and host contexts to filter out irrelevant attacks. The prototype system is evaluated on the KDD 1999 intrusion detection dataset, where the experimental results have shown competitive precision and recall values of the system compared with previous approaches.

Keywords: Context, Context-aware Cyber Security, Semantic Networks.

1 Introduction

Internet communications and distributed network environments have become rich media for electronic data transfer. Due to huge amounts of data transmission, it becomes vital to build effective security policies and threat-detection systems that are capable of analyzing network data. There has been a significant amount of work performed in the area of intrusion detection; however, most of the recent approaches focus on processing alerts with no semantics or context aware capabilities. Providing semantic capabilities on the top of Intrusion Detection Systems (IDS) can have a great benefit in inferring possible relationships between network attacks. Additionally, context aware capabilities can be utilized to filter out most of false positive alerts about possible attacks. The current IDSs are broadly categorized as host-based or network-based. In host-based IDSs, the system is only aware of the host environment. By contrast, network-based IDSs have a better knowledge of events happening at the network level. Both network-based and host-based IDSs utilize rule engines to disseminate alerts, with a significant number of such alerts being false alarms due to the lack of contextual information concerning hosts and network events. To address this problem, we propose a layered attack detection system that utilizes semantics and context. The

L. Uden et al. (Eds.): 7th International Conference on KMO, AISC 172, pp. 431–442.
springerlink.com

system is designed to detect all relevant attacks occurring under particular contexts. It consists of one semantic layer and two context layers. The semantic layer is represented by semantic networks [1, 2] which are graphs with nodes representing attacks and edges modeling semantic relationships between attacks; we use semantic networks to infer all relevant attacks in a sequence of suspicious network connections events. By contrast, context layers are applied on top of semantic networks to filter out irrelevant attacks by matching them with contextual information about attack profiles and network hosts. The contributions of this paper are three-fold: First, we utilize an automatic approach to create a semantic network of relevant attacks using the KDD 1999 dataset [3]. Second, we apply the Conditional Entropy Theory [4] to create attack context profiles based on attack features. Context profiles are then used to filter out all non-relevant predictions made by the semantic network. Finally, we create host context filters consisting of contextual facts about hosts and attacks. We employ a logical based approach to filter out attacks based on host-attack contexts. The context filters are applied in a layered manner to filter semantic network predictions. The reminder of this paper is organized as follows: The next section describes the approach used in creating system semantics and contexts layers. Section 3 reports our experiment with the KDD 1999 intrusion detection dataset. Section 4 presents the related work with an emphasis on various techniques and methods which are related to semantics and context in the cyber security domain. Section 5 provides conclusions and points out future work.

2 The Approach

In this section, we provide an overview of our approach and a description of the prototype system. Figure 1 illustrates the major components in our system. The detection process occurs in two phases: First, the attack detection models are built using historical audit network data, and then we use these models in a layered manner to simulate the actual detection on network evaluation data. We use a network data repository (Audit Network Data in Fig. 1) to store a set of events collected by network sensors, as well as their characteristics. We use the publicly available intrusion detection dataset (KDD 1999) [3] as the network data repository. Although some researchers have identified drawbacks in KDD 1999 [5, 6], this dataset is nevertheless one of the most widely datasets in evaluating intrusion detection systems. The KDD 1999 is formatted as a set of network connection records labeled as normal activities or attacks. Each connection record has 41 features, which characterize that particular connection in terms of connection protocols, the services requested, etc. The attacks found in the KDD 1999 dataset are also categorized in one of four categories: 1. User to root (U2R), 2. Remote to local (R2L), 3. Denial of service (DOS), and 4. Probe Attacks. Before using our system we create the semantic network model, the simple Bayesian network model, and the attack context profiles, as are explained next.

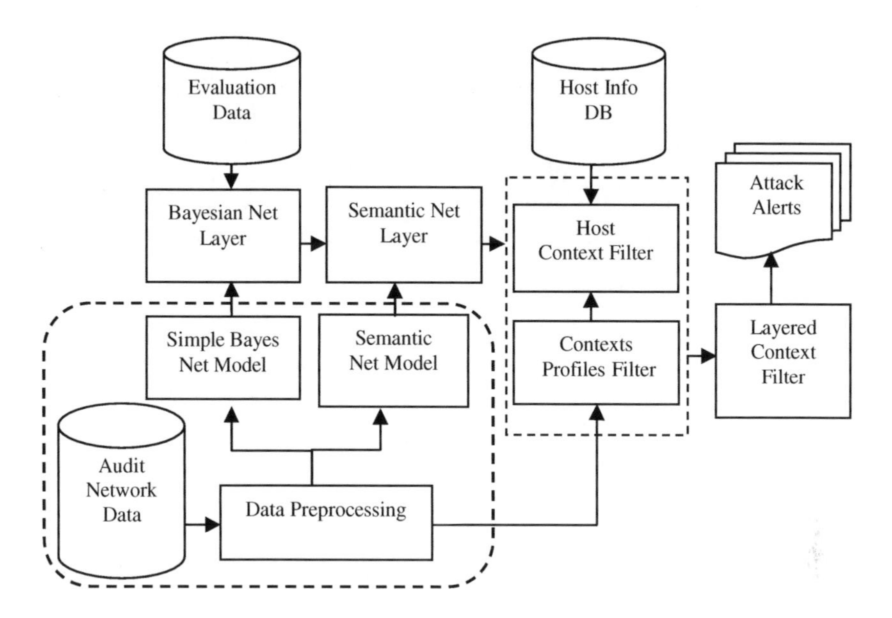

Fig. 1 Major System Components

2.1 Semantic Network Creation Process

The semantic network we create is a graph with nodes representing attacks and edges connecting relevant attacks. Each node in the created semantic network represents one of the 22 attacks in KDD 1999 dataset. Normal activity is also treated as a node (although it is not an attack). The semantic network is created based on similarities of attack features. The numbers on edges represent the degree of relevance between nodes. A semantic network is described as follows [7]:

Definition (1) **[semantic network]:** Let $X = \{x_1 \dots, x_n\}$ be the set of attacks where each $x_i \in X$ has a binary feature vector of values $f_i = \{f_{i1} \dots, f_{im}\}$. The semantic network SN^{sn} (V^{sn}, E^{sn}) is a directed graph where V^{sn} is set of nodes and E^{sn} is a set of edges, and $V^{sn} \subseteq X$ and $|V^{SN}| \leq |X|$ and each edge links two relevant attacks $< v_i, v_j >$ and has a relevant score $w(v_i, v_j)$ where $0 < w(v_i, v_j) \leq 1$.

The semantic network is created as follows: All numerical features in the KDD 1999 dataset are discretized. Then, a universal feature vector is generated as the union of all unique dataset feature values. This is denoted by $U_f = \cup_{(1 \leq i \leq n)} < f_{i1} \dots, f_{iv} >$ where n is the number of unique feature values in the dataset, and v is the number of unique values in feature f_i domain. The universal feature vector is then used to build the attack feature frequency vector for all attacks. Each attack feature frequency vector consists of the frequency value for each feature that occurs in a specific attack. We automatically assign relevance scores between attacks on semantic networks using the Anderberg similarity coefficient measure [8], which calculates similarity between two binary vectors. Since Anderberg is a binary similarity measure, we convert the attack feature frequency vectors to binary vectors of zeros and ones using a cutoff data transformation method [9]. For instance, the *teardrop* attack usually occurs when the connection protocol is User Datagram Protocol (UDP), therefore, the binary vector of *teardrop* attack includes the value 1 for the entry (Protocol *is* UDP, attack *is teardrop*), whereas if the connection protocol is the Transmission Control Protocol (TCP), then the binary vector of *teardrop* attack includes the value 0 for the entry (Protocol *is* TCP, attack *is teardrop*).

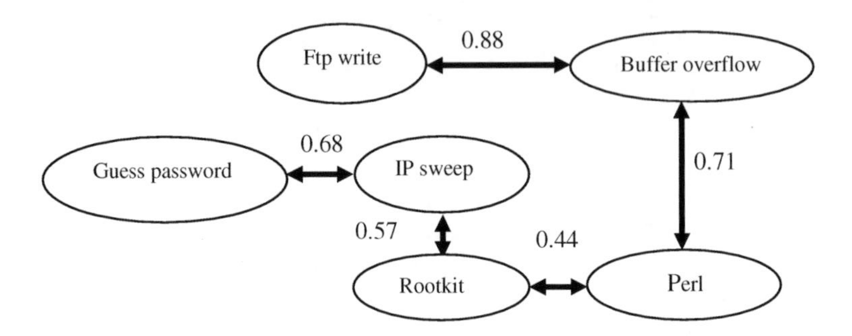

Fig. 2 A partial semantic network for some KDD 1999 Attacks

The binary feature vectors are then used to compute the Anderberg similarity between pairs of attacks, and to further calculate the relevance scores on the edges of the semantic network. Figure 2 illustrates a partial semantic network for some attacks in KDD 1999 dataset. The path between "Buffer overflow", which is in the U2R category, and "Guess password", which is a R2L category of attack, is used in calculating the relevance score between these two attacks by multiplying all scores on that path $0.68 \times 0.57 \times 0.44 \times 0.71 = 0.125$. This represents one possible relevance score, however, we should infer the most accurate semantic relationship between these two attacks, which is the maximum relevance score using all possible paths between these two attacks[7].

2.2 Simple Bayesian Network Prediction

Our system requires an initial node to start searching in the semantic network to identify relevant attacks. To provide such node we create a simple Bayesian network model which takes as input the connection records in the database, and provides as output the attack with the highest probability for the corresponding connection record (see Fig. 1).We extract the conditional probabilities from the simple Bayesian network model and then we calculate the combined probability of any attack a_i as follows.

$$P(a_i|F) = \frac{P(f_1|a_i) \cdot P(f_2|a_i) \ldots, (f_n|a_i)}{P(F)} \tag{1}$$

Where $P(a_i|F)$ is the probability of attack a_i given the feature vector F with n features, under the assumption that such features are independent. The semantic network identifies all attacks relevant to the initial input node, based on the relevance score bounded by a user-defined threshold identified (a lower threshold results in more attacks retrieved by the semantic network, and a higher threshold results in fewer attacks). However, not all the relevant attacks have the same context; therefore, context filters are applied to allow attacks that are both relevant and within context.

2.3 Context Filters Module

The main purpose of the context module is to identify attacks that are relevant and specific to a particular context, and at the same time to filter out irrelevant attacks reducing the amount of false positives alerts. In our system the context module recognizes two types of context: attack context profiles, and host contexts. The attack context profiles are used to discard attacks which do not match a given set of context profiles. The context profile of each attack consists of features which minimize the conditional entropy [4], and thus the uncertainty about that attack by observing specific values of these features. The second type of context is the host context which is used to discard attacks which do not match the target host context in terms of host vulnerabilities, application types, running services, and the affected platforms. The following two sections describe both context types in detail.

2.3.1 Conditional Entropy Attack Context Profiles

Conditional entropy calculates the degree of uncertainty of one random variable given another [4]. The conditional entropy is strongly related to information and probability theories. In the context of cyber security, conditional entropy can be defined as the amount of information needed to infer the degree of uncertainty about one event based on the occurrence of another. We used the KDD 1999 connection record features to build the context profiles of different attacks in the dataset. The conditional entropies are calculated by utilizing the conditional probabilities of attacks given the values of their features, and the probability distribution of the feature values in the data set. Thus, we first calculate the conditional entropy of attack y in the dataset given a specific observed value of feature x. Then we calculate the sum of conditional entropies of all attacks given a specific feature, this value is called *global conditional* entropy. The conditional entropy is described as follows [4]:

Definition (2) **[Conditional entropy]:** The conditional entropy (H) of variable y given variable x is the sum of entropies for all values of variable y given values of variable x. It is denoted by $H(Y|X) = -\sum_{i=1}^{n}\sum_{j=1}^{m} pXY(x_i, y_j)\log_2 p_{Y|X}(y_j|x_i)$.

An important factor in building attack context profiles is the individual contribution of a specific feature on the occurrence of a particular attack. To address this issue, we calculate the sum of conditional entropies of a specific attack y with respect to all values of a specific feature x. This sum is called the *local conditional entropy* of a specific attack y given feature *x*. The lower the value of local conditional entropy of attack *y* given feature *x*, the lower the degree of uncertainty about the context of occurrence of attack *y* given specific values for *x*. Features which produce smaller global and local conditional entropy values convey better quality information; consequently, they are used in building attack context profiles. For any feature *x* selected in the context profile of attack y, the values of *x* which give a conditional probability value greater than 0 of attack *y* are used to build one context profile rule for attack *y*, where the context profile rule is a set of feature values which are used to minimize the uncertainty about the context of each attack. To build the context profile of attack A_i, we select the set of features which give the lowest *m* values of the global conditional entropy, and the set of features which give the lowest *n* values of the local conditional entropy of attack A_i. The features which are common between these two sets are used to build the context profiles rules for attack A_i. The intersection of such features in a a sequence of events, minimizes the uncertainty about the context of attack A_i, due to its low entropy. The value *m* is selected based on global features selection on KDD 1999 dataset. The first 14 highly ranked features are used in several works [10]. By contrast, the value *n* is selected based on the local feature selection method used in [11] to select the best features for each of the KDD attacks categories. On average, 7 features were selected by domain experts for each category. Thus, the common features between the first 14 features which gives low global conditional entropy values and the first 7 features which gives low local conditional entropy values (for specific attack) are the set of features used to create each attack context profile. We created 15 unique context profiles. Some context profiles match more than one attack as some attacks have the same context, thus each of the 22 attacks has one context profile. We did not create any context profiles for the normal category, since it does not correspond to any attacks. The context profiles are then used to automatically filter out any attack which does not match its corresponding profile. Any connection record that has no matching context profile is treated as a normal activity and hence no further context filtering is performed on that connection.

2.3.2 Host Context

The entropy-based context profiles, which form the first context filter, deal with features of attacks and they are not aware of contextual information about network hosts such as a host having a patch against a predicted attack. We address this situation by incorporating as much contextual information about hosts and their environment as possible to further filter out the potential attacks that were generated by semantic networks and survived the entropy-based attack context profile filters. Host contexts contain facts about a specific host; each fact is of a specific type *t*, which can be about host vulnerabilities, host operating system, the installed applications, the running services, the host patches, etc. The facts in the host contexts are modeled as predicates, and implemented using a relational database. The definition of host contextual facts is given below.

***Definition* (3) [Host context]:** *The context of host h denoted by c (h), is a set of predicates*$\{p_1 \ldots p_n\}$. *Each predicate is represented as a host context fact f of type t, where* $\{f_{1(h)} \ldots, f_{i(h)}\}$ *represent the facts about host h such, the expression* $isT(h, c)$ *is true, that is, (h) has the context (c) iff* $(\forall f_{d(1 \le d \le i)(h)} \in \{t\} \ (f_{d(h)})) = True$.

Each predicate is represented as a logical fact *d* of type t. In addition to host context facts, we create a set of attack context facts; these facts represent the conditions that should exist on a target host *h,* that can lead to specific attack *y* and they are different from the conditional entropy based facts. The format of attack facts is similar to host context facts. We use the context fact predicates in a disjunctive normal form ($\vee$) to match all host context facts with attack facts of the same type. The conjunctive normal form ($\wedge$) is also used to match host context facts of different types with attack facts. The context of an attack *A* on host *h* matches the context *c(h)* of host *h,* when at least one host context fact of each type *t* matches one or more fact of the attack *A* for the same type *t* as defined below:

***Definition*(4) [Context Matching]:** *Given all context facts* $\{f_{1(h)} \ldots, f_{i(h)}\} \in \{t\}$ *of of host h and all context facts* $\{f_{1(Al)} \ldots, f_{m(Al)}\}$ *of attack* $A_{l(1 \le l \le n)}$ *detected on host host h by a semantic network, the attack* (A_l) *matches the context c(h) of host h iff* $\forall f_{e(1 \le e \le m)(Al)} \in \{t\} (\exists f_{d(1 \le d \le i)(h)} \in \{t\} \mid f_{e(Al)} = f_{d(h)})$.

The following example shows a scenario of a *Perl* attack predicted by semantic network on host *h*. Perl is one of the U2R attacks which can lead to denial of service on the target host. The context facts about operating system, vulnerabilities, and running applications on a target host *h* are as follows: *hasOS(h, redhat linux 5.0), hasVulnerability(h, CVE-1999-1386), has Application (h, Apache server 1.3.1).* In addition, the following facts represent the context facts required for a *Perl* attack on target host *h*: *isOS(h, redhat linux5.0), existVulnerability(h, CVE-1999-1386),* given these facts about host *h*, the perl attack is in the context of host *h*, due to the fact that the conjunctive normal expression evaluates to True by matching context facts about perl attack with those about host *h*. We used the layered context filtering as a final step in our system. The layered context filtering is performed by applying host context filter on the output of the entropy based contexts profiles filter. The layered context filtering is performed as follows: If A_l is an attack included in the results of entropy context profiles filter and if it matches host *h* context *c(h)*, then attack A_l will be included in the final attacks alert list.

3 Experiments and Evaluation

In order to evaluate our system, we performed two types of experiments using the KDD 1999 dataset. The first type of experiments evaluated the quality of the answers of semantic networks using several relevance thresholds denoted by *t*, and then utilizing only the entropy context profiles. The second experiment evaluated the layered attack detection approach, which consisted of using simple Bayesian networks layer to make initial detection, and then semantic network layer to retrieve relevant attacks, and applying both attacks contexts profiles and host context filters. We developed our system in an Oracle database. Additionally, we used Weka [12] and Knime [13] data mining tools to perform data preprocessing tasks and to extract the probability distributions of the simple Bayesian network model. The KDD 1999 dataset was divided into two parts, the training part and the evaluation one: The first part (training one) was used in three tasks, first, to train simple the simple Bayesian network, and then to create semantic network and context profiles. The second part (the evaluation one) was used in our experiments. We used the 10% version of KDD dataset, which consists of 494,020 connections, 97,278 of which are normal connections, and 396,744 are attacks distributed among all 4 categories (DOS, R2L, U2R, and PROBE).We selected 75% of the dataset connection records to be in the training part, representing about 370,515 connection records, and 123,505 connections representing about 25% of connection records to be in evaluation part. Both experiments were performed on a desktop machine with Intel(R) core ™ 2 CPU 2.4 GHZ and 2-GBYTES RAM. We used precision, recall, and F- measure as our metrics defined below in formulas (2), (3), and (4) respectively:

$$P = \frac{TP}{TP+FP} \quad (2) \qquad R = \frac{TP}{TP+FN} \quad (3) \qquad F = \frac{2 \times P \times R}{(P+R)} \quad (4)$$

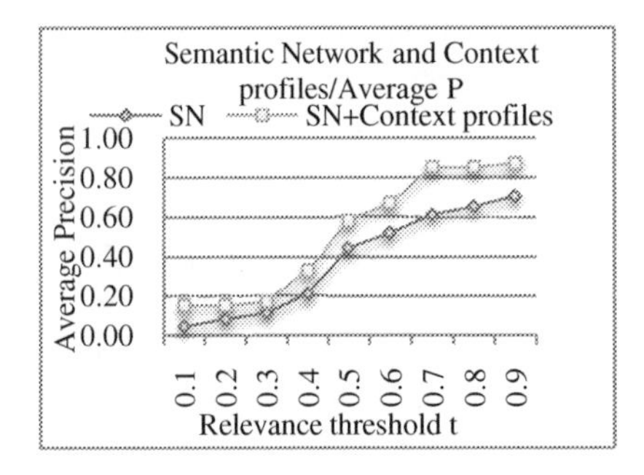

Fig. 3 Average P (SN and context profiles)

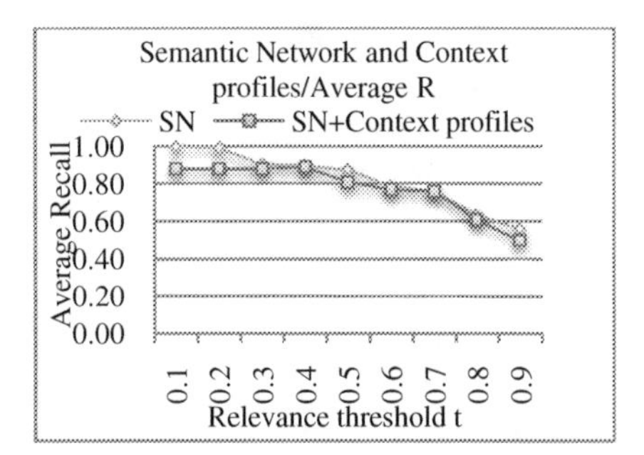

Fig. 4 Average R (SN and context profiles)

TP, FP, and FN are the true positives, false positives, and false negatives respectively. The true positives are the number of attacks retrieved by system layers which are considered relevant to connection record under evaluation. For normal connection records (i.e., no attack), the number of relevant predictions is expected to be 1, as the relevant answer for normal connections should be normal only. The false positive rate is the number of irrelevant attack predictions for the connection record under evaluation. The first type of experiments measured the usefulness of applying context profiles as filters on the related attacks that were recommended by the semantic networks. The semantic network retrieves the relevant attacks based on the relevance score threshold *t*. Lower relevance thresholds result in more retrieved attacks; however, some false positives are expected. Higher relevance thresholds retrieve fewer but very relevant attacks; however some relevant ones may be missed. Figures 3 and 4 show the precision and recall results of using both semantic networks and attack context profiles for detecting attacks which are supposed to be relevant based on semantics and context of occurrence. As observed in figure 3, the precision values are higher (better) when semantic networks and context profiles are used, compared with precision values of using only semantic networks without context profiles. At the 0.7 relevance threshold, semantic networks with no context profiles achieve 0.61 precision compared to 0.85 when the contexts profiles filter is used to filter semantic network results. It can be observed from the results that the context profiles filter removes most of false positives (normal activities), which are included in semantic network results. The semantic network has a superior average recall, which is almost 1 at the 0.1 and 0.2 semantic relevance thresholds. Recall results reported on figure 4 shows that the context profile filters missed some relevant attacks, beyond the 0.7 relevance threshold, when the recall values of the semantic network and context profiles filters start to decline. However, both recall lines are convergent beyond the 0.7 threshold. This decline shows that the semantic network starts missing some relevant attacks beyond the 0.7 threshold.

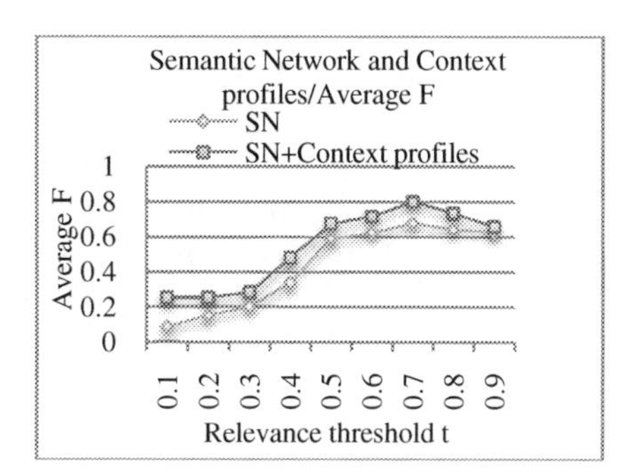

Fig. 5 Average F measure (SN with context profiles)

Figure 5 shows the F measure values for the first experiment. The observed results show that using the semantic network with context profiles, has a better F measure values at all semantic relevance thresholds between 0.1 and 0.9. The best F measure value when semantic network and context profiles are used together is 0.8; this value is obtained at 0.7 relevance threshold. By contrast, this value is approximately 0.68 when the semantic network is used without contexts profiles filter. To The second type of experiment was performed to compare the performance of (1) Semantic network without any context filters, (2) Semantic network with host context filter, and (3) Semantic network with both context filters (context profiles, and host context). The semantic network predictions are used as inputs to contexts profiles filter. The contexts profiles filter checks these predictions, and filters out attacks which do not match any context profile. The outputs of the context profiles filter are then used as inputs to the host context filter. In the current implementation, we collected about 100 vulnerabilities to create host contexts. The collected vulnerabilities cover all attacks in the KDD 1999 dataset confirmed by several vulnerability databases [14, 15]. In total, 22 host contexts were created, all attacks predicted by context profiles which do not match host context are discarded by host context filter.

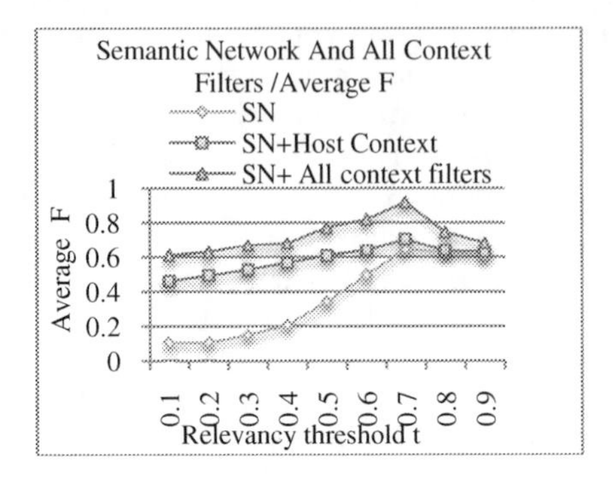

Fig. 6 Average F measure (SN, and all context Filters)

Figure 6 shows the results of F- measure values for the second experiment. When both context filters are used after semantic network, the best F value obtained at the 0.7 relevance threshold is 0.92. As observed in figure 6, the best F value at 0.7 relevance score obtained of applying the host context filter on the results of semantic network is about 0.71 which is lower than when both context filters are applied. The host context filter showed some limitations in handling normal connections when no context profiling was used. When both context filters were used together, a superior precision was achieved, resulting in a better F-value (see figure 6).

4 Related Work

Several research articles are relevant to this work. In [17] a four layer semantic schema was presented to extract attack scenarios and attack knowledge; however, the presented approach did not utilize any host-based context matching. The authors of [18] proposed a web attack detection system using semantics of application headers. The described system was designed to detect header and payload based attacks. In [19] the authors developed a prototype to detect suspicious activities in wireless sensor networks by correlating data in time and spatial domains. In [20] the latent semantic analysis was utilized for security attacks detection at run time. The approach achieved high accuracy; however, the process requires selecting features at run time using latent semantic analysis, which is computationally expensive. By contrast, we used pre-calculated semantic networks to minimize run time computation. The authors of [21] used contextual information about hosts and their vulnerabilities with the goal of minimizing false alarm rate for intrusion detection. Based on their results, the system minimized the false alarm rate. In [22] the authors studied the effect of correlating IDS attack signatures with static and dynamic network information to derive network context. In [23] vulnerability analysis was used as a context information source to minimize false alarm rates. Event correlation was used in [24] to infer events within the same context. The authors of [25] used entropy based measures to build normal and anomaly contexts.

5 Conclusions and Future Work

We designed, implemented, and evaluated a new contextual semantics approach for detecting cyber-attacks. The semantic layer is represented by semantic network which models semantic relationships between attacks in order to infer all relevant attacks given in a sequence of events. The context layer consists of two filters that can be applied on the top of semantic network predictions, to filter out irrelevant attacks by matching them with contextual information on attack profiles and network hosts. We found that our layered approach, which consists of the semantic network and both context filters, can achieve more than 99% precision on the KDD 1999 dataset. The value of F-measure was more than 92% as well. The proposed approach is promising in cyber-attack prediction problem. We are planning to evaluate our system using different datasets. Currently, we are working on building a complete context information repository in our system by adding more contextual information filters.

Acknowledgments. This work was partially supported by a grant from Northrop-Grumman Corporation, USA.

References

1. Sowa, J.: Semantic Networks. In: Shapiro, S.C. (ed.) Encyclopedia of Artificial Intelligence, pp. 1493–1511. Wiley, New York (1992)
2. Sowa, J.: Semantic Networks, `http://www.jfsowa.com/pubs/semnet.htm`
3. Knowledge discovery in databases DARPA archive. Task Description, `http://www.kdd.ics.uci.edu/databases/kddcup99/task.html`
4. Shannon, C.: The Mathematical Theory of Communication. University of Illinois Press (1949)
5. McHugh, J.: Testing intrusion detection systems: A critique of the 1998 and 1999 DARPA intrusion detection system evaluations as performed by Lincoln Laboratory. ACM Transactions on Information and System Security 3(4), 262–294 (2001)
6. Kayacik, G., Zincir, A.: Analysis of Three Intrusion Detection System Benchmark Datasets Using Machine Learning Algorithms. In: IEEE Intelligence and Security Informatics, Atlanta, USA (2005)
7. Karabatis, G., Chen, Z., Janeja, V.P., Lobo, T., Advani, M., Lindvall, M., Feldmann, R.L.: Using Semantic Networks and Context in Search for Relevant Software Engineering Artifacts. In: Spaccapietra, S., Delcambre, L. (eds.) Journal on Data Semantics XIV. LNCS, vol. 5880, pp. 74–104. Springer, Heidelberg (2009)

8. Duarte, J., Dos, S., Melo, L.: Comparison of Similarity Coefficients Based On Rapid Markers In The Common Bean. Genetics and Molecular Biology 22(3), 427–432 (1999)
9. Pensa, R., Leschi, C., Besson, J., Boulicaut, J.: Assessment of Discretization Techniques For Relevant Pattern Discovery From Gene Expression Data. In: 4th Workshop on Data Mining in Bioinformatics (2004)
10. Güneş, A., Nur, Z., Malcolm, I.: Selecting Features for Intrusion Detection: A Feature Relevance Analysis on KDD 99. In: Third Annual Conference on Privacy, Security and Trust, PST, Canada (2005)
11. Kumar, K., Nath, B., Kotagiri, R.: Layered Approach Using Conditional Random Fields for Intrusion Detection. IEEE Transactions on Dependable and Secure Computing 7(1), 35–49 (2010)
12. Weka Data mining and machine learning software, `http://www.cs.waikato.ac.nz/ml/weka/`
13. Konstanz Information Miner, `http://www.knime.org/`
14. IBM Internet Security Systems, `http://xforce.iss.net/xforce/xfdb/588`
15. National Vulnerability Database, `http://web.nvd.nist.gov/view/vuln/search?execution=e2s1`
16. Mrutyunjaya, P., Manas, R.: A Comparative Study of Data Mining Algorithms for Network Intrusion Detection. In: First International Conference on Emerging Trends in Engineering and Technology, pp. 505–507 (2008)
17. Wei, Y.: Semantic Approach for Attack Knowledge Extraction in Intrusion Detection Systems. In: 29th Annual IEEE International Conference on Local Computer Networks (2004)
18. Vaidehil, V., Srinivasan, N., Anand, P., Balajil, A., Prashanthl, V., Sangeethal, S.: A Semantics Based Application Level Intrusion Detection System. In: International Conference on Signal Processing, Communications and Networking (2007)
19. Ganesh, K., Sekar, M., Vaidehi, V.: Semantic Intrusion Detection System Using Pattern Matching and State Transition Analysis. In: International Conference in Recent Trends in Information Technology (2011)
20. Lassez, J., Rossi, R., Sheel, S., Mukkamala, S.: Signature Based Intrusion Detection using Latent Semantic Analysis. In: IEEE International Joint Conference on Computational Intelligence, pp. 1068–1074 (2008)
21. Lexi, P., Benedikt, W., Volker, W.: A Context Aware Network-IDS. In: 13th Nordic Workshop on Secure IT Systems, NordSec Copenhagen, Denmark (2008)
22. Frédéric, M., Mathieu, C., Lionel, B., Yvan, L.: Context-Based Intrusion Detection Using Snort, Nessus and Bugtraq Databases. In: Third Annual Conference on Privacy, Security and Trust, Fredericton, New Brunswick, Canada (2005)
23. Liu, X., Xiao, D.: Using Vulnerability Analysis to Model Attack Scenario for Collaborative Intrusion Detection. In: 10th International Conference on Advanced Communication Technology, pp. 1273–1277 (2008)
24. Zhou, J., Heckman, M., Reynolds, B., Carlson, A., Bishop, M.: Modeling Network Intrusion Detection Alerts For Correlation. ACM Transactions and Information System Security 10(1), 1–31 (2007)
25. Gu, Y., McCallum, A., Towsley, D.: Detecting anomalies in network using maximum entropy estimation. In: ACM SIG-COMM Conference on Internet Measurement, pp. 345–351 (2005)

Collaborative Network Development for an Embedded Framework

Jonathan Bar-Magen Numhauser, Antonio Garcia-Cabot, Eva Garcia, Luis de-Marcos, and Jose Antonio Gutierrez de Mesa

UAH - University of Alcala
CC - Department of Computer Science
Alcala de Henares, Madrid, Spain

Abstract. In the following paper we will describe the work developed on a collaborative and component or modular based methodology for native multi-platform development. This investigation work is the continuation of a previous version, for which we adapted to support development of web intelligent projects for various devices as well as using native context development in each device.

Keywords: Collaborative, Framework, Components, Methodology, Embedded Software Development.

1 Introduction

Software project development has been substantially affected by online communities. The influence applied by members of these communities and the development implications has changed the matter in which projects are being developed for web usage.

Collaborative based development has to go hand in hand with suitable methodologies and technological elements that may ensure a fluent and successful result.

Part of these methodologies have been studied by our investigation group and tested on CMS frameworks, in specific Joomla. As an inevitable result the progress of the collaborative methodology was improved during the last few months with the objective of obtaining a more flexible and optimized work methodology that will allow us to implement new technological functionalities, as well as increase significantly the feedback of the collaborative communities by introducing intelligent agents.

As a result we obtained an improved methodology, which is able to apply intelligent feedback to new breakthrough projects that are based on native software development, respecting the limitations of the components that form part of the final project and allowing the users of our methodology the opportunity to adapt, structure and obtain the most of the proposed work method.

In the following paper the specific details of the working methodology for collaborative and modular development oriented to native based software development will be exposed and explain.

L. Uden et al. (Eds.): 7th International Conference on KMO, AISC 172, pp. 443–453.
springerlink.com

2 Environmental Requisites

The essence of a Framework is its ability to solve the largest amount of issues posed to it in a specific context, as well as comply with the requisites of this environment.

As an initial step of the adaptation of the original collaborative methodology, we pursued the possibility to increase its functional capabilities, translated in a better and optimized Framework, adapted to the most common requisites of a network based development methodology.

The following section describes the context requisites and the means used to solve some of the most common issues in a methodological development process as well as the development of a strongly dynamic Framework.

2.1 Native Context

First we will like to point to the fact that the meaning of "Native" can be interpreted in many ways. In our case, following the long experience acquired in mobile projects development, by native we refer to the fact of developing a project in the basic requirements and available tools that offers a specific hardware element. E.g. the case of project development targeted to mobile devices, can be approached as a unified and adaptable project in which the developers dedicate their main efforts to create a single version targeted to various devices, or it can be approached as a distributed and pre-processed development in which the developer creates a specific original version, and from that he dedicates most of the efforts to adapt this original version to the rest of targeted devices, considering the limitations and potentials of each device, and creating a specific version for each.

For our investigation work we decided to apply the second alternative, and so we named it Native development, as it requires the creation of a native version of the main project on each device it is targeted to.

The resulting methodology is dedicated to offer a context in which the working process will be oriented to native development, and make a complete and optimized use of all the resources in project development to ensure the best work process

2.1.1 Methodological Implications

As part of the overall methodology, the native development has important implication in some of the basic aspects of project development. The most common project development steps are known to be Analysis, Design, Development, Implementation and Maintenance.

Native development will affect most of the steps described. The decision to develop in these terms influences the overall aspects by placing a higher importance in the design of a proper project that will easily be adapted to new hardware requirements, resulting in a dedicated development of projects prepared for this future adaptations, as well as a number of Maintenance steps that will allow the improvements and repair of the new out coming versions. The implementation step will be the most influenced by this working method during the testing and execution of the resulting version.

2.1.2 Native or Embedded Analysis and Design

The first and second step of any methodology for project development is the analysis of the requisites for the project and the design of the components that will form the out coming product.

Native development will force the analyst and the designer to consider the limitations in versioning, as well as the differences in each device for a proper 100% utilization of the device resources.

Experience in each step is crucial, and in these two more than in any other, as the blueprints for the future project is laid down, from cost measurement to time dedication, functionalities and optimizations, all these aspects will be planned in these steps.

2.1.3 Native Development

Development of native based projects will represent in most cases the step in which many of the efforts will be applied to. Depending on the previous steps, the efforts may vary, but it will always represent the core work of the our methodology.

Native development requires the developer to be aware, at parallel times, of the limitations of the main device in which he is developing the Gold version of the project, as well as the overall limitations that he may encounter in the devices for which the analysts planned to implement the resulting program.

The use of a proper Framework, together with an elaborate feedback database will determine the level of experience the developer will need to have for a proper completion of the project.

Eventually, when we later describe the collaborative nature of our methodology, it will show how in all the development steps it will experiment a significant reduction in work effort.

2.1.4 Native Implementation and Maintenance

The last two general steps of the methodology is the implementation and maintenance. Once completed the Golden version of the project, the adaptation process begins by implementing and testing the version in all targeted devices.

By an exhaustive process of modifications, improvements, feedback with the Framework, and work optimization, the resulting versions for each device are obtained. Maintenance is a necessary step in any software project, and in these cases there is no difference. Depending on the methods used for development of the variety of versions, modifications to the project may be extremely cheap or extremely costly. If there was a proper use of the proposed Framework, as well as a rich API database the modifications to the Gold version may easily be propagated to the other native versions. For our methodology we have been using a pre-process xml compiling for real time modification of the source code before compilation and deployment.

These precompiled steps will significantly affect the time cost of the maintenance process. Applied to component based development, each component will be dependable of this type of maintenance.

2.2 *XML Pre-processed Build*

ANT and Maven are well known compiling engines that allow a powerful set of administration steps in implementation and deployment.

In this section we will describe the use of a specialized step in ANT compilations, which has been used to allow the developers in our working methodology to implement modifications of various numbers of versions in the original *golden* version, without the need to create new independent versions.

2.2.1 Build.xml

For the purpose of pre-compiling the projects, we opted for the use of Ant calls of standard build.xml files. In contrary with the overall Ant calls, in this case a program has been elaborated to facilitate the analysis of the code, to build the new source files from the base code files, and compile the source files once they are filtered.

The process of filtering the base source code and obtaining new source code for further compiling will be executed by a specific program created for that purpose.

2.2.2 Tags

The program that filters the source code makes use of specific Tags that will guide him through the filtering of determined sections of the source code.

The Tags enclose sections of the source code which are known to differ by device, so eventually when a Tag is selected for compilation, the code section in that area is used, while code sections that their Tags haven't been called will not be compiled, allowing a variety of versions to be compiled from a single source code.

2.2.3 Methodological Implications

By using a single source code for all the targeted versions, and allowing such tool to be used on all levels, from the specific source code of the project, to the API used by the project, the tedious work of creating a single version for every targeted device won't be needed anymore, reducing the development cost in time.

On the other hand, the implementation and maintenance will be significantly influenced by this method as the testing and modification to a problem may be propagated to all versions with a single change.

The implications in time cost are significant in the test cases obtained from our investigations.

2.3 API

By adopting the use of a native based project development, the team had to consider the aggregation of a specific API to the workflow.

The API pretends to cover all acting aspects of the development process for which may be technical aspects.

As it was explained in the previous section, pre-processed tag build tasks of the source code may be applied on a higher level, and in our case it will be to the API.

The dependency on this element is of such importance that it may define the viability for the development of projects.

2.3.1 Basic Aspects

The API will consist of a number of generic components from which the developer may extract the best functionalities. As they are submitted to pre-processing from the build tool, there will be a need to have several adaptability considerations, and use the appropriate tags to control the critical code zones.

Apart from the pre-processing considerations, the API will follow the overall behavior of other common API libraries, and will offer a valuable feedback to the developing step, as well as the implementation and maintenance.

2.3.2 Technical Requisites

Making use of a developing API requires that we address a number of issues. The first is the technological context. For each native context, there should be an associated API, which implies that there may be more than one API, as a project may cover a number of native environments.

Another issue is the compiling process. As it was mentioned before, building of a native version for the project will enforce the need to have a pre-process support for the version. It is not mandatory, but is recommended to ensure a better adaptation step, as it was mentioned in sections before.

The final and maybe most crucial issue are the components and modular structure of the API. The methodology that we have been working on is a collaborative and component based methodology, which requires the users of this methodology to orient every aspect of the project development to component based concepts. The API is no exception, and it should allow the developers to add and modified existing components for future improvements. The concept of component development will be treated in the following section.

3 Collaborative Framework

In the previous section we laid out all the possible environmental requisites of a project that may follow the steps determined by our methodology. In this section we will describe the collaborative nature of the methodology, and how its adaptation from a component based methodology together with the community characteristic resulted in an improvement of the standard development process.

3.1 Components Methodology

Our investigation follows the work that took place by our team on the possible use of a more adaptable methodology for CMS projects, in concrete Joomla. From this work we extracted some of the most useful characteristics that we successfully acquired from exhaustive ROI studies from practical work.

Components were one of the main aspects that we extracted from our previous work. The components based methodologies define a number of steps in which to follow and ensure a proper component recycle for future projects. Its implications on collaborative projects have been substantial and can be further studied.

In our case, we adapted this concept to a Native project environment, as it was explained in the previous section. But the question is how it differs from a non native environment.

We considered that project should be targeted on various environments regardless the technological difference of each, and that the solution that we chose to implement is the native based development. In effect this method differs from our previous practical work, as CMS uses generic script based encoding, which in turn can be read by much of the available devices in the market. But as we argued before, the downside is the generalization of the resulting product, and the high probability of not using to the fullest of each device potential.

Components methodology applied to native development changes the working method of a project life cycle. The need for motile native contexts as well as the limitations of each device will require that the components created, for future reuse, comply with much of the characteristics of each native technology, forcing a developer team to acquire enough expertise so it can target a larger group of devices.

The cost is greater in the beginning for a developing team, but with time a team that makes a constant use of this methodology, will find that the developing cost is reduced exponentially, as more components are available in each native technology, and the tedious work of creating new components subsides.

3.1.1 Community and Collaboration

Many of the open source CMS frameworks are strongly dependable of collaboration. The collaboration is generally applied to the development of components, and its sharing.

Some sharing of these components are for profit while others not, but in any case it represents an interesting tendency to elaborate project basic elements in a larger context.

The collaborative elaboration of components has been studied in our last work, and in this case we have to analyze its potential to modify and optimize the native development of projects.

It is fundamental to state that an active community is necessary to habilitate this working method, and a certain organization, and a local and global level is in order as well. As part of our previous work, we established a couple of fundamental roles to the methodology, which will be further used and developed in this case.

3.1.2 Collaborative Roles

The roles that were mentioned in the previous section are the Global Project Manager (GPM) and the International Communications Manager (ICM). These two roles we originally created to solve some of the problems that we encountered when implementing the methodology on practical cases.

For the evolution of our methodology, we obviously have to consider the implications of adding the Native requisite to the formula. Collaborative component sharing on Native components for each different technological context requires roles to allow a fluent transition of knowledge from one global team to other global teams, and ensure the health of a consistent component database.

We translate the entities that played a role in CMS development, and adapt it to multi-platform (multi-version) projects that will need this approach to improve the development process.

As a solution for part of the difficulties that we encounter, we decided to add a new role to the overall methodology, and describe its impact.

On the developing level, professionals with experience on multi-platform development should have a better communication with developers in other global groups. Thus we introduce a role similar to the GPM, but on a more technical level, Components Technological Manager (CTM). This new role will have a greater technological level, and will be in constant collaboration with the other two global roles, for which main role is to maintain a global consistency of the components.

By distributing the work, and creating a specific role for components managing, we charge more collaborative work on the GPM and ICM, and more development work on the CTM.

The main concept is that the creation and maintenance of a component can take place in any of the global groups, but that the managing of the component, which include version control, documentation control and etc., should be under the supervision of one of the groups, allowing for a change of responsibility in case the group wants to resign from it.

This main rule ensures a better quality control, and allows an improve feedback.

3.2 Collaborative Native Component Development Process

By now all necessary elements of our proposed methodology have been described, and in this section we will expose the fundamental steps to execute this methodology, as well as some results from our practical studies.

3.2.1 Practical Application

The use of this proposed methodology is to improve the resulting work process, and allow developing teams to adopt and adapt this work method to their needs.

This is why the initial motivation for the continuation of the first investigation work was to solve a number of issues that could be found in a multi-platform development process, and which are not common in CMS development.

The success of the practical application of the first version of this methodology in CMS development proved to us that an adaptation is in order.

3.2.2 Analysis and Design

The application of this methodology is directed to project development. As any methodology, it covers all aspects of a project life cycle. The initial step is to establish the requisites, in the analysis step. By making use of the traditional roles as well as the new roles described in the previous section, the analysis step will be formed by the basic definition of project development as well as the establishment of the possible existing components that will be used in each targeted device or platform.

The targeted devices will be also a main issue to study, as it will define the work needed to cover the project expectation on various versions.

3.2.3 Development

The development step will result in creation or recycle of not only the basic components for its execution, but also the possible distribution of this developed components if they do not exist in the general database.

In the development process there will be a dependency on the behavior of external agents, developing groups and components that may be found in the collaborative community.

Depending on the GPM and ICM experience and abilities the developing process will produce good or bad results, and their good communication with the CTM will be fundamental for the recycle and creation of components.

3.2.4 Implementation and Maintenance

The implementation and maintenance step will mainly be dependable of the local group, and occasionally the CTM will intervene to ensure that possible solutions to problems in maintenance can be found in the community, instead of developing from scratch.

3.2.5 Results

As a consequence of applying this proposed and improved methodology on a number of development projects, we obtained results regarding costs, organization, functionality and quality.

Generally the most valuable constant to be studied and judged is the time cost value. By this variable it is decided if the methodology is worth the effort.

The results confirm that our methodology had obtained a number of variable favorable to our interests.

First is the overall project development process. As it was stated before, our prediction was that the first cycles would cost more than the following, and that these cycles will affect the developers experience as well as the tools functionality.

With each cycle we observed on the developers' level an improvement in much of the common tasks, as it is the acquisition of requisites for the project, the design towards targeted devices, the development technological issues and even in later cycles the implementation and maintenance was influenced positively.

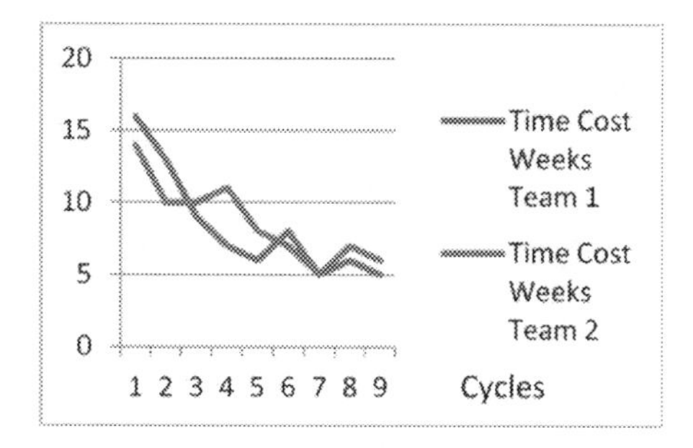

Fig. 1 Project development time cost two teams

On the framework level, the feedback obtained in the initial cycles from the communities on components was crucial for a smooth starting, and with time the information started to flow on both channels, from the community and to the community.

The local and global components database was enriched with every cycle, reducing the developing cost in time.

In organization techniques, we noticed a difficulty to adapt the existing developing teams to the new working environment. The aggregation of a new role, in some cases three for those teams with no experience with the previous methodology, resulted in the first cycles a fall in productivity.

After a few cycles, the teams were able to put aside much of the issues encountered in other areas, and dedicate more time to organization issues. It is then when we started to notice an improvement in the functionality of the team, and the resulting project.

Quality and functionality of the resulting product was always part of our concern, mainly because a methodology is planned on those necessary requisites.

As it was stated on the very first cycle there was no improvement in quality and functionality of the resulting product. There was even a downgrade in functionality as the users of the methodology sorted themselves to adapt into the new Framework.

Once products were created, the results started to differ. Many quality issues were immediately solved by using collaborative components from the communities, and other developers group. There was a case of mutual feedback, as a component was modified for better results.

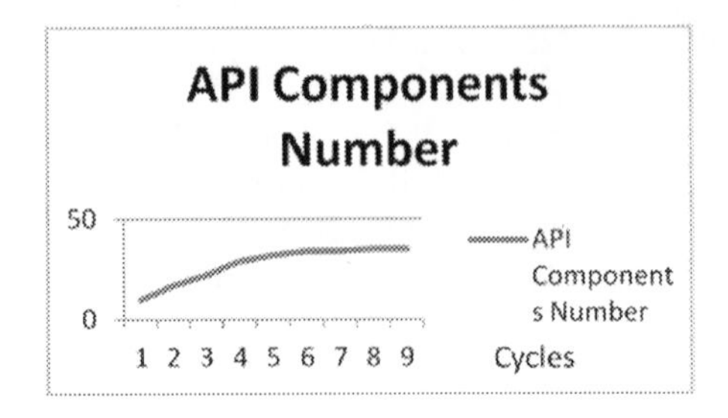

Fig. 2 Components database evolution

Eventually on later projects the process became stable and improvements were reduced to a constant, but the quality improved significantly compared to previous versions.

4 Conclusions

Considering the outcome of the practical application that took place on a variety of projects oriented to a variety of devices, we concluded the following.

A possible adaptation of the first version of the collaborative methodology was possible, with the aggregation of new elements to adapt the steps to native development context.

The use of new tools, as may be the code pre-processing or the API allowed the teams subjected to this methodology to initiate a feedback database that will contribute to the global collaborative database.

Finally the results obtained from the improvements were satisfactory, allowing us to continue with the research of this new alternative methodology.

Acknowledgments. This research is co-funded by:

(1) The University of Alcalá FPI research staff education program,

(2) The Spanish Ministry of Industry, Tourism and Commerce AVANZA I+D program (grant : TSI-020301-2009-31).

Authors also want to acknowledge support from the TIFyC research group.

References

1. Bar-Magen, J., Gutierrez Martinez, J., De Marcos, L., Gutierrez de Mesa, J.: Colaborative and Component Based Software Development Methodology. In: Proceedings of the IADIS International Conference Applied Computing, Rumania, pp. 287–290 (2010)
2. Yildiz, U., Godart, C.: Designing Decentralized Service Compositions: Challenges and Solutions. In: Web Information System and Technologies, Spain, pp. 60–71 (2007)
3. Hornung, G., Simon, K., Lausen, G.: Mashups over the Deep Web. In: Web Information System and Technologies, Portugal, pp. 228–241 (2008)

4. Boella, T., Remondino, M.: Collaboration and Human Factor as Drivers for Reputation System Effectiveness. In: Web Information System and Technologies, Portugal, pp. 3–16 (2009)
5. LeBlanc, J.: Learning Joomla 1.5 Extension Development. Packt Publishing, Birmingham (2008)
6. Griss, M.L.: Software Reuse Architecture, Process, and Organization for Business Success. In: Proceedings of the Eighth Israeli Conference on Computer Systems and Software Engineering, Israel, pp. 86–98 (1997)
7. Lakhani, K., Von Hippel, E.: How open source software works: "free" user-to-user assistance, Cambridge, MA 02142, USA (2002)
8. Murphy, C.: Adaptive Project Management Using Scrum. Methods & Tools Global knowledge Source for Software Development Professionals 12(4), 10–23 (2004)
9. Ghosh, G.: Agile, Multidisciplinary Teamwork, Oslo, Norway (2004)
10. Wake, W.C.: Extreme Programming as Nested Conversations. Methods & Tools Global Knowledge Source for Software Development Professionals 10(4), 2–13 (2002)
11. Wake, F.M.: How to Select a QA Collaboration Tool. Methods & Tools Global Knowledge Source for Software Development Professionals 11(1), 26–31 (2003)
12. Ponder, M.: Component-Based ethodology and Development Framework for Virtual and Augmented Reality Systems Unified. École Polytechnique Fédérale De Lausanne, Rumbaugh (2004)
13. Dabbs Halloway, S.: Component Development for the Java (TM) Platform. Addison-Wesley Longman Publishing Co., Inc., Boston (2002)
14. Hightower, R.: JSF for nonbelievers: JSF component development. IBM developerWorks (2005)
15. McDirmid, S., Matthew, F., Hsieh, W.: Java Component Development in Jiazzi. School of Computing University of Utah, USA

A User-Centric Approach for Developing Mobile Applications

Aleš Černezel and Marjan Heričko

University of Maribor, Faculty of Electrical Engineering,
Computer and Information Science, Institute of Informatics,
Smetanova 17, Maribor, Slovenia

Abstract. Creating mobile applications that are accepted by users is critical for a successful business. This article presents a user-centric approach for developing applications. In the theoretical part, we will present some background information about the three analytic tools that were used with this approach. In the practical part, these analytic tools will be practically applied on a case study for a mobile application. First, the three analytic tools are practically applied and after identifying use cases, the application will be implemented. The application consists of two parts: server and client. The server is based on Java EE technology and the client is based on the Android platform.

Keywords: Mobile applications, User-centric design, Personas, Scenarios.

1 Introduction

Mobile devices are becoming more popular every year, which naturally implies a greater development of applications for such devices [1]. There are a variety of platforms and operating systems for mobile devices, each offering its own specific features. This paper will focus on developing a mobile application, with a brief description of the features that the application will offer. The development process will employ a user-centric approach. The main focus of this approach is to focus on the end users of the application and consider their needs, desires, and limitations. User requirements are iteratively tested, thus such applications are more suitable for users.

This paper is organised in the following way: The second chapter describes user-centric design, why users are important, how to learn about them, and a detailed description of three analytic tools, which were used for user-centric development: personas, scenarios, and use cases. Section 3 describes a case study of a mobile application for foreign exchange students. First, the analytic tools are used in practice and after generating user requirements, the developing of the application is briefly described.

L. Uden et al. (Eds.): 7th International Conference on KMO, AISC 172, pp. 455–465.
springerlink.com

2 User-Centric Design

User-centric design is a design philosophy where the primary focus is on the users. While designing and building the application, one must consider all the users' desires, needs and limitations, in order to make the application more convenient to end users. The application must fit the users and not vice versa [2], [3]. Design is a multi-stage problem where certain tasks must be performed, as seen in Figure 1.

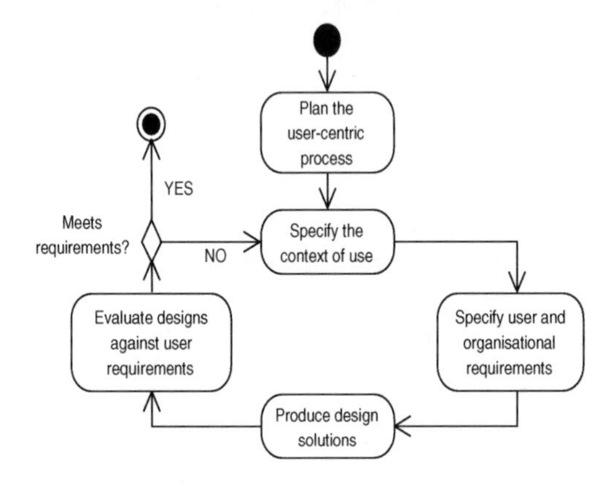

Fig. 1 The iterative user-centric design process [4]

User-centric design seeks answers to questions such as [3]:

- Who are the users of our application?
- What are the users' tasks and goals?
- What do users expect from our application?
- What is the context in which the application will be used?
- What are the users' experience levels?
- What information do the users need? And in what form?

Answering these questions can improve both usefulness and usability. Usefulness relates to relevance – does the application match the users' needs? Usability relates to ease of use – it shouldn't be difficult for users to use the application, even without any instructions. Usability can be further improved by following these guidelines [3]:

- **Visibility.** Visibility helps users predict the effects of their actions. Any function performing elements should be highly visible. Users should be able to tell what they can and cannot do in any situation.
- **Accessibility.** Users need to find important information quickly and easily. There should not be too many ways to find information as this can confuse users. Large pieces of information (e.g. text, tables, and charts) should be chunked and organized into smaller pieces that are meaningful to the user.

- **Feedback.** Users should receive immediate feedback after performing an action. They should also be able to tell if the action is finished or still processing.
- **Errors.** Users can make incorrect actions, such as submitting an empty form. The focus should be on minimizing user error, but if errors do occur, users should be spared technical details and enjoy the support of a recovery mechanism. No data should be lost.

2.1 Knowing Your Users

The previously mentioned guidelines are general and are insufficient for making a specific application usable. It is important to discover how they interact with the said application by involving users from the very beginning and including them as an integral part of the development process. Observing them at work can help analyse tasks, workflows, and goals.

User feedback is important and can be improved by using various methods such as walk-throughs, card sorting, paper prototypes, think-aloud sessions, and surveys. Explore different designs and approaches before making any final decisions and consult with users.

Assumption validation should be done by means of usability testing. Usability testing is an iterative process, as can be seen in Figure 1, and should be done throughout the development cycle. It is the only way to assure if the application meets the users' needs.

2.2 Analytic Tools

Analytic tools help with the design process. There are a few tools that are used in the analysis of user-centred design, namely: personas, scenarios, and use cases. In the next few chapters, we will describe all three of the mentioned analytic tools.

2.2.1 Persona

The word persona is derived from Latin, where it originally referred to a theatrical mask [5]. It is a fictional model of a user that focuses on the individual's goals when using an artefact. It resembles classic user profiles with one key difference. It is not a description of an average user, but rather combines various patterns of users' behaviour, desires, needs, and goals into a single individual. They do not represent user roles and are therefore different from actors found in use case diagrams [6].

Personas are created after studying the primary group of users, which typically involves an interview or a survey. The gathered information is used to form one or more personas – usually more than one, as it would be difficult and impractical to fit all the details into one persona. Ideally, the behaviour of the personas should not overlap, so that the number of personas can be kept to a minimum. All personas combined should contain all of the users' characteristics. If any other persona can be identified in the future (containing some non-overlapping characteristics), then there was something missing in the first place [6].

Each persona is usually described in one or two pages, consisting of goals, skills, attitudes, and their environment. In order to make the persona more alive to the development team, some made-up personal details are added to the description. The number of personal details must be balanced – too much gets in the way and too little can turn the persona into a generic user [6].

There are three types of personas [7]:

- **Primary** personas describe users for whom the design is for. If the design fails for the primary persona, then the design fails entirely.
- **Secondary** personas are not the main target, but they should be satisfied if at all possible.
- **Anti-**personas are the types of users for whom you are not designing.

Each persona is derived from the users' goals and it is therefore important to focus on them when studying users. One should be aware of the difference between tasks and goals. Goals are end states, whereas tasks are merely steps that help users reach their goals. A very common mistake, which can significantly reduce the usability of an application, is focusing on users' tasks instead of their goals.

Goals can be categorized into four groups [6]: (1) personal goals, which are simple, universal and, of course, personal. For example: do not make mistakes. (2) Corporate goals, which are the goals of the users' work organisation. They are important as they increase the success of an organisation. For example: increase the market share of a product. (3) Practical goals, which are the bridge between personal and corporate goals. For example: satisfying a client connects the corporate goal of "improving profit" with the personal goal of "being productive". (4) False goals, which are goals that are not relevant for the user. They are usually objectives that support the technical details of a system, for example: save memory or run in a browser.

Given the above classifications, certain facts must be taken into account. If different goals result in a conflict, personal goals usually take precedence. This is because goals that are closer to the user are the most important. Therefore, one should not only focus on corporate goals, but also focus on personal goals and improve efficiency through them [6].

2.2.2 Scenarios

Carroll [8] gives a simple definition of scenarios: "*Scenarios are stories. They are stories about people and their activities.*" Scenarios are fictional narrative stories, describing the sequence of events in a persona's daily life. The story should be concise and focus on relevant details with a length of around one page [7]. Even though scenarios do not seem to be more than ordinary stories, they do hold various details about events and tasks, which give details about the specific personas – even emotional and physical ones [2]. Scenarios are a great tool for communication with other people, because stories are easier to follow [8].

Scenarios contain characteristic elements. They include, or presume, a *setting*: a starting state of the story, which mentions the persons and objects that are involved. Scenarios also include *agents* or *actors*, usually personas that were defined in the previous step. Each of them has its own *goals* or *objectives* that he or she wants to achieve, given the circumstances of the setting. Every scenario has at least one agent and at least one goal. Multiple goals are not equally important. One goal is usually the defining goal and typically answers the question: "Why did this story happen?" Similarly, one agent might be the principal actor and provides an answer to the question: "Who is this story about?" The plot of the scenario includes a sequence of *actions* and *events*. Those can facilitate, obstruct, or even be irrelevant to given goals [8].

There are different types of scenarios. Depending on the content, one can identify [9]: (1) problem scenarios, which describe what users can do in a certain situation; (2) activity scenarios, which propose an improvement for current practices; (3) information scenarios, which describe users' understanding, perception, and interpretation of information; and (4) interaction scenarios, which describe physical actions and system responses that enact and respond to the users' goals and needs.

Depending on the outcome of the story, we can identify [2], [7]: (1) a best case scenario, where the outcome of the story is positive, (2) a worst case scenario, where the main agent does not accomplish the main goal, and (3) an average case scenario, where nothing extraordinary occurs and the outcome is neutral.

Carroll [8] states five reasons for scenario-based design [9]:

- Vivid descriptions of user experiences evoke reflections about design issues.
- Scenarios concretely fix an interpretation and a solution, but are open-ended and easily revised.
- Scenarios can be written at multiple levels, from many perspectives and for many purposes.
- Scenarios can be abstracted and categorized.
- Scenarios anchor design discussion in work, supporting participation among stakeholders and appropriate design outcomes.

And what are the weaknesses of the scenarios? With scenarios it is difficult and impractical to describe specific low-level tasks. They are also based on assumptions that are not always true to life [7].

2.2.3 Use Cases

Use cases describe interactions between the actor and the world. They are represented as a sequence of simple steps in order to achieve the actor's goal. Steps are described at an intermediate level of detail. Each task is often described with more than one use case [7]. Exception handling is also possible: the normal flow is defined first and in case of an exception, the alternate flow is defined [10].

There are various ways to describe a use case [7], [10]: (1) a structured text, where the use case is similar to a scenario; (2) graphical notation for modelling flows, such as a UML activity diagram; and (3) a two-column table, where the first column consists of steps performed by the actor, and the second describes world feedback.

UML use case diagrams can be used to model multiple use cases on a single diagram. The diagram consists of a system boundary, one or more actors, and one or more use cases. Actors and use cases are connected according to their interactions. Each use case is modelled as an ellipse, containing a short description. A more detailed description is done separately, usually in one of the three previously mentioned ways [11], [12].

The main advantage of use cases is that they decompose complex tasks into several smaller units that are easier to follow and manage. Their weakness is that they are not so detailed and that they can sometimes simplify or even omit key steps [7].

3 Case Study: Mobile Application for Foreign Students

The practical part of this paper presents a case study in the form of an application designed for foreign exchange students (Erasmus) visiting the University of Maribor, Slovenia. The application should offer useful information for foreign students, especially information that is unrelated with the study. A user-centric approach will be used to further identify user specifications for the application.

In the next few chapters we will describe how to identify user requirements for the application in a user-centric way. Firstly, we will identify our users and create personas. Secondly, those personas will be used in various scenarios. Thirdly, use cases will be derived from scenarios.

3.1 Identifying Users and Creating Personas

A brief description of the application shows that the target group includes foreign students. A typical Erasmus exchange student is European, 20 to 25 years old, computer literate, understands English, likes to socialize, and goes to parties in their free time. He or she is unfamiliar with the city of Maribor and its people and does not understand the local language. The duration of the exchange varies from 3 months to one year. As a student, he or she is can benefit from government subsidized meals and other discounts.

This group of users (foreign students) becomes our primary persona. No other primary personas can be identified. Tourists and local students can be identified as a secondary persona, and a non-English speaking person can be identified as an anti-persona. In this case study, we will only briefly describe the primary persona.

Helga Eriksson, foreign student: *Helga is 20 years old and studies computer science in Uppsala, Sweden. She decided to participate in the Erasmus exchange programme because she wants to see places and meet new people. During her one year exchange at the University of Maribor, she wants to complete the 2nd year of her studies.*

She has a big desire to learn about foreign places because she has never been abroad. She is an open person and has no trouble making new friends. She is staying in a student dormitory and will eat subsidized meals. She has no means of transportation, so she will use public transportation. Like all Erasmus students, she will also receive a scholarship.

In her free time she likes to meet with friends, goes to the movies, concerts, and clubs. In addition to entertainment, she is also interested in natural and cultural attractions. Her hobbies are: cycling, reading books, and participating in social networks.

She expects that the exchange will offer her new personal and professional experience as well as new acquaintances that may help her in the future.

By examining a persona, one can identify various interests, lifestyles, and tasks that help generate system requirements. For example, the given persona has an interest in entertainment, meeting people, and exercising. She does not need to arrange a place to stay and does not own a transportation vehicle. The amount of free time she has is unknown.

3.2 Writing Scenarios

The next step after identifying personas is using them in various scenarios with daily life events. The scenarios will focus on the primary persona, as it is the only one. Each scenario will be analysed according to the given theoretical background.

Scenario 1: *Helga and her two friends are going to a concert on a Friday night. They are torn between two options and must make a decision first. Both performers are local groups, which neither of the girls is familiar with. They decide to search the Web for more information. They find lots of pictures, but no user feedback. Despite insufficient data, they decide to go to the first concert.*

The *setting* of the given scenario can be found in the first sentence: a primary persona with two friends on a Friday night. All three *actors* have the same *goal*: going to a concert. Solving the dilemma could be identified as a *sub-goal*. Searching the Web for more information is an *action* to complete the sub-goal, which is necessary to complete the primary *goal*.

Scenario 2: *Helga has finished early with her lectures. She is hungry and wants to eat something. As she only has half an hour of free time, she starts to look for a fast food restaurant that offers subsidized meals. This significantly narrows her options, so she wishes she had a list or a map of providers of subsidized meals. She asks a passer-by for help, but has trouble with communication as he does not speak English. The second passer-by has never heard of the term "subsidized meal" as he was not a student. She quickly realizes that she must ask people her age. She gets lucky with the third person and manages to return to her lectures in time.*

The *setting* of this scenario is: the primary persona is hungry and has half an hour of free time. Her main *goal* is to eat. Searching for a restaurant and asking people for information are *actions* that support reaching the *goal*. The fact that she had to ask three people to find useful information tells us that the situation did not go as planned. The scenario also shows how clever and adaptable the main persona is when she finds out that she needs to ask people her age. The need for a list or a map of restaurants will help with identifying requirements for the application.

Scenario 3: *During her exam period Helga wants to take a break and go to a short trip to Pohorje (a nearby mountain range). She decides to take the bus, because the distance is too much. She prints the bus timetable and goes to sleep. Overwhelmed with excitement and busy preparing her things, she forgets to check the location of the bus station. The next morning she has only 15 minutes to get there and because she does not know exactly where the bus station is, she misses her bus. She is very angry and disappointed, but still does not want to cancel her trip. Unfortunately, it is a Sunday, when the buses do not drive as often. She then waits one hour for the next bus in a nearby coffee shop.*

The third scenario is a mixture of a worst-case and average-case scenario, because she partly fails her primary *goal* – at first she fails when she misses the bus, but then decides to create a new, similar, *goal* in which she succeeds. The *setting* of this scenario is: a Sunday during the exam period, when the primary actor is planning a short trip. The printed bus timetable can be identified as an *object*. The scenario also shows that the primary persona can sometimes forget key things – in this case the location of the bus station.

3.3 Identifying Use Cases

After writing scenarios, the next step is to derive use cases from them. All scenarios include a desire for information that would make certain steps easier. For this case, the metaphor *activity* was created. An activity can be: a club that people can go to; a concert or other type of event; a restaurant that offers subsidized meals; the theatre/opera/cinema; a museum, sightseeing or short trips; shopping etc.

The attribute set for an activity is the following: category, title, description, contact information, and location. If the activity is an event, it can also include a date and time of the event. Each activity can have multiple tags for easier classification.

Use case 1: *activity rating*. In scenario 1, the primary persona found it difficult to choose the concert she wanted to attend. She wanted to get some user feedback. This could be done by allowing users to rate activities and sort them by popularity. Users should also be given the option to change their vote.

Use case 2: *list of activities*. Users will be able to view a list of activities with an additional filter and sort option. Categories and tags are used for filtering, and user rating and time of occurrence (for events) can be used for sorting. More details about an activity will be given in the detail screen.

User case 3: *quick search*. Scenario 2 shows the need to quickly search for a restaurant with subsidized meals. Predefined short questions and answers will be used to build a filter by tags and categories. Answering the last question will reveal results, which will then allow even further filtering. An example quick search for the 2nd scenario: (1) what do you want to do? I am hungry; (2) what do you want to eat? Fast food; and (3) filter by tag "subsidized".

Use case 4: *displaying activities on a map*. This use case is an upgrade of the second use case. It is an alternative way to display a list of activities so that the user can see which activities are closer. The primary persona can use this feature in the 2nd scenario to find the nearest fast food restaurant.

Figure 2 shows all four identified use cases on a UML use case diagram.

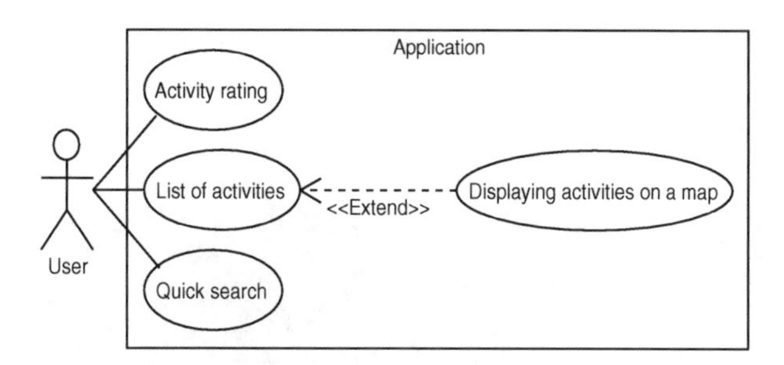

Fig. 2 A UML use case diagram for identified use cases

3.4 Creating the Application

The specified use cases were used as a basis for creating the application. The application consists of two parts. The first part is the web server, hosting representational state transfer (REST) services. The second part is the mobile client application, which consumes these services. The platform Java EE was chosen for the web server and the platform Android was chosen for the mobile client application. The focal point of this paper is on the client side, therefore only the mobile application will be described.

Android version 2.1 was chosen in order to support as many devices as possible. In order to achieve offline operations, the mobile application has its own data storage, which is regularly updated by connecting to the web server. An SQLite database was chosen for this operation. The main function of the web server is to provide clients with activities of various kinds and sources via REST services. As mentioned in the previous chapter, an activity is a very broad and universal metaphor and can be used for all kinds of things. If there is a future need to add a certain type of activity that is not yet supported (e.g. new sport activities, restaurants, exhibitions, events etc.), it must only be implemented on the server side and the clients will be automatically updated.

The user interface is designed to be clean, simple and intuitive. All the main functionalities are at the top level. The core functionality of the application is listing activities in various ways: sorting by popularity (top rated), browsing by category, upcoming events, and a quick search. Only the most important data is shown in the list of activities: category, title, tags, rating, and number of votes. More information, such as a detailed description, address, contact information, opening hours etc. can be found in the detail screen, along with the possibility to vote or re-vote. Every list of activities on the screen can also be viewed in map mode using Google Maps. Each activity is shown as a pin on the map on which the user can click in order to discover more details. This way the user can view the geographical layout of the activities and see which ones are nearby.

The quick search feature is implemented according to the description of the use case – predefined questions and answers lead to a list of activities. Users can use the quick search to find for more specific activities quickly and simply. For example: only three steps are required if the user wishes to find pizza restaurants. A demonstration of this feature is shown in Figure 3. The second and third screen of Figure 3 show the questions and answers, while the last screen displays the resulting list of activities.

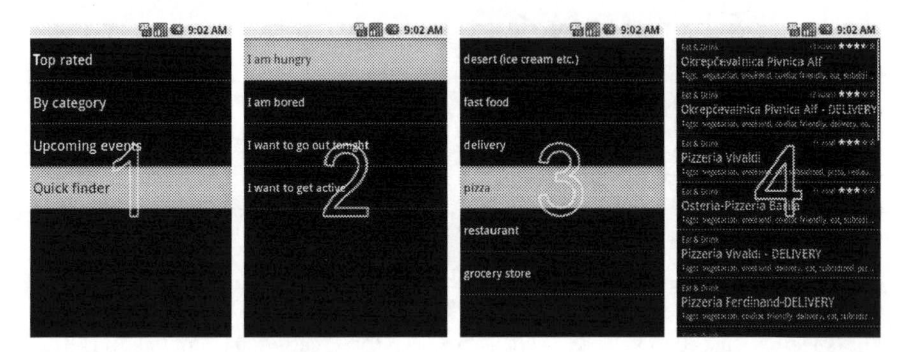

Fig. 3 Demonstration of the quick search feature

4 Conclusion

Developing applications for mobile devices is currently a trending topic. As the number of applications continues to grow rapidly, the importance of making useful, and user-accepted applications is significant. Therefore, focusing on the user is important. One of the ways that this can be achieved is to perform a user-centric design, which was presented in this paper. The case study showed how easily this type of design can be performed and how the results can be interpreted and implemented. The use of personas can help identify groups of end users for our application. These personas are then used in scenarios where real-life examples occur and help us analyse use cases, which are a fundamental user specification. We plan on continuing this research in several directions. More than one persona will be taken into account and will be more thoroughly described. Also, more detailed scenarios will be written, and consequently, more use cases can be identified.

Readers are advised to interpret the results of this paper by considering the following limitations. Only one primary persona was identified. If more personas were identified, including non-primary personas, the user requirements of this application could be different. This also applies to writing scenarios – if more personas were identified, more scenarios would be identified. Also, more scenarios contain more details and can provide more use cases, which represent end user requirements.

References

1. Mobile application development popularity and increasing demand of mobile phones. Article Alley (August 26, 2010), http://www.articlealley.com/article_1718458_11.html
2. User-centered design. Wikipedia, http://en.wikipedia.org/wiki/User-centered_design
3. Katz-Haas, R.: User-Centered Design and Web Development (July 01, 1998), http://www.stcsig.org/usability/topics/articles/ucd%20_web_devel.html
4. Bevan, N.: Usability Net Methods for User Centred Design. In: Hci International 2003 Proceedings: Human-Computer Interaction: Theory and Practice (part 1), vol. 1, pp. 434–436 (2003)
5. Persona. Wikipedia, http://en.wikipedia.org/wiki/Persona
6. Blomkvist, S.: Persona – an overview, pp. 1–8 (October 2002)
7. Massimi, M.: Fieldwork 2: analysis, requirements (February 01, 2011)
8. Carroll, J.M.: Five reasons for scenario-based design. Interacting with Computers 13(1), 43–60 (2000)
9. Põldoja, H.: Scenario-Based Design, http://www.slideshare.net/hanspoldoja/scenariobased-design
10. Writing effective Use Cases and User Story Examples, http://www.gatherspace.com/static/use_case_example.html
11. Kim, S.-K., Carrington, D.: Integrating use-case analysis and task analysis for interactive systems. In: APSEC 2002 Proceedings of the Ninth Asia-Pacific Software Engineering Conference, pp. 12–21 (2002)
12. Introduction to UML 2 Use Case Diagrams. Agile Modeling, http://www.agilemodeling.com/artifacts/useCaseDiagram.htm

A Novel Agent-Based Framework in Bridge-Mode Hypervisors of Cloud Security

Maziar Janbeglou and WeiQi Yan

AUT University, Auckland, New Zealand

Abstract. Cloud computing has been introduced as a tool for improving IT proficiency and business responsiveness for organizations as it delivers flexible hardware and software services as well as providing an array of fundamentally systematized IT processes. Despite its many advantages, cloud computing security has been a major concern for organizations that are making the transition towards usage of this technology. In this paper, we focus on improving cloud computing security by managing and isolating shared network resources in bridge-mode hypervisors.

Keywords: cloud computing, security, virtual machine, virtualization, virtual networking, hypervisor.

1 Introduction

Every organization actively seeks for new ways of delivering business needs to their customers. With increasing demands on usage of computer network, the means of fulfilling this demand should be based on competitiveness and lower cost. Cloud computing has then been introduced as a tool for obtaining this goal.

Cloud computing is a broad expression that includes delivering hosted services such as computation, storage, and IO/network over the Internet [1,2]. There are many definitions for cloud computing [3]. Gartner [4] defines cloud computing as a style of computing where massively scalable IT-enabled capabilities are delivered 'as a service' to external customers using Internet technologies. According to NIST (National Institute of Standards and Technology), "Cloud computing is a model for enabling convenient, on-demand network access to a shared pool of configurable computing resources (i.e. networks, servers, storage, applications, and services) that can be rapidly provisioned and released with minimal management effort or service provider interaction" [5]. The advantages of using cloud computing make organizations pay particular attention to it because it is on-demand, self-service, location independent, elastic, accessible by network from everywhere, and measurable to be billed[6-8].

The reasons why companies are likely to transition into cloud computing IT solutions [9-11] are:

L. Uden et al. (Eds.): 7th International Conference on KMO, AISC 172, pp. 467–479.
springerlink.com

- Using on-demand resource allocating [12] that enables users to consume computing capabilities (i.e. applications, server time, network storage) when required
- Applying multi-tenant Virtual Machine (VM) and resource pooling systems to use diverse computing resources (i.e. hardware, software, processing, servers, network bandwidth) to serve multiple consumers – those resources are dynamically assigned [13]
- Using scalable and elastic resource allocation based on demands[14]
- Automatically controlling, optimizing, and measuring resource allocation for charging purpose, as well as allowing easy monitoring, controlling and reporting [15]

Generally, the cloud environment consists of three core components [16]:

- Software as a Service (SaaS): In SaaS, whole software or application processes run on physical server and become available to consumers over the Internet [17-20].
- Platform as a Service (PaaS): PaaS is software and product development tools as well as programming languages which are leased by providers so that clients can build and deploy their own applications for their specific use[21].
- Infrastructure as a Service (IaaS): IaaS delivers computing services, storage, and network, typically in the form of VMs [22] running on hosts

There are also four types of deployment models[23]:

- Private Cloud: This model is for usage of individual organization and not shared among other organizations.
- Community Cloud: This model is shared with several set of organizations.
- Public Cloud: The most common type of cloud infrastructure that is made to be available for public
- Hybrid Cloud: This model is made by combining two or more deployment models (private, community, or public).

Although usage of cloud computing has significantly decrease the total cost of delivering services and provides scalability and high quality of service for organizations[24-26], it has also introduced many new and unknown security risks[27]. Hence, more consideration is needed to identify potential security risks, possible solutions of overcoming the risks, and evaluation on the feasibility of the proposed solutions.

In order to deliver a secure cloud with minimized risks, every aspect of cloud service methods and deployment models should be carefully considered. However, in this paper, we will only be focusing on the security of IaaS service method and its application in all the mentioned deployment models. Basically, IaaS deals with virtualized resources such as VM, storage, and database service. Because cloud computing is inherently adopted from traditional computer networks, security issues in traditional computer networks should be reassessed in relation to IaaS in cloud computing.

2 Virtualization

Virtualization is responsible for splitting resources on a single physical machine into multiple VMs. In other words, it provides the ability of installing multiple Operating Systems (OS) on different VMs on a same physical machine and as a result it increases the machine utilization. It is the main factor of reducing cost in cloud computing because it supports separating and allocating resources among multiple tenants. Virtualization helps Cloud Service Providers (CSP) to solve the complexity issues in delivering services, managing shared resources and utilizations, isolating VMs, and providing security.

The risks towards virtualization of IaaS model are studied from four different aspects. They are Virtualized systems risks, Hypervisor risks, Virtual machine risks, and Virtual network risks.

2.1 Virtualized System Risks

Using virtualized systems does bring some negative consequences. This technology makes the security management more complex inasmuch as it needs more controlling and monitoring of the shared resources. Besides, larger security threats arise when many VMs are combined into a physical machine. Any compromised VM could lead to security risks for the rest of VMs. Furthermore, some virtualized systems use easy ways of sharing information among VMs. It can be an opportunity for performing attacks if they are not carefully monitored. Since these systems are dynamic and flexible to changes, defining security boundaries will be complicated[5].

From data communication point of view, there is a similar relationship between security concerns in physical machines running on traditional networks and VMs running on virtualized systems since they both use a shared medium to communicate [28]. Because computing resources are shared with companies in public model, gaining control over physical security is under serious risks. In other words, companies do not have any control on how the resources are being shared and managed among different companies[29]. This is the main reason of increasing security risks in virtualized systems. Although some security remedies have been applied to strengthen the security of virtualized systems, there is no assurance of protecting VMs intercommunications on the same server[30].

2.2 Hypervisor Risks

The hypervisor is a main part of any virtualized systems. It is the most important empowering component of splitting hardware from the OS. This layer provides physical server resources sharing and VM/host isolation. Therefore, the capability of the hypervisor is essential to provide the necessary isolation when an attack happens[31]. Because hypervisor is a software program, it is inherently vulnerable to the growth of volume and complexity of application codes [32]. Basically, software codes running on a VM should not be able to affect the code running on the physical machine itself and within different VMs. However, existence of some probable bugs in the software or limitations of implementing virtualized systems may put this isolation at risk. Major inherent vulnerabilities in current hypervisors are Rogue Hypervisors, External Modification of the Hypervisor, VM Escape, and Denial-of-Service.

Fig. 1 illustrates the VMs use of the shared resources in VMWare ESX server architecture. As we mentioned above, using shared resources is the main reason of threats towards VMs.

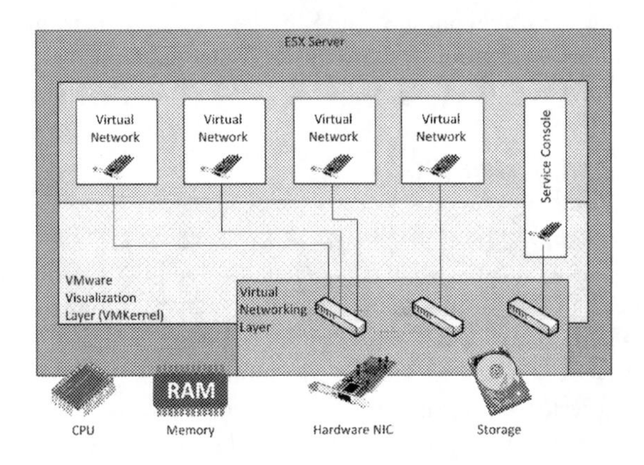

Fig. 1 VMWare ESX server architecture

2.3 Virtual Machine Risks

VM techniques, such as Xen [33], VMware [34] and Hyper-V, offer on-demand virtualized IT infrastructures [35]. VM instances use the shared resources on a physical server to deliver business needs. While they are working on a same physical machine and using the shared resources, security threats will become a general problem for all of the VMs.

Some threats are common towards all computerized systems. Denial-of-Service is one example of these kinds of threats. Other types of threats are only related to VMs. These attacks, which are inherently destructive for all of the VMs in a virtualized server, are listed [36-39]:

- Shared clipboard attack: because the memory is shared among VMs, the attacks through shared clipboard are done by moving clipboard information between malicious programs in VMs of different security realms.
- Keystroke logging attack: some VM technologies provide keystroke logging function and capture screen updates to be transferred across virtual terminals in the VM.

- Monitoring VMs from an infected host: because all network packets will be transferred via a host, the host may be able to affect the VMs by the following functions:

- Providing full control of VMs such as starting, stopping, pausing, and restarting function of the VMs.
- Providing full controlling and monitoring resources available to the VMs, including CPU, memory, storage, and network usage of VMs
- Manipulating shared resources such as adjusting the number of CPUs, amount of memory, amount and number of virtual disks, and number of virtual network interfaces available to a VM
- Providing full access to monitoring the applications running inside the VM
- Manipulating data stored on the VM's virtual disks
- Monitoring VMs from another VM: basically, VMs do not have direct access to one another's virtual disks on the host. However, if the VM technology uses a virtual hub or switch to connect the VMs to the host, intruders can use hacking techniques such as Address Resolution Protocol (ARP) poisoning and network packets redirection [40] to redirect packets going to or from the other VM.
- VM backdoors: through backdoor, communication channels are opened between the guests and hosts and this allows intruders to potentially perform dangerous operations.

Some traditional protection methods such as Intrusion Detection and Prevention Systems (IDS and IPS), Firewalls, and log monitoring [41] are applicable as software services on VMs to increase safety and upholding servers integrity [42]. However, the existence of new and unknown security risks in cloud computing demands more considerations then relying only in traditional protection tools.

2.4 *Virtual Network Risks*

In virtual networks, almost all the physical network threats such as IP/MAC spoofing [43,44], ARP spoofing [45], and packet sniffing[46] are likely to happen. In physical networks, firewall and encryptions mechanisms are the main tools for applying security. Isolation does a similar function in virtual networks.

Some hypervisors offer three choices for users to configure virtual network: bridge mode, Network Address Translation (NAT) mode, and host only mode [47,48]. Depends on which model is being used by a CSP, different security mechanisms should be considered.

In bridge mode, all the VMs' network interfaces are directly attached to a physical server's Network Interface Card (NIC). This state is equivalent to that of several computers being connected to a virtual hub in physical networks.

In NAT mode, a simulated NAT server is implemented inside the hypervisor. It is responsible for forwarding the VMs network traffics to the physical NIC on the host. In this mode, the VMs are not directly connected to the physical NIC.

Therefore, hypervisor should be equipped with some preconfigured MAC/IP addresses for respective VMs. By this, hypervisor creates a point-to-point link between each VM and its virtual network interfaces with preset MAC and IP addresses [49].

Host only mode is used to provide intercommunicating among VMs and, theoretically, with the host.

3 Agent-Based Proposed Virtual Network Model

The main step of performing any computer network related attacks is the identification of potential victims. Attackers use network scanner applications in order to identify available VMs residing in a physical machine, discover open ports, and uncover services on ports. As highlighted earlier in this paper, usage of shared resources has led to security weaknesses in virtualized systems. This paper proposes a model to improve the IaaS security on the shared network resources by making VMs invisible from attackers, and as a result, preventing them from performing the key step of network-related attacks.

Isolation plays a very important role to provide security in virtualized systems and it is considered for all types of shared resources. Among these resources, the focus of this study is limited to the shared networking resources. Hypervisors use virtual network layer as a concept to isolate VMs running on a physical server. As it is illustrated in Fig. 1, virtual networking layer is responsible for providing VMs' intercommunications in a secured and isolated manner.

The proposed model introduces an agent, which itself is a VM with a Linux OS to provide a centralized virtual network management for all of the VMs residing in a physical server. Application of the agent in VM will ensure no performance downturn as the packets do not hit any physical interface. The aim of this agent is to help the hypervisor to provide security by confining the visibility and accessibility of the VMs network resources.

In this model, as illustrated in Fig. 2, the agent has two Virtual Network Interfaces (VNI). One of them is directly connected to the hypervisor (VNI:0) and the other one is for serving VMs (VNI:1). Services running on the agent are listed as below:

- Dynamic Host Configuration Protocol (DHCP): DHCP server is responsible for generating random Internet Protocols (IP) for VMs. The IPs will eventually be assigned to respective VMs. Most network scanners use IP boundaries to look for available resources between the defined beginning and end of boundary. The usage of random IP assigning in this model will counteract network scanners in locating available VMs.
- Point-to-Point-Tunneling-Protocol (PPTP) applied in Virtual Private Network (VPN): It provides encryption for VMs intercommunications. The usage of encryption prevents network sniffer applications such as "Wireshark" to get any information related to the existence of VMs and also the infrastructure of the whole system.

- Packet-filter: It is in charge of approving/rejecting any request being sent towards the virtual network interfaces.

Application of this model consists of the following steps:

- Generating network-sub-interfaces and random IP address configurations
- Generating PPTP configurations
- Customizing the packet-filter

3.1 Generating Network Sub-interfaces and Random IP Address Configurations

Network sub-interfaces refers to sub-interfaces that are created from VNI:1. Each of the sub-interfaces will be assigned to corresponding VMs. The sub-interfaces serve as communication channel for VMs to communicate with the agent. As such, the number of network sub-interfaces should be equal to the number of available VMs.

In this model, randomly generated IP address configurations are assigned to the VMs by a DHCP service running on the agent. The advantage of using random IPs is to cause difficulty in guessing and identifying active VMs by hackers using network scanner applications. Normally, DHCP is set to assign IPs from a large range of IPs for example from 192.168.0.100/24 to 192.168.0.200/24. However, in

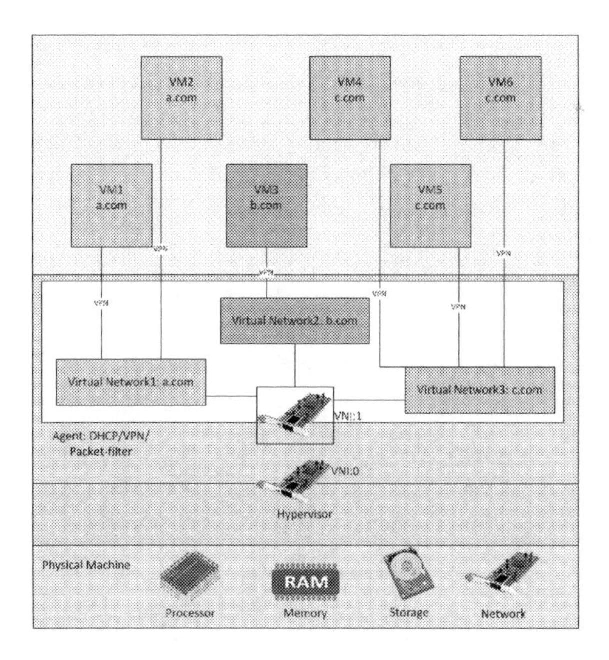

Fig. 2 An example of applying the proposed model in a public cloud

this model, DHCP deals with limited numbers of random and unique pools in a very small sub-network from one specified IP class to reduce the boundary of broadcast in the whole network. The proposed subnet for VMs is 255.255.255.252 (/30). In this subnet, besides Broadcast-IP and Network-ID, only two IPs are available. One is assigned to a VM and the other is assigned to the network sub-interface on the agent to become the default-gateway for the VMs. Therefore, VMs are only able to send their packets to the agent or the gateway on VNI:0.

Table 1 shows an example of randomly generated and assigned IP addresses for each VM.

Table 1 Example of randomly generated network configurations

Virtual Machine	Network IP Address	VM IP Address	Broad-Casting IP	Network-sub-interface IP
1	10.107.237.160	10.107.237.162	10.107.237.163	10.107.237.161
2	10.166.168.24	10.166.168.26	10.166.168.27	10.166.168.25
3	10.252.221.108	10.252.221.110	10.252.221.111	10.252.221.109
4	10.57.27.116	10.57.27.118	10.57.27.119	10.57.27.117
5	10.97.116.128	10.97.116.130	10.97.116.131	10.97.116.129
6	10.99.213.100	10.99.213.102	10.99.213.103	10.99.213.101

3.2 Generating PPTP Configurations

Encryption is the main part of every security model. PPTP is used to apply encryption in the proposed model. After VMs get the IP from DHCP pool, they make a VPN connection to the agent. Therefore, all the packets between VMs and the agent would be transferred securely. This will be an added security layer especially for VMs belonging to the same company (i.e. VM1-2 from a.com, and VM4-6 from c.com) as they both will have trusted sharing resources. In this state, VMs are assigned IPs from a same subnet so that they will be able to use their shared resources. The following steps should be applied for generating IP addresses in PPTP service:

1. Counting the number of domains available (3 domains in our example: a.com, b.com, and c.com)
2. Counting the maximum number of VMs in each domain (two VMs in a.com, one VM in b.com, and three VMs in c.com)
3. Calculating appropriate subnet for each domain. The appropriate subnet is the smallest possible subnet to cover a group of VMs working in a domain. (Appropriate subnet for domain a.com is 255.255.255.248, for domain b.com is 255.255.255.252, and for domain c.com is 255.255.255.248)
4. Generating and assign IPs to each VM and network-sub-interface

According to our example illustrated in Fig. 2, Table 2 displays IP address configurations for each VM that is being assigned by PPTP service.

Table 2 An example of assinged IP addresses by PPTP in our proposed model

Virtual Machine	domain	VM IP address Assigned by PPTP	subnet	Network-sub-interface IP
1	a.com	172.25.50.118	255.255.255.248	172.25.50.117
2	a.com	172.25.50.119	255.255.255.248	172.25.50.117
3	b.com	172.27.33.122	255.255.255.252	172.27.33.121
4	c.com	172.29.159.2	255.255.255.248	172.29.159.1
5	c.com	172.29.159.3	255.255.255.248	172.29.159.1
6	c.com	172.29.159.4	255.255.255.248	172.29.159.1

Virtual network in our model is a conceptual definition for VMs belonging to the same domain.

3.3 Customizing the Packet-Filtering

In addition to PPTP and the safety it provides, packet-filtering, which acts as a centralized firewall on the agent, also improves the security of the whole system by confining any internal-communications via dropping packets originating from internal VMs residing in different domains.

When the agent receives a packet, it may be originated from either a genuine VM or an attacker's VM. If the source IP address is equal to any IP addresses in the DHCP pool, it is most likely from a genuine VM. Otherwise, the packet is suspected to be originated from the attacker. Since in this method the attacker is not able to guess or find the available IPs in DHCP pool, they have to assign the IP address manually and as a result, the packets being sent by the attackers will be easily identified and dropped by the packet-filtering service. In addition, if the attacker gets an IP address from the DHCP server, they can only communicate with the default-gateway, not the other VMs. If the packet could satisfy these two criteria, it eventually will be proceed to the hypervisor.

Hackers using a compromised VM may provide any of the ARP-related and port scanning attacks to get more information about the available VMs [50]. They might find some VMs' IP addresses on broadcasted packets originating from the virtual network (if bridge mode is selected on the hypervisor). However, because the packet-filtering bans any VMs intercommunications within different domains, the VMs' IP addresses remain invisible for attackers. As a result VMs would be safe from any ARP-related attacks.

3.4 Evaluation

There is no one-size-fits-all solution for securing cloud computing. Different CSPs have applied different security architectures to meet the variety of security requirements of their customers. But, what is common in all cloud security architectures is that VMs residing in a physical server will be using the shared networking resources. Although it provides an easier way of sharing information among VMs, it can also cause numerous security risks. The available solutions to overcome this issue cannot ensure total protection of the security requirements. With this being said, another security layer is needed to cover the vulnerabilities of the existing models. The proposed model suggests VMs invisibility as a solution for this problem.

An important characteristic of cloud computing is that unnecessary details are invisible[51]. This model, similarly, considers the information of VMs (such as IP address) as unnecessary details that should be invisible from cloud users.

The proposed model, as an add-on security layer, is matched with Integrity key aspect of a computer related security system (Confidentiality, Integrity, and Availability) or CIA. Integrity in CIA model refers to prevention of unauthorized modification or alternation of information whereby VMs are kept hidden from one another and as a result a compromised VM will be prevented from doing any modification of available shared network resources on the physical machine. Prevention of unauthorized disclosure of information is another impact of applying the proposed model. Hiding and limiting public access towards VMs prevent unapproved access from available VMs. This definition is consistent with the confidentiality aspect of CIA model. This model also fulfills the availability of CIA security system by making a specific VM available for its genuine user. As it is stated earlier, in this model users are only able to access and manipulate their own VMs.

Availability in this model goes hand in hand with authorization and accountability of AAA (Authentication, Authorization, and Accountability) security framework. Authentication, in this model, refers to limiting and controlling the use of shared networking resources. The proposed agent is in charge of managing, monitoring, and controlling any access towards VMs. Each VM in this model is only able to communicate with the agent. Therefore, any possible intercommunication would be possible via the agent. Accountability of this model is being done by monitoring the agent packet-filtering's log. Any illegal access is being logged by the agent and is sent to the CSP administrators for taking appropriate actions.

The whole idea of the proposed model is to hide VMs via isolation of shared networking resources. As private cloud is known to be a non-shared deployment model, application of any additional security models will only decrease the system performance. The usage of the proposed model in this case will be redundant. The proposed model is better to be implemented in community, hybrid, and public cloud because the invisibility of VMs is more crucial.

4 Conclusion

Security has been a major concern for organizations that are applying cloud computing. In this paper the security risks towards each virtualized system component have been stated and then a proposed model was introduced. This model suggests invisibility as a key concept to secure cloud computing. It can be achieved by applying random IP address configurations, encryption, and packet-filtering services in the model. Finally, the evaluation of proposed model was stated to identify how this system matches the current security architecture.

References

1. Wu, X., Wang, W., Lin, B., Miao, K.: Composable IO: A Novel Resource Sharing Platform in Personal Clouds. In: Jaatun, M.G., Zhao, G., Rong, C. (eds.) CloudCom 2009. LNCS, vol. 5931, pp. 232–242. Springer, Heidelberg (2009)
2. Weiss, A.: Computing in the clouds. netWorker, 16–25 (2007), doi:10.1145/1327512.1327513.
3. Yeh, J.T.: The Many Colors and Shapes of Cloud. In: Jaatun, M.G., Zhao, G., Rong, C. (eds.) CloudCom 2009. LNCS, vol. 5931, p. 1. Springer, Heidelberg (2009)
4. David, W.C.: Cloud computing: Key initiative overview. Gartner (2010), doi:EUKEINCLCOOVRW012110
5. Mell, P., Grance, T.: The National institute of standards and technology (NIST) definition of cloud computing (2009)
6. Onwubiko, C.: Security issues to cloud computing, pp. 271–288. Springer, London (2010), doi:10.1007/978-1-84996-241-4_16
7. Abramson, D., Buyya, R., Giddy, J.: A computational economy for grid computing and its implementation in the nimrod-g resource broker. Future Generation Computer Systems (FGCS) 18(8), 1061–1074 (2002)
8. Kouzes, R.T., Anderson, G.A., Elbert, S.T., Gorton, I., Gracio, D.K.: The changing paradigm of data-intensive computing. Computer 42, 26–34 (2009)
9. Mather, T., Kumaraswamy, S., Latif, S.: Cloud security and privacy: An enterprise perspective on risks and compliance. O'Reilly Media (2009)
10. Zhang, Q., Cheng, L., Boutaba, R.: Cloud computing: State-of-the-art and research challenges. Journal of Internet Services and Applications, 7–18 (2010), doi:10.1007/s13174-010-0007-6
11. Gourley, B.: Cloud computing and cyber defense. A white paper provided to the national security council and homeland security council as input to the White House review of communications and information infrastructure (2009)
12. Yang, H., Wu, G., Zhang, J.-z.: On-Demand Resource Allocation for Service Level Guarantee in Grid Environment. In: Zhuge, H., Fox, G.C. (eds.) GCC 2005. LNCS, vol. 3795, pp. 678–689. Springer, Heidelberg (2005)
13. Oh, T.H., Lim, S., Choi, Y.B., Park, K.-R., Lee, H., Choi, H.: State of the Art of Network Security Perspectives in Cloud Computing. In: Kim, T.-h., Stoica, A., Chang, R.-S. (eds.) SUComS 2010. CCIS, vol. 78, pp. 629–637. Springer, Heidelberg (2010)
14. Espadas, J., Molina, A., Jimenez, G., Molina, M., Ramirez, R., Concha, D.: A tenant-based resource allocation model for scaling Software-as-a-Service applications over cloud computing infrastructures (2011), doi:10.1016/j.future.2011.10.013
15. Mahmood, Z.: Cloud computing for enterprise architectures: concepts, principles and approaches. In: Cloud Computing for Enterprise Architectures, pp. 3–10. Springer (2011)
16. Dawoud, W., Takouna, I., Meinel, C.: Infrastructure as a service security: Challenges and solutions. In: 2010 7th International Conference on Informatics and Systems, INFOS 2010, Cairo, Egypt, March 28-30 (2010)
17. Dave, T.: Enabling application agility - Software as a Service, cloud computing and dynamic languages. Journal of Object Technology, 29–32 (2008)
18. Frederick, C., Gianpaolo, C.: Architecture strategies for catching the long tail. Microsoft Corporation (2006)

19. Gillett, F.E.: The new tech ecosystems of cloud, cloud services, and cloud computing. Forrester Research (2008)
20. Turner, M., Budgen, D., Brereton, P.: Turning software into a service. Computer 36(10), 38–44 (2003), doi:10.1109/mc.2003.1236470
21. Lawton, G.: Developing software online with Platform-as-a-Service technology. Computer, 13–15 (2008), doi:10.1109/mc.2008.185
22. Buyya, R., Yeo, C.S., Venugopal, S., Broberg, J., Brandic, I.: Cloud computing and emerging IT platforms: Vision, hype, and reality for delivering computing as the 5th utility. Future Generation Computer Systems, 599–616 (2009), doi:10.1016/j.future.2008.12.001
23. Dillon, T., Chen, W., Chang, E.: Cloud Computing: Issues and Challenges. In: 24th IEEE International Conference on Advanced Information Networking and Applications, AINA, pp. 27–33 (2010)
24. Wang, C., Wang, Q., Ren, K., Lou, W.: Ensuring data storage security in cloud computing. Cryptology ePrint archive, report 2009/081 (2009)
25. Grossman, R.L.: The case for cloud computing. IT Professional 11, 23–27 (2009)
26. Grossman, R.L., Gu, Y.: On the varieties of clouds for data intensive computing (2009)
27. Pearson, S.: Taking account of privacy when designing cloud computing services. In: 2009 ICSE Workshop on Software Engineering Challenges of Cloud Computing, CLOUD 2009, Vancouver, BC, Canada, May 23 (2009)
28. Anthony, T.V., Toby, J.V., Robert, E.: Cloud computing: A practical approach. McGraw-Hill (2010)
29. Llanos, D.R.: Review of grid computing security by anirban chakrabarti, pp. 45–45. Springer (2007) ISBN: 3540444920.45-45, doi:10.1145/1317394.1317406
30. Szefer, J., Keller, E., Lee, R.B., Rexford, J.: Eliminating the hypervisor attack surface for a more secure cloud. In: 18th ACM Conference on Computer and Communications Security, CCS 2011, Chicago, IL, United states, October 17-21 (2011)
31. Tolnai, A., Von Solms, S.H.: Securing the cloud's core virtual infrastructure. In: 5th International Conference on Broadband Wireless Computing, Communication and Applications, BWCCA 2010, Fukuoka, Japan, November 4-6 (2010)
32. Klein, G., Elphinstone, K., Heiser, G., Andronick, J., Cock, D., Derrin, P., Elkaduwe, D., Engelhardt, K., Kolanski, R., Norrish, M., Sewell, T., Tuch, H., Winwood, S.: Formal verification of an OS kernel. In: 22nd ACM SIGOPS Symposium on Operating Systems Principles, SOSP 2009, Big Sky, MT, United states, October 11-14 (2009)
33. Barham, P., Dragovic, B., Fraser, K., Hand, S., Harris, T., Ho, A., Neugebauer, R., Pratt, I., Warfield, A.: Xen and the art of virtualization. In: SOSP 2003: Proceedings of the 19th ACM Symposium on Operating Systems Principles, Lake George, NY, United states, October 19-22 (2003)
34. Nishikiori, M.: Server virtualization with VMware vSphere 4. Fujitsu Scientific and Technical Journal, 356–361 (2011)
35. Muthu, R.: Component-based development for cloud computing architectures. In: Cloud Computing for Enterprise Architectures, pp. 91–113. Springer (2011)
36. dos Santos Ramos, J.C.C.: Security challenges with virtualization. universidade de lisboa (2009)
37. Larry, D.: Virtualization: What are the security risks? ZDnet (2008), http://www.zdnet.com/blog/security/virtualization-what-are-the-security-risks/821 (accessed January 22)

38. Garfinkel, T., Pfaff, B., Chow, J., Rosenblum, M., Boneh, D.: Terra: a virtual machine-based platform for trusted computing. In: Proceedings of the Nineteenth ACM Symposium on Operating Systems Principles, Bolton Landing, NY, USA (2003)
39. Joel, K.: Virtual machine security guidelines (September 2007)
40. Janbeglou, M., Zamani, M., Ibrahim, S.: Redirecting outgoing DNS requests toward a fake DNS server in a LAN. In: 2010 IEEE International Conference on Software Engineering and Service Sciences, ICSESS 2010, Beijing, China, July 16-18 (2010)
41. Nourian, A., Maheswaran, M.: Privacy and security requirements of data intensive computing in clouds. In: Handbook of Data Intensive Computing. Springer Science and Business Media (2011), doi:10.1007/978-1-4614-1415-5 19
42. Rittinghouse, J.W., Ransome, J.F.: Cloud computing: implementation, management, and security. CRC Press (2009)
43. Basak, D., Toshniwal, R., Maskalik, S., Sequeira, A.: Virtualizing networking and security in the cloud, pp. 86–94 (2010), doi:10.1145/1899928.1899939
44. Ferguson, P., Senie, D.: Network ingress filtering: Defeating denial of service attacks which employ IP source address spoofing. RFC 2827 (2000)
45. Ramachandran, V., Nandi, S.: Detecting ARP Spoofing: An Active Technique. In: Jajodia, S., Mazumdar, C. (eds.) ICISS 2005. LNCS, vol. 3803, pp. 239–250. Springer, Heidelberg (2005)
46. Thawatchai, C.: Sniffing packets on LAN without ARP spoofing. In: 3rd International Conference on Convergence and Hybrid Information Technology, ICCIT 2008, Busan, Republic of Korea, November 11-13 (2008)
47. Pu, X., Liu, M., Jin, J., Cao, Y.: A modeling of network I/O efficiency in Xen virtualized clouds. In: International Conference on Electronics, Communications and Control, ICECC 2011, Ningbo, China, September 9-11 (2011)
48. Schoo, P., Fusenig, V., Souza, V., Melo, M., Murray, P., Debar, H., Medhioub, H., Zeghlache, D.: Challenges for cloud networking security mobile networks and management, pp. 298–313. Springer, Heidelberg (2011), doi:10.1007/978-3-642-21444-8_26
49. Wu, H., Ding, Y., Yao, L., Winer, C.: Network security for virtual machine in cloud computing. In: 5th International Conference on Computer Sciences and Convergence Information Technology, ICCIT 2010, Seoul, Republic of Korea, November 30-December 2 (2010)
50. Man, N.D., Huh, E.-N.: A collaborative intrusion detection system framework for cloud computing. In: International Conference on IT Convergence and Security 2011, ICITCS 2011, Suwon, Republic of Korea, December 14-16 (2012)
51. Masayuki, O., Tetsuo, S., Takuya, S.: Security architectures for cloud computing, fujitso (2010)

Points or Discount for Better Retailer Services – Agent-Based Simulation Analysis

Yuji Tanaka, Takashi Yamada, Gaku Yamamoto,
Atsushi Yoshikawa, and Takao Terano

Department of Computational Intelligence and Systems Science,
Tokyo Institute of Technology

Abstract. Service management at commodity goods retailers requires various kinds of strategic knowledge. This paper focuses on the (dis)advantages of mileage point and discount services. To uncover the characteristics of the two strategies, we are developing an agent-based simulator to analyze the behaviors of competing retailing stores and their customers. The retailer agents adaptively increase or decrease sales promotion of mileage point service and discounting based on their experiences and strategies to acquire customers. Customer agents, on the other hand, make a decision to choose one of the two retailers to repeatedly purchase daily commodities based on customer types and utilities. To explore the better strategies of retailers, we have conducted intensive simulation experiments. Our computational results have shown that the emergence of retailors' cooperative behaviors and their stable relations strongly depends on to what extent the retailers decide the discounting rather than mileage point strategy at the very first stages.

Keywords: Mileage points services, Discount services, Agent-based simulation, Sales promotion of commodity goods.

1 Introduction

Mileage point services or frequent shoppers programs have been widespread among various retailing services such as airline companies, credit cards, hotel chains, and so on. As so many retailers, even retailing stores for daily commodities such as supermarkets, have introduced such services, it becomes hard for them to differentiate their marketing promotion strategies. On the other hand, discount services are a traditional method to enclose customers to specified retailers. From the point of view of knowledge management, it is worth to analyze which strategies of mileage points or discount services are better for the retailers. So far, however, we do not have easy-to-use and effective methods to uncover the benefits and/or defects of such service systems.

To cope with the issues, in this paper, we apply Agent-Based Modeling (ABM) to analyzing strategic behaviors of competing retailers and their customers in order to explore co-existing conditions of plural retailers. ABM is a cutting-edge

L. Uden et al. (Eds.): 7th International Conference on KMO, AISC 172, pp. 481–491.
springerlink.com

technique to understand various social and economic phenomena [1, 2, 3]. ABM focuses from global phenomena to individuals in the model and tries to observe how individuals with individual characteristics or 'agents' will behave as a group. In our simulation model of the paper, the model consists of the two kinds of agents: retailers and customers. The retailer agents adaptively increase or decrease sales promotion of mileage point service and discounting based on their experiences and strategies to acquire customers. Customer agents, on the other hand, repeatedly change the decisions to choose one of the two retailers to repeatedly purchase daily commodities based on customer types and utilities.

The rest of the paper is organized as follows: Section 2 discusses related work on the literature and defining the problems to cope with in the paper. Section 3 describes the principles and design of the agent-based simulator. Section 4 explains the experiments, results, and discussion. Section 5 gives some concluding remarks.

2 Related Works and Problem Description

The mileage points services give customers certain amount of points in proportion to the amount of money they spend to have goods or services. The services are considered to be beneficial for both customers and retailers. The customers have direct benefits, and the retailors have the information on the customers' behaviors about the goods or services [4].

In the marketing literatures, they have investigated the roles of mileage services or frequent shopper programs (FSPs), Loyalty Programs (LPs) [4, 5, 6, 7, 8]. For example, in [8], Yi and Jeon conduct questionnaire survey research to customers about the effects of loyalty on the retailer and the program. However, such survey research is hard to measure the effects of complex competing situations, because of the limit of available data and surveys. In [5], Berman categorizes various FSPs and LPs, and then discusses their characteristics. Based on his research, we have focused on the competitive situation of commodity good retailers.

From the literature, they usually focus on the usage of the services or economic roles. There are very few researches on the mechanisms in the competing retailors, which are very hard to investigate in the case studies.

Compared with the research in the literature, in this paper, we will observe emergent dynamic phenomena, in which two competing retailors learn to make two kinds of decisions on their customers: points or discount. The basic idea of the learning behaviors comes from multi-agent learning mechanisms, for example, in [9, 10]. The mechanisms are similar to the one in the literature of Game Theory [11], however, we believe that the dynamic properties cannot be analyzed without ABM techniques.

The originality of our study is summarized in the following two points:

1. Base on our previous studies [2, 12], we propose a new ABM, which focuses on analyzing competing situations of two learning retailors among the same groups of customers. The retailors periodically changes their promotion

strategies based on the previous rewards. The customers purchase one kind of goods in each period from the either of the two retailers.
2. From intensive simulation experiments, although they are very abstract ones compared with case studies, we are able to find stable and/or unstable conditions of the decision strategies, which explain the stylized facts on the real world problems in [5, 7, 13].

3 Agent-Based Simulation Model for Competing Retailors and Customers

The overview of the proposed ABM is shown in Figure 1. In the model, there exist two competing retailer shops (Red and Blue), who sell one kind of goods. At every step, groups of customer agents purchase it from either the two retailer shops they would like to. The decisions depend on retailors' promotion strategies: points or discount services.

At every simulation step, the both retailer and customer agents make their decisions and behaviors as follows:

1. Customer agents select a retailer Red or Blue, then purchase one good at one time, we consider one step is corresponding to 1 day;
2. At every 100 steps or one term, retailer agents Red and Blue make decisions about promotion costs based on their experience on the sales values; and
3. Repeat customer step 1) for 100 steps), then execute retailer step 2), If the retailer steps reached at a terminal condition (e.g., 1,000 terms), then terminate.

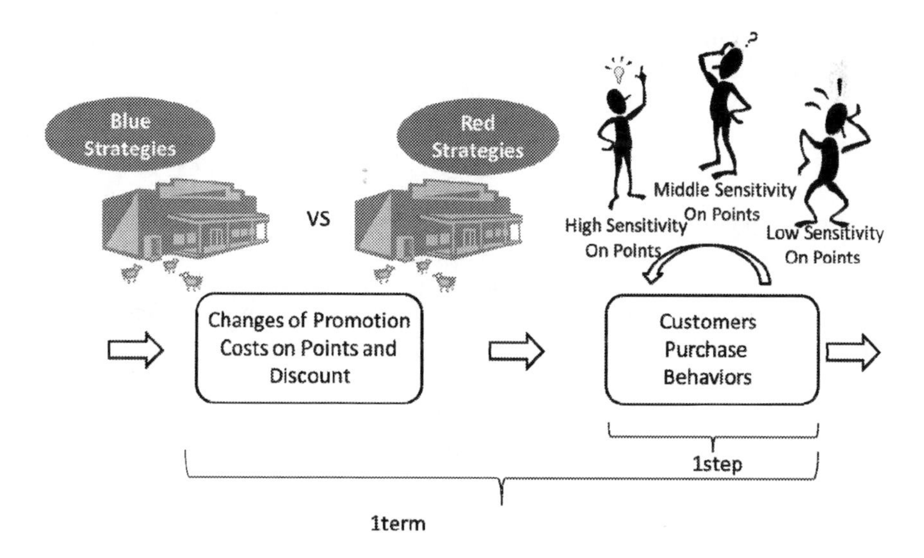

Fig. 1 Overview of the Simulation Model

3.1 Description of Retailer Agents

Retailer agents (Blue and Red) make decisions on discount or points based on their current sales amounts and the observation on the previous decisions of the opposite side. Each retailer agent has the evaluation value of the strategy $V(s,a)$, where s means the strategy at the previous term of the opposite side, and a means the own strategy of the current term. $V(s,a)$ is determined by the equation (1):

$$V(s,a) \leftarrow V(s,a) + \alpha[r_t - V(s,a)] \tag{1}$$

where

$$s = \{Increase, Decrease, Keep\_Level\} \tag{2}$$
$$a = \{Increase, Decrease, Keep\_Level\} \tag{3}$$
$$r_t = P_{(t-1)} - P_t \tag{4}$$

Pt denotes the benefit of term t, and a retailer agent learns to change the behavior based on the difference of terms t and $(t-1)$. A retailer agent selects the action a, which maximizes $V(s,a)$. After determining the action, the retailer agent changes the promotion values *unitPr* based on equations (5) (when *Increase*), (6) (when *Decrease*), or (7) (when *Keep-Level*):

$$Pr_{i,t} = Pr_{i,(t-1)} + unitPr \tag{5}$$

$$Pr_{i,t} = Pr_{i,(t-1)} - unitPr \tag{6}$$

$$Pr_{i,t} = Pr_{i,(t-1)} \tag{7}$$

Assignment of the ratio of promotion values depends on the strategies of each retailer agent. The strategy of retailer i (Blue or Red) is denoted $strategy_i$, which are represented in equation (8). Each real value element runs between 0.0 and 1.0 and the sum is equal to 1.0.

$$strategy_i = \{discount_{all,i}, discount_{frequency,i}, point_{all,i}, point_{frequency,i}\} \tag{8}$$

Discount and point services are given to every customer agent, however, the value is different whether the customer is categorized to be excellent or not. When the customer agent is the more excellent, the discount or point services are added the more. The values $discount_{i,t}$ and $point_{i,t}$ of retailer agent i at term t are determined by equations (9) and (10) for ordinal customers and (11) and (12) for excellent customers:

$$discount_{i,t} = Pr_{i,t} * discount_{all,} \tag{9}$$

$$point_{i,t} = Pr_{i,t} * point_{all,i} \tag{10}$$

$$discount_{i,t} = Pr_{i,t} * (discount_{all,i} + discount_{frequency,i}) \tag{11}$$

$$point_{i,t} = Pr_{i,t} * (point_{all,i} + point_{frequency,i}) \tag{12}$$

The profit of the retailer agent i is calculated by equations (13), (14), (15), and (16). If the value of $S_{i,t}$ becomes zero or minus, then the agent becomes bankruptcy stage and the simulation stops.

$$Sl_{i,t}=P_i * N_{i,t} \tag{13}$$

$$B_{i,t}=Sl_{i,t} - VC_{i,t} - CC \tag{14}$$

$$VC_{i,t}=Pr_{i,t} *N_{i,t} *(discount_{all,i} + point_{all,i})+ Pr_{i,t} *Nfrequency_{i,t} *(discount_{frequency,i} + point_{frequency,i})+PC*N_{i,t} \tag{15}$$

$$S_{i,t}=S_{i,(t-1)}+B_{i,t} \tag{16}$$

where $Sl_{i,t}$ is the sales amount, $B_{i,t}$ is the benefit, P_i means the price of the good, $N_{i,t}$ is the number of sales, $Nfrequency_{i,t}$ means the number of sales for excellent customers, $Pr_{i,t}$ is the promotion cost, $VC_{i,t}$ is the varying cost, and $S_{i,t}$ is the surplus of the retailer agent i at term t. Also, CC means the fixed cost and PC means the prime cost.

3.2 Description of Customer Agents

There are plural customer agents in the model. We assume that one customer agent represents a group of 100 individual customers with the same properties on the purchase decisions and behaviors. They select one of the retailer agents based on discount and point services and purchase one good at each step. The more customer agents' purchase from one retailer, the utility about the points becomes the larger. The mechanism is inspired by the research literature [6, 13]. Each customer agent consumes their point at the end of the term and the points are expired.

The selection of the retailer agents is based on the roulette selection method with the utility of the customer obtained from the retailer. The probability of customer agent k for retailer agent i , $probability_{i,k}$ is given by equation (17):

$$probability_{i,k}=U_{i,k}/(\textstyle\sum^{n}_{j=1} U_{j,k}) \tag{17}$$

where, n is the number of retailers in the market (In this case, n=2, Blue and Red), $U_{i,k}$ is the utility of customer agent k, who purchase the good from retailer i. The value of $U_{i,k}$ is determined by the promotion cost of retailer agent i and its promotion strategies as follows:

$$U_{i,k}=Pr_{i,t} * (p_{discount}*U_{i,d} +p_{point}\cdot U_{i,p}) \tag{18}$$

$p_{discount}$, and p_{point} will change whether the customer agents are excellent or not. When normal (resp., excellent), the value is determined by equations (19) and (20) (resp., (21) and (22)):

$$p_{discount}=discount_{all,i} \tag{19}$$

$$p_{point}=point_{all,i} \tag{20}$$

$$p_{discount}=discount_{all,i} +discount_{frequency,i} \tag{21}$$

$$p_{point}=point_{all,i} +point_{frequency,i} \tag{22}$$

where, $U_{i,d}$ and $U_{i,p}$ respectively mean the utilities of discount and point services of retailer agent *i*. The value will be changed by the number of purchase at *i*. They are given by equations (23) and (24):

$$U_{i,d}=1 - U_{i,p} \tag{23}$$

$$U_{i,p}=purchase*sensitivity+intercept \tag{24}$$

where, *purchase-sensitivity* means the sensitivity of customers' behaviors about the point service and *intercept* is the initial utility value of the point service. The customer agents are categorized as *High*, *Middle*, and *Low* on the behaviors about point services. Based on prior turning of the simulation, we set the values shown in Table 1.

Table 1 Parameters of Customer Agent Sensitivity on Point Service

Purchase-Sensitivity	Sensitivity Value	Purchase ratio at Blue
High	0.020	70%
Middle	0.009	50%
Low	0.000	30%

3.3 Implementation of the Simulator

The simulator does not intend to get realistic results, but aims at understanding stylized facts on service strategies of competing retailers, so that we have simply implemented the specifications described in the previous sub-sections. The simulator is developed from full scratch with JAVA language. The simulator runs on a usual Window 7 PC with 16MB main-memories. A standard result of one run is obtained within 0.01 CPU seconds. It is enough simple that we are able to test various scenarios changing agent and environmental parameters. The major results are described in the following section.

4 Experiments and Discussions

4.1 Experimental Settings

Using the agent-based simulator explained in the previous section, we conduct intensive experiments to explore the co-existence conditions or cooperative situations of the two competing retailers Blue and Red. We utilize the following parameter set for the simulation shown in Table 2. In the series of the experiments, we set 100 customer agents and 1,000 terms. These settings mean that there are 10,000 individuals in the market, and that we analyze the changes of the market for about 3 years. We also change the 35 strategies patterns of Blue and Red shown in Table 3.

Table 2 Parameter Settings of the Experiments

Number of Terms	1,000
Number of Steps of Each Term	100
Number of Trials of Each Scenario	50
Number of Customer Agents	100
High: Middle: Low Sensitivity Ratio	20:60:20
Number of Retailer Agents	Blue and Red
Learning Ratio of Retailer Agents α	0.8
Initial Promotion Price Pr_0	150
Initial Surplus S_0	6,000,000
Fixed Value FC	600,000
Price of the Good P	1,000
Prime Cost of the Good PC	700
Strategy of Blue	Eq. (25)
Strategy of Red	Eq. (25)

As a result, we set 35*35 = 1,225 scenarios in the experiments. In each scenario, we execute 50 trials and summarize the results.

Table 3 List of Strategies of a Retailer Agent

Strategy #	$discount_{all}$	$point_{all}$	$discount_{frequency}$	$point_{frequency}$
1	1	0	0	0
2	0.75	0.25	0	0
3	0.75	0	0.25	0
4	0.75	0	0	0.25
5	0.5	0.5	0	0
6	0.5	0	0.5	0
7	0.5	0	0	0.5
8	0.5	0	0.25	0.25
9	0.5	0.25	0	0.25
10	0.5	0.25	0.25	0
11	0.25	0.75	0	0
12	0.25	0	0.75	0
13	0.25	0	0	0.75
14	0.25	0	0.25	0.5
15	0.25	0	0.5	0.25
16	0.25	0.25	0	0.5
17	0.25	0.5	0	0.25
18	0.25	0.25	0.5	0

Table 3 (*continued*)

19	0.25	0.5	0.25	0
20	0.25	0.25	0.25	0.25
21	0	1	0	0
22	0	0.75	0	0.25
23	0	0.75	0.25	0
24	0	0.5	0	0.5
25	0	0.5	0.5	0
26	0	0.5	0.25	0.25
27	0	0.25	0	0.75
28	0	0.25	0.75	0
29	0	0.25	0.5	0.25
30	0	0.25	0.25	0.5
31	0	0	1	0
32	0	0	0.75	0.25
33	0	0	0.5	0.5
34	0	0	0.25	0.75
35	0	0	0	1

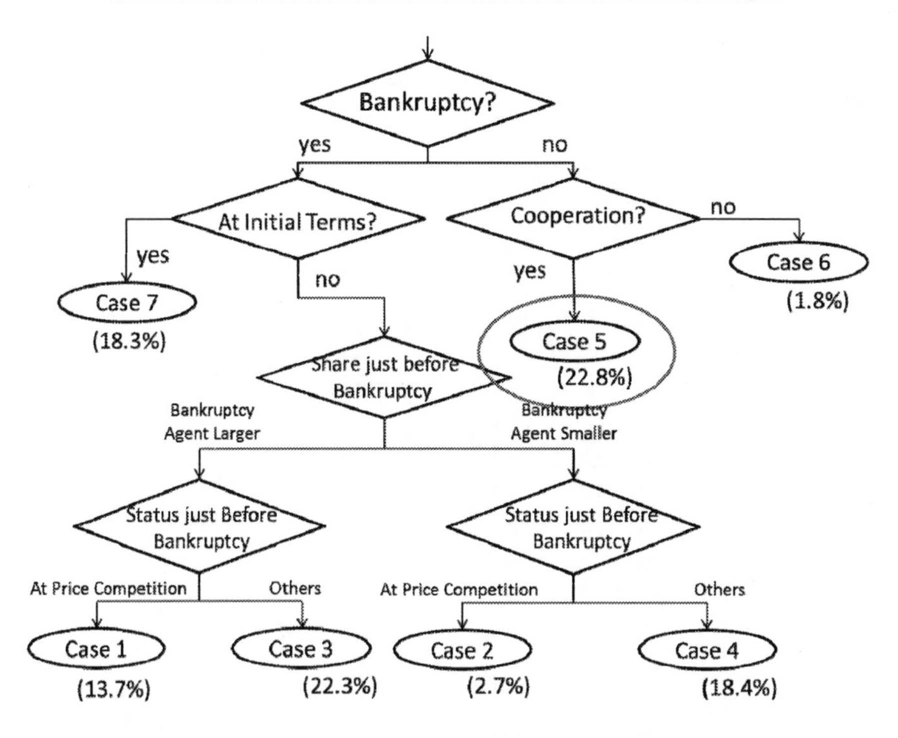

Fig. 2 Summary of Experimental Scenarios

4.2 Results and Discussions

The results of the experiments are summarized in Figure 2. Among the simulation experiments, the cases of the bankruptcy of either Blue or Red agents has become 78.2%. The key events of the simulation terms are: Bankruptcy occurrence, Market share is high/low just before the bankruptcy, Price competition status just before the bankruptcy, and The Emergence of cooperative behaviors of retailer agents. From Figure 2, we consider that in our model, the bankruptcy is usual, however, some scenarios, stable cooperative behaviors occur as stylized facts. One of such typical results of the stable cooperation level are depicted in Figure 3.

In Figure 3, (A), (B), (C), and (D) respectively represent the changes of retailer agents status on profits, number of sales, promotion costs, and surplus. Form the Figure 3, in this simulation case, we have observed the following results:

1. In the initial terms during 1 and 100, both Blue and Red retailers increase the promotion costs, as a results, both of them lose the profits. The competitiveness is very hard, because of the rapid changes of the number of sales. Furthermore, the surplus of Blue and Red rapidly decreases.
2. In the middle term around 200, however, the both retailers spontaneously decrease the promotion costs. As a result, they are able to increase their surplus. From the detailed analysis on the decisions and behaviors of Blue and Red, during the terms, they are changing their decisions from defeat to cooperate. This situation is very similar to a typical case of the strategy changes in the repeated prisoners' dilemma game [11]. Although the design and principle of the agent-based simulator is so different from the game theoretical approaches, the emergence of the spontaneous cooperative behaviors is very interesting.

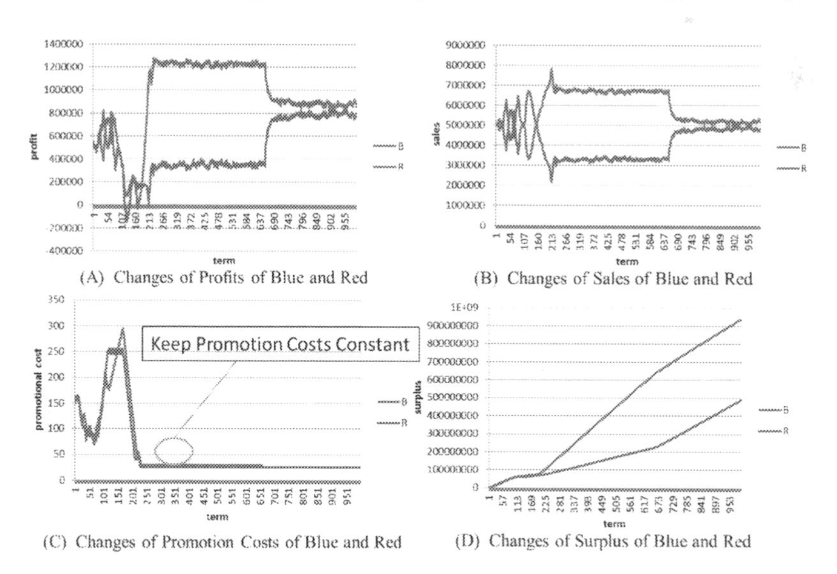

(A) Changes of Profits of Blue and Red

(B) Changes of Sales of Blue and Red

(C) Changes of Promotion Costs of Blue and Red

(D) Changes of Surplus of Blue and Red

Fig. 3 Typical Case of Cooperation Emergence

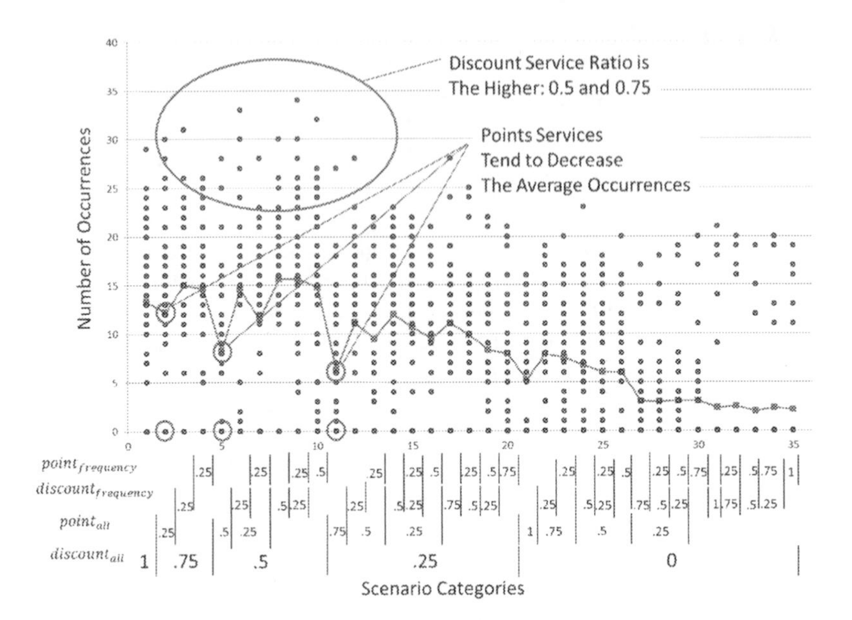

Fig. 4 Scenario Categories and Number of Corporative Behavior Emergence

We have further analyzed all the cases focusing on the case 5 in Figure 2, in which the cooperative behaviors have been frequently observed. The results are shown in Figure 4.

The x-axis of Figure 4 represents the characteristics of strategies shown in Table 3. The y-axis represents the number of occurrence of cooperative behaviors. Please note that when the value of $discount_{all}$ in both Blue and Red agents are equal to 0.25 or 0.50, the cooperative behaviors tend to occur. Especially, strategies between 5 and 15 are remarkable. Also, there are no other cases where the cooperative behaviors often occur. This suggests that the emphasis of point service strategies might be wrong and that the discount service strategies are better in committing markets, if all the retailers would adopt point service systems in their services in the early stages of promotions.

5 Concluding Remarks

This paper has proposed a novel agent-based model (ABM) for analyzing service management at commodity goods retailers. The ABM has especially focused on the (dis)advantages of mileage point and discount services. To uncover the characteristics of the two service strategies, we have implemented a simulator to analyze the behaviors of competing retailing stores and their customers. Then, we have conducted intensive computer experiments to uncover the characteristics of point and discount service systems among the retailing shops.

The main contribution of the paper to knowledge management in organization literature is summarized as follows:

1. The proposed agent-based model reveals the effectiveness of the issues, which are hard to examine in the real situations.
2. From intensive simulation experiments, although they are very abstract ones compared with case studies, we are able to find stable and/or unstable conditions of the decision strategies.
3. About the decision making in competing marketing conditions, the experimental results suggest that discount services strategies combined with the other point service strategies would work well, when every participant would adopt the point services.

Our future work related to the ABM include the analyses of point service systems as a whole with more retailer and customer participations and the work to ground the ABM with real world data.

References

1. Toriyama, M., Kikuchi, T., Yang, C., Yamada, T., Terano, T.: Who is a Key Person to Transfer Knowledge in a Business Firm - Agent-Based Simulation Approach. In: Proc. 5th Knowledge Management in Organizations, KMO 2010, pp. 41–51 (2010)
2. Kobayashi, T., Takahashi, S., Kunigami, M., Yoshikawa, A., Terano, T.: The Effect of Management Style Formalization with Growth of Organization - An Agent-Based Model. In: Proc. 6th Knowledge Management in Organizations, KMO 2011, p. 20 (2011)
3. Chen, S.-H., Terano, T., Yamamoto, R. (eds.): Agent-Based Approaches in Economic and Social Complex Systems. In: VI -Post-Proceedings of the AESCS (2011)
4. International Workshop 2009. ABSS, vol. 8. Springer (2009)
5. Dowling, G.R., Uncles, M.: Do Customer Loyalty Programs Really Work? MIT Sloan Management Review, 16 pages (1997)
6. Berman, B.: Developing an Effective Customer Loyalty Program. California Management Review 49(1), 123–148 (2006)
7. Liu, Y.: The Long-Term Impact of Loyalty Programs on Consumer Purchase Behavior and Loyalty. Journal of Marketing 71(4), 19–35 (2007)
8. Humby, C., Hunt, T., Phillips, T.: Scoring Points How Tesco Continues to Win Customer Loyalty. Kogan Page Ltd. (2008)
9. Yi, Y., Jeon, H.: Effects of loyalty programs on value perception, program loyalty, and brand loyalty. Journal of the Academy of Marketing Science 31(3), 229–240 (2003)
10. Kutschinski, E., Uthmann, T., Polani, D.: Learning Competitive Pricing Strategies by Multi-Agent Reinforcement Learning. Journal of Economic Dynamics and Control 27(11-12), 2207–2218 (2003)
11. DiMicco, J.M., Greenwald, A., Maes, P.: Learning Curve: A Simulation-Based Approach to Dynamic Pricing. Electronic Commerce Research 3, 245–276 (2003)
12. Gibbons, R.: Game Theory for Applied Economists. Princeton University Press (1992)
13. Kobayashi, M., Kunigami, M., Yamadera, S., Yamada, T., Terano, T.: How a Major Mileage Point Emerges through Agent Interactions using Doubly Structural Network Model. In: WEIN 2009 (Workshop on Emergent Intelligence on Netwoked Agents) (2009)
14. Doan, R.J., Simon, H.: Power Pricing. Free Press (1997)
15. Taylor, G.A., Neslin, S.A.: The Current and Future Sales Impact of a Retail Frequency Reward Program. Journal of Retailing 81(4), 293–305 (2005)

Cloud-IO: Cloud Computing Platform for the Fast Deployment of Services over Wireless Sensor Networks

Dante I. Tapia[1], Ricardo S. Alonso[1], Óscar García[1], Fernando de la Prieta[2], and Belén Pérez-Lancho[2]

[1] R&D Department, Nebusens, S.L., Scientific Park of the University of Salamanca, Edificio M2, Calle Adaja, s/n, 37185, Villamayor de la Armuña, Salamanca, Spain
[2] Department of Computer Science and Automation, University of Salamanca, Plaza de la Merced, s/n, 37008, Salamanca, Spain
{dante.tapia,ricardo.alonso,oscar.garcia}@nebusens.com, {fer,lancho}@usal.es

Abstract. In the recent years, a new computing model, known as Cloud Computing, has emerged to react to the explosive growth of the number of devices connected to Internet. Cloud Computing is centered on the user and offers an efficient, secure and elastically scalable way of providing and acquiring services. Likewise, Ambient Intelligence (AmI) is also an emerging paradigm based on ubiquitous computing that proposes new ways of interaction between humans and machines, making technology adapt to the users' necessities. One of the most important aspects in AmI is the use of context-aware technologies such as Wireless Sensor Networks (WSN) to perceive stimuli from both the users and the environment. In this regard, this paper presents Cloud-IO, a Cloud Computing platform for the fast integration and deployment of services over WSNs.

Keywords: Cloud Computing, Ambient Intelligence, Wireless Sensor Networks, Multi-Agent Systems, Real-Time Locating Systems.

1 Introduction

Nowadays the society is currently immersed in an innovator dynamics that has encouraged the development of intuitive interfaces and systems with a certain intelligence level that are capable of recognizing the users' necessities and responding to them in a discreet way, often even imperceptible [1]. Thus, is necessary to consider people as the center of the developments in order to build intelligent and technologically complex scenarios in a wide range of different areas such as medicine, domestic, academic or public administration, where technology adapts to people's necessities and their environment, and not on the contrary. Here is where concepts such as Ambient Intelligence (AmI) [2], Cloud Computing [3], agent technology [3] [4] and innovative visualization interfaces

L. Uden et al. (Eds.): 7th International Conference on KMO, AISC 172, pp. 493–504.
springerlink.com

and techniques [5] come into play in order to achieve an improved human-machine interaction.

Systems based on Ambient Intelligence encourage technology to adapt to people, detecting their necessities and preferences and allowing them to enjoying services without need of being conscious or learning complicated operations. In this sense, agent and multi-agent systems provide these AmI-based systems with the necessary intelligence for responding to the users' preferences and needs [6]. The Cloud Computing model is centered on the user and offers a highly effective way of acquiring and supplying services. Cloud Computing defines a new economic model based on a new way of consuming services. Furthermore, all these features are performed in a reliable and secure way, with an elastic scalability which is capable of attending to sharp and a priori unpredictable changes on the demand, and at the expense of a very low increase in the management costs [7]. Likewise, Wireless Sensor Networks (WSNs) provide an infrastructure capable of supporting distributed communication needs and increase the mobility, the flexibility and the efficiency of users [8]. Specifically, Wireless Sensor Networks allow gathering information about the environment and reacting physically to it, thus expanding users' capacities and automating daily tasks [9].

There are three key functionalities that would make an AmI-based platform be a powerful tool for supporting a wide variety of situations: locating, telemonitoring and personal communication. Locating systems allow knowing the exact position of users and objects of interest [10]. This way, the platform could identify each user, know where he is, as well as provide him with services in an automatic and intelligent way, without need of him to start the interaction. Telemonitoring (or sensing) allows obtaining information about users and their environment (e.g., temperature, humidity, etc.), taking it into account when offering them customized services in keeping with the environment status [11]. Finally, personal communication, by means of voice or data, allows users to communicate among them through the platform, no matter their location or the communication media they choose. In order to an AmI-based platform can be successful regarding versatility, it should be capable of integrating these three services.

In this regard, the Cloud-IO platform is designed to provide solutions for the mentioned necessities in different environments. Thus, Cloud-IO is intended not only for covering the essential needs of society, but also for supporting the development of innovative and very specific technological solutions.

The rest of this paper is organized as follows. The next section depicts the main objectives that the development of the Cloud-IO platform intends to address. Then, the main components and services that will be integrated into the Cloud-IO platform are described. Finally, the conclusions and future lines of work are depicted.

2 Background and Problem Description

Currently, there are two main computing models that still dominate information technologies. On the one hand, the centralized computing model, which is typical in super-powerful mainframe systems with multiple terminals connected to them. On the other hand, the distributed computing model, whose the most widespread example is the client-server model, with largely demonstrated efficiency [12]. During the last years, a new model, known as Cloud Computing, has emerged with the aim of giving support to the explosive growth of the number of devices connected to Internet and complementing the increasing presence of technology in people's daily lives and, more specifically, in business environments [3] [7] [12].

The Cloud-IO platform is aimed to offer locating, telemonitoring (sensing) and voice/data transmission (personal communication) services using a unique wireless infrastructure and supported by a Cloud Computing model. Furthermore, the platform is focused on supporting applications based on Ambient Intelligence paradigm and that are conscious of the location of the users and the environment state, as well as allowing them to keep intercommunicated through the platform.

Wireless Sensor Networks are identified as one of the most promising technologies by different technological analysts and specialized journals [8] [9] [11], because of they give support to the current requirements related to the deployment of networks that cover the communication needs flexibly in time, space and autonomy, without need of a fixed structure. The user identification is a key aspect for an adequate services customization and environment interaction [13]. In this sense, knowing the exact geographic location of people and objects can be very useful in a wide range of application areas, such as industry or services [10]. The advantage of knowing and visualize the location of all the resources in a company and how these interact and collaborate in the different productive processes is a clear example of the demand for a platform as Cloud-IO. Other good example of this demand are all those emergence situations where is required to locate people, such as forest fires or nuclear disasters. Whether in home automation, healthcare telemonitoring systems or hospitals, WSNs are used for collecting information from the users and their environment [9]. Nonetheless, the development of software for remote telemonitoring that integrates different subsystems demands the creation of complex and flexible applications. As the complexity of an application increases, it needs to be divided into modules with different functionalities. There are several telemonitoring developments based on WSNs [14]. However, they do not take into account their integration with other architectures and are difficult to adapt to new scenarios [15]. Furthermore, even though there are some existing approaches that integrates personal communications, such as voice, over WSNs [16], current wireless sensor technologies have interoperability, manageability and mobility deficiencies [17]. In this regard, all these issues can be addressed developing a new platform that integrates locating, telemonitoring and personal communication services in a unique wireless sensor infrastructure.

In the last years, several approaches aimed at merging WSNs and the Cloud model have been proposed [18] [19] [20]. Nevertheless, many of these proposals are not realistic or scalable enough, which results in not much practical approaches. Moreover, many of the research done until the date do not consider desirable features such as fault-tolerance, security and the reduction of the communication response times, and, when they are considered, they are usually achieved at the expense of energy efficiency. Furthermore, these approaches do not consider the integration of locating, telemonitoring and personal communication in the same infrastructure.

3 The Cloud-IO Platform

This section describes the main components of the Cloud-IO platform. First, the basic functioning of the platform is depicted. After that, the different modules that make up the platform are described. As this is a research work that will be finished in Q4 2013, this paper presents a preliminary description that will be extended and published further on.

The main objective of this work is to revolutionize the interoperability, accessibility and usability of systems based on WSNs through a new platform based on Cloud Computing services. Therefore, the Cloud-IO project consists of the development of a modular, flexible and scalable platform that integrates, on the one hand, identification/locating, telemonitoring and personal communication services in a unique network infrastructure, and, on the other hand, the creation of interactive and customized services managed from a Cloud infrastructure. In addition, most of its multiple functionalities must be accessible through 3D virtual scenarios generated automatically by a new graphical environment integrated in the own platform.

It is worth mentioning that this project is not only aimed at integrating pre-existing technology, but also designing and developing innovative hardware and software in order to build a unique platform. Therefore, it will be developed a new specific hardware, both for the network infrastructure and the services implemented over it. Moreover, it will be developed a new middleware that manages all functionalities of the platform, making use of techniques and algorithms based on Artificial Intelligence [21] [22] on practically each of these functionalities: dynamic routing, real-time locating [10], 3D virtual model automatic generation, as well as self-configuring and error-recovery, among many others. This way, it will be achieved a platform that will not only make use of existing technology, but also perform important and relevant improvements on the current state of the art of different technological areas. The basic architecture of the Cloud-IO platform, shown in Figure 1, is formed by the next main components:

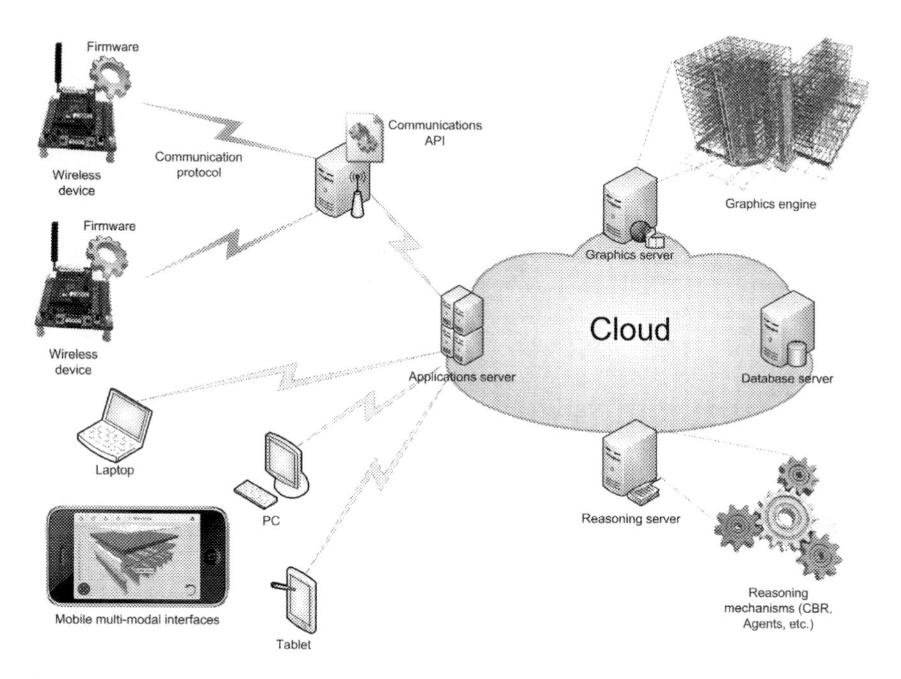

Fig. 1 Basic architecture of the Cloud-IO platform

- **Wireless sensor infrastructure.** The main components and services of the Cloud-IO platform are based on this infrastructure. This infrastructure consists of a set of wireless physical devices over which part of the low-level middleware, described below, will run. This low-level middleware allows the rest of the platform to access to their functionalities.
  - **Wireless devices.** They form the hardware layer of the wireless sensor infrastructure. In this sense, a set of wireless devices with reduced energy consumption and physical size must be developed. Each of them should share a common architecture made up of a microcontroller, a radio transceiver and a set of physical interfaces for the data exchange between the device itself and the sensors and actuators connected to those interfaces [15]. According to the application, these devices must be able to be supplied by batteries or an extern source. Moreover, and also according to each application or system to which they can belong, these devices should include Digital Signal Processors (DSP) [23], for the personal communication service, or physical interfaces intended for the intercommunication with sensors and actuators, for the telemonitoring service [15].
  - **Low-level middleware.** Over the wireless sensor infrastructure it must be developed a low-level middleware formed by different layers aimed at allowing the data exchange among the wireless devices and the rest of the components of the Cloud-IO platform. This low-level middleware is

formed by the firmware embedded in the wireless devices, a communications API (Application Programming Interface) for accessing the features of the devices from any extern machine (e.g., a PC), as well as a communications protocol for the data exchange between the firmware and the communications API.

  - **Firmware.** It consists of code that can run on each of the wireless devices. By means of this firmware the devices can connect automatically among them, thus building a network topology for the data exchange among devices. Likewise, the firmware must implement the required functionalities for each of the three main services of the platform (i.e., telemonitoring, identification and locating, personal communication) and respond to the received commands remotely from the communications API by means of the communications protocol. Regarding the devices intended for the personal communication service, it must be also developed the firmware specific to be run over the Digital Signal Processors [23] in order to implement the voice codecs required for that service [16].
  - **Communications protocol.** It must be developed a communications specific protocol that allows transporting the required data among the devices, as well as between the devices and the communications API. This protocol must allow transporting data with different quality-of-service, network latency and data throughput rate. This way, the different necessities of the main three services (i.e., telemonitoring, locating and personal communication) must be covered in an optimal way.
  - **Communications API.** The communications API allows the rest of the components of the Cloud-IO platform to interact with the wireless devices in order to collect information from the sensors connected to them, as well as send and receive voice/data streams. In addition, this API must incorporate a set of innovative locating algorithms [10] in order to provide the rest of the components of the platform with accurate and real-time position of people and objects that carry the wireless devices focused on being used by the identification and locating service. The communications API should run over a server in the entity that acts as consumer of the Cloud-IO services.

- **Cloud Computing infrastructure.** This infrastructure is, along with the wireless sensor infrastructure, the other main pillar of the Cloud-IO platform. Thanks to this, the tasks requiring an elevated computational cost can be moved to remote powerful machines, as these tasks are very difficult to be implemented in the machines and devices, often mobile, used by the client users of the platform. This infrastructure should be basically formed by a set of high-performance hardware machines, as well as a data communications network infrastructure for the interconnection between the client machines and the remote high-performance machines.

- **Hardware.** The hardware infrastructure that must support the Cloud Computing components should be formed by a set of high-performance machines that will be sited, generally, remotely with regard to the client machines infrastructure (i.e., the wireless infrastructure connected to the client machine, as well as the set of multimodal interfaces [1]). This set of high-performance machines are responsible for executing the tasks with elevated computational cost that, otherwise, could not be feasible to be done in the client machines, which usually have a low or moderate performance.
- **Data communication networks.** Consist of a set of data communication networks, either public, private or public-private partnered, aimed at the data transmission between client machines and the remote high-performance machines where the software of the Cloud Computing infrastructure will run. These networks could be from a public network based on cable, ADSL or 3G, until a local intranet based on Ethernet or Wi-Fi, or even a mixed solution (e.g., a Virtual Public Network or VPN). On the one hand, the information gathered by the wireless sensor infrastructure are transmitted from the client machines to the remote high-performance machines. On the other hand, all the information processed by the reasoning mechanisms, as well as the rendered tridimensional images, are transmitted from the remote high-performance machines to the client machines, including the multimodal interfaces. This way, client machines are freed from a great part of the computational load.
- **Software components of the Cloud Computing infrastructure.** These components make up the software layer that run over the Cloud Computing hardware infrastructure and are responsible for performing remotely a great part of the high-cost computational load, which, otherwise, would be difficult to be done by the client machines, especially mobile multimodal interfaces.
  - **High-level middleware.** This middleware layer must make possible the data exchange among the different Cloud software components, as well as the interaction among those components and the client machines, such as the mobile multimodal interfaces [1]. This middleware should offer its functionalities through protocols similar to Web Services [24]. This will allow the remote access to them from practically any kind of device which includes an adequate web browser.
  - **Graphics engine.** The graphics engine is an essential part of the Cloud-IO platform. It should be based on a set of certain APIs and a well-defined list of functionalities. It must run over the hardware infrastructure mentioned before and allow, remotely, performing the tridimensional objects modeling, the layout and animation of them, and, lastly, the rendering of the scenarios and the objects. This graphics engine is a high-complexity component that should

contemplate multiple and different technologies, such as techniques for lighting, shadowing, culling (i.e., saving time when rendering objects hidden by the presence of other objects in the scenarios, as, for example, the use of Binary Space Partitioning or BSP [25]), etc.

– **Reasoning mechanisms.** Integrated with the Cloud Computing infrastructure itself, it will be implemented a set of complex and innovative reasoning mechanisms based on agent technology [4]. This way, all the tasks related to alerts management, tracking and patter recognition can run on the Cloud itself, thus freeing the client machines, and especially the generally light multimodal interfaces, from performing these tasks. These reasoning mechanisms include the algorithms intended for managing the call routing and queuing required by the personal communication service.
– **Databases.** Optionally, these databases could be remotely stored, so that all the information and the important data required for the different services (i.e., telemonitoring, locating and personal communication) in the Cloud-IO platform could be provided by machines more secure and powerful than those that could have the clients of the platform.

- **Multimodal interfaces.** Last but not least, the platform will include a set of innovative multimodal interfaces [2] that should allow showing in an enhanced way all the information provided by the platform from the data received from the Cloud Computing services. This way, all the information and services of the platform will be able to be accessed by almost any device that can execute a web browser. Specifically, the design and development are focused on those mobile multimodal interfaces that can be implemented in devices so light as a mobile phone or a tablet PC.

3.1 Telemonitoring, Locating and Communication Services

Therefore, three main differentiated services must be developed over the Cloud-IO platform. These services must be fully integrated among them and work simultaneously over the platform. These services (telemonitoring, locating and personal communication) are shown in Figures 2, 3 and 4, respectively:

1. **Telemonitoring service.** This service allows gathering information about the context and the users in the defined application scenarios, as well as acting consequently managing alerts and other relevant situations. In order to do this, the design and development of this service should make use of specific functionalities of the wireless devices, which, through different physical interfaces (hardware) could be connected to different home automation sensors (e.g., temperature, humidity, gases, smokes, etc.) [15], biomedical sensors (e.g., breath rhythm, body temperature, fall detection, etc.) [14], biometric sensors (e.g., recognition of faces, eyes or fingerprints) and actuators (e.g., light alarms, sound alarms, data delivery through the platform, etc.).

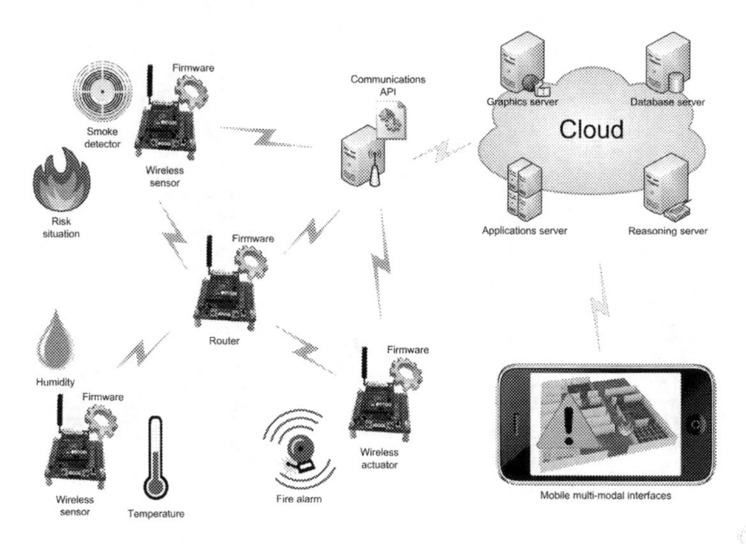

Fig. 2 Telemonitoring service in the Cloud-IO platform

2. **Real-time identification and locating service.** This service allows identifying and estimating the position of users and objects in every moment. As the other services, this service must also allow utilizing algorithms for managing alerts derived from the position of the users and objects according to the area where they are, they permissions, etc. In this sense, the service must make use of an innovative set of locating algorithms [10] that will be developed. The main objective of the development of these algorithms is to achieve a much better accuracy than with existing locating systems when locating people and objects, as well as allow its adequate operation both outdoors and indoors.
3. **Personal communication service.** This service allows the transmission of voice and instant messaging through the wireless devices in the platform. To do that, it must be used an innovative architecture of wireless devices that puts together microcontrollers, transceivers and Digital Signal Processors [23], along with the implementation in the devices and the platform of voice codecs [16] and reasoning mechanisms based on agent technology [4] and focused on the personal communications routing, as well as the implementation of optimized queuing algorithms.

This way, thanks to the integration of these three services in a unique wireless infrastructure of fast deployment, the Cloud-IO platform will be able to deploy applications as the next example. Let's imagine a scenario when a natural disaster has happened, such as a possible leak in a nuclear power station. The Cloud-IO platform will allow facing this situation in the next way. As a previous stage, operators will place a set of wireless devices along the perimeter of the zone. These devices will act as basic communications infrastructure and as beacons when locating the operators themselves throughout the environment. This infrastructure will not require previous calibrating and will be deployed in a few minutes. Simultaneously, these beacons will incorporate different environmental

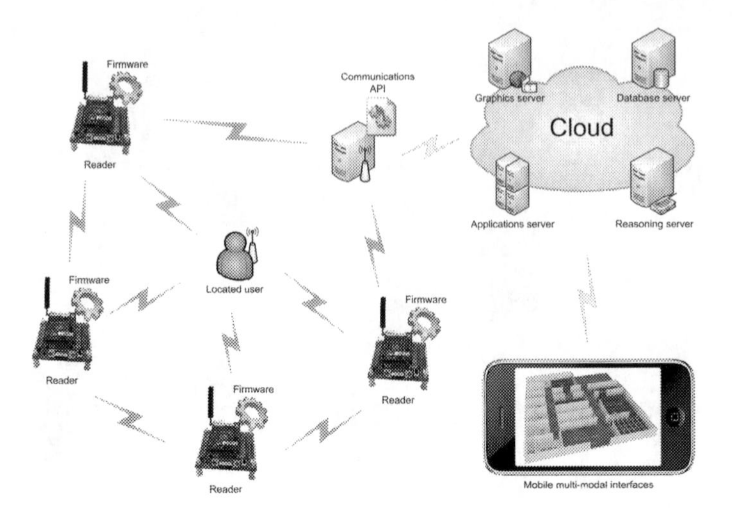

Fig. 3 Real-time identification and locating service in the Cloud-IO platform

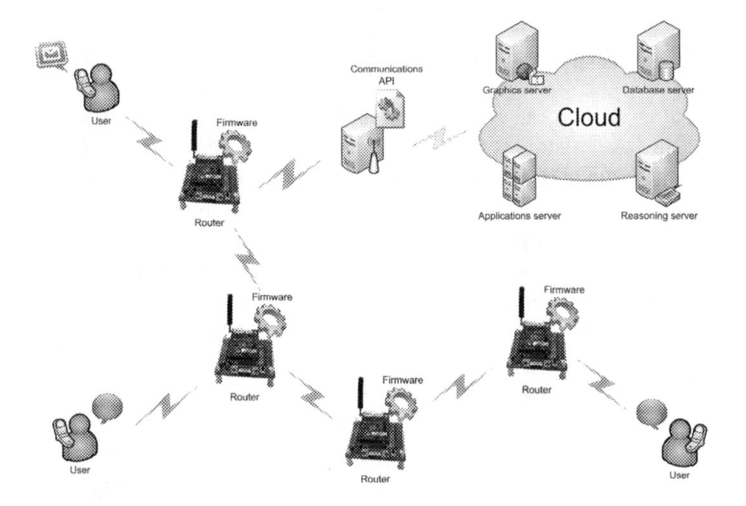

Fig. 4 Personal communication service in the Cloud-IO platform

sensors for measuring humidity, temperature, radiation, wind direction, and will be connected to a GPS receiver in order to have a global reference. Operators will carry small wireless devices (i.e., tags) to be identified and located throughout the environment by the beacon devices, as well as communicate among them via voice.

This way, deploying a unique infrastructure in a few minutes, the services in the platform will provide a complete information about the emergence situation: real-time location of operators, environmental and radiation measurements, as well as the information that operators will transmit among them via voice in real-time. All the information will be collected by the platform, which, thanks to different techniques of object modeling and tridimensional representation [25], will show it to those authorized users that want to watch this information in a mobile device, a PC or a remote control center, all of this through Internet.

4 Conclusions and Future Work

This work presents Cloud-IO, a Cloud Computing platform for the fast integration and deployment of services over Wireless Sensor Networks. The Cloud-IO platform integrates in a unique wireless infrastructure three main services that are key for the implementation of AmI-based applications: telemonitoring (sensing), locating and personal communication (voice and data transmission among users). Furthermore, the platform makes use of agent technology for implementing reasoning mechanisms aimed at routing and queuing data and voice transmissions, as well as a new graphics engine and multimodal interfaces in order to enhance the interaction between users and the applications deployed using this platform.

Future work includes the full development of all the projected functionalities. This includes the production of the first hardware prototypes for the wireless devices. At the software level, the firmware embedded in the devices, the communications API and the high-level middleware in the Cloud infrastructure will be developed. Then, the graphics engine and the agent-based reasoning mechanisms will be designed and developed. At a further stage, the three main services (telemonitoring, locating and personal communication) will be implemented over the platform. Finally, the system will be implemented in two real scenarios related to elderly care and a fire department in order to test its actual performance.

Acknowledgments. This project has been supported by the Spanish Ministry of Economy and Competitiveness. Project IDI-20111471.

References

1. Aarts, E., de Ruyter, B.: New research perspectives on Ambient Intelligence. J. Ambient Intell. Smart Environ. 1, 5–14 (2009)
2. Sadri, F.: Ambient intelligence: A survey. ACM Comput. Surv. 43, 36:1–36:66 (2011)
3. Rodríguez, S., Tapia, D.I., Sanz, E., Zato, C., de la Prieta, F., Gil, O.: Cloud Computing Integrated into Service-Oriented Multi-Agent Architecture. In: Ortiz, Á., Franco, R.D., Gasquet, P.G. (eds.) BASYS 2010. IFIP AICT, vol. 322, pp. 251–259. Springer, Heidelberg (2010)
4. Wooldridge, M.: An Introduction to MultiAgent Systems, 2nd edn. Wiley (2009)
5. Cuřín, J., Kleindienst, J.: SitCom: Virtual Smart-Room Environment for Multi-modal Perceptual Systems. In: Nguyen, N.T., Borzemski, L., Grzech, A., Ali, M. (eds.) IEA/AIE 2008. LNCS (LNAI), vol. 5027, pp. 476–485. Springer, Heidelberg (2008)
6. Corchado, J.M., Bajo, J., de Paz, Y., Tapia, D.I.: Intelligent environment for monitoring Alzheimer patients, agent technology for health care. Decision Support Systems 44, 382–396 (2008)

7. Marston, S., Li, Z., Bandyopadhyay, S., et al.: Cloud computing — The business perspective. Decision Support Systems 51, 176–189 (2011)
8. Baronti, P., Pillai, P., Chook, V.W.C., Chessa, S., Gotta, A., Hu, Y.F.: Wireless sensor networks: A survey on the state of the art and the 802.15.4 and ZigBee standards. Comput. Commun. 30(7), 1655–1695 (2007)
9. Sarangapani, J.: Wireless Ad hoc and Sensor Networks: Protocols, Performance, and Control, 1st edn. CRC (2007)
10. Liu, H., Darabi, H., Banerjee, P., Liu, J.: Survey of Wireless Indoor Positioning Techniques and Systems. IEEE Transactions on Systems, Man, and Cybernetics, Part C: Applications and Reviews 37, 1067–1080 (2007)
11. Corchado, J.M., Bajo, J., Tapia, D.I., Abraham, A.: Using Heterogeneous Wireless Sensor Networks in a Telemonitoring System for Healthcare. IEEE Transactions on Information Technology in Biomedicine 14, 234–240 (2010)
12. Zissis, D., Lekkas, D.: Addressing cloud computing security issues. Future Generation Computer Systems 28, 583–592 (2012)
13. Vera, R., Ochoa, S.F., Aldunate, R.G.: EDIPS: an Easy to Deploy Indoor Positioning System to support loosely coupled mobile work. Personal and Ubiquitous Computing 15, 365–376 (2011)
14. Jurik, A.D., Weaver, A.C.: Remote Medical Monitoring. Computer 41, 96–99 (2008)
15. Tapia, D.I., Alonso, R.S., De Paz, J.F., Zato, C., de la Prieta, F.: A Telemonitoring System for Healthcare Using Heterogeneous Wireless Sensor Networks. International Journal of Artificial Intelligence 6, 112–128 (2011)
16. Wang, C., Sohraby, K., Jana, R., et al.: Voice communications over zigbee networks. IEEE Communications Magazine 46, 121–127 (2008)
17. Arbanowski, S., Lange, L., Magedanz, T., Thiem, L.: The Dynamic Composition of Personal Network Services for Service Delivery Platforms. In: 4th IEEE International Conference on Circuits and Systems for Communications, ICCSC 2008, pp. 455–460 (2008)
18. Kolli, S., Zawodniok, M.: A dynamic programming approach: Improving the performance of wireless networks. Journal of Parallel and Distributed Computing 71, 1447–1459 (2011)
19. Khattak, A.M., Truc, P.T.H., Hung, L.X., et al.: Towards Smart Homes Using Low Level Sensory Data. Sensors (Basel) 11, 11581–11604 (2011)
20. Patel, S.V., Pandey, K.: Design of SOA Based Framework for Collaborative Cloud Computing in Wireless Sensor Networks. International Journal of Grid and High Performance Computing 2, 60–73 (2010)
21. Russell, S.J., Norvig, P., Canny, J.F., et al.: Artificial intelligence: a modern approach. Prentice Hall, Englewood Cliffs (1995)
22. Ahmad, I., Kamruzzaman, J., Habibi, D.: Application of artificial intelligence to improve quality of service in computer networks. Neural Computing and Applications 21, 81–90 (2011)
23. Masten, M.K., Panahi, I.: Digital signal processors for modern control systems. Control Engineering Practice 5, 449–458 (1997)
24. Cerami, E.: Web Services Essentials: Distributed Applications with XML-RPC, SOAP, UDDI & WSDL, 1st edn. O'Reilly Media, Inc. (2002)
25. Fryazinov, O., Pasko, A., Adzhiev, V.: BSP-fields: An exact representation of polygonal objects by differentiable scalar fields based on binary space partitioning. Computer-Aided Design 43, 265–277 (2011)

Personalization of the Workplace through a Proximity Detection System Using User's Profiles

Carolina Zato, Alejandro Sánchez, Gabriel Villarrubia, Javier Bajo, Sara Rodríguez, and Juan F. De Paz

Departamento Informática y Automática, Universidad de Salamanca, Salamanca, Spain
{carol_zato,asanchezyu,gvg,jbajope,srg,fcofds}@usal.es

Abstract. This article presents a proximity detection prototype that will be included in the future in an integral system primarily oriented to facilitate the labor integration of people with disabilities. The main goal of the prototype is to detect the proximity of a person to a computer using ZigBee technology and then, to personalize its workplace according to his user's profile. The system has been developed as an open MultiAgent System architecture using the agent's platform PANGEA, a Platform for Automatic coNstruction of orGanizations of intElligent Agents.

Keywords: proximity detection, Zigbee, RTLS, open MAS, agent platform, personalization, user's profiles, disabled people.

1 Introduction

Due to the advance of the technologies and communications, intelligent systems have become an integral part of many people's lives and the available products and services become more varied and capable, users expect to be able to personalize a product or service to meet their individual needs and will no longer accept "one size fits all". Personalization can range from simple cosmetic factors such as custom ring-tones to the complex tailoring of the presentation of a shopping web site to a user's personal interests and their previous purchasing behaviour [15]. It is expected to expand these innovative techniques to a wide range of fields. One of the segments of the population, which will benefit with the advance of personalized systems, will be people with disabilities [16], contributing to improve their quality of life [17]. Considering the near future, public and private companies will be provided with intelligent systems specifically designed to facilitate the interaction with the human users. These intelligent systems will be able to personalize the services offered to the users, depending on their concrete profiles. It is necessary to improve the services' supply, as well as the way to offer them [18]. Technologies such as Multiagent Systems and Ambient Intelligence based on mobile devices have been recently explored as a system of interaction with the

L. Uden et al. (Eds.): 7th International Conference on KMO, AISC 172, pp. 505–513.
springerlink.com

dependent people [19]. These systems can provide support in the daily lives of dependent people [20].

This article presents a multi-agent based proximity detection prototype, specifically developed for a work environment, which can facilitate tasks such as activating and personalizing the work environment; these apparently simple tasks are in reality extremely complicated for some people with disabilities.

The rest of the paper is structured as follows: The next section introduces the detection proximity prototype, the technology used and how it works. Section 3 presents MAS in which the prototype is included. Section 4 explains the case study and finally, in section 5 some conclusions and future work are presented.

2 Detection Proximity Prototype

In this section we revise the proposed proximity detection prototype, focusing on the technology used and on the functioning of the prototype.

2.1 Technology Used

ZigBee sensors are used to deploy the detection prototype. ZigBee is a low cost, low power consumption, two-way wireless communication standard that was developed by the ZigBee Alliance [5]. It is based on the IEEE 802.15.4 protocol [2], and operates on the ISM (Industrial, Scientific and Medical) band at 868/915MHz and a 2.4GHz spectrum. Due to this frequency of operation among devices, it is possible to transfer materials used in residential or office buildings while only minimally affecting system performance [1]. Although this system can operate at the same frequency as Wi-Fi devices, the possibility that it will be affected by their presence is practically null, even in very noise environments (electromagnetic interference). ZigBee is designed to be embedded in consumer electronics, home and building automation, industrial controls, PC peripherals, medical sensor applications, toys and games, and is intended for home, building and industrial automation purposes, addressing the needs of monitoring, control and sensory network applications [5]. ZigBee allows star, tree or mesh topologies. Devices can be configured to act as network coordinator (control all devices), router/repeater (send/receive/resend data to/from coordinator or end devices), and end device (send/receive data to/from coordinator) [6]. One of the main advantages of this system is that, as opposed to GPS type systems, it is capable of functioning both inside and out with the same infrastructure, which can quickly and easily adapt to practically any applied environment.

Our prototype must allow performing efficient indoor locating in terms of precision because computers are very close one each other, for these reason the Real-Time Locating Systems (RTLS) Model was chosen. The infrastructure of a Real-Time Locating System contains a network of reference nodes called *readers* [12] and mobile nodes, known as *tags* [12][13]. Tags send a broadcast signal which includes a unique identifier associated to each tag. Then, readers obtain the identifier, as well as specific measurements of the signal. These measurements give information about the power of the received signal (*e.g.*, RSSI), its quality (*e.g.*, LQI, Link Quality Indicator), the Signal to Noise Ratio (SNR) or the Angle of Arrival (AoA) to the reader, amongst many others. These signals are gathered and processed in order to calculate the position of each tag.

RTLS can be categorized by the kind of its wireless sensor infrastructure and by the locating techniques used to calculate the position of the tags. This way, there is a range of several wireless technologies, such as RFID, Wi-Fi, UWB (Ultra Wide Band), Bluetooth and ZigBee, and also a wide range of locating techniques that can be used for determining the position of the tags [14].

2.2 How the Prototype Works

The proposed proximity detection system is based on the detection of presence by a localized sensor called the control point (where the ZigBeeReaderAgent is deployed), which has a permanent and known location. Once the Zigbee tag carried by the person has been detected and identified, its location is delimited within the proximity of the sensor that identified it. Consequently, the location is based on criteria of presence and proximity, according to the precision of the system and the number of control points displayed.

The parameter used to carry out the detection of proximity is the RSSI (Received Signal Strength Indication), a parameter that indicates the strength of the received signal. This force is normally indicated in mW or using logarithmic units (dBm). 0 dBm is equivalent to 1mW. Positive values indicate a signal strength greater than 1mW, while negative values indicate a signal strength less than 1mW.

Under normal conditions, the distance between transmitter and receiver is inversely proportional to the RSSI value measured in the receiver; in other words, the greater the distance, the lower the signal strength received. This is the most commonly used parameter among RTLS.

RSSI levels provide an appropriate parameter for allowing our system to function properly. However, variations in both the signal transmission and the environment require us to define an efficient algorithm that will allow us to carry out our proposal. This algorithm is based on the use of a steps or measurement levels (5 levels were used), so that when the user enters the range or proximity indicated by a RSSI level of -50, the levels are activated. While the values received are less than the given range, each measurement of the system activates a level. However, if the values received fall outside the range, the level is deactivated. When the maximum number of levels has been activated, the system interprets this to mean that the user is within the proximity distance of detection and wants to use the computer equipment. Consequently, the mechanisms are activated to remotely switch on both the computer and the profile specific to the user's disability.

The system is composed of 5 levels. The tags default to level 0. When a user is detected close to a reader, the level is increased one unit. The perceptible zone in the range of proximity gives an approximate RSSI value of -50. If the user moves away from the proximity area, the RSSI value is less than -50, resulting in a reduction in the level. When a greater level if reached, it is possible to conclude that the user has remained close to the marker, and the computer will be turned on.

On the other hand, reaching an initial level of 0 means that the user has moved a significant distance away from the workspace, and the computer is turned off.

The system uses a LAN infrastructure that uses the wake-on-LAN protocol for the remote switching on and off of equipment. Wake-on-LAN/WAN is a technology that allows a computer to be turned on remotely by a software call. It can be implemented in both local area networks (LAN) and wide area networks (WAN) [4]. It has many uses, including turning on a Web/FTP server, remotely accessing files stored on a machine, telecommuting, and in this case, turning on a computer even when the user's computer is turned off [7].

3 System Architecture

This proximity detection prototype is integrated within a open MAS that includes all the agents and information needed to create an integral system for helping disabled people in the workplace.

The open MAS has been created using PANGEA. There are many different platforms available for creating multiagent systems that facilitate the work with agents [8][9][10][11]; however our aim is to have a tool that allows users to create an increasingly open and dynamic multiagent system (MAS). PANGEA is a service oriented platform that allows to the implemented MAS to take the maximum advantage of the distribution of resources. To this end, all services are implemented as Web Services.

The own agents of the platform are implemented with Java, nevertheless the agents of the detection prototype are implemented in .NET and nesC.

Using PANGEA, the platform will automatically launch the following agents:

- OrganizationManager: the agent is responsible for the actual management of organizations and suborganizations. It is responsible for verifying the entry and exit of agents, and for assigning roles. To carry out these tasks, it works with the OrganizationAgent, which is a specialized version of this agent.
- InformationAgent: the agent is responsible for accessing the database containing all pertinent system information.
- ServiceAgent: the agent is responsible for recording and controlling the operation of services offered by the agents.
- NormAgent: the agent that ensures compliance with all the refined norms in the organization.
- CommunicationAgent: the agent is responsible for controlling communication among agents, and for recording the interaction between agents and organizations.
- Sniffer: manages the message history and filters information by controlling communication initiated by queries.

These agents interact with the specific agents of the detection prototype:

- ZigbeeManagerAgent: it manages communication and events and is deployed in the server machine.
- UsersProfileAgent: it is responsible for managing user profiles and is also deployed in the server machine.
- ClientComputerAgent: these are user agents located in the client computer and are responsible for detecting the user's presence with ZigBee technology, and for sending the user's identification to the ZigbeeManager Agent. These agents are responsible for requesting the profile role adapted for the user to the ProfileManagerAgent.
- DatabaseAgent: the detection proximity system uses a database, which stores data related to the users, sensors, computer equipment and status, and user profiles. It can also communicate with the InformationAgent of PANGEA.
- ZigBeeCoordinatorAgent: it is an agent included in a ZigBee device responsible for coordinating the other ZigBee devices in the office. It is connected to the server by a serial port, and receives signals from each of the ZigBee tags in the system.
- ZigBeeReaderAgent: these agents are included in several ZigBee devices that are used to detect the presence of a user. Each ZigBeeReaderAgent is located in a piece of office equipment (computer).

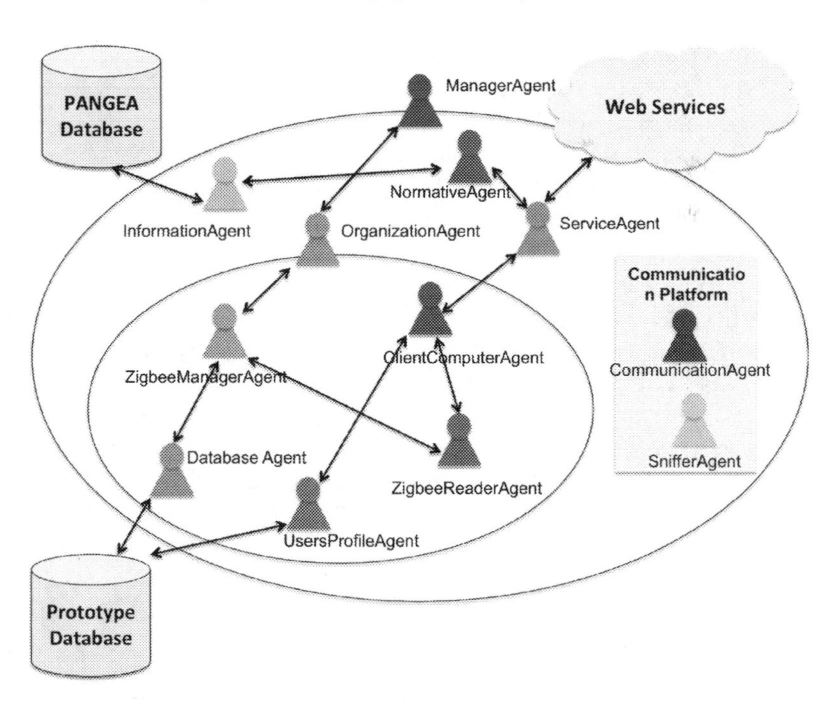

Fig. 1 System architecture

Every user in the proposed system carries a Zigbee tag, which is detected by a ZigBeeReaderAgent located in each system terminal and continuously in communication with the ClientComputerAgent. Thus, when a user tag is sufficiently close to a specific terminal (within a range defined according to the strength of the signal), the ZigBeeReaderAgent can detect the user tag and immediately send a message to the ClientComputerAgent. Next, this agent communicates the tag identification to the UsersProfileAgent, which consults the database to create the xml file that is returned to the ClientComputerAgent. After,, the ClientComputerAgent interacts with the ServiceAgent to invoke the Web Services needed to personalize the computer according to his profile.

4 Case Study

This paper presents a proximity detection system that is used by people with disabilities to facilitate their integration in the workplace. The main goal of the system is to detect the proximity of a person to a computer using ZigBee technology. This allows an individual to be identified, and for different actions to be performed on the computer, thus facilitating workplace integration: automatic switch on/off of the computer, identifying user profile, launching applications, and adapting the job to the specific needs of the user. Thanks to the Zigbee technology the prototype is notably superior to existing technologies using Bluetooth, infrareds or radiofrequencies, and is highly efficient with regards to detection and distance. Additionally, different types of situations in a work environment were taken into account, including nearby computers, shared computers, etc.

In our Case Study we have a distribution of computers and laptops in a real office environment, separated by a distance of 2 meters. The activation zone is approximately 90cm, a distance considered close enough to be able to initiate the activation process. It should be noted that there is a “Sensitive Area” in which it is unknown exactly which computer should be switched on; this is because two computers in close proximity may impede the system’s efficiency from switching on the desired computer. Tests demonstrate that the optimal distance separating two computers should be at least 40cm.

Figure 2 shows two tools that the system provides in the main server, where the ZigbeeManagerAgent is running. The screen shown above allows tracking the flow of events and controlling which computers are on and who are the users, identifying their tags and consulting the UsersProfileManagerAgent. Moreover, this tool allows executing applications or programs remotely. Figure 3 shows the screen that appears in the user’s computer when someone with a tag is close enough to it. The ClientComputerAgent running in this computer detects his presence and then, the computer is switched on with all the personal configurations.

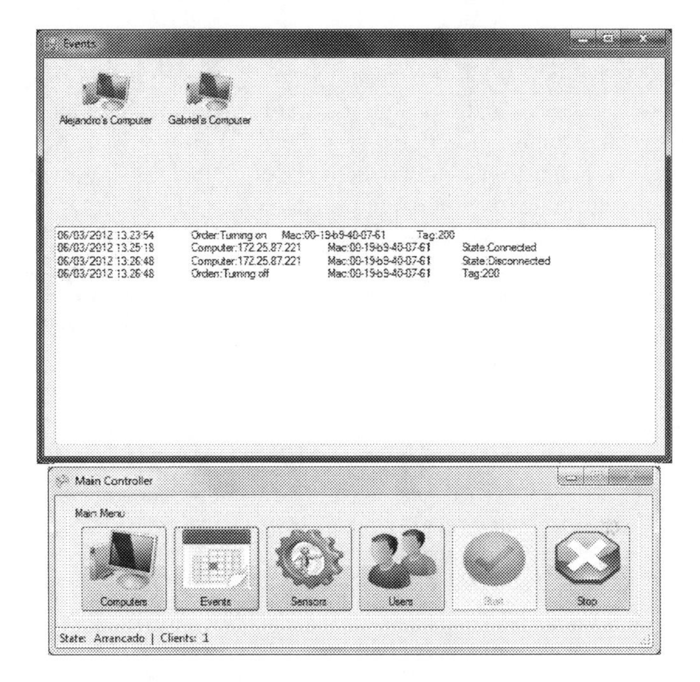

Fig. 2 Screenshot of the prototype in the main server

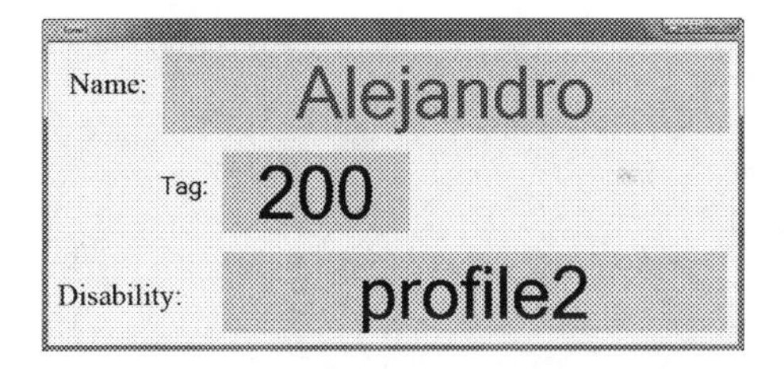

Fig. 3 Screenshot of the prototype in the main server

5 Conclusions

This prototype allows the detection and identification of a user making possible to detect any special needs, and for the computer to be automatically adapted for its use. This allows the system to define and manage the different profiles of people with disabilities, facilitating their job assimilation by automatically switching on or off the computer upon detecting the user's presence, or initiating a procedure that automatically adapts the computer to the personal needs of the user. This prototype is specifically oriented to facilitate the integration of people with disabilities into the workplace.

The prototype is part of complete and global project in which different tools for helping disabled people will be included. Using the PANGEA, that models all the services as Web Services and promotes scalability, the addition in the future of all those services that conform the global project will be easier. Some of these future services include pointer services, predictive writing mechanisms, adaptation for alternative peripheral, virtual interprets in language of signs, identification of objects by means of RFID, etc.

Acknowledgements. This project has been supported by the Spanish CDTI. Proyecto de Cooperación Interempresas. IDI-20110343, IDI-20110344, IDI-20110345, and the MICINN TIN 2009-13839-C03-03 project. Project supported by FEDER funds.

References

1. Huang, Y., Pang, A.: A Comprehensive Study of Low-power Operation in IEEE 802.15.4. In: Proceeding of the 10th ACM Symposium on Modeling, Analysis and Simulation of Wireless and Mobile Systems, s.n., Chaina (2007)
2. Singh, C.K., et al.: Performance evaluation of an IEEE 802.15.4 Sensor Network with a Star Topology (2008)
3. Universidad Pontificia de Salamanca. [En línea] (2011), http://www.youtube.com/watch?v=9iYX-xney6E
4. Lieberman, P.: Wake on LAN Technology, White paper (2011), http://www.liebsoft.com/pdfs/Wake_On_LAN.pdf
5. ZigBee Standards Organization: ZigBee Specification Document 053474r13. ZigBee Alliance (2006)
6. Tapia, D.I., De Paz, Y., Bajo, J.: Ambient Intelligence Based Architecture for Automated Dynamic Environments. In: Borrajo, D., Juan, L.C., Corchado, M. (eds.) CAEPIA 2007, vol. 2, pp. 151–180 (2011)
7. Nedevschi, S., Chandrashekar, J., Liu, J., Nordman, B., Ratnasamy, S., Taft, N.: Skilled in the art of being idle: reducing energy waste in networked systems. In: Proceedings of the 6th USENIX Symposium on Networked Systems Design and Implementation, Boston, Massachusetts, April 22-24, pp. 381–394 (2009)
8. Agent Oriented Software Pty Ltd. JACK™ Intelligent Agents Teams Manual. s.l.: Agent Oriented Software Pty. Ltd. (2005)
9. Hübner, J.F.: J -Moise+ Programming organisational agents with Moise+ & Jason. Technical Fora Group at EUMAS 2007 (2007)
10. Giret, A., Julián, V., Rebollo, M., Argente, E., Carrascosa, C., Botti, V.: An Open Architecture for Service-Oriented Virtual Organizations. In: Braubach, L., Briot, J.-P., Thangarajah, J. (eds.) ProMAS 2009. LNCS, vol. 5919, pp. 118–132. Springer, Heidelberg (2010)
11. Galland, S.: JANUS: Another Yet General-Purpose Multiagent Platform. Seventh AOSE Technical Forum, Paris (2010)
12. Liu, H., Darabi, H., Banerjee, P., Liu, J.: Survey of Wireless Indoor Positioning Techniques and Systems. IEEE Transactions on Systems, Man, and Cybernetics, Part C: Applications and Reviews 37(6), 1067–1080 (2007)

13. Tapia, D.I., De Paz, J.F., Rodríguez, S., Bajo, J., Corchado, J.M.: Multi-Agent System for Security Control on Industrial Environments. International Transactions on System Science and Applications Journal 4(3), 222–226 (2008)
14. Tapia, D.I., Bajo, J., De Paz, J.F., Alonso, R.S., Rodríguez, S., Corchado, J.M.: Corchado Using Multi-Layer Perceptrons to Enhance the Performance of Indoor RTLS. In: Progress in Artificial Intelligence - EPIA 2011. Workshop: Ambient Intelligence Environmets (2011)
15. Pluke, M., Petersen, F., Brown, W.: Personalization and User Profile Management for Public Internet Access Points (PIAPs). CiteSeerX Scientific Literature Digital Library. Online Resource (July 2009), doi: 10.1.1.117.8111
16. Carretero, N., Bermejo, A.B.: Inteligencia Ambiental. CEDITEC: Centro de Difusión de Tecnologías, Universidad Politécnica de Madrid, España (2005)
17. Corchado, J.M., Bajo, J., Abraham, A.: GERAmI: Improving the delivery of health care. IEEE Intelligent Systems 23(2), 19–25 (2008)
18. Macarro, A., Bajo, J., Jiménez, A., de la Prieta, F., Corchado, J.M.: Learning System to Facilitate Integration through Ligthweigth Devices. In: Proceedings FUSION 2011, Chicago, US (2011) ISBN: 978-0-9824438-1-1
19. Anastasopoulos, M., Niebuhr, D., Bartelt, C., Koch, J., Rausch, A.: Towards a Reference Middleware Architecture for Ambient Intelligence Systems. In: ACM Conference on Object-Oriented Programming, Systems, Languages, and Applications (2005)
20. Ranganathan, V.K., Siemionow, V., Sahgal, V., Yue, G.H.: Effects of aging on hand function. Journal of the American Geriatrics Society 49, 1478–1484 (2001)

A Visualization Tool for Heuristic Algorithms Analysis

Laura Cruz-Reyes[1], Claudia Gómez-Santillán[1], Norberto Castillo-García[1], Marcela Quiroz[1], Alberto Ochoa[2], and Paula Hernández-Hernández[1]

[1] Instituto Tecnológico de Ciudad Madero, División de Estudios de Posgrado e Investigación. Juventino Rosas y Jesús Urueta s/n, Col. Los mangos, C.P. 89440, Cd.Madero, Tamaulipas, México
lcruzreyes@prodigy.net.mx, {cggs71,norberto_castillo15, paulahdz314}@hotmail.com, qc.marcela@gmail.com,
[2] Instituto de Ingeniería y Tecnología, Universidad Autónoma de Ciudad Juárez. Henry Dunant 4016, Zona Pronaf, C.P. 32310, Cd. Juárez, Chihuahua, México
doctor_albertoochoa@hotmail.com

Abstract. The performance of the algorithms is determined by two elements: efficiency and effectiveness. In order to improve these elements, statistical information and visualization are key features to analyze and understand the significant factors that affect the algorithm performance. However, the development of automated tools for this purpose is difficult. In this paper a visual diagnosis tool named VisTHAA, which provides researchers statistical and visual information about instances and algorithms, is proposed. Besides, VisTHAA allows researchers to introduce characterization measurements, make algorithm comparisons with a non-parametric test and visualize the search space ruggedness and the behavior of the algorithm. Due to the above, in this study we used VisTHAA as a tool for improving the efficiency of a reference algorithm in the literature. In the study case, the experimental results showed that through visual diagnosis it was possible to increment the algorithm's efficiency to 93% without a considerable loss of effectiveness.

Keywords: Algorithm's Performance, Visualization, VisTHAA.

1 Introduction

The objective of experimental algorithmics is to promote that the experiments are relevant, accurate and reproducible in order to produce knowledge to improve the design of algorithms and their performance accordingly [1]. Nowadays, it is not enough to know that an algorithm is the best in solving a particular set of instances, it is also important to understand and explain formally its behavior. The performance of heuristic algorithms depends on many factors including design quality; thus, a poor design can lead to a poor performance. There are no rules or guidelines widely accepted to design properly heuristics. Particularly, researchers

L. Uden et al. (Eds.): 7th International Conference on KMO, AISC 172, pp. 515–524.
springerlink.com

consume too much time to fine-tune an algorithm; usually they adjust their control parameters manually by trial and error requiring a considerable time.

In this paper VisTHAA (Visualization Tool for Heuristic Algorithm Design) is introduced, which is a modular visual diagnostic tool that will allow other researchers extend its features and functionalities thorough modules integrated into an architecture to facilitate the analysis of heuristics. The current version of VisTHAA has the following modules available: Input and Pre-processing data, characterization of the instances, visualization of the instances and the algorithm's behavior, visualization of the three-dimensional search space, and the algorithm comparative analysis.

In the literature there are other tools similar to VisTHAA [2, 3] however none of them are able to allow the researchers to introduce their own characterization measurements or to visualize the search space in three dimensions. In the study case we use some features of VisTHAA, which make our tool different from the other ones, to solve the optimization problem under study: the Bin-Packing Problem (BPP).

The remaining paper is organized as follows. In Section 2 we show a detailed description of the proposed tool, emphasizing their main modules and the interaction among them. In Section 3 a study case is presented, the used methodology aims to identify areas of improvement, which allows the algorithm adjustment. Section 4 shows the analysis of the experimental results, using the well-known Wilcoxon Test to validate statistically the results. Finally, we discuss the most important remarks and some future work in Section 5.

2 VisTHAA

VisTHAA is a visual tool applied to the analysis of heuristic algorithms. Figure 1 shows only the modules considered for this initial release. The most innovative part are the characterization modules, which can be applied to the problem,

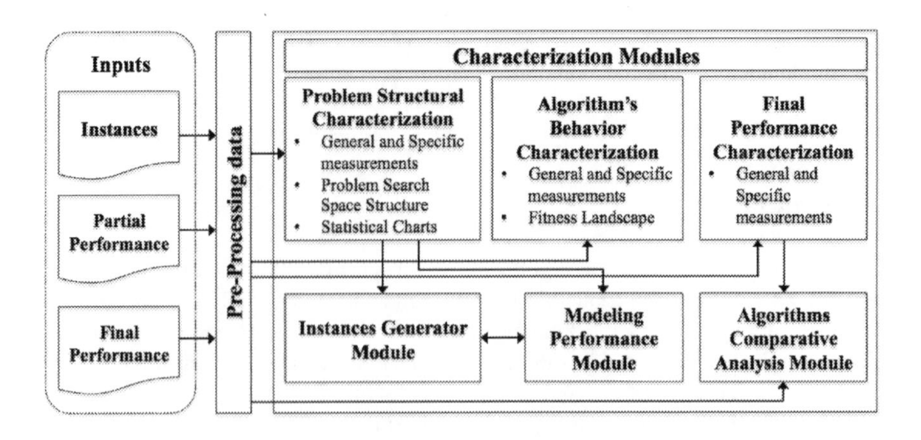

Fig. 1 Architecture of VisTHAA

algorithm, or final results, and contribute at every phase of the heuristic optimization process by identifying which factors have a significant influence on performance. This tool is available from https://sth-se.diino.com/lauracruzreyes/VisTHAA/ExecutableCode.

2.1 Input and Pre-processing Data Module

Researchers use their own formats for reading and processing data stored in files. When they want to analyze their data, they need to adjust their files to the input formats required for the available tool. It is complicated for them to change the format of theirs files. In the design of VisTHAA, the inputs may represent instances of a problem, algorithm data or performance data. In order to simplify the researcher's work, we design a preprocessing module. In the case of the input of the instances, different formats and different problems can be read.

2.2 Characterization Module

This module includes a set of options to characterize and quantify the factors that define the problem structure as well as partial and final algorithm performance. As we can see in Figure 1, in each characterization option exists a set of general-purpose measurements, in order to identify which factors impact on the heuristic optimization process, i.e., *mean*, *variance* and *mode*. Another contribution that is closely linked to the characterization module is the use of a calculator, which shows the instance parameter and general-purpose measurements. From this stored information the researcher is allowed to introduce new characterization measurements in infix notation. This module uses the shunting yard algorithm [4]. In the final performance module are available measurements proposed by Quiroz and Cruz [5, 10] to quantify the quality of the solutions found by the algorithm.

a) **Search Space Structure:** Represent the objective function domain elements that are created randomly [6]. Search space structure is obtained by a random walk algorithm, which gives exactly the same probability of choosing to each possible solution of the instance, this is because random walk is a blind search technique, so it does not prefer search space areas whit the greatest potential of finding the optimum, in [7] a detailed explanation of random walk can be found. Representing the search space through three-dimensional visualization techniques, helps the researcher to analyze how the instance structure is, that is, it is possible to see if the instance is rough or not, and make decisions about intensification and diversification strategies to be applied to each instance.
b) **Statistical Charts:** Through them it is possible to visualize characteristics that describe the problem factors and in some cases identify patterns in them. VisTHAA allows the researchers use frequency charts, which characterize the distribution of values in a dataset. The scatter plots display the correlation between a pair of variables.
c) **Algorithm's Behavior Fitness Landscape:** This kind of graphic shows the algorithm's behavior during its execution, displaying the fitness values obtained in each iteration giving the researcher visual arguments to identify if

the algorithm is improving or not, and then adjust the diversification and intensification strategies. A formal definition of fitness landscape can be found in [8].

d) Algorithms Comparative Analysis Module: In this module the performance of two algorithms is evaluated. This evaluation is made through the Wilcoxon test, which is a non-parametric procedure that determines whether or not the differences between two datasets are statistically significant [9].

3 Study Case

The study case has a three-main-phases methodology: 1) data input, 2) instance characterization, and 3) instance visualization. Once these phases are concluded, if areas of improvement are observed, a redesign phase is performed. Finally the results are validated using the Wilcoxon test, which allows verifying if the results are better statistically.

The optimization problem under study is the Bin Packing Problem (BPP), which is one of the most studied optimization problems. BPP is classified as a NP-hard combinatorial optimization problem [10]. BPP consists in finding an objects distribution that minimizes the number of bins used. In practice, this kind of problems has been solved by heuristic approaches; the Weight Annealing Heuristic (WABP) proposed by Loh et al. [11], is one of them. Basically, WABP has four steps. The first step initializes the parameters; then, the second step constructs and initial solution using the FFD procedure; the third step is the main, it consists in trying to improve iterative the current solution through a set of swaps between some items from different bins; finally, the last step outputs the results.

3.1 *Phase 1: Introducing the Instances to VisTHAA*

In order to introduce the instances to VisTHAA, two additional files are required. The first one is named *metainstance*, which describes the format of the instances. The second one is named *logbook*, which is used to specify the number of instances to be introduced, the name of the file that describes the instances, and the name of each one of them. Figure 2 shows a BPP instance, the *metainstance* and the *logbook* files.

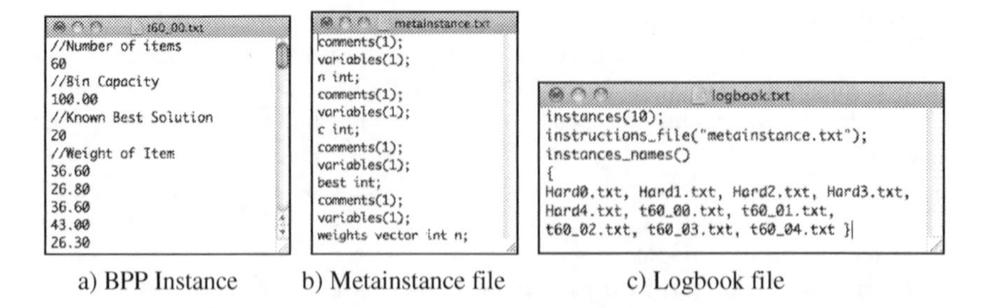

a) BPP Instance b) Metainstance file c) Logbook file

Fig. 2 Example of description of instances files for their correct introduction to VisTHAA

3.2 Phase 2: Characterization of BPP

After the instances are loaded, the researcher can perform the characterization of BPP instances. The statistical characterization is made through measurements. So far, it is possible to select between basic statistics or specialized measurements for BPP reported in literature. VisTHAA has implemented some of the specialized measurements for BPP [5, 10], which can be utilized to gain knowledge of the instances. Nevertheless, VisTHAA gives the researchers the opportunity to introduce their own measurements from variables already defined in the metainstance file.

Figure 3 shows an example of how to introduce a new measurement, in this case, the *normalized_range*. Once a set of measurements is introduced, the researcher can decide which of them to consider for generating the attributes matrix over the set of selected instances; the purpose is to obtain information of the calculated values over a set of instances.

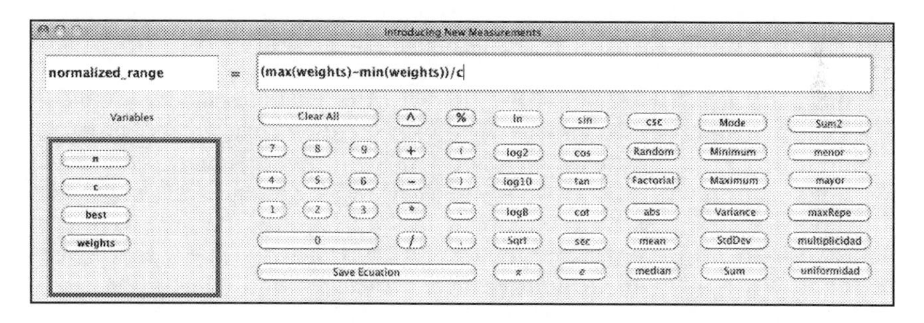

Fig. 3 Module to introduce new measurements to VisTHAA

Table 1 shows the attributes matrix over the loaded instances. It is observed that, for the measured attributes, there are significant differences among the analyzed instances; this is because the instances belong to a two different datasets. The most evident attribute is *uniformity* (attribute 5 in Table 1), which measure the level of uniformity of the weights distribution. In average, the instances belonging to the Hard dataset are highly uniform, with a value of 0.926 (92.6%), whereas the instances belonging to the T dataset are not so uniform, only a 0.473 (47.3%).

Table 1 Attributes matrix over the instances to be optimized

Instance	*Att1*	*Att2*	*Att3*	*Att4*	*Att5*	*Att6*	*Att7*	*Att8*
Hard0	0.201	0.349	1.005	2	0.929	0.272	0.042	0.148
Hard1	0.200	0.349	1.000	1	0.950	0.276	0.043	0.149
Hard2	0.200	0.349	1.005	2	0.890	0.277	0.044	0.149
Hard3	0.200	0.347	1.015	2	0.940	0.272	0.042	0.147
Hard4	0.201	0.350	1.010	3	0.920	0.277	0.041	0.148
t60_00	0.251	0.495	1.200	3	0.466	0.333	0.072	0.244
t60_01	0.251	0.475	1.071	2	0.550	0.333	0.070	0.224
t60_02	0.250	0.498	1.250	3	0.466	0.333	0.081	0.248
t60_03	0.250	0.495	1.224	2	0.450	0.333	0.079	0.245
t60_04	0.250	0.498	1.250	3	0.433	0.333	0.078	0.248

From Table 1, the symbolism *Att* of the measured attributes represents: 1) *minor*, 2) *mayor*, 3) *multiplicity*, 4) *max repetitions*, 5) *uniformity*, 6) *normalized mean*, 7) *normalized standard deviation*, 8) *normalized range*. In the experiments of the next two sections only is considered the characteristic of uniformity.

3.3 Phase 3: Visualization of Instances and Algorithm Behavior

All of the experiments in this section where performed under the next hardware specifications: Intel Atom processor at 1.67 GHz and 1 GB of RAM. In order to analize WABP heuristic and two variants, instances from recognized benchamrks were used, these instances can be found in a well-known Internet sites [12, 13].

The algorithm behavior of WABP is analized primarily with visual support. The initial configuration of this algorithm is: scaling = 0.05; iteration number = 50; initial temperature = 1; and the temperature reduction = 0.05.

The visualization of the search space is another way to characterize the problem structure. The literature states that the ruggedness of an instance is related with its difficulty [5].

To build a graph of the search space, first we extract a representative sample of the different solutions of the instance with their corresponding fitness, through a random walk. The resulting one-dimensional data set (fitness values) is converted into a two-dimensional set through performing "K" partitions of length "I". Finally, for each solution, we get the value of the X-axis (the position relative to K), the value of the Y-axis (the position relative to I) and the value of the Z-axis (the fitness of the solution). With these three variables can be graphed the search space in three dimensions. This simplified graph shows the neighborhood structure partially.

Figure 4 shows two graphics of the fitness landscapes that correspond to two types of instances used in the experiment. As we can see, Figure 4a is not rugged, which implies that the solutions of that instance are very similar to each other. On the other hand, Figure 4b represents a rugged instance. Moreover, the instances belonging to the Hard dataset have similar fitness landscapes graphics; and it is the same with the T dataset instances. Then the Hard dataset is uniform and produces a not rugged surface, while The T dataset is not uniform and produces a rugged surface.

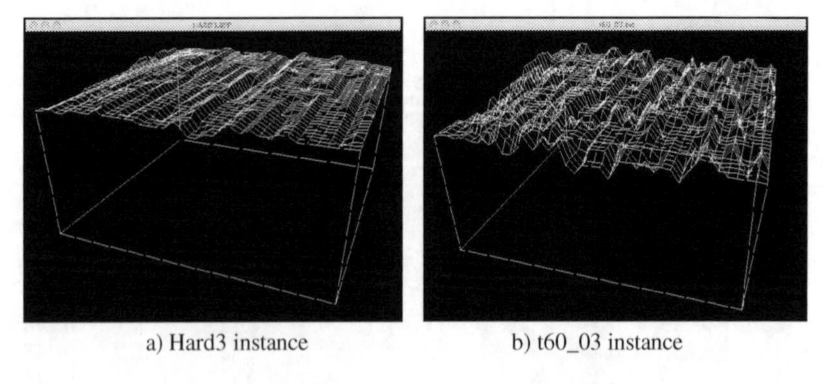

a) Hard3 instance b) t60_03 instance

Fig. 4 Fitness landscapes visualization of two instances in VisTHAA

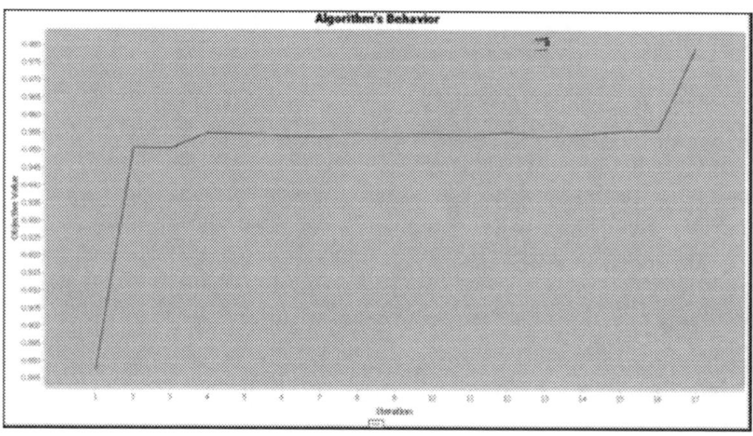

a) Hard3 instance

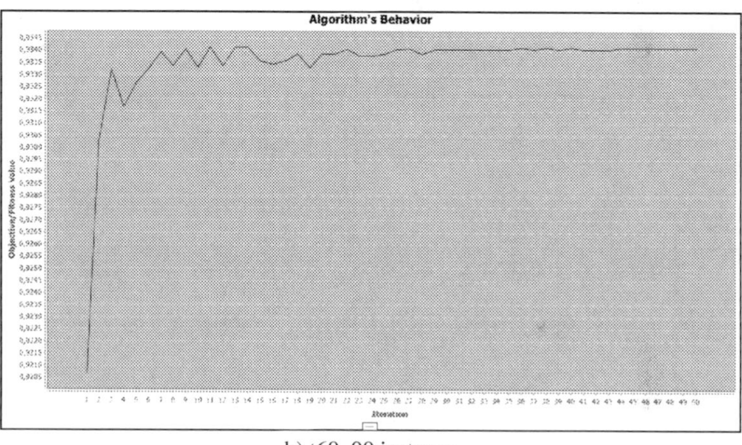

b) t60_00 instance

Fig. 5 Behavior of WABP Algorithm with two instances

3.4 Algorithm Redesign Phase

According to the visual analysis of the number of iterations without improvement, for the algorithm under study, two possible values for *nloop* parameter were considered; where *nloop* is the number of iterations of the algorithm. The first one was obtained by recognizing the following pattern: except for hard3, the remaining instances converge in early iterations. Moreover, Hard3 instance is the last to converge (17 iterations), which is why the maximum limit number of 17 iterations was set for the first reconfiguration. The second number of maximal iterations was obtained by computing the average of the number of iteration where each instance achieves its last improvement, which was 3.1 iterations; consequently, a maximum limit number of 3 iterations was used for the second reconfiguration.

4 Experimental Results

Table 2 shows that with the initial algorithm configuration, WABP finds four of ten optimum values, which add up to 381 used bins. With the first algorithm reconfiguration it can be seen that WABP achieves a considerable saving of 297 iterations, reflecting an improvement of 63% in computer time, without a loss in the quality of the solutions, that is, WABP find the same objective values for every instance. Analyzing the algorithm's performance with the second reconfiguration, it can be noted that a bigger saving of 437 iterations, which represents the 93% in computer time, is achieved. However, there is a slight loss of quality of 0.26%.

In order to analyze the obtained results and to validate statistically the forcefulness in the improvement of the algorithm performance, the Wilcoxon test was performed. VisTHAA has available this non parametric test.

Table 2 Experimental results decreasing the number of maximal iterations

		***nloop*=50**		***nloop*=17**		***nloop*=3**	
Instances	*Optimum*	*Bins*	*Iterations*	*Bins*	*Iterations*	*Bins*	*Iterations*
Hard0	56	56	50	56	17	56	3
Hard1	57	57	50	57	17	57	3
Hard2	56	57	50	57	17	57	3
Hard3	55	55	17	55	17	56	3
Hard4	57	57	50	57	17	57	3
t60_00	20	21	50	21	17	21	3
t60_01	20	21	50	21	17	21	3
t60_02	20	21	50	21	17	21	3
t60_03	20	21	50	21	17	21	3
t60_04	20	21	50	21	17	21	3
TOTAL	381	387	467	387	170	388	30

Due to three experiments were conducted (*nloop*=50, *nloop*=17, *nloop*=3) and the Wilcoxon test only deals with two samples, it was necessary do the test twice, the first one with the results of the initial algorithm configuration against the first reconfiguration, and the second one with the results of the first reconfiguration against the second reconfiguration. Both, the first and the second Wilcoxon tests performed revealed that there are significant differences with a 99.5% of confidence.

In addition, Table 2 shows that the reference algorithm WABP, over the instances of the T dataset, could not find the optimum value for any. On the other hand, the instances belonging to the Hard dataset were a little easier to solve by WABP. This could be explained by the instance structure, that is, the fitness landscapes graphics of the instances belonging to the T dataset, in general, are more rugged than the other ones corresponding to the Hard dataset. According to the literature, the ruggedness of an instance is related with its difficulty, in our study, we could observe that characteristic, since the T dataset instances were the more rugged and the most difficult to solve by WABP.

5 Conclusions and Future Work

VisTHAA is a tool for analyzing the behavior of heuristic algorithms during the optimization process. Although VisTHAA is in its initial stage, includes relevant facilities for handling generic instances, statistical and visual characterization of the optimization process and statistical comparison of heuristic strategies.

By analyzing WABP algorithm, the feasibility of VisTHAA was validated, finding that after to characterize in different ways the instances and the algorithm behavior, the visual diagnosis was important to identify what kind of strategy we must follow in order to improve the algorithm performance. The graphics provide extra information than can be complemented with the attributes matrix. Specifically, analyzing the behavior graphic we realized that WABP could improve its performance by reconfiguring one of its control parameter. Two reconfigurations were performed, which were better than its predecessor in efficiency. The first adjustment allowed the significant improvement of the initial configuration of the 63%, without a loss in the quality of the solutions. The second adjustment allowed the significant improvement of the 93%, with an insignificant loss of 0.26%.

As future work, experimentation with different parameter settings is considered, as well as tests other heuristics and try to identify other factors that are crucial for the algorithm performance. Besides, it is important to identify which characteristics of the instances make them difficult to solve for WABP, once identified them, it could be possible develop improvement strategies. Preliminary, instances with high uniformity causes a not rugged search surface for WABP.

We have also considered the expansion of VisTHAA with new functionality, including on-line operation. Currently the process is off-line by requiring researches provide all the data for the optimization process. With the on-line operation, the algorithms can be run on the tool.

References

1. McGeoch, C.C.: Experimental Analysis of Algorithms. In: Pardalos, P.M., Romeijn, H.E. (eds.) Handbook of Global Optimization, vol. 2, pp. 489–513 (2000)
2. Chuin, H., Chong, W., Halim, S.: Tuning Tabu Search Strategies via Visual Diagnosis. In: MIC 2005 The 6th Metaheuristics International Conference. School of Information Systems, Singapore Management University (2005)
3. Jin, M.V.: Visualizing Swarm algorithms. Master's Thesis. Faculteit Toegepaste Wetenschappen (2004)
4. Dijkstra, E.W.: ALGOL-60 Translation. Stichting Mathematisch Centrum. 2e Boerhaavestraat 49, Amsterdam, Rakenafdeling (1961)
5. Quiroz, M.: Caracterización de Factores de Desempeño de Algoritmos de Solución de BPP. Master's thesis. Instituto Tecnológico de Cd. Madero, México (2009)
6. Bartz-Beielstein, T., Chiarandini, M., Paquete, L. (eds.): Experimental Methods for the Analysis of Optimization Algorithms, 1st edn., 457p. Springer, Heidelberg (2010)
7. Gómez, C.G., Cruz-Reyes, L., Meza, E., Schaeffer, E., Castilla, G.: A Self-Adaptative Ant Colony System for Semantic Query Routing Problem in P2P Networks. Computación y Sistemas 13(4), 433–448 (2010)

8. Gendreau, M., Potvin, J.Y.: Handbook of Metaheuristics, 2nd edn. International Series in Operations Research & Management Science, vol. 146. Springer (2010) ISBN: 978-1-4419-1663-1, e-ISBN: 978-1-4419-1665-5
9. Mendenhall, W., Sincich. T.: Statistics for Engineering and the Science, 4th edn. Prentice-Hall (1997)
10. Cruz-Reyes, L.: Clasificación de Algoritmos Heurísticos para la solución de Problemas de Bin Packin. PhD thesis, Centro Nacional de Investigación y Desarrollo Tecnológico, Cuernavaca, Morelos, México (2004)
11. Loh, K.H., Golden, B., Wasil, E.: Solving the one-dimensional bin packing problem with a weight annealing heuristic. Computers and Operations Research 35(2008), 2283–2291 (2006)
12. Klein, R., Scholl, A.: Bin Packing, Allowed in, `http://www.wiwi.uni-jena.de/Entscheidung/binpp/`
13. Beasley, J.E.: OR-library: Distributing test problems by electronic mail. Journal of the Operational Research Society 41, 1069–1072, Allowed in, `http://people.brunel.ac.uk/~mastjjb/jeb/orlib/binpackinfo.html`

QuPreSS: A Service-Oriented Framework for Predictive Services Quality Assessment

Silverio Martínez-Fernández[1], Jesús Bisbal[2], and Xavier Franch[1]

[1] GESSI, Universitat Politècnica de Catalunya (UPC), C/Jordi Girona 1-3, 08034, Barcelona, Spain
{smartinez,franch}@essi.upc.edu
[2] Dept. ICT, Universitat Pompeu Fabra (UPF), C/Tánger 122-140, 08018, Barcelona, Spain
jesus.bisbal@upf.edu

Abstract. Nowadays there are lots of predictive services for several domains such as stock market and bookmakers. The value delivered by these services relies on the quality of their predictions. This paper presents QuPreSS, a general framework which measures predictive service quality and guides the selection of the most accurate predictive service. To do so, services are monitored and their predictions are compared over time by means of forecast verification with observations. A systematic literature review was performed to design a service-oriented framework architecture that fits into the current body of knowledge. The service-oriented nature of the framework makes it extensible and interoperable, being able to integrate existing services regardless their heterogeneity of platforms and languages. Finally, we also present an instantiation of the generic framework architecture for the weather forecast domain, freely available at http://gessi.lsi.upc.edu/qupress/.

1 Introduction

A service system is a dynamic configuration of people, technologies, organisations and shared information that create and deliver value to customers and other stakeholders [6]. Examples of customers receiving a service are: a person taking a bus to go somewhere or going to a restaurant to have a meal, or an information technology (IT) company subcontracting the information backup process to another company in order to save costs and time.

Service-oriented architecture (SOA) has become in the last years the most usual strategy to implement service systems. Basically, this emerging development paradigm allows service providers to offer loosely coupled services. These services are normally owned only by the providers. As a result, the service customer (or client or user) does not need to worry about the development, maintenance, infrastructure, or any other issue related to the service operation. Instead, the customer just has to find and choose the most appropriate service.

L. Uden et al. (Eds.): 7th International Conference on KMO, AISC 172, pp. 525–536.
springerlink.com

Due to the success of service technology, there is a big offer of services covering many domains. Therefore it is needed to assess which service is the most appropriate for the customer's needs. Each customer has different needs, which require to be satisfied by the provider: quality, reputation, cost, security, personalisation, locality and so on.

Among all kind of services, we concentrate on predictive (or forecasting) services. We define *predictive services* as those services whose main functionality is to show in advance a condition or occurrence about the future. Predictive services emerge in a lot of domains to predict: stock market prices, results in bookmakers, election polls, sales forecasting and so on. We are quite used to see this functionality offered by websites that provide forecasts over specific data [15], but they are not so often exposed as web services (WS).

An example that is really familiar to all of us is weather forecast. Weather conditions affect our decisions in daily routines such as deciding what to wear first thing in the morning or planning a trip. To make these decisions, different services like the weather forecast section on TV news or specialised web sites (such as forecastadvisor.com and forecastwatch.com) are consulted. However, sometimes their predictions do not agree and therefore a fundamental question arises:

> Given a portfolio of candidate predictive services, which one is expected to be the most accurate to satisfy some given user needs?

In order to answer the question above, two related activities become necessary. On the one hand, *service monitoring* by using a monitoring infrastructure that observes the behaviour of such predictive services [3]. On the other hand, *forecast verification*, defined as the process of assessing the quality of a forecast by comparing it with its corresponding observation [14]. Both activities need to exist together: forecast verification cannot be implemented without monitoring real data, and it makes no sense to establish a monitoring infrastructure if quality analysis is not done. Therefore, a general framework that performs both activities is needed in order to properly answer the question above.

This paper presents an approach to evaluate the quality of predictive services. It has four objectives, which are respectively covered in the following four sections, that are: (1) to understand the state-of-the-art in predictive services quality assessment; (2) to identify gaps in current research about this problem; (3) to propose a generic framework software architecture to support predictive service monitoring, verification and selection; and (4) to develop an instantiation of the generic framework software architecture as a proof-of-concept in the weather forecast domain.

2 State of the Art in Predictive Services

To make a thorough and unbiased state-of-the-art about forecast verification for predictive services, we have performed a Systematic Literature Review (SLR) following the guidelines of [9]. In Section 2.1, an excerpt of the systematic review protocol is presented. A summary of the results is reported in Section 2.2. To see all the details and the rest of the SLR, the reader is referred to [11].

2.1 Systematic Literature Review Protocol

The protocol consists of three main stages: identifying the research questions that the review is intended to answer, formulating the strategy to search for primary studies, and determining the study selection criteria and procedures.

Research questions (RQ). The aim of this review consists of identifying the existing research on predictive services in order to measure how accurate their predictions are. It addresses the following questions:

RQ1. Are there existing SOA frameworks which measure predictive service quality?
RQ2. In which domains such frameworks are being applied?
RQ3. Which are the main criteria used to evaluate predictive services?

Search strategy. Our primary interest is finding service-oriented frameworks for forecasting. We started then with a search string with two parts: different synonymous for "forecast" looking for the functionality, and three different keywords for service orientation (the general term "service-oriented", the architecture "SOA" and the technology "web service") covering the target platform:

1. (forecast OR forecasting OR foresight OR foretell OR foretelling OR forethought OR predict OR prediction OR predictive OR predictability OR predicting OR prognosis OR prognosticate OR prognostication OR prevision OR anticipation OR outlook) AND ("service-oriented" OR SOA OR "web service")

However, the systematic search with this string was not promising. We understood that the search could be too specific so we decided to complement this search string with another one more general, looking for state-of-the-art papers on service orientation that can include thus forecasting as part of the results:

2. (SOA OR "service-oriented") AND ("systematic review" OR survey OR "state-of-the-art")

Study selection. We rigorously identified 213 studies from the literature with the search strategy, of which 18 passed the selection criteria (which is based on three stages processing title, abstract and full text, respectively) and quality assessment. The included studies, which are presented in Table 1, covered three topics: SOA, monitoring and prediction in SOA systems. The column document type indicates if the paper is a journal article (JA), a conference paper (CP), a technical report (TR), or a master thesis (MT).

Table 1 Primary studies considered in the systematic literature review

Authors	Year	Type	Doc. Type	Ref.
L.B.R. de Oliveira et al.	2010	SOA	JA	[16]
M.H. Valipour et al.	2009	SOA	CP	[21]
M.P. Papazoglou et al.	2007	SOA	JA	[19]
U. Zdun et al.	2006	SOA	JA	[23]
G. Canfora et al.	2009	Monitoring	CP	[2]
M. Oriol	2009	Monitoring	MT	[17]
A.T. Endo et al.	2010	Monitoring	TR	[5]
A. Bertolino	2007	Monitoring	CP	[1]
N. Delgado et al.	2004	Monitoring	JA	[3]
R. Guha	2008	Prediction	JA	[7]
S. Punitha et al.	2008	Prediction	CP	[20]
M. Marzolla et al.	2007	Prediction	CP	[12]
H.N. Meng et al.	2007	Prediction	CP	[13]
D.M. Han et al.	2006	Prediction	CP	[8]
C.B.C. Latha et al.	2010	Prediction	JA	[10]
N. Xiao et al.	2008	Prediction	JA	[22]
A.H. Murphy	1993	Prediction	JA	[14]
B. Domenico	2007	Prediction	TR	[4]

2.2 Review Results

The data extraction and synthesis of knowledge arisen from the SLR are discussed below in relation to the three research questions.

RQ1: SOA Frameworks which measure predictive service quality. The literature review found only one service-oriented framework which measures predictive service quality [4]. In this work, geosciences web services are used to integrate sources of data. Once the data is collected, they perform forecast verification with observations.

Nevertheless, the SLR located other service-oriented frameworks that monitor and/or predict but do not consider forecast quality.

There are numerous monitoring frameworks to get quality of services (QoS) and to alert when a service level agreement (SLA) between a provider and a customer is predicted to be violated (for details, the reader is referred to the related work section of [18]). Some of these monitoring frameworks have their own prediction models to predict the performance of services in order to balance the workload in a composite SOA. Other studies predict the performance of services either to help users during the selection process of a provider [12] or to improve the reliability of services using these performance predictions (e.g., to detect bottlenecks at design time [20], to promote efficiency [22] or to avoid a service crash because of exceeding memory usage [13]).

On the other hand, predictions models which require huge amounts of data from different sources and vast computing power are starting to use SOA due to its benefits. We found examples in the weather forecast domain [10], macroeconomic domain [8] and drug design domain [7].

RQ2: Predictive domains. The only domain in which we have found existing SOA frameworks that measure predictive services quality is weather forecast [4].

Moreover, we found other composite SOA frameworks and simple WS which do not measure how accurate predictive (web) services are, but perform predictions. Their domains were: weather forecast [10], macroeconomic analysis [8], drug design [7] and QoS prediction with the final goal of improving composition of services and service's reliability [22]. It is worth to note other relevant predictive domains that we found in the excluded studies but whose systems are not service-oriented [11]: business and economic forecasting, automotive forecasting, prediction of flight delays, prediction of protein's issues in medicine, sales forecasting for the calculation of safety stock and results in betting shops.

RQ3: Criteria to evaluate predictive services. There are three criteria that determine if a forecast is good (or bad), namely consistency, quality and value [14]. Consistency is the correspondence between judgments and forecasts. There exists a difference between judgments (forecaster's internal assessment which is only recorded in his/her mind) and forecasts (forecaster's external spoken/written statements regarding future conditions). Quality is the correspondence between forecasts and observations (i.e., the accuracy of the forecasts). A forecast has high quality if it predicts the observed conditions well according to some aspect like accuracy and skill. Value refers to the incremental benefits of forecasts to users. A forecast has value if it helps the user to make a better decision.

In this paper, we focus on quality because of three reasons. First, it is useful in the forecast verification process. Second, consistency could not be entirely projected onto a software prototype, since it contains an element of uncertainty and subjectivity by definition. Third, value requires knowledge about the impact of predictions which is out of the IT scope of this paper.

SLR Conclusion. As a main result, we can conclude that forecast verification for predictive services has not received sufficient attention and it justifies the focus of further research. More effort is necessary to integrate methods for forecast verification in SOA monitoring frameworks. This has motivated our work.

3 Predictive Services Quality

As we mentioned in Section 1, given a portfolio of candidate predictive services, we aim to find the most accurate to satisfy some given user needs. To this end, we need the following four inputs, which are summarised in Fig. 1:

- Predictive services. The set of sources that give predictions.
- Ground truth service. It refers to the information that is collected on location (i.e., the real observations gathered by sensors once they happen). This source is trusted and reliable, hence only one is needed.
- Predictive context. It consists of relevant context conditions for predictions (like current date, location, etc).
- Customer query. The desired predictions in a specific domain.

Among all available predictive services, the one with the highest likelihood to make right predictions should be used. By performing forecast verification with the above inputs, the predictive service with the highest quality is selected.

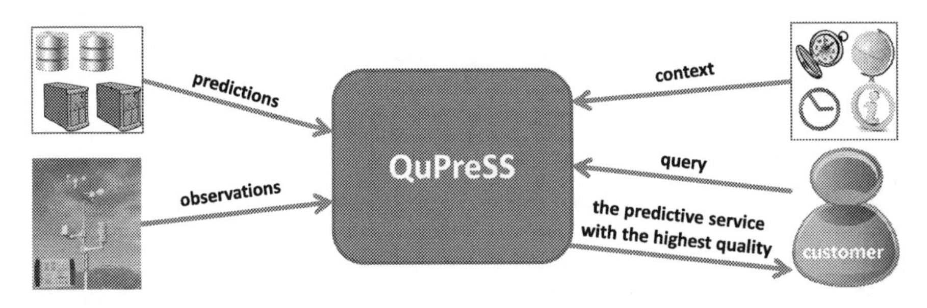

Fig. 1 Inputs and outputs of the QuPreSS (QUality for PREdictive Service Selection) framework

More formally, the problem targeted by the QuPreSS framework can be announced as follows.

The prediction problem. Let $PS = \{PS_1, PS_2, \ldots, PS_n\}$ be a set of predictive services in a given domain, *GT* the ground truth for this domain, *C* the context of the prediction, and *Q* the customer query as follows:

- $PS_i : Time \times D_0 \rightarrow D_I$
- $GT : Time \times D_0 \rightarrow D_I$
- $C \subseteq \langle Y_1, Y_2, \ldots Y_m \rangle$
- $Q \subseteq \langle Time \times D_0 \rangle$

where *Time* is the moment for which a prediction or observation is wanted, D_0 the parameters to every available query dimension that define the conditions or occurrences to be predicted or observed, D_I the prediction or observation, and Y_i the value of every relevant context condition.

Then the prediction problem can be defined as: to find a prediction function $pred(PS, GT, C, Q) \rightarrow PS$ such that:

$$\forall i: 1 \leq i \leq n: quality(pred(PS, GT, C, Q), GT, C, Q) \geq quality(PS_i, GT, C, Q)$$

where *quality* is the function that measures the forecast quality of a predictive service PS_i for the query *Q*, given the context *C* and the data gathered from the ground truth *GT*.

Examples of quality measures appear in Section 5. There are two ways to calculate quality: measured-oriented calculation focuses on one or two overall aspects of forecast quality (e.g., accuracy and skill) whereas distribution-oriented calculation constitutes forecast quality in its fullest sense. For example, the distribution-oriented approach allows comparing two or more sets of forecasts [14].

Forecast quality is inherently multifaceted in nature. Various aspects of forecast quality are: bias, association, accuracy, skill, reliability, resolution, sharpness, discrimination and uncertainty. To see their definitions and relevant distributions, the reader is referred to [14].

4 Framework Architecture

In this section we present a software architecture that implements the prediction problem as announced in the previous sections. The most important functional requirement of our architecture is to select the most appropriate predictive service to satisfy a customer's request according to some evolvable forecasting model. Subordinated to this main requirement, we have identified others: show the prediction of the highest ranked service, monitor registered predictive services, save their predictions, and get and save the real behaviour according to some trusted ground truth service. Besides, it should offer data to external systems and services. Among the non-functional requirements, the platform shall be: extensible to add new predictive services, adhered to standards to facilitate interoperability, able to process continuous data flows, compatible with existing components when possible, aligned with legal statements of external sources, and robust.

In order to fulfil the previous requirements, and following Section 3 above, the architecture presented in Fig. 2 was designed. Its components are:

- External sources: the Ground Truth service and a portfolio of predictive services. They are external and heterogeneous. They can be implemented by WS or other technologies. To integrate all these heterogeneous sources, we chose an SOA. If an external source is not exposed as a WS, it is wrapped into a WS proxy. Predictive services have to provide forecasts following a pre-defined document format to be easily integrated, or otherwise they are also wrapped into WS proxies to homogenise these formats.
- A Monitor service to save in a systematic manner the QoS of each predictive service and the response for every periodical request to the WS. These requests contain the predictions.
- The Forecasting Data Collector service, to collect both the ground truth and the predictions gathered by the Monitor service.
- Two database management systems (DBMS): one in charge of saving observations (Ground Truth Database) and another one to save predictions (Forecast Data Database). In those prediction domains where predictions change very frequently, it would be better to use data stream management systems (DSMS) instead of DBMS to process continuous data flows.
- The QuPreSS web application and the QuPreSS service which show/offer the results of the QuPreSS framework. The QuPreSS web application is designed

for human customers whereas external services can use the QuPreSS service. They verify predictions quality by means of the Forecast Verifier component. Both of them serve to support decisions about choosing between different potential predictive services and to directly redirect to the most accurate one (with the help of the Invocator component).

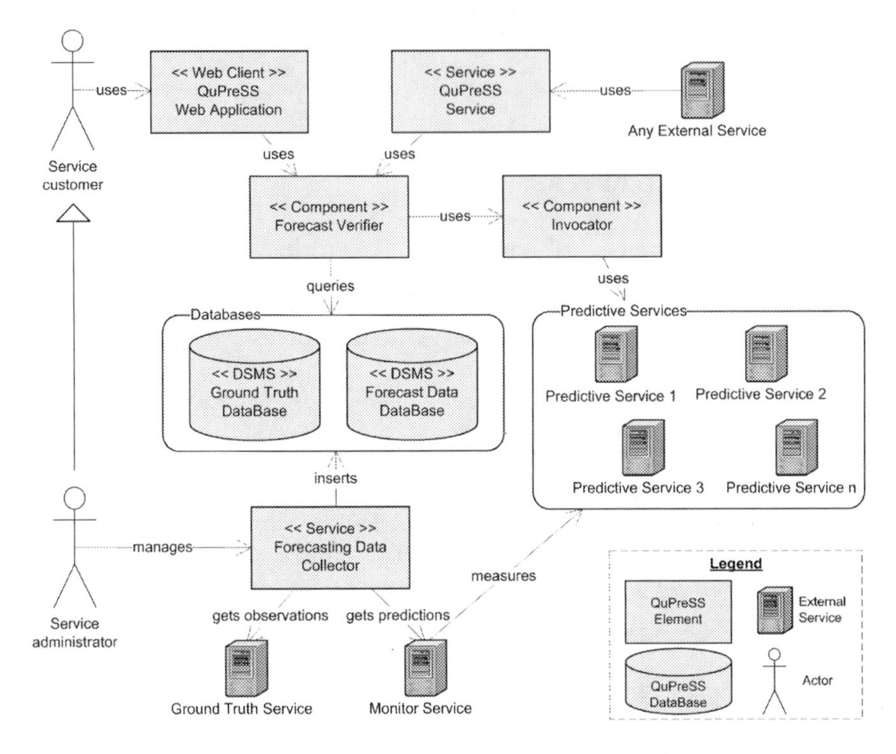

Fig. 2 QuPreSS framework architecture

Fig. 2 shows two human actors: the Service Customer and the Service Administrator. The Service Administrator manages the forecasting data collector service. After configuring this service, the system is set up to read, parse and save prediction data from registered predictive services, as observations from the ground truth service. On the other hand, either a service customer or external services get observations, predictions and services quality assessment from the QuPreSS framework.

5 An Exemplar Implementation for the Weather Forecast Case

In this section, the development of a prototype is discussed as a proof-of-concept. It fulfils the architecture requirements defined in the previous section, since it is an instantiation of the QuPreSS framework with all its elements for the weather forecast domain. These elements have been implemented using WS technologies,

MySQL databases and Java Web technologies. Besides, it uses the existing SALMon monitor [18], invoking the services that its SOA offers.

During the selection process of predictive services [11], we found that most of the WS which provide weather forecasts for cities all over the world are not free. However, there are other kinds of sources that without being WS (e.g., XML files), provide free weather forecasts. Thus, we deployed WS proxies which use them as a backend. In order to facilitate the integration of new services, the format of the response exposed by these proxies is the same for all of them. As a result, with this approach we resolved both the lack of free WS and the integration of heterogeneous sources.

The ground truth is obtained from AEMET (the Spanish Meteorological Agency) weather stations[1]. On the other hand, the SALMon monitor service saves in systematic manner predictions for four Catalan cities. These predictions come from the following predictive services: AEMET forecasts[2], open data Meteocat[3] (the Catalan Meteorological Agency), and Yahoo! Weather RSS feed[4].

In this proof-of-concept, the accuracy parameter taken into account was how precise the forecasts of high and low temperatures are. To verify the correctness of predictive services, we used the mean-squared error (MSE) and the approximation error (AE). Both of them quantify the difference between values implied by an estimator and the true values.

The formula to calculate the MSE of a predictive service *PS* is:

$$MSE_{PS} = \frac{\sqrt{\sum_{i=1}^{n}\left(\mu_{T_{GT}} - T_{PS_i}\right)^2}}{n} \tag{1}$$

where μ_{TGT} is the average of temperatures (high or low) of the corresponding observations (ground truth), T_{PSi} is a predicted temperature (high or low) from the *PS*, *i* is an index (which refers to a date), and *n* is the total amount of observations from the *GT* and predictions of the *PS*.

The AE measures the discrepancy between an observation and its prediction:

$$AE_{PS} = \frac{\sum_{i=1}^{n} |T_{GT_i} - T_{PS_i}|}{\sum_{i=1}^{n} |T_{GT_i}|} \tag{2}$$

where *T* indicates a high or low temperature that is either a prediction from the *PS* or its corresponding observation from the ground truth (*GT*), *i* is an index (which refers to a date), and *n* is the total amount of observations/predictions compared.

We had a limitation regarding the availability of several different observations that would correspond to a given prediction, since predictions refer to an area (e.g., Barcelona), but observations are taken in very specific locations (e.g., Sagrada Familia). In this case, we used the most centric (in a geographic sense) observation. Another limitation is to find correspondent observations to specific

---

[1] ftp://ftpdatos.aemet.es/datos_observacion/resumenes_diarios/

[2] http://www.aemet.es/es/eltiempo/prediccion/municipios

[3] http://dadesobertes.gencat.cat/ca/dades-obertes/prediccions.html

[4] http://developer.yahoo.com/weather/

predictions (e.g., if an observation says that it rained just a drop, can we consider right a prediction that said that is was going to rain?).

Results. High and low temperatures of four cities (Barcelona, Girona, Lleida and Tarragona) have been monitored since July 2011. Every day both real observations from AEMET weather stations and forecasts from three predictive services (AEMET, Meteocat and Yahoo!) are saved in our DBMS.

The Forecast Verifier component calculates prediction errors (MSE and AE) of the three predictive services in a period of time (by default the last two weeks) for a given city. Then, it returns either a ranking of predictive services ordered by quality (so that customers may select themselves, usually the first in the ranking), or the actual forecast of the predictive service with most quality. In the second case, the Invocator component is responsible to directly offer forecasts from the most accurate predictive service, hiding any analysis and redundant forecasts.

The presentation layer has been implemented by a web client application for customers and a WS for external services. This tool is freely available on-line at: http://gessi.lsi.upc.edu/qupress/.

6 Conclusions and Future Work

Forecast quality is of obvious importance to users since forecasts guide their decisions, and also to providers because it affects their reputation. Nevertheless, as we have shown in the SLR results, we can conclude that forecast verification for predictive services has not received sufficient attention and it justifies the focus of further research. More effort is necessary to integrate methods for forecast verification in SOA monitoring frameworks.

The main contribution of this paper is the presentation and development of QuPreSS, a generic SOA framework which performs forecast verification with observations in predictive domains. This architecture is scalable and easy to integrate with other systems and services. The goal of the architecture is to determine which predictive service provides better predictions by assessing the forecast quality of all of them. This assessment assists customers in making better decisions and gives them an edge over the competition. Besides, providers of these services can use this evaluation to improve their predictions.

We have identified general issues that need to be solved. Different predictive services from the same domain may provide predictions with different characteristics, which make them not easily comparable. For instance, when a service provides hail probability, other provides wind speed. Another example, when one service provides forecasts two days in advance, and another does so one week in advance. Likewise, it is needed to know the quality aspect or error that behaves better in a specific case.

At present, more important future work relates to: (1) increasing the amount of monitored services and collect more data about predictions, (2) applying this architecture to other domains apart from the weather forecast domain, (3) identifying the current knowledge about parameters that determine predictive service quality, and (4) studying the forecast value criteria of predictions.

Acknowledgments. This work has been supported by the Spanish project TIN2010-19130-C02-01. We would also like to thank Marc Oriol and Marc Rodríguez for their friendly cooperation while integrating QuPreSS with SALMon.

References

1. Bertolino, A.: Software testing research: Achievements, challenges, dreams. In: Proc. of the Future of Software Engineering Symposium, FOSE, pp. 85–103. IEEE-CS Press (2007)
2. Canfora, G., Di Penta, M.: Service-Oriented Architectures Testing: A Survey. In: De Lucia, A., Ferrucci, F. (eds.) ISSSE 2006-2008. LNCS, vol. 5413, pp. 78–105. Springer, Heidelberg (2009)
3. Delgado, N., Gates, A.Q., Roach, S.: A taxonomy and catalog of runtime software-fault monitoring tools. IEEE Transactions on Software Engineering 30(12), 859–872 (2004)
4. Domenico, B.: Forecast Verification with Observations: Use Case for Geosciences Web Services, `http://www.unidata.ucar.edu/projects/THREDDS/GALEON/Phase2Connections/VerificationUseCase.html` (cited February 19, 2012)
5. Endo, A.T., Simao, A.S.: Formal Testing Approaches for Service-Oriented Architectures and Web Services: a Systematic Review. Technical Report Instituto de Ciências Matemáticas e de Computaçao 348 (2010) ISSN-0103-2569
6. Gregory, M., Spohrer, J.: Succeeding through service innovation: A service perspective for education, research, business and government. IBM IfM - University of Cambridge Institute for Manufacturing (2008) ISBN: 978-1-902546-65-0
7. Guha, R.: Flexible Web Service Infrastructure for the Development and Deployment of Predictive Models. J. Chem. Inf. Model 48(2), 456–464 (2008)
8. Han, D., Huang, H., Cao, H., Cui, C., Jia, C.: A Service-Oriented Architecture Based Macroeconomic Analysis & Forecasting System. In: Zhou, X., Li, J., Shen, H.T., Kitsuregawa, M., Zhang, Y. (eds.) APWeb 2006. LNCS, vol. 3841, pp. 1107–1117. Springer, Heidelberg (2006)
9. Kitchenham, B.: Procedures for Performing Systematic Literature Reviews. Keele University Technical Report, TR/RE-0401 (2004)
10. Latha, C.B.C., Paul, S., Kirubakaran, E., Sathianarayanan: A Service Oriented Architecture for Weather Forecasting Using Data Mining. Int. J. of Advanced Networking and Applications 2(2), 608–613 (2010)
11. Martínez-Fernández, S.: Accuracy Assessment of Forecasting Services. Master Thesis Universitat Politècnica de Catalunya (2011)
12. Marzolla, M., Mirandola, R.: Performance Prediction of Web Service Workflows. In: Overhage, S., Ren, X.-M., Reussner, R., Stafford, J.A. (eds.) QoSA 2007. LNCS, vol. 4880, pp. 127–144. Springer, Heidelberg (2008)
13. Meng, H.N., Qi, Y., Hou, D., Zhang, Y., Liu, L.: Forecasting software aging of service-oriented application server based on wavelet network with adaptive genetic algorithm. In: Proc. of 3rd International Conference on Natural Computation, ICNC, pp. 353–357 (2007)
14. Murphy, A.H.: What is a good forecast? An essay on the nature of goodness in weather forecasting. Weather Forecasting 8, 281–293 (1993)

15. Nikolopoulos, K., Metaxiotis, K., Assimakopoulos, V., Tavanidou, E.: A first approach to e-forecasting: a survey of forecasting Web services. Information Management & Computer Security 11(3), 146–152 (2003)
16. de Oliveira, L.B.R., Romero Felizardo, K., Feitosa, D., Nakagawa, E.Y.: Reference Models and Reference Architectures Based on Service-Oriented Architecture: A Systematic Review. In: Babar, M.A., Gorton, I. (eds.) ECSA 2010. LNCS, vol. 6285, pp. 360–367. Springer, Heidelberg (2010)
17. Oriol, M.: Quality of Service (QoS) in SOA Systems: A Systematic Review. Master Thesis Universitat Politècnica de Catalunya (2009)
18. Oriol, M., Franch, X., Marco, J.: SALMon: A SOA System for Monitoring Service Level Agreements. Universitat Politècnica de Catalunya Technical Report, LSI-10-18-R (2010)
19. Papazoglou, M.P., Traverso, P., Dustdar, S., Leymann, F.: Service-Oriented Computing: State of the Art and Research Challenges. Computer 40(11), 38–45 (2007)
20. Punitha, S., Babu, C.: Performance Prediction Model for Service Oriented Applications. In: Proc. of 10th IEEE International Conference on High Performance Computing and Communications, HPC, pp. 995–1000 (2008)
21. Valipour, M.H., Amirzafari, B., Maleki, K.N., Daneshpour, N.: A brief survey of software architecture concepts and service oriented architecture. In: Proc. of 2nd IEEE International Conference on Computer Science and Information Technology, ICCSIT, pp. 34–38 (2009)
22. Xiao, N., Fu, W., Chen, T., Huang, Q.: A service-oriented monitoring system with a forecasting algorithm of a time sequence-based hybrid model. International Journal of Parallel, Emergent and Distributed Systems 23(2), 137–151 (2008)
23. Zdun, U., Hentrich, C., van der Aalst, W.M.P.: A survey of patterns for service-oriented architectures. International Journal of Internet Protocol Technology 1(3), 132–143 (2006)

A Knowledge Management Model Applied to Health Tourism in Colombia

Luz Andrea Rodríguez Rojas[1], Giovanny M. Tarazona Bermudez[2], and Alexis Adamy Ortiz Morales[2]

[1] Fundación Universitaria los Libertadores, Bogotá – Colombia
larodriguezr1@libertadores.edu.co
[2] Universidad Distrital "Francisco José de Caldas", Bogotá - Colombia
{gtarazona,aortiz}@libertadores.edu.co

Abstract. Health tourism is an emerging sector in many countries, where developing economies find an opportunity to grow by providing quality medical services at competitive prices. The proposed model is applied to Knowledge Management and is also based on Medical Tourism from the perspective of Information and Communications Technology – ICT. This model represents a fundamental tool for achieving competitive tourism worldwide and for improving decision-making in the industry, since the fundamental weaknesses of such an industry are related to information-wise aspects, value and reliability when transferring knowledge. This management model provides a guide to achieve these objectives through the establishment of tools for its implementation.

1 Introduction

Human beings have always wanted to improve their quality of life and keep their welfare. Developments of social cultural and economic aspects of human health have occurred recently, leading to other kinds of developments in areas like medicine, technology, industry and tourism. Due to these developments and also due to factors like cost, transport, quality and medical security – all these related to surgical and medical procedures – there has been an increase in the number of potential patients that are willing to travel to different countries where the conditions to receive suitable medical treatments are a lot more favorable than in their own countries. This social phenomenon has been called "health tourism", and constitutes an activity that is now considered to be of great economic value within the services industry worldwide, having regional leading countries in the field, such as India, Thailand, Mexico and Brazil.

Health tourism, also known as medical tourism, is still not as popular in Colombia, but its potential growth is significant. This is why official bodies like the Ministry of Commerce, Industry and Tourism and also the regional Chambers

L. Uden et al. (Eds.): 7th International Conference on KMO, AISC 172, pp. 537–546.
springerlink.com

of Commerce – among others – are drawing their attention to the growth of this new market, encouraging investment projects that allow consolidation and creation of competitive levels of service with respect to other leading countries in this sector.

The National Planning Department issued an official document titled CONPES 3678 [1] in 2010, where the Productive Transformation Program (PTP) was established as a public-and-private strategy to consolidate eight world-class sectors and promote economic growth. Among these eight world-class sectors, both health and tourism were included.

The absence of a Knowledge Management model limits the transfer of knowledge and experiences between the people involved in the development of Health Tourism. The lack of a suitable model also hinders decision-making and prevents participants from easily achieving highly competitive levels of service that compare with those offered in world-leading countries.

2 Health Tourism General Aspects

An internationally accepted definition of health tourism states that it is "the process in which a person travels abroad for the purpose of receiving either health or welfare-associated services" [2].

Factors like high cost, long waiting lists in the local hospitals, access to new technologies and skills in the destination countries combined with low-cots transport and easy to access Internet-based marketing, have played a major role in the development of health tourism. Various Asian countries have taken the lead in this new market, but most of the other countries have also tried to participate [3].

2.1 Different Categories of Health Tourism

2.1.1 Welfare

In this category we find every activity that is aimed at improving people's quality of life, achieving high levels of self-satisfaction in terms of physical, mental, emotional, spiritual and social health, regardless of any medical condition or vulnerability of the human body. These activities consider welfare from a health tourism viewpoint, which includes physical and recreational activities as well as specific practices that fulfill the same purpose.

2.1.2 Aesthetic Medicine

This category refers to all medical specialized practices that involve treatment and surgery in order to restore, maintain, and promote the beauty of the human body. Through aesthetic medicine, it is possible to prevent and treat all aesthetic pathologies. The purpose is to enhance physical features making them aesthetic and more beautiful; this also attempts to retard aging processes and promotes both physical and mental health of people.

2.1.3 Preventive Medicine

In this category, we find medical practices that are intended to avoid illnesses by keeping track and control over the possible causes of people's pains and sufferings. Preventive medicine includes health promotion and protection activities once a particular illness or disease has appeared.

2.1.4 Curative-Care Medicine

This category deals with the diagnosis and immediate treatment of illnesses. It includes pharmaceutical treatment and treatment at a hospital every time there is presence of vulnerability symptoms in the body that suggests health deteriorating conditions of specific organic functions.

2.2 Health Tourism in Colombia

In general, tourist-related activities in Colombia have shown a significant increase in the period 2002-2007, which can be seen in an annual growth of 15%, according to the data from Proexport, and also according to figures from the Productive Transformation Program run by the Ministry of Commerce Industry and Tourism. In 2008, the percentage of health tourists that visited Colombia accounted for about 2.2% of the whole (1`222.966 tourists), as reported by DAS border control agents and international airport tourist control. These figures are significant in spite of the restrictions imposed on tourism by the global economic crisis; this had a positive impact on the Colombian economy in terms of a new economic alternative, which represented revenues of 126 million dollars.

This perspective leads to setting objectives towards achieving revenues over 6 thousand million dollars by 2032 for Colombia [4], which will be represented by the sums of foreign currency spent on medical treatments and welfare-related activities. These revenues will also be represented by all other expenses associated to the visits of all these health tourists, aiming at having a bigger share of the market and also a higher impact.

3 Knowledge Management Model Applied to Health Tourism

Recently, there has been a lot of debate about knowledge Management as an instrument to model and measure intangible assets. Some of the best-known and widely-implemented models in this context are KPMG Consulting, Andersen, KMAT, and the knowledge creation model.

Knowledge is the result of the social actions of organizations, and it is much more that just the result of individual learning [5]. Knowledge circulation, both codified and tacit, represents a key element in the learning process associated to the development of innovative activities and, therefore, associated to the creation of competitive advantages. The dissemination of information and communication technologies develops a set of activities whose main input is knowledge-based information.

Knowledge regional spaces gather all the knowledge created by all shorts of institutions, namely public, private, productive, educational, and research organizations. An example of this is the productive promotions sponsored by the specialized regional and national entrepreneurial associations [6].

It was not long ago that Colombian tourism started to be regarded as one of the most appropriate instruments to articulate local development processes, where participants and their relationships become essential elements [7].

3.1 Advances in Colombia

Colombia has not ignored the social phenomenon of health tourism nor the widespread offers on this matter. Hence, many Colombian cities have already started different projects aimed at forming and consolidating health clusters throughout the country. In Bogota and Cundinamarca, the Chamber of Commerce of Bogota together with the city council and Proexport lead the development of a programme called Capital Health (Salud Capital). In Medellin, Proantioquia leads a programme called Health with no Borders (Salud sin Fronteras). In Cali, Universidad Javeriana encourages the development of a project based on what they call Health Competitive Capabilities for the region, also known as Health Valley (Valle de la Salud); while in the region of Santander, an organization called CARCE has promoted the development of a Health Export project through a proposal called Santander Health (Salud Santander). Apart from all these schemes and programmes, there are other health-care organizations that carry out projects independently in order to export health services from Viejo Caldas, Barranquilla and Cartagena.

Factors like lower cost, shorter waiting lists, a variety of services, and the opportunity to combine tourism with health-care activities, join to shape this new industry; additionally, these conditions open the path to potentially create new jobs and wealth in a region, while at the same time provide ill people with high-quality treatment at affordable prices [8].

3.2 The Participants Involved

Table 1 Participants involved in health tourism. Source: Cárdenas, F. & González, C. (2011) [9].

Name	Category
Support Service	Administration
	Accommodation
	Airlines
	Financial Services
	Food
	Interpreters and guides
	Local Transport
	Tourist Attractions

Table 1 (*continued*)

Health Instructions	Clinics
	Hospitals
	Health Centers
Customers	Health Tourists
	Third Party Payers
Providers	Medicines
	Medical Resources and Equipment Clinical Laboratories
Thechnology and Science	Educational Centers
	Research Centers
	Accreditation Organizations
	Chamber of Commerce
State Entities	Ministries
	Proexport
	Superintendents
	DANE

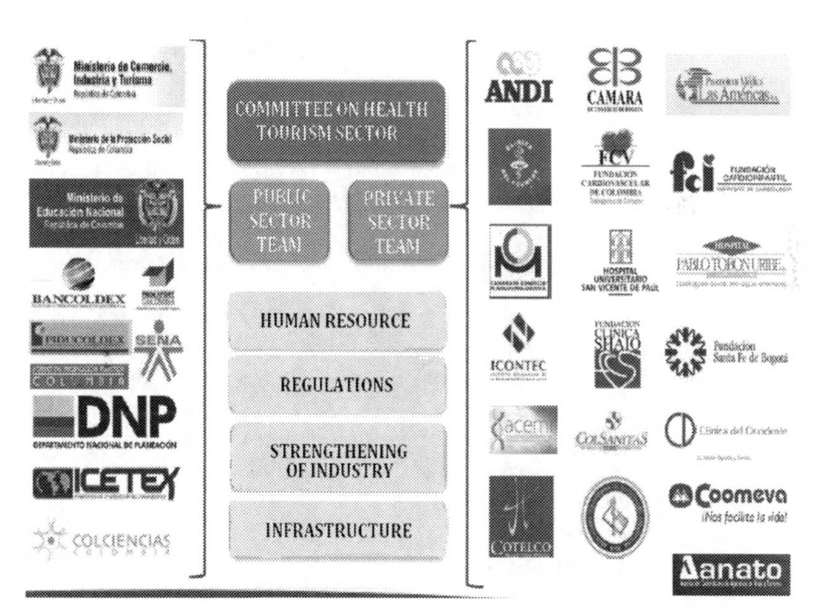

Fig. 1 The Participants involved

3.3 *Proposed Model*

The knowledge management model for Health Tourism in Colombia is based on five development pillars intended to gather the various fundamental ideas that

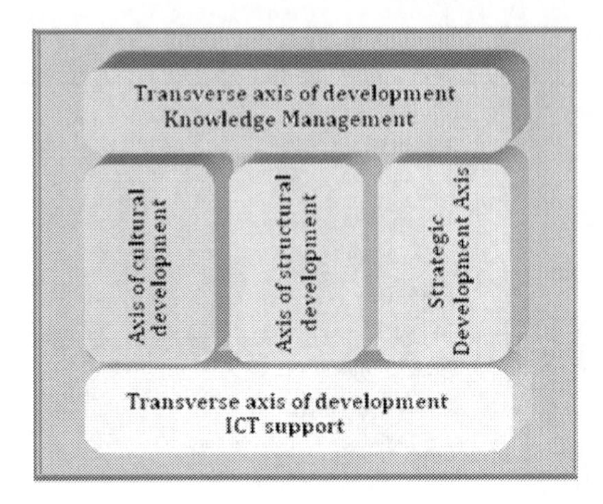

Fig. 2 Knowledge management model for health tourism. Source: The authors.

support the whole project, which will permit attaining integrating goals to ensure the creation of a corporate network, sustainable in time. The two pillars associated to Information and Communication Technologies (ICTs), and also associated to Knowledge Management, respectively, are easily distinguishable for being cross-sectional.

The three remaining pillars correspond to the following aspects of the proposal: culture, structure, and strategy. From these three perspectives, the idea is to provide appropriate tools that support the proposal's implementation and application in order to strengthen health tourism.

3.3.1 Cross-Sectional Development Pillar: ICT Technical Support

This development pillar of the model, which is connected to the technical support provided by Information and Communication Technologies, represents a cross-sectional element intended to be bedrock upon which the other pillars develop. It should allow the easy administration of the network by using elements and techniques that capture, record, convert, store, present, manage, and transmit all sorts of information and knowledge, which ultimately become technical management media for health tourism.

3.3.2 Cultural Development Pillar

Culture, as one of the pillars for development, is intended to create an image that strengthens an associative-wise project for all the parties through the identification and commitment to cultural elements previously defined. This in turn will permit the existence of cooperation bonds between the parties, since there will be common awareness of the collective need to develop an entrepreneurial network.

3.3.3 Structural Development Pillar

The structure, as a development pillar, is intended to create a series of regulations, both legal y constituent, that become a type of guidance, defining the position and the procedures to be followed in particular situations. This involves the establishment of clear rules and general conditions for the proper development of all activities.

3.3.4 Strategic Development Pillar

The administration of a strategic-wise development pillar attempts to consolidate a series of tactical moves oriented to the fulfillment of the system's objective, namely health tourism services being offered according to high-quality standards in Colombia

3.3.5 Cross-Sectional Development Pillar

Knowledge Management: the creation of a knowledge-management model proposal, which permits strengthening health tourism by implementing virtual activities, pursues the creation of intangible products such as a knowledge transfer scheme, proper information management, and the development of technology-assisted activities. These objectives will represent better tangible results for the parties involved in the system, whether they are institutions, companies or customers. This model is intended to offer a series of fundamental ideas that will allow the construction of a proposal based on several viewpoints regarding the activities of the different participants of the cluster. This also permits interaction between the different parties of the system through technological tools that attain competitive development and at the same time guarantee joint management to achieve the main goals of the entrepreneurial network. This network must spin around the services being offered, namely high quality medical tourism services. This pillar offers integral tools like the following:

A *Knowledge Management Portal for Health Tourism*: The creation of a health tourism portal represents the most important activity of technological implementation to strengthen the health tourism cluster through knowledge management. This is because this portal constitutes an effective communication tool that can be used by all the participants of the system as a platform for the development of any management-associated activity. This Knowledge Management portal, intended exclusively for health tourism, can be run in several ways according to the functions to be performed:

- Informative portal: The portal plays an informative role whenever advertising purposes are necessary to promote health tourism and its associated products. Likewise, it serves to advertise the advantages and benefits offered together with the constituent members in charge of such products and/or services. The idea is to reinforce all this information by publishing articles of interest that are closely related to medical tourism. Other articles or publications can be cited, mentioned or linked to this web site.

- Interactive portal: The portal plays an interactive role when it allows users to register and actively participate by using the site itself as a health tourism platform. There are various benefits to this, namely access to certain types of contents and applications that can be aimed at customers by including elements such as forums, chats, and virtual communities. This may also include participation in customer-oriented events, patient diagnosis, and service quotes, among other advantages.
- Transactional portal: The health tourism portal plays a transactional role whenever it is used to allow user registration for people who intend to consolidate commercial bonds and therefore their participation is not only for communication purposes but for management purposes. This type of management requires the parties to exchange information and money, which implies the use of higher levels of cyber security; hence the portal may also represent a hyper-link to access extranets, allowing reliable specific transactions to be made.

Customer Service Center: The creation of a customer service center leads to having the health tourism cluster equipped with an information mechanism intended for users. Such an information mechanism will also be intended for people or companies that show commercial interest in the system. In that sense, the customer service center serves as a communication center that receives and makes calls, and also issues messages through mechanisms like e-mail, fax, telephone calls, and video-conferences. The purpose of these activities is to provide users with information and technical support.

Suplí Network: The creation of a supply network begins by identifying the main links or parties of the cluster that represent the center of all economic activity in the system. In the case of the health tourism cluster, the main parties are health-care service institutions; which means that the supply-associated activities must be around the needs of such institutions. By managing an application that allows controlling and coordinating supply activities among raw-material providers, medical service providers, institutions and customers, the purpose is to create proper means of communication, negotiation and acquisition between the parties. This must ensure the fulfillment of particular needs in terms of medicine supply, surgical and medical material supply as well as the supply of equipment and hospital services.

Knowledge Community: The creation of a knowledge community around a health tourism cluster is intended to take advantage of the boom that such a communication tool had among all people surfing the Internet. This community might have a great impact on potential clients. One of the advantages offered by social networks is that they gather the people interested in medical topics and this type of participation attracts more people, resulting in an ever-growing network of potential users within the health tourism services cluster. This represents great advantages in terms of communication and cooperation, which is even more significant from a user viewpoint since the information and support provided in the portal will come from peer users who can give advice and recommendations about procedures and services based on their own experience. This may additionally have a positive impact in terms of promotion and advertising.

Virtual Learning System: The implementation of a virtual learning system pursues the development of activities aimed at producing knowledge, training people and improving the processes involved in health tourism practices. This particular tool is founded on the following principles: Every organization is made of people and is aimed at serving end users, who are also people; therefore, the management of all learning and training activities at all levels implies improving the conditions for the health tourism cluster on the whole. The idea is to strengthen aspects such as quality, regulation compliance, customer service, second language acquisition, accreditations and certifications, the use of technological tools, and the creation of innovation and research processes. In that respect, Colombia has a technological platform already (the platform of a governmental education agency called SENA). In recent years, this platform has offered virtual education to many people around the country; hence it already provides the necessary technical support to carry out health tourism projects. Thus, it would be important to organize the contents as well as the logistics of the actual implementation.

Virtual medical records: The creation of virtual medical records is a technological application intended to implement a jointly-administrated programme by means of an extranet made up of all medical institutions participating in the system. The network is to serve and be managed by authorized medical personnel in order to look up and feed the medical records of the different patients; thus allowing health-care professionals to know patients condition at any time. This will help to stipulate the kind of treatment, medications, procedures, tests and diagnosis that patients have undertaken in the past. These activities are intended to consolidate the medical information of patients, creating a clinical data store to easily identify health tourists before, during, and after any medical procedures. In the long run, this information can be used when the same tourists wish to undergo another procedure.

Health tourist agenda: The creation of a health tourist agenda as a technological tool attempts to manage the logistics behind patients' activities from the moment they travel to the destination country to the moment they leave hospital and go back to their countries. These conditions must be extended to include monitoring patients for a period of time after leaving. The health tourist agenda is a tool created from the moment health-tourism service conditions are agreed, and must contain all specifications to offer both medical and support services regarding flight tickets, accommodation bookings, medical appointment scheduling, clinical tests, and surgical procedures. Additional specifications may include scheduling of other complementary services like local transport services, tourist guides, interpreters, and access to tourist attractions and places of general interest for tourists.

4 Conclusions

Knowledge management has various application fields. It seems that any human activity is likely to be continuously improved by implementing knowledge management, and so health tourism in not the exception.

A region's image is an influential factor associated to the notion of security and it constitutes an attractive feature for health tourists since it has a direct impact on their decisions when choosing their medical tourism destinations.

The human resource component happens to be an essential factor in terms of competitiveness within the medical tourist supply in emergent markets like the Colombian market.

In Colombia, aspect like development and research represent weaknesses for the health tourism sector due to its low management indices within institutions, despite the existing proposals to attain proper implementation of its services.

In Colombia, there are projects in progress aimed at extending the scope of the health services offered as well as the scope of the corresponding infrastructure. The resulting benefits of such projects can be used by health tourism clusters.

Colombia is compelled to develop strategies so that the institutions that intend to export health services obtain certifications in order to offer international-standard services.

Knowledge management provides essential tools to make the Health Tourism sector world-class. Thus, it is possible to fulfill the expectations officially written by the government in its Productive Transformation Policy.

References

1. Departamento Nacional de Planeación.: Documento CONPES 3678 (2010)
2. Ministerio de Comercio, Industria y Turismo de Colombia.: Turismo & salud, Una dupla ganadora. Recuperado el, (Octubre 30, 2011) (2009), https://www.mincomercio.gov.co/ptp/descargar.php?id=40519
3. Connell, J.: Medical tourism: Sea, sun, sand and y surgery. Revista Tourism Management 27, 1093–1100 (2006)
4. Ministerio de Comercio, Industria y Turismo de Colombia.: Desarrollando sectores de clasemundial en Colombia. Recuperado el, (Octubre 30, 2011) (2009), https://www.mincomercio.gov.co/ptp/descargar.php?id=40518
5. Nonaka, I., Takeuchi, H.: The Knowledge-Creating Company: How Japanese Companies Create the Dynamics of Innovation. Oxford University Press, Londres (1995)
6. Casas, R.: La Formación de Redes de Conocimiento. Una perspectiva regional desde México, IIS- UNAM. Anthropos, Barcelona (2001)
7. Merinero, R.: Desarrollo local y Análisis de RedesSociales: el valor de las relaciones como factor del desarrollo socioeconómico. [Versión electrónica] Recuperado de (2010), http://revista-redes.rediris.es/pdf-vol18/vol18_11.pdf
8. Reisman, D.: Health tourism: social welfare through international trade. MPG Books Group, UK (2010)
9. Cárdenas, F., González, C.: Propuesta de un modelo de gestiónsustentada en el comercio electrónico como herramienta de fortalecimiento del cluster de turismo de saludpara Bogotá. Trabajo de Grado, Universidad Distrital, Colombia (2011)

Adopting a Knowledge Management Concept in Securing the Privacy of Electronic Medical Record Systems

Suhaila Samsuri, Zuraini Ismail, and Rabiah Ahmad

Abstract. As the enhancement of Information Technology (IT) in various management fields escalates, the realization of knowledge management (KM) concept is considered as significant. This concept can be applied in considering information privacy as a component in designing a computerized system. Consequently, this paper proposes the adoption of explicit and tacit knowledge concepts in identifying the privacy of information components in order to construct a secured electronic medical record (EMR) system. A preliminary investigation involving interviews with a selected group of hospital information system (HIS) developers has been conducted earlier on. The findings of this study revealed four important components to be factored in any privacy preservation framework for HIS. They are namely, legislation, ethical, technology and cultural. By applying the KM concept into these four components, it can further derive more systematic guidelines in order to achieve the privacy preservation of information. Nevertheless, further research must be developed as to yield a more inclusive guideline in designing an EMR system that is reliable, specifically in addressing the need of securing patient's personal medical information privacy.

Keywords: Knowledge management, explicit knowledge, tacit knowledge, information privacy, electronic medical record, hospital information system.

1 Introduction

Knowledge management (KM) is the most fundamental concept to be considered in securing the privacy of any information management system. Currently, KM concept has been incorporated into numerous advanced computer systems and adopted into many system designs [1]. The norms in information technology (IT) usually involve with the activities of documentation, sharing of information, exchanging ideas with others by sending and receiving e-mails or using other computer mediated interaction, such as teleconferencing and virtual meeting [2]. In the health care industry particularly, the development of a Hospital Information System (HIS) and the transmission from paper-based to paperless-based record system has encouraged the advancement in health data management and technologies, such as the digitization of medical records, creation of central record systems and the development of healthcare data warehouse [3]. These initiatives

L. Uden et al. (Eds.): 7th International Conference on KMO, AISC 172, pp. 547–558.
springerlink.com

have entailed yet another concern in IT field which is the personal medical information privacy issue [4]. Various studies have been carried out in efforts to mitigate the personal medical information privacy issues since it involves many stakeholders such as the government, system developers and the public. It also warrants the latest and most innovative security mechanisms to be embedded in the existing technology system. Eventually, the health care providers must also address all issues with regards to the skills, knowledge and qualification required of the users of the system in order to operate the system efficiently. Therefore, the sole purpose of this study is to examine the KM role in identifying the components which are appropriate for enhancing the EMR system with built-in privacy preservation elements.

This paper is divided into several sections, firstly is the introduction. Secondly, the previous studies section will elaborate on the important terms conceptually including KM, explicit and tacit knowledge as well as the information privacy. The third section will detail out the research methodology utilized in this study. It will be followed by a discussion in the fourth section and finally, the summary of this study along with suggestions for future research in the conclusion section.

2 Previous Studies

According to several studies, knowledge management (KM) can be simplified as a formal, directed process of determining what information a company has, that could benefit others in the company and then devising ways to make it easily available [5][6][8]. It involves with the step of knowing how the knowledge is captured, evaluated, cleansed, stored, provided and used [7][8]. Smuts et al [2] defined KM as a class of information system (IS) which is applied to managing organizational knowledge. The use of technology will definitely add value by reducing time, effort and cost in allowing people to share knowledge and information. The KM approach requires the integration of the people, processes and technology involved in designing, capturing and implementing the intellectual infrastructure of an organization, including the necessary changes in management attitudes, organizational behaviour and policy [9][10]. Technologically, intranets, groupware, data warehouse, networks, bulletin boards video conferencing are some of the key tools for storing and distributing this intelligence [11].

In IT, KM is described as the various processes such as acquisition, creation, renewal, archival, dissemination and application (conversion of new knowledge into action or behaviour modification) of knowledge. Shami et al [12] generally explained the KM concept as managing the corporation's knowledge through a systematically and organizationally specified process for acquiring, organizing, sustaining, applying, sharing and renewing both the tacit and explicit knowledge of employees to enhance organizational performance and value creation. For the purpose of this research, the KM concept shall be referred as the knowledge of the employees or users of the system in handling the daily tasks of the electronic medical record (EMR) system which are mainly related with the patients' medical records. The underlying challenges however remain in the competency of the users in transforming their skills from manual to automation. It is said that the pursuit of tacit, explicit, self-knowledge, self-renewal and innovation are timeless, endless and relentless [8].

Knowledge management concept can be divided into two approaches; explicit and tacit knowledge approaches. The explicit knowledge can be transmitted across an organization through documentations, drawings, standard operating procedures, manuals of best practices and the like. Information system is said to be playing a central role in facilitating the dissemination of explicit knowledge assets over company intranets or between organizations via the internet [13]. Explicit knowledge is found in databases, memos, notes and documents, which are usually represented as user manuals for using a particular computer system like an EMR. Similarly, it can also be in the form of a set of training handouts provided for staffs and users on the guidelines in using the information system. A comprehensive policies and code of conducts on the information security and privacy procedures need to be formulated and explained clearly to all parties involved.

In contrast, tacit knowledge seems to be more complicated due to its nature as intuitive, hard to define knowledge that is mainly experienced based. It is a kind of knowledge of "Intellectual Capability" that is not easily catalogued. It is completely incorporated in the individual [14]. Tacit knowledge is an accumulative knowledge that is embodied in the individual, escapes definition and quantitative analysis and is learned through trial and error analysis and transferred through socialization, demonstration and imitation [15]. It is suggested that the success and effectiveness of the implementation of a new electronic medical record system in the HIS entirely dependable on the experience and skills of the system users. The users could originate from various fields of educational background, namely doctors, nurses, IT officers or administration staffs.

Smith [8] has clarified on the differences between explicit and tacit knowledge of daily tasks in workplace, refer to Table 1. This classification is useful in providing a KM guideline towards the information privacy protective usage and maintenance of an EMR system in the hospital.

Table 1 Use of the explicit and tacit knowledge in the workplace [8]

Explicit knowledge – academic knowledge or "know-what" that is described in formal language, print or electronic media, often based on established work processes, use people-to-documents approach	**Tacit knowledge** – practical, action-oriented knowledge or "know-how" based on practice, acquired by personal experience, seldom expressed openly, often resembles intuition
Work process – organized tasks, routine, orchestrated, assumes a predictable environment, linear, reuse codified knowledge, create knowledge objects	**Work practice** – spontaneous, improvised, web-like, responds to a changing, unpredictable environment, channels individual expertise, create knowledge
Learn – on the job, trial-and-error, self-directed in areas of greatest expertise, meet work goals and objectives set by organization	**Learn** – supervisor or team leader facilitates and reinforces openness and trust to increase sharing of knowledge and business judgment

Table 1 (*continued*)

Teach – trainer designs using syllabus, uses formats selected by organization, based on goals and needs of the organization, may be outsourced	**Teach** – one-on-one, mentor, internships, coach, on-the-job training, apprenticeships, competency based, brainstorm, people to people
Type of thinking – logical, based on facts, use proven methods, primarily convergent thinking	**Type of thinking** – creative, flexible, unchartered, leads to divergent thinking, develop insights
Share knowledge – extract knowledge from person, code, store and reuse as needed for customers, e-mail, electronic discussions, forums	**Share knowledge** – altruistic sharing, networking, face-to-face contact, videoconferencing, chatting, storytelling, personalize knowledge
Motivation – often based on need to perform, to meet specific goals	**Motivation** – inspire through leadership, vision and frequent personal contact with employees
Reward – tied to business goals, competitive within workplace, compete for scarce rewards, may not be rewarded for information sharing	**Reward** – incorporate intrinsic or non-monetary motivators and rewards for sharing information directly, recognize creativity and innovation
Relationship – may be top-down from supervisor to subordinate or team leader to team members	**Relationship** – open, friendly, unstructured, based on open, spontaneous sharing of knowledge
Technology – related to job, based on availability and cost, invest heavily in IT to develop professional library with hierarchy of databases using existing knowledge	**Technology** – tool to select personalized information, facilitate conversations, exchange tacit knowledge, invest moderately in the framework of IT, enable people to find one another
Evaluation – based on tangible work accomplishments, not necessarily on creativity and knowledge sharing	**Evaluation** – based on demonstrated performance, ongoing, spontaneous evaluation

Currently, the study on the integration of Smith KM workplace concept with the electronic healthcare tasks particularly on the preservation of information privacy is not available. It is appropriate to look at the definition of information privacy as this is the objective in designing an effective EMR system before proceeding into the adoption of KM in the EMR system during its implementation stage. Basically, information privacy is understood as the ability of an individual to control the collection, retention, and distribution of information about an individual. It is also the right of data owner to determine for what purposes their personal information is stored and used [16]. In medical practices, information privacy is recognized as a careful and restricted communication of information about one's condition and medical history to one's caregivers. An individual expects that access to it will be carefully restricted [17]. In this context, privacy can be restricted or controlled by limiting what personal data is made available to others. In other words, it is how personal data is disclosed and shared with others [18]. The following three previous models have formed the foundation towards designing a secured information system as well as sustaining the information privacy preservation.

The first model that was proposed by Fischer-Hübner [19][20] comprised of, namely;

i. Protection enacted by the government through laws
ii. Protection through the employment of privacy-enhancing technologies
iii. Self-regulation for a fair information practice based on codes of conduct which are promoted by business players
iv. Educating privacy among the consumers and IT professionals

Fischer-Hübner was one of the researchers who were actively involved in the information security and privacy technology research. The Fischer-Hübner's model seems to be conveniently more acceptable by many countries since it correlates with the privacy protection laws that were enacted by their respective government. The second model was raised by Peter Swire who was highly involved in privacy and business law research. His presentation on 'Privacy Today' [21] highlighted that privacy protection can be achieved through the following:

i. Technology: the adoption of security and privacy mechanism in the system
ii. Law: the necessary rules and regulations in controlling the information accessibility
iii. Market: business entities and commercial organizations do provide some degree of security and privacy control to protect their customers, but still subjected to the their own requirements
iv. Choice of individual: it is entirely up to the individual to choose any business or commercial organizations that promote best or poor privacy protection.

Besides focusing on technology and legal aspects, Swire has included market into the list mainly because most of information privacy cases were directly linked to commercial and business dealings over the internet. His focus of attention was primarily on the public's interest rather than the personal information privacy.

The third model was suggested by Holvast [22] which outlined four components of privacy protection. These components were initially suggested by the Electronic Privacy Information Center (EPIC), a public interest research center which was established in 1994 in Washington, D.C. The components were;

i. Comprehensive laws: using general law that governs the collection, use and dissemination of personal information by both public and private sector. A neutral party is required to oversee the implementation to ensure compliance
ii. Sectoral laws: there are specific laws which govern only specific technical applications or specific areas, such as the financial privacy
iii. Self-regulation: using a code of conduct or practice established by companies and industries. In other words, basically self-policing
iv. Technology of privacy: using available technology-based systems such as privacy-enhancing technologies that are able to provide several privacy protection mechanisms

These components seem to be tailor designed for the use of system developers in the United States (US). It is due to the fact that the US government differentiates between the comprehensive and sectoral law, as currently being practiced.

A thorough examination and consideration was made on these three models and the various components associated. This is to ensure the relevance of the models adoption and adaptation in incorporating privacy protection of personal medical information as well as being the foundation for this study.

3 Methodologies

The literature review on previous studies, including the function of KM in protecting the information privacy and the adopted privacy models has provided the background of this study. The research methodology involved with interviews session as a preliminary investigation of this study. The purpose of the interviews was to gauge the existing procedures of personal information management and privacy mechanism embraced by the Malaysian hospitals. The interviews served as a cognitive assessment in verifying the feasibility of achieving the research objectives. The interviewees were identified under strict scrutiny in order to ensure the reliability of their input. This is primarily essential because the final conclusion of this study will certainly aid in further understanding of the required components in designing and developing a comprehensive privacy protection guideline that is skewed appropriately for the Malaysian environment.

Three important criteria were formulated in justifying the interview process. They include,

i. Semi-structured interviews that were conducted based on the models, which the analysis were then referred to the current Personal Data Protection Act.
ii. Semi-structured interviews that were selectively conducted with authorized personnel whom were highly involved in policy making in data protection in the government, have hands-on experience in matters related to hospital privacy and data protection policy.
iii. Interviews were also conducted to determine the practitioners' perspective, whereby the interviewees were chosen based on their knowledge in privacy policy and also their strong contribution towards the development of the EMR system in Malaysia.

The interviewees were diligently selected across the Peninsular Malaysia, whom has a wide range of responsibilities in the development of hospital information system. They were designated as the Information Technology Director and Officer, Electronic Medical Record Officer, doctor and magistrate, as depicted in Table 2. The interviewees were mostly selected due to their contribution and involvement in the establishment of HIS and belonged to the Ministry of Health (MOH) monitoring committee of HIS. They were contacted through phone calls in order to secure a face-to-face appointment. Both government and semi-government hospitals personnel were included.

Table 2 Pilot Study Interviewee

No	Interviewee	Designation	Organization
1	A	Information Technology Director	Ministry of Health
2	B	Information Technology Officer	Ministry of Health
3	C	Electronic Medical Record Officer	Government Hospital
4	D	Electronic Medical Record Officer	Semi-government Hospital
5	E	Doctor	Government Hospital
6	F	Magistrate	Malaysian Magistrate Court

There were six (6) interviews conducted between November 2009 and March 2010 period. Each session took 45 to 50 minutes to complete. The findings from these interviews were then compared with the existing privacy protection models. The findings have been useful in devising a new set of models as a guideline in designing a capable electronic medical record system that protects the personal medical information privacy.

4 Findings

Based on the preliminary investigation, this study has successfully listed the proposed components necessary for the information privacy protection. For reference purpose, the transcripts of the interviews are available in text and the crux of the discussions and points were also listed. Unlike the models which were proposed by EPIC, the comprehensive components mentioned below need to be incorporated simultaneously during the early stage of system development to ensure the efficiency of the system. The four components identified for the proposed models are as follows, accompanied with comments by interviewees.

A. Legislation

Legislation or Law is the most decisive factor in protecting an individual's information privacy. The law (whatever rules enacted by the government that refer to the controlling and the manipulation of personal information) is significant in assuring individual's right over his/her personal information. As mentioned by a legal expert, "personal health/medical information in Malaysia is categorized under 'sensitive information' in personal data protection act" (Interviewee F). The Malaysia Personal Data Protection Act 2010 for example, enlisted seven principles in any act of deliberation and manipulation of an individual's personal information; general principle, notice & choice principle, disclosure principle, security principle, retention principle, data integrity principle and access principle.

B. Ethics

Ethics is a self-regulation that is basically referred to several codes of ethics that are being practiced by certain organizations or institutions such as the hospitals. For countries without any privacy or data protection law, the people can only cling on to their 'codes of ethics' in their pursuit of protecting individual's rights. The code of ethics therefore must also be stated in their service policies. In a hospital scenario, the principles of ethics are usually conformed to the international standard of services provided in Medical Act or Nurses Act, which also covers the policy of individual privacy. One of the officers from the Health Ministry of Malaysia remarked, "since the personal data protection act has yet to be enforced, our employees including the doctors, nurses and other staffs must strictly adhere to the ethical code of conducts" (Interviewee B).

C. Technology

Technology must also be incorporated in the privacy protection initiatives due to the transformation of information management in the hospital into paperless-based system. The embarkation to the paperless-based system has made the usage of privacy mechanism technology apparently inevitable. One of the interviewees expressed that, "we are developing our own health information system in order to improve security and privacy control over health information management with our very own privacy mechanism technologies. Patients will have the opportunity to access their own personal medical information" (Interviewee C).

D. Culture

Culture is the most crucial model to be considered and has been a distinctive factor from the previous studies. Most of the interviewees have cited that the culture-based privacy represents as a key model, which its practice differ from one country to another. This model replicates Moor's perception on privacy which stated that, 'the concept of privacy has distinct cultural aspect which goes beyond the core values. Some cultures may value privacy and some may not' [23]. Western countries should not expect similar privacy practices in the Asian countries, which value privacy much less than them. Karat said, 'privacy means different things to different people' [24]. As stated by one of the interviewees "Malaysia is perhaps not fully ready for the full implementation of privacy protection by using the western framework and orientation. We are one unique society where distinction gap in social cultures and values co-exist, unless a mandatory modification is done for the privacy concept to suit us" (Interviewee D). It is the instilled culture of the people that determines the extent of privacy preservation of personal information. However, it should be noted that privacy of information is still commonly regarded as a basic human right. Therefore, in order to avoid any conflicts with regards to privacy and its implementation, adequate measures of precaution must be in place.

Both interviewees A and E have mentioned a general interest within the community which entails the individual's concern on personal medical information privacy. Nonetheless, this is not categorized in any of the components identified.

5 Discussion: Towards a Human-Centered System Design

The personal information privacy models listed above; technology, legislation, ethics, and culture are the obligatory components to be factored in designing an EMR system. In developing a personal information management system with built in security and privacy protection, the balance of combination between the concept of human-centered and technology-centered design needs to be taken into account to warrant the effectiveness of the system. Human-centered design embraces the physical and psychological needs of the human user, enabling the user to function at the highest level possible. The technology-centered model is also vital in providing the current and latest security mechanisms to be embedded into the system, making it capable to achieve the health care provider's goal in its usability and usefulness. Legislation or law is also a salient aspect to be included when dealing with the issues of information privacy and security. Inevitably, new technology and system designs must also adhere to the rules and guidelines in the privacy or data protection act legislated by the government. Due to its nature, the researcher has sorted both technology and legislation as technology-centered development, whereas ethics and culture components are categorized as human-centered design. The justification on this classification is further explained in the next paragraph. The role of KM is also required in providing a systematic guideline to assist the system users in managing the system effectively.

The ethics component in computer and technology field can be described as the study of those behavioural actions of IT professionals that will benefit all of society [25]. Several scholars [19][20] preferred to refer the term as self-regulations provided by an organization in its 'code of conduct' or 'code of ethics'. As ethics or self-regulations usually have its own set of recorded rules to be distributed to the users or employees, the researcher has classified this model as a type of explicit knowledge. Thus, whenever it is applied in a HIS environment especially in the EMR unit in the hospital, it can be referred as explicit task. By adapting it into the Smith [8] study, explicit task can be illustrated as a kind of comprehensive manual for handling an EMR system with security and privacy mechanism attached. The manual shall include all the modules listed by Smith such as how to operate the system (work process), a complete training syllabus on EMR operation (teach), security and privacy etiquette (motivation) and so on.

Additionally, the culture component can be viewed as what a group learns over a period of time as the group solves its problems of survival in an external environment and its problems of internal integration [26]. Such learning is simultaneously a behavioural, cognitive and emotional process [27]. A task using tacit knowledge is accomplished through specific individual involvement and requires the individual to make changes to their existing behaviour [28]. Thus, the cultural component is suitable for tacit knowledge to be applied, in providing the employees with excellent attitude towards protecting information privacy while they handle an EMR system. This study summarized the adaptation of Smith's study called Human-Centered Design for EMR System in the Table 3 below;

Table 3 Human-Centered Design for EMR System

Ethic Component	Culture Component
Explicit Tasks – a comprehensive manual for handling an EMR system with security and privacy mechanism and concern, follow the hospital 'code of conduct'	**Tacit Tasks** – based on employees past experiences and awareness in handling patient's personal information
Work process – understand the manual and know how to operate the EMR system	**Work practice** – experienced and expert in operating an EMR system
Learn – on the job, trial-and-error, handling system based on employees' 'code of ethics'	**Learn** – senior staffs or nurses to facilitate and reinforce the task with openness and trust in sharing their knowledge
Teach – IT officer as trainer, provides a set of comprehensive modules for syllabus EMR training	**Teach** – IT officer provide hands-on training for the hospital staffs, one-on-one, mentoring, internships and coaching
Share knowledge – distribute skills through e-mail, electronic discussions, forums	**Share knowledge** – distribute knowledge through altruistic sharing, networking, videoconferencing and personalized knowledge
Motivation – towards HIS mission in securing patient personal information privacy	**Motivation** – inspired through reliable and trusted organizational culture
Reward – gaining the trust and confidence from patients	**Reward** – 'excellent service award' for staff as a kind of recognition
Relationship – senior doctors or nurses to junior doctors or nurses	**Relationship** – spontaneous sharing of knowledge among any hospital staff assigned in handling the EMR
Evaluation – does the EMR technology effective in securing patient information privacy?	**Evaluation** – does the EMR system achieve its usability goal?

6 Conclusion

This study emphasizes on the importance of knowledge management (KM) role in ensuring the components that are relevant to any electronic medical record (EMR) system development. Through interviews administered, four components have emerged as mentioned by the HIS experts. This on-going study intends to adopt the legislation, ethics, technology and culture in securing the privacy of electronic medical record system. Essentially, the application of the system does not depend on the technology-centered development alone. It must also incorporate the explicit and tacit knowledge concept into the ethics and culture model during the system's implementation and maintenance. The incorporation of those concepts will definitely assist in generating a system which is more of human-centered state-of-the-art. It is hoped that this preliminary investigation will be useful in integrating the concepts of KM with computerization and information privacy aspect, particularly in the industry of health care. Nonetheless, further studies need to be conducted in developing a comprehensive guideline in adopting a systematic EMR work tasks which also incorporate the KM concepts. A proper and comprehensive training needs to be provided that includes information privacy awareness in handling patient's information. This may ensures the confidentiality, integrity and availability of the information is consistently preserved. Future works of this study may consider other aspect of qualitative data collection such as observation or case study. However, both explicit knowledge and tacit knowledge has played a significant role towards the establishment of privacy preservation for HIS.

Acknowledgments. Suhaila Samsuri is a PhD candidate, currently a staff of International Islamic University Malaysia. Both Dr Zuraini Ismail and Dr Rabiah Ahmad are Associate Professors at their respective universities. This study is funded by the Research University Grant (RUG) from Universiti Teknologi Malaysia (UTM) and Ministry of Higher Education (MOHE) Malaysia with project number Q.K 130000.2138.01H98. The authors would like to thank all the interviewees for their cooperation and support in this research.

References

1. Samsuri, S., Ismail, Z., Ahmad, R.: User-Centered Evaluation of Privacy Models for Protecting Personal Medical Information. In: Abd Manaf, A., et al. (eds.) ICIEIS 2011, Part I. CCIS, vol. 251, pp. 301–309. Springer, Heidelberg (2011)
2. Smuts, H., van der Merwe, A., Loock, M., Kotzé, P.: A Framework and Methodology for Knowledge Management System Implementation. In: SAICSIT 2009, pp. 70–79 (2009)
3. Xiong, L., Xia, Y.: Report on ACM Workshop on Health Information and Knowledge Management (HIKM 2006). SIGMOD Record 36(1), 39–42 (2007)
4. Samsuri, S., Ismail, Z., Ahmad, R.: Privacy models for protecting personal medical information: A preliminary study. In: International Conference of Research and Innovation in Information Systems, ICRIIS 2011, pp. 1–5 (2011)
5. Liss, K.: Do we know how to do that? Understanding knowledge management. Harvard Management Update, 1–4 (1999)
6. Kankanhalli, A., Tanudidjaja, F., Sutanto, J., Tan, B.C.Y.: The Role of IT in Successful Knowledge Management Initiatives. Communications of the ACM 46(9), 69–73 (2003)
7. Chait, L.: Creating a successful knowledge management system. Prism (second quarter 1998)
8. Smith, E.A.: The role of tacit and explicit knowledge in the workplace. Journal of Knowledge Management 5(4), 311–321 (2001)
9. Petrides, L.A.: Knowledge Management, Information Systems and Organizations. EDUCAUSE Center for Applied Research (20) (2004)
10. Anthens, G.H.: A Step Beyond A Database. Computerworld 25(9), 28 (1991)
11. Maglitta, J.: Smarten Up! Computerworld 29(23), 84–86 (1995)
12. Shami Zanjani, M., Mehrasa, S., Modiri, M.: Organizational dimensions as determinant factors of KM approaches in SMEs. World Acad. Sci. Engin. Technol. 45 (2008)
13. Sanchez, R.: "Tacit Knowledge" versus "Explicit Knowledge" Approaches to Knowledge Management Practice (2005), `http://www.knowledgeboard.com/download/3512/Tacit-vs-Explicit.pdf` (accessed on January 2012)

14. Tuzhilin, A.: Knowledge Management Revisited: Old Dogs, New Tricks. ACM Transactions on Management Information Systems 2(3), 13 (2011)
15. Murphy, F., Stapleton, L., Smith, D.: Tacit Knowledge and Human Centred Systems: The Key to Managing the Social Impact of Technology. In: International Multitrack Conference of Advances in Control Systems (2004)
16. Sidek, Z.M., Abdul Ghani, N.: Utilizing Hippocratic Database for Personal Information Privacy Protection. Jurnal Teknologi Maklumat 20(3) (2008)
17. Blumenthal, D., Tavenner, R.N.: The "Meaningful Use" Regulation for Electronic Health Records. The New England Journal of Medicine 363(6), 501–504 (2010)
18. Whitley, E.A.: Informational privacy, consent and the "control" of personal data. Information Security Technical Report 14, 154–159 (2009)
19. Fischer-Hübner, S., Thomas, D.: Privacy and Security at risk in the Global Information Society. In: Loade, B. (ed.) Cybercrime. Routledge, London (2000)
20. Fischer-Hübner, S.: Privacy-Enhancing Technologies (PET), Course Description. Karlstad University Division for Information Technology, Karlstad (2001)
21. Swire, P.: Privacy Today, International Association of Privacy Professionals (2008), `http://www.privacyassociation.org` (accessed on March 2010)
22. Holvast, J.: History of Privacy. In: Matyáš, V., Fischer-Hübner, S., Cvrček, D., Švenda, P. (eds.) IFIP WG 9.2, 9.6/11.6, 11.7/FIDIS. IFIP AICT, vol. 298, pp. 13–42. Springer, Heidelberg (2009)
23. Moor, J.H.: Towards a Theory of Privacy in the Information Age. In: CEPE 1997 Computers and Society, pp. 27–32 (1997)
24. Karat, J., Karat, C.M., Broodie, C.: Human Computer Interaction viewed from the intersection of privacy, security and trust. In: Sears, A., Jacko, J.A. (eds.) The Human Computer Interaction Handbook: Fundamentals, Evolving Technologies and Emerging Applications, 2nd edn. Lawrence Erlbaum Associates, New York (2008)
25. De Ridder, C., Pretorius, L., Barnard, A.: Towards teaching Computer Ethics. Technical Reports (2001), `http://osprey.unisa.ac.za/TechnicalReports/UNISA-TR-2001-8.pdf`
26. MacIntosh, E.W., Doherty, A.: The influence of organizational culture on job satisfaction and intention to leave. Sport Management Review 13, 106–117 (2010)
27. Scheres, H., Rhodes, C.: Between cultures: Values, training and identity in a manufacturing firm. Journal of Organizational Change Management 19, 223–236 (2006)
28. Hicks, R.C., Dattero, R., Galup, S.D.: A metaphor for knowledge management: explicit islands in a tacit sea. Journal of Knowledge Management 11(1), 5–16 (2007)

An Overview on the Structure and Applications for Business Intelligence and Data Mining in Cloud Computing

A. Fernández[1], S. del Río[2], F. Herrera[2], and J.M. Benítez[2]

[1] Dept. of Computer Science, University of Jaén, Jaén, Spain
alberto.fernandez@ujaen.es

[2] Dept. of Computer Science and Artificial Intelligence,
CITIC-UGR (Research Center on Information and Communications Technology),
University of Granada, 18071 Granada, Spain
saradelriogarcia@gmail.com, {herrera,jmbs}@decsai.ugr.es

Abstract. Cloud Computing is a new computational paradigm which has attracted a lot of interest within the business and research community. Its objective is to integrate a wide amount of heterogeneous resources in an online way to provide services under demand to different types of users, which are liberated from the details of the inner infrastructure, just concentrating on their request of resources over the net. Its main features include an elastic resource configuration and therefore a suitable framework for addressing scalability in an optimal way. From the different scenarios in which Cloud Computing could be applied, its use in Business Intelligence and Data Mining in enterprises delivers the highest expectations. The main aim is to extract knowledge of the current working of the business, and therefore to be able to anticipate certain critical operations, such as those based on sales data, fraud detection or the analysis of the clients' behavior. In this work, we give an overview of the current state of the structure of Cloud Computing for applications on Business Intelligence and Data Mining. We provide details of the layers that are needed to develop such a system in different levels of abstraction, that is, from the underlying hardware platforms to the software resources available to implement the applications. Finally, we present some examples of approaches from the field of Data Mining that had been migrated to the Cloud Computing paradigm.

1 Introduction

The Cloud Computing infrastructure has its origins in the concept of grid computing, which has the aim of reducing computational costs and to increase the flexibility and reliability of the systems. However, the difference between the two lies in the way the tasks are computed in each respective environment. In a computational grid, one large job is divided into many small portions and executed on multiple

L. Uden et al. (Eds.): 7th International Conference on KMO, AISC 172, pp. 559–570.
springerlink.com

machines, offering a similar facility for computing power. On the other hand, the computing cloud is intended to allow the user to obtain various services without investing in the underlying architecture and therefore is not so restrictive and can offer many different services, from web hosting, right down to word processing [3].

A Service Oriented Architecture (SOA) [27] is one of the basis of Cloud Computing. This type of system is designed to allow developers to overcome many distributed enterprise computing challenges including application integration, transaction management and security policies, while allowing multiple platforms and protocols and leveraging numerous access devices and legacy systems [2]. We can find some different services that a Cloud Computing infrastructure can provide over the Internet, such as a storage cloud, a data management service, or a computational cloud. All these services are given to the user without requiring them to know the location and other details of the computing infrastructure [11].

One of the successful areas of application for Cloud Computing is the one related to Business Intelligence (BI) [1, 18]. From a general perspective, this topic refers to decision support systems, which combines data gathering, data storage, and knowledge management with analysis to provide input to the decision process [24] integrating data warehousing, Data Mining (DM) and data visualization, which help organizing historical information in the hands of business analysts to generate reporting that informs executives and senior departmental managers of strategic and tactical trends and opportunities.

No need to say that the processing of a high amount of data from an enterprize in a short period of time has a great computational cost. As stated above, with these constraints a new challenge for the research community comes out with the necessity to adapt the systems to a Cloud Computing architecture [26]. The main aim is to parallelize the effort, enable fault tolerance and allowing the information management systems to answer several queries in a wide search environment, both in the level of quantity of information as for the computational time.

Along this contribution, we will first study the common structure of this type of models in order to face DM problems. Specifically, we will describe a standard infrastructure from the Cloud Computing scenario, presenting the different layers that must be taken into account to implement the management of the data, the parallel execution engine and the query language [4, 8].

Apart from the specific application of BI, the possibilities that the paradigm of Cloud Computing offers to DM processes with the aid of cloud virtualization is quite significative. This issue grows in relevance from the point of view of the parallelization of high computational cost algorithms, for example those methodologies based on evolutionary models [22, 32, 31]. Cloud Computing platforms such as MapReduce [7] and Hadoop (`http://hadoop.apache.org`) are two programming models which helps the developers to include their algorithms into a cloud environment. Both systems have been designed with two important restrictions: first, clouds have assumed that all the nodes in the cloud are co-located, i.e. within one data centre, or that there is relatively small bandwidth available between the geographically distributed clusters containing the data. Second, these clouds have assumed that individual inputs and outputs to the cloud are relatively small, although the aggregate data managed and processed are very large.

In the last years, many standard DM algorithms have been migrated to the cloud paradigm. In this work we will present several existing proposals that can be found in the literature that adapt standard algorithms to the cloud for making them high efficient and scalable. The growing number of applications in real engineering problems, such as computer vision [34], recommendation systems [19] or Health-care systems [28] show the high significance of this approach.

This contribution is arranged as follows. In Section 2 we introduce the main concepts on Cloud Computing, including its infrastructure and main layers. Next, Section 3 presents the architecture approaches to develop BI solutions on a Cloud Computing platform. The programming models for implementing DM algorithms within this paradigm, together with some examples, is shown in Section 4. Finally, the main concluding remarks are given in Section 5.

2 Basic Concepts on Cloud Computing

We may define an SOA [27] as an integration platform based on the combination of a logical and technological architecture oriented to support and integrate all kind of services. In general, a "Service" in the framework of Cloud Computing is a task that has been encapsulated in a way that it can be automated and supplied to the clients in a consistent and constant way.

The philosophy of Cloud Computing mainly implies a change in the way of solving the problems by using computers. The design of the applications is based upon the use and combination of services. On the contrary that occurs in more traditional approaches, i.e. grid computing, the provision of the functionality relays on this use and combination of services rather than the concept of process or algorithm.

Clearly, this brings advantages in different aspects, for example the scalability, reliability, and so on, where an application, in the presence of a peak of resources' demand, because of an increase of users or an increase of the data that those provide, can still give an answer in real time since it can get more instances of a determinate service; the same occurs in the case of a fall of the demand, for which it can liberate resources, all of these actions in a transparent way to the user.

The main features of this architecture are its loose coupling, high inter-operativity and to have some interfaces that isolate the service from the implementation and the platform. In an SOA, the services tend to be organized in a general way in layers or levels (not necessarily with strict divisions) where normally, some modules use the services that are provided by the lower levels to offer other services to the superior levels. Furthermore, those levels may have different organization structure, a different architecture, etc.

There exists different categories in which the service oriented systems can be clustered. One of the most used criteria to group these systems is the abstraction level that offers to the system user. In this manner, three different levels are often distinguished , as we can observe in Figure 1. In the remainder of this section, we

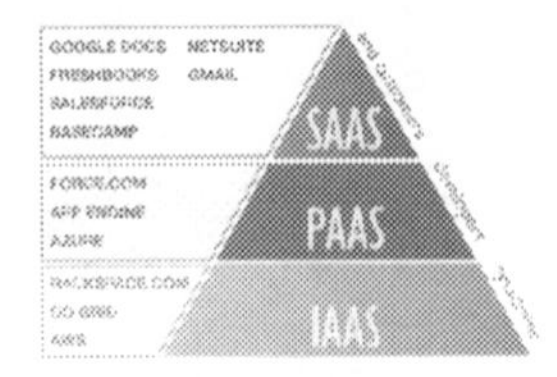

Fig. 1 Illustration of the layers for the Services Oriented Architecture

will first describe each one of these three levels, providing the features that defines each one of them and some examples of the most known systems of each type. Next we will present some technological challenges that must be taken into account for the development of a Cloud Computing system.

2.1 Infrastructure as a Service (IaaS)

IaaS is the supply of hardware as a service, that is, servers, net technology, storage or computation, as well as basic characteristics such as Operating Systems and virtualization of hardware resources [16]. Making an analogy with a monocomputer system, the IaaS will correspond to the hardware of such a computer together with the Operating System that take care of the management of the hardware resources and ease the access to them.

The IaaS client rents the computational resources instead of buying and installing its own data center. The service usually is billed based in its actual usage, so it has the advantage that the client pays for what it uses and it uses what he needs in each moment. Also, according to the dynamical scaling associated to Cloud Computing, in the case of low loads of work it uses (and pays for) less resources, and in the presence of a higher resources demand, IaaS can provide them to attend the punctual necessities of that client, being this task done in real time. It is also frequent that the contract of the service includes a maximum that the user cannot exceed.

One kind of typical IaaS clients are scientific researchers and technicians, which thanks to the IaaS and the wide volume of infrastructure that they offer as a service, they can develop tests and analysis of the data in a level that could not be possible without the access to the this big scale computational infrastructure.

2.2 Platform as a Service (PaaS)

At the PaaS level, the provider supplies more than just infrastructure, i.e. an integrated set of software with all the stuff that a developer needs to build applications, both for the developing and for the execution stages. In this manner, a PaaS provider does not provide the infrastructure directly, but making use of the services of an IaaS it presents the tools that a developer needs to, having an indirect access to the IaaS services and, consequently, to the infrastructure [16].

If we follow the analogy set out in the previous subsection relative to a monocomputer system, the PaaS will correspond to a software layer that enables to develop components for applications, as well as applications themselves. It will be an Integrated Developer Environment, or a set of independent tools, that allows to develop an engineering software problem in all its stages, from the analysis and model of the problem, the design of the solution, its implementation and the necessary tests before carrying out the stage of deployment and exploitation. In the same manner, a programming language that counts with compilers and libraries for the different Operating Systems will allow that the same application can be deployed in different systems without the necessity of rewrite any piece of code.

2.3 *Software as a Service (SaaS)*

In the last level we may find the SaaS, i.e. to offer software as a service. This was one of the first implementations of the Cloud services, along with the gaining in importance of the Internet usage. It has its origins in the host operations carried out by the Application Service Providers, from which some enterprises offered to others the applications known as Customer Relationship Managements [9].

Throughout the time, this offer has evolved to a wide range of possibilities, both for enterprises and for particular users. Regarding the net support, although these services are performed through the Internet, as it provides the geographical mobility and the flexibility needed, a simple exchange of data in this manner will not assure the privacy of them. For this reason, Virtual Private Networks are often employed for this aim, as they allow to transmit data through the Internet in an encrypted way, maintaining the privacy and security in the information exchange between the client application of the user and the SaaS application store in the cloud.

2.4 *Technological Challenges in Cloud Computing*

Cloud computing has shown to be a very effective paradigm according to its features such as on-demand self-service since the custermors are able to provision computing capabilities without requiring any human interaction; broad network access from heterogeneous client platforms; resource pooling to serve multiple consumers; rapid elasticity as the capabilities appear to be unlimited from the consumer's point of view; and a measured service allowing a pay-per-use business model. However, in order to offer such a advantageous platform, there are some weak points that are needed to take into account. Next, we present some of these issues:

- Security, privacy and confidence: Since the data can be distributed on different servers, and "out of the control" of the customer, there is a necessity of managing hardware for computation with encoding data by using robust and efficient methods. Also, in order to increase the confidence of the user, several audits and certifications of the security must be performed.

- Availability, fault tolerance and recovery: to guarantee a permanent service (24x7) with the use of redundant systems and to avoid net traffic overflow.
- Scalability: In order to adapt the necessary resources under changing demands of the user by providing an intelligent resource management, an effective monitorization can be used by identifying a priori the usage patterns and to predict the load in order to optimize the scheduling.
- Energy efficiency: It is also important to reduce the electric charge by using microprocessors with a lower energy consumption and adaptable to their use.

3 Cloud Computing for Business Intelligence Processes

As it was stated in the introduction of this work, the philosophy of Business Intelligence systems is to satisfy the necessity of analyzing large amounts of data in a short period of time, usually in just a matter of seconds or minutes. This high volume of data may come as a result of different applications such as prediction of loan concessions and credit policies to clients based on risks, classification or clustering of clients for beam marketing, product recommendations and extraction of patterns from commercial transactions among others.

In the remainder of this section, we will first present an architecture to develop BI solutions on a Cloud Computing platform. Then, we will stress the goodness on the use a Cloud Computing with respect to other similar technologies.

3.1 Organization of the Cloud Computing Environment

To address the goals stated in the beginning of this section, in [4, 8] the authors revisited a basic model and process for analyzing structured and unstructured user generated content in a business warehouse. This data management organization architecture based on clouds follows a four layer architecture, although there exists several similar approaches [21, 23]. We must point out that the three first tiers are common for DM and BI approaches, whereas the last one is specifically designed for data warehousing and On-Line Analytical Processing (OLAP) applications. The description of these components is enumerated below:

- The first level is the infrastructure tier based on Cloud Computing. It follows the structure introduced in Section 2, that is, it includes a network architecture with many loosely coupled computer nodes for providing a good scalability and a fault tolerance scheme. As suggested, the system must take into account a dynamic/elastic scheme such that the performance, cost and energy consumption of the node machines is managed at run-time.
- The second layer is devoted to the parallel data storage. The relational data bases may have difficulties when processing the data along a big number of servers, so that there is a necessity of supporting new data base management systems based on the storage and retrieval of key/value pairs (as opposed to the relational model

based on foreign-key/primary-key relationships). Some examples of systems that are optimized for this purpose are Google BigTable [5] or Sector [13].

- The third level is the execution environment. Due to the large number of nodes, Cloud Computing is especially applicable for distributed computing tasks working on elementary operations. The most known example of cloud computing execution environment is probably Google MapReduce [7] and its open source version Hadoop [10], but other projects can be found as feasible alternatives [17, 33]. All these environments aim at providing elasticity by allowing to adjust resources according to the application, handling errors transparently and ensuring the scalability of the system.
- The last tier is the high querying language tier, which is oriented towards OLAP, and Query/Reporting Data-Warehousing tools. This layer is the interface to the user and it provides the transparency to the other tiers of the architecture. Some query languages have been proposed like Map-Reduce-Merge [36] or the Pig Latin language [25] which has been designed to propose a trade-off between the declarative style of SQL, and the low-level, procedural style of MapReduce.

3.2 *On the Suitability of Cloud Computing for Business Intelligence and Data Mining*

Traditionally, when having a high amount of data to be processed in a short period of time, a grid computing environment was the most suitable solution in order to reduce computational costs and to increase the flexibility of the system. The similarities with Cloud Computing are evident, since both of them are composed of loosely coupled, heterogeneous, and geographically dispersed nodes. However, the main difference between the two lies in the way the tasks are computed in each respective environment. In a computational grid, one large job is divided into many small portions and executed on multiple machines, offering a similar facility for computing power. On the other hand, the computing cloud is intended to allow the user to obtain various services without investing in the underlying architecture and therefore is not so restrictive and can offer many different services, from web hosting, right down to word processing [3].

Additionally, the advantages of this new computational paradigm with respect to other competing technologies are clear. First, Cloud application providers strive to give the same or better service and performance as if the software programs were installed locally on end-user computers, so the users do not need to spend money buying a complete hardware equipment for the software to be used, i.e. a simple PDA device is enough to run the programs on the Cloud.

Second, this type of environment for the data storage and the computing schemes allows enterprizes to get their applications up and running faster, with a lower necessity of maintenance from the IT department since it automatically manages the business demand by assigning more or less IT resources (servers, storage and/or networking) depending on the computational load in real time [30]. Finally, this inherent elasticity of this system makes the billing of the infrastructure to be done according to the former fact.

4 Adaptation of Global Data Mining Tasks within a Cloud Computing Environment

In this section we aim at pointing out the promising future that is foreseen for the Cloud Computing paradigm regarding the implementation of DM algorithms in order to deal with very large data-sets for which it has been prohibitively expensive until this moment.

The idea behind all the proposals we will introduce is always distributing the execution of the data among all the nodes of the cloud and to transfer the least volume of information as possible to make the applications highly scalable and efficient, but always maintaining the integrity and privacy of the data [14, 29].

As introduced in Section 3, it is necessary to transform the data stored into multidimensional arrays to "Pig data" [25] to be able to carry out an online analysis process by using the MapReduce/Hadoop scheme or related approaches (such as Sector/Sphere [13]), also reducing the storage costs [8]. In this programming model, users specify the computation in terms of a map and a reduce function, and the underlying runtime system automatically parallelizes the computation across large-scale clusters of machines, handles machine failures, and schedules inter-machine communication to make efficient use of the network and disks.

Map, written by the user, takes an input pair and produces a set of intermediate key/value pairs. The MapReduce library groups together all intermediate values associated with the same intermediate key I and passes them to the reduce function. The reduce function accepts an intermediate key I and a set of values for that key. It merges these values together to form a possibly smaller set of values. The intermediate values are supplied to the user's reduce function via an iterator. This allows us to handle lists of values that are too large to fit in memory.

The map invocations are distributed across multiple machines by automatically partitioning the input data into a set of M splits. The input splits can be processed in parallel by different machines. Reduce invocations are distributed by partitioning the intermediate key space into R pieces using a partitioning function (e.g., hash(key) mod R). The number of partitions (R) and the partitioning function are specified by the user (see Figure 2).

In brief, these systems are oriented to distribute the data sets along the cloud and to distribute the computation among the clusters, i.e. instead of moving the data among the machines, they define mapping functions to create intermediary tuples of <key, value> and the use of reduction functions for this special processing. As an example, in a program aimed at counting the number of occurrences of each word in a large collection of documents, the map function will emit each word plus an associated count of occurrences whereas the reduce function sums together all counts emitted for a particular word.

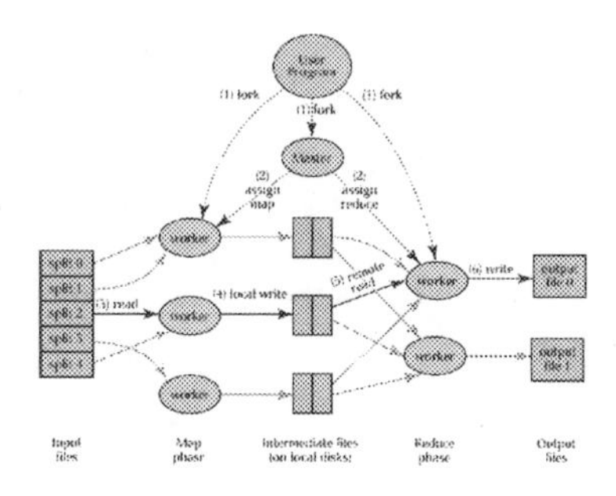

Fig. 2 Overall flow of execution a MapReduce operation

Although this is a relatively new framework, many DM algorithms have been already implemented following the guidelines of this programming model. For example, in [6] the authors present a classification model which tries to find an intermediate model between Bayes and K-nearest neighbor techniques by using a certain kind of subtree to represent each cluster which is obtained by clustering on training set by applying a minimum spanning tree MapReduce implementation, and then perform the classification using idea similar to KNN.

Another approach that is developed using the MapReduce model aims at addressing the progressive sequential pattern mining problem [15], which intrinsically suffers from the scalability problem. Two Map/Reduce jobs are designed; the candidate computing job computes candidate sequential patterns of all sequences and updates the summary of each sequence for the future computation. Then, using all candidate sequential patterns as input data, the support assembling job accumulates the occurrence frequencies of candidate sequential patterns in the current period of interest and reports frequent sequential patterns to users.

Gao et al. introduces in [12] an experimental analysis using a Random Decision Tree algorithm under a cloud computing environment by considering two different schemes in order to implement the parallelization of the learning stage. The first approach was that each node built up one or more classifiers with its local data concurrently and all classifiers are reported to a central node. Then the central node will use all classifiers together to do predictions. The second option was that each node works on a subtask of one or more classifiers and reports its result to a central node, then the central node combines work from all local nodes to generate the final classifiers and use them for prediction.

Other works are based on different cloud computing environments, a Particle Swarm Optimization was designed for the Amazon Elastic Compute Cloud (`http://aws.amazon.com`), where the candidate solutions are presented by the set of task-service pairs, having each particle to learn from different exemplars, but to learn the other feasible pairs for different dimensions. The constructive position building procedure guarantees each position was shown to be feasible and this scheme greatly reduces the search space and enhances the algorithm performance [35].

Lin and Luo proposed a novel DM method named FD-Mine [20] that is able to efficiently utilize the cloud nodes to fast discover frequent patterns in cloud computing environments with data privacy preserved. Through empirical evaluations on various simulation conditions, the proposed FD-Mine showed to deliver excellent performance in terms of scalability and execution time.

Finally, it is important to point out that there exist open source DM libraries from which the users can use the techniques implemented under these software platform. We may stress the Mahout library (`http://mahout.apache.org/`) which is mainly based on clustering approaches, but also parallel frequent pattern mining; and the Pentaho Business Analytics (`http://www.pentaho.com/big-data/`) which offers unmatched native support for the most popular big data sources including Hadoop, NoSQL (not relational data base models) and analytic databases. Additionally, the relevance in this area can be stressed by the vast amount of available commercial SaaS products such as Actuate (`http://www.actuate.com/`), ComSCI (`http://www.comsci.com/`) or FPX (`http://www.fpx.com/`); however, for most of them it is a bit unclear what they truly offer to the user in terms of which techniques they may implement for managing and mining the data, i.e. what they include in their toolbox to enable analysts to create reports and custom analyzes.

5 Concluding Remarks

In this work we have presented an overview on BI and DM applications within the Cloud Computing environment. In particular, we aimed at stressing the significance and great possibilities of this topic in the near future, since it offers an scalable framework for those high dimensional problems which are hard to overcome with the standard technologies.

Taking this into account, we have first introduced the main features of the Cloud Computing infrastructure, i.e. the different levels of abstraction which allow us to understand its nature. Then, we described its advantages with respect to other classical technologies such as grid computing and therefore how DM applications obtains higher benefits from this scheme.

Next, we have described the specific architecture that is needed in order to develop BI and DM applications within a cloud framework. Specifically, a four layer structure is suggested following the recommendations given in the specialized literature. This structure is composed of a cloud infrastructure to provide the service nodes and the communication network at the lowest level, a parallel data storage to distribute the information across the cloud, a execution environment which must take advantage of the characteristics offered by the cloud and finally the use of a high query language to fully exploit the features of the parallel execution tier.

To end with, we have presented several DM algorithms that have been migrated to Cloud Computing, examining the particularities of their implementation which are mainly based on the MapReduce/Hadoop scheme, which is currently the most important execution environment to allow an efficient parallelization of the data processing within the cloud.

References

1. Abelló, A., Romero, O.: Service-Oriented Business Intelligence. In: Aufaure, M.-A., Zimányi, E. (eds.) eBISS 2011. LNBIP, vol. 96, pp. 156–185. Springer, Heidelberg (2012)
2. Alonso, G., Casati, F., Kuno, H., Machiraju, V.: Web Services: Concepts, Architectures and Applications. Springer, Heidelberg (2004)
3. Buyya, R., Yeo, C., Venugopal, S., Broberg, J., Brandic, I.: Cloud computing and emerging it platforms: Vision, hype, and reality for delivering computing as the 5th utility. Future Generation Computer Systems 25(6), 599–616 (2009)
4. Castellanos, M., Dayal, U., Sellis, T., Aalst, W., Mylopoulos, J., Rosemann, M., Shaw, M.J., Szyperski, C. (eds.): Optimization Techniques 1974. LNBIP, vol. 27. Springer, Berlin (1975)
5. Chang, F., Dean, J., Ghemawat, S., Hsieh, W.C., Wallach, D.A., Burrows, M., Chandra, T., Fikes, A., Gruber, R.E.: Bigtable: A distributed storage system for structured data. ACM Trans. Comput. Syst. 26(2) (2008)
6. Chang, J., Luo, J., Huang, J.Z., Feng, S., Fan, J.: Minimum spanning tree based classification model for massive data with mapreduce implementation. In: Fan, W., Hsu, W., Webb, G.I., Liu, B., Zhang, C., Gunopulos, D., Wu, X. (eds.) ICDM Workshops, pp. 129–137. IEEE Computer Society (2010)
7. Dean, J., Ghemawat, S.: MapReduce: simplified data processing on large clusters. Communications of the ACM 51(1), 107–113 (2008)
8. d'Orazio, L., Bimonte, S.: Multidimensional Arrays for Warehousing Data on Clouds. In: Hameurlain, A., Morvan, F., Tjoa, A.M. (eds.) Globe 2010. LNCS, vol. 6265, pp. 26–37. Springer, Heidelberg (2010)
9. Duer, W.: CRM, Customer Relationship Management. MP editions (2003)
10. Foundation, T.A.S.: Hadoop, an open source implementing of mapreduce and GFS (2012), http://hadoop.apache.org
11. Furht, B., Escalante, A. (eds.): Handbook of Cloud Computing. Springer, US (2010)
12. Gao, W., Grossman, R.L., Yu, P.S., Gu, Y.: Why naive ensembles do not work in cloud computing. In: Saygin, Y., Yu, J.X., Kargupta, H., Wang, W., Ranka, S., Yu, P.S., Wu, X. (eds.) ICDM Workshops, pp. 282–289. IEEE Computer Society (2009)
13. Grossman, R.L., Gu, Y., Sabala, M., Zhang, W.: Compute and storage clouds using wide area high performance networks. Future Generation Comp. Syst. 25(2), 179–183 (2009)
14. Gupta, V., Saxena, A.: Privacy Layer for Business Intelligence. In: Meghanathan, N., Boumerdassi, S., Chaki, N., Nagamalai, D. (eds.) CNSA 2010. CCIS, vol. 89, pp. 323–330. Springer, Heidelberg (2010)
15. Huang, J.-W., Lin, S.-C., Chen, M.-S.: DPSP: Distributed Progressive Sequential Pattern Mining on the Cloud. In: Zaki, M.J., Yu, J.X., Ravindran, B., Pudi, V. (eds.) PAKDD 2010, Part II. LNCS, vol. 6119, pp. 27–34. Springer, Heidelberg (2010)
16. Hurwitz, J., Bloor, R., Kaufman, M., Halper, F.: Cloud Computing for Dummies. Wiley (2010)
17. Isard, M., Budiu, M., Yu, Y., Birrell, A., Fetterly, D.: Dryad: Distributed Data-parallel Programs from Sequential Building Blocks. In: 2nd ACM SIGOPS/EuroSys European Conference on Computer Systems, EuroSys 2007, pp. 59–72. ACM, New York (2007)
18. Jun, L., Jun, W.: Cloud computing based solution to decision making. Procedia Engineering 15, 1822–1826 (2011)
19. Lai, C.F., Chang, J.H., Hu, C.C., Huang, Y.M., Chao, H.C.: Cprs: A cloud-based program recommendation system for digital tv platforms. Future Generation Comp. Syst. 27(6), 823–835 (2011)

20. Lin, K.W., Luo, Y.C.: A fast parallel algorithm for discovering frequent patterns. In: GrC, pp. 398–403. IEEE (2009)
21. Liyang, T., Zhiwei, N., Zhangjun, W., Li, W.: A conceptual framework for business intelligence as a service (saas bi). In: Fourth International Conference on Intelligent Computation Technology and Automation, ICICTA 2011, pp. 1025–1028. IEEE Computer Society, Washington, DC (2011)
22. McNabb, A.W., Monson, C.K., Seppi, K.D.: Parallel pso using mapreduce. In: IEEE Congress on Evolutionary Computation, pp. 7–14. IEEE (2007)
23. Mircea, M., Ghilic-Micu, B., Stoica, M.: Combining business intelligence with cloud computing to delivery agility in actual economy. Journal of Economic Computation and Economic Cybernetics Studies (in press, 2012)
24. Negash, S., Gray, P.: Business intelligence. Communications of the Association for Information Systems 13, 177–195 (2004)
25. Olston, C., Reed, B., Srivastava, U., Kumar, R., Tomkins, A.: Pig latin: a not-so-foreign language for data processing. In: SIGMOD 2008: Proceedings of the 2008 ACM SIGMOD International Conference on Management of Data, pp. 1099–1110. ACM (2008)
26. Ouf, S., Nasr, M.: Business intelligence in the cloud. In: IEEE 3rd International Conference on Communication Software and Networks, ICCSN 2011, pp. 650–655 (2011)
27. Papazoglou, M., Van Den Heuvel, W.J.: Service oriented architectures: Approaches, technologies and research issues. VLDB Journal 16(3), 389–415 (2007)
28. Shen, C.P., Jigjidsuren, C., Dorjgochoo, S., Chen, C.H., Chen, W.H., Hsu, C.K., Wu, J.M., Hsueh, C.W., Lai, M.S., Tan, C.T., Altangerel, E., Lai, F.: A data-mining framework for transnational healthcare system. Journal of Medical Systems, 1–11 (2011)
29. Upmanyu, M., Namboodiri, A.M., Srinathan, K., Jawahar, C.V.: Efficient Privacy Preserving K-Means Clustering. In: Chen, H., Chau, M., Li, S.-h., Urs, S., Srinivasa, S., Wang, G.A. (eds.) PAISI 2010. LNCS, vol. 6122, pp. 154–166. Springer, Heidelberg (2010)
30. Velte, A.T., Velte, T.J., Elsenpeter, R. (eds.): Cloud Computing: A Practical Approach. McGraw Hill (2010)
31. Verma, A., Llor, X., Goldberg, D.E., Campbell, R.H.: Scaling genetic algorithms using mapreduce. In: ISDA, pp. 13–18. IEEE Computer Society (2009)
32. Wang, J., Liu, Z.: Parallel data mining optimal algorithm of virtual cluster. In: Ma, J., Yin, Y., Yu, J., Zhou, S. (eds.) FSKD (5), pp. 358–362. IEEE Computer Society (2008)
33. Warneke, D., Kao, O.: Exploiting dynamic resource allocation for efficient parallel data processing in the cloud. IEEE Transactions on Parallel Distributed Systems 22(6), 985–997 (2011)
34. White, B., Yeh, T., Lin, J., Davis, L.: Web-scale computer vision using mapreduce for multimedia data mining. In: Proceedings of the Tenth International Workshop on Multimedia Data Mining, MDMKDD 2010, pp. 9:1–9:10. ACM, New York (2010)
35. Wu, Z., Ni, Z., Gu, L., Liu, X.: A revised discrete particle swarm optimization for cloud workflow scheduling. In: Liu, M., Wang, Y., Guo, P. (eds.) CIS, pp. 184–188. IEEE (2010)
36. Yang, H., Dasdan, A., Hsiao, R.L., Parker, D.S.: Map-reduce-merge: simplified relational data processing on large clusters. In: SIGMOD 2007: Proceedings of the 2007 ACM SIGMOD International Conference on Management of Data, pp. 1029–1040. ACM (2007)

RESTful Triple Space Management of Cloud Architectures

Antonio Garrote Hernández and María N. Moreno García

Universidad de Salamanca
agarrote@usal.es, mmg@usal.es

Abstract. In this paper we present a job coordination service for distributed applications being executed in a Hadoop cluster based on the use of a RDF backed triple space and a RESTful web services interface. The system provides an efficient and simple coordination mechanism to resolve data dependencies between applications and it can be used at the same time as a rich source of information about the state and activity of the cluster that can be processed to build additional services and resources.

1 Introduction

Cloud infrastructure and distributed data processing frameworks like Apache's Hadoop[1], where storage and processing capacity can be dynamically expanded, have made it easier to build applications capable of processing sheer amounts of data in a scalable way. These applications are built from a large collection of distributed applications that must coordinate their execution in order to process incoming data usually stored as plain data files in a distributed file system that can be easily processed by Hadoop's line oriented data interface. The final output of this process is refined information that can also be stored in the distributed file system, in a relational database system or any other data store. The drawback of this architecture is the increasing complexity of the application coordination task as well as the increasing number of unstructured data sources that must be tracked and warehoused. In this paper we present a triple space [7] coordination system[2] that uses a RDF graph as a blackboard system with a RESTful [8] interface that applications in

[1] http://hadoop.apache.org/

[2] http://github.com/antoniogarrote/clusterspace

L. Uden et al. (Eds.): 7th International Conference on KMO, AISC 172, pp. 571–579.
springerlink.com

the cluster can use to track their data dependencies and publish application's execution state information using using HTTP standard methods to assert and retract this information from the RDF graph. The use of RDF and a shared ontology provides a snapshot of the distributed system execution that can be queried and processed to provide higher level cluster management services on top of the basic coordination system.

1.1 Problem Description

Complex data processing scenarios on a distributed system using frameworks like Apache's Hadoop often involve building a graph of individual applications with data dependencies among them. Raw data is inserted on the boundaries of the application graph and then loaded into the HDFS file system, either periodically or on demand. These data are processed by different applications following a data path through the application graph where intermediary data outputs are stored in temporary locations of the HDFS file system. Finally the multiple possible outputs generated must also be stored in the distributed file system. Managing this complex network of data dependencies is the main goal of the system presented in this paper, this task is accomplished capturing the state of the distributed application graph as RDF graph available through a HTTP RESTful interface. Managing dependencies is only one aspect to be considered in distributed data processing systems. Errors in application execution must be tracked and all kind of notifications must be generated, for example, notifying users that the output of a data processing path has finished its computation. The system here described addresses this problem using the RDF graph as a triple space, a blackboard style system where queries over the graph can be registered and will be automatically triggered where certain conditions are met. Distributed systems must also be easy to extend. Applications must be as loosely coupled as possible, so the configuration of the data processing graph can be rearranged, adding new applications or removing existing ones. The use of a plain HTTP interface, where RESTful semantics for HTTP methods are observed, along with the extensible capabilities of RDF make it possible for applications in the data processing graph to only share a small RDF ontology to describe data dependencies while additional RDF terms can be used by groups of application for more complex coordination tasks specific to that subset of applications. Finally, an important problem in distributed data processing systems is managing the knowledge generated and stored in the system. Big distributed systems evolve along time with different teams building new applications, adding data sources and consuming output from other applications whose authors might be unaware of this fact. Using a single RDF graph as a coordination mechanism allows to capture an important portion of the implicit knowledge about the system, like data sources, data dependencies, information not being recently accessed, etc. All this information can be used to build a layer of high level services on top of the coordination system that can be used to effectively manage the execution of the data processing graph.

2 Components

The term "cloud architecture" [3] is a broad term that can refer to very different systems. In this section we will narrow its scope describing in a precise way the different components in the architecture where the triple space coordination system is embedded and the additional services built on top of it. The figure 1 shows the relation between these building blocks.

2.1 Distributed File System

The distributed file system is the main data repository in the cluster, available for all applications being executed. In our case Hadoop Distributed File System (HDFS)[3] [4] is the underlying implementation used. HDFS offers a reliable storage mechanism implementing tasks like data replication and distribution. It also offers an scalable solution whose storage capacity can be expanded as required adding more nodes to the cluster.

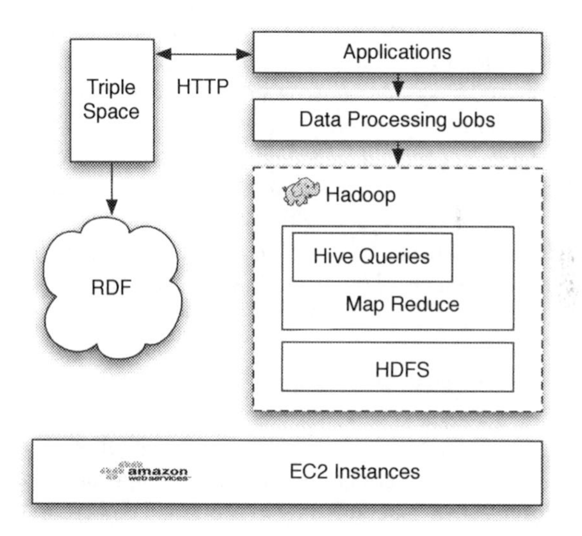

Fig. 1 Components of the distributed system and how they are related to the triple space coordination service

2.2 Data Assets

Every data unit stored in the cluster that is available for applications being executed is referred to as a data asset. The most generic data asset in the cluster are files stored

[3] http://hadoop.apache.org/hdfs/

in the distributed file system. These files are imported as raw unstructured data by applications and then transformed into different data assets.

2.3 Data Processing Jobs

Data processing jobs are groups of batch operations being executed by applications in the cluster, transforming input data assets and generating new data assets as results. In our case, data processing jobs consist of Hadoop map-reduce [5] jobs processing files available in the Hadoop distributed file system. Job operations with a higher level of abstraction, like Apache Hive's[4] SQL-like queries that are finally executed as a collection of map-reduce jobs are also counted as data processing jobs. Data processing jobs may have dependencies on other jobs that must be met before the job can begin its execution.

2.4 Applications

Applications designate every unit of business logic being executed in the cluster. The main task performed by applications is the execution of data processing jobs. Applications in our model are autonomous and direct communication between applications is avoided. However, applications are addressable using the HTTP protocol by the triple space coordination service. The triple space coordination service makes use of the HTTP protocol to notify applications when a certain event in the triple space have taken place. These notifications can be used to trigger data processing jobs when their dependencies are met. Additionally, applications have an associated state that is maintained by the triple space coordination service.

3 Triple Space Coordination Service

The triple space coordination service is an implementation of a blackboard style tuple space communication system backed by a RDF graph and accessible through a RESTful HTTP interfaces mapping classic tuple space operations [8] to HTTP protocol methods. Applications in the cluster can request three main operations in the coordination service:

- Application registration
- Assertion of data as RDF triples
- Data hook registration

Applications must first register into the service providing some application details as structured RDF information, like the accessible location of the application as an URL or the repository where the application source code is stored. This application information is transformed into a set of RDF triple assertions and then added to the

[4] http://hive.apache.org/

RDF graph. Once registration is finished, the RDF sub-graph encoding the application information will be accessible as an HTTP resource that can be read or updated by applications in the cluster using regular HTTP verbs for resource manipulation.

During its execution, applications may add arbitrary new data assertions to the RDF graph mapped as HTTP resources. Additionally, applications can add RDF assertions to the triple space using specific HTTP entry points for data assets. Triples added through the data assets HTTP interface are validated by the triple space service and additional information is added, including time-stamps, RDF type assertions and a updated execution state predicate. Applications in the cluster can obtain the RDF graph associated with a data asset dereferencing the associated resource URI. Applications updating any resource in the HTTP API or adding assertions to the triple space RDF graph must identify themselves using the HTTP User-Agent header and the URI of the API HTTP resource created by the application when registering into the service. This information is used by the service to track changes on the RDF graph.

The last kind of triple space operation is the registration of data hooks. Data hooks make possible for applications to be notified by the triple space coordination system when changes in the RDF data graph occur. Data hooks are created as a different kind of HTTP resource in the triple space HTTP interface. They are described using a small ontology including predicates to specify a SPARQL [6] query and a callback URL. The triple space service registers the SPARQL query associated to every data hook created in the system and evaluates the query every time new data added to the RDF query may change the result set returned by the query. When the query is evaluated results are sent back to the application using a HTTP PUT request to the callback URL associated to the data hook. The application can use this notification to trigger a data processing being hold awaiting for its dependencies to be fulfilled or to update its internal state.

3.1 Cluster Ontology and RDF Encoding

Communication between applications and the triple space consist in the exchange of two types of RDF graphs: RDF data encoding the representation of the HTTP API services like applications, data hooks and data assets, and arbitrary RDF triples added by applications to the coordination service graph. In the case of RDF data containing the representation of a HTTP API resource, a shared ontology is used by applications to ensure easy inter-operability. This ontology includes a small set of around 20 RDF properties used to describe certain types of information:

- State of applications, resources, etc.
- Distributed file system paths.
- Time format, creation and update time-stamps.
- Resource RDF types.

Correct use of this vocabulary is enforced by the triple space coordination service running validations on the received RDF data. This vocabulary is also extended by certain types of applications and resources. For example, data assets consisting of Hive tables declare and additional sub-type of the `cs:DataAsset` RDF type and information about the location of the HDFS file, table name and Hive partitions.

On the other hand, use of differnt RDF ontologies is possible on arbitrary RDF data added to the graph by applications. This allows for subsets of applications to coordinate their execution through the triple space service and to extend the coordination service through the use of new RDF vocabularies.

RDF data in HTTP requests and responses are encoded using JSON-LD. JSON is a popular format for web APIs and in contrast with other RDF serializations, numerous libraries supporting the format exist for most programming languages. It also makes it possible for application developers using the HTTP API of the coordination service to interact in most cases with the service using familiar JSON objects without having to deal with the complexity of the RDF data model.

3.2 Implementation Details

The triple space service is implemented as a JavaScript application being executed in the Node.js V8 JavaScript platform[5]. Node.js offers an excellent platform for developing scalable web services thanks to its implementation of an asynchronous evented HTTP server. Specific support for Node.js is also available in different cloud infrastructure providers.

RDF and SPARQL implementations are provided by RDFStore-js[6], a pure JavaScript implementation of a RDF and SPARQL database we have developed. This library provides an events API for RDF graphs that is used to by the coordination service to perform the evaluation of SPARQL queries associated to data-hooks. RDFStore-js can also be executed in client applications using node, Java applications using Mozilla's Rhino JavaScript implementation as well as in the browser. This offers a convenient mechanism for client applications to process RDF data received from the coordination service when processing JSON-LD [1] encoded RDF as plain JSON objects is not enough for their functionality.

In the triple store coordination service the underlying persistent storage layer for RDFStore-js is a MongoDB[7] cluster where RDF triple assertions are stored as JSON documents. A replica set of two MongoDB instances to provide better availability to the system on the event of the coordination service node failure.

4 Higher Level Services

The RDF graph maintained by the triple space coordination service can be understood as a snapshot of the cluster state at any given moment of time, including

[5] http://nodejs.org/

[6] http://github.com/antoniogarrote/rdfstore-js

[7] http://www.mongodb.org/

information such as available data sources, the dependency graph between applications and data assets or cluster health information. The use of an extensible semantic technology like RDF makes it possible to store this information provide an makes it possible to develop an additional layer of services processing this information implementing higher level cluster coordination services.

4.1 Application Monitoring Service

The application monitoring service registers a data hook in the triples space requesting the value of the `cs:state` property for all `cs:Application` RDF resources registered in the graph. Every time an application associated to one of these resources is updated, the application monitoring service is notified by the triple space coordination application with the updated list of states for the applications in the cluster. The application monitoring service uses this information to provide a web interface where the state of cluster applications can be browsed, implementing functionality like email alerts or the required logic to restart failed applications.

4.2 Hive Data Catalog Service

`cs:HiveAssets` RDF resources are specialized data assets storing information about Hive tables and partitions. They also include temporal information about the date when a particular Hive table and partition have been modified. New tables being created and partitions being removed are also translated into HTTP requests adding and removing RDF assertions.

The Hive data catalog application registers a data hook in the coordination service consisting of a SPARQL query requesting the value in the graph of the `cs:tableName` RDF property associated to `cs:HiveAssets` whose state is updated by an application at a given time-stamp. Using this information as well as Hive's data definition language, information for those tables is retrieved and stored.

This information is used by the Hive data catalog service to build a description of the Hive tables available in the system and offer a web interface to users where the available tables as well as the applications modifying each table can be browsed. The information can also be used to determine applications affected by administrative changes in Hive tables as well as to look for data tables no longer being used.

4.3 Cluster Log Service

The cluster log service is a module of the triple space coordination service logging individual assertion and removal of RDF triples in the RDF graph. Every time an application request triggers the insertion or deletion of RDF data from the graph, the RDF data is serialized and logged to HDFS using Apache's Flume[8]. The format

[8] https://cwiki.apache.org/FLUME/

of the logged data used is a variation of the Turtle syntax [2] where a time-stamp and the URI identifying the application in the User-Agent of the HTTP triggering the RDF graph modification are added. This line oriented log format makes it easier the log processing using map reduce Hadoop jobs. This information is used to audit the performance of the cluster and compute usage statistics.

5 Conclusions and Future Work

In this paper we have presented a coordination service for distributed data processing applications based on the use of RDF and RESTful HTTP services. The main goal of the service is to coordinate the execution of data driven application work flows. At the same time, it can be used as an structured data source about the state of the cluster, including application's execution state, functional dependencies between applications and data sources. This information can be used to implement higher level services like the data catalog and application monitoring services.

The extensible design of RDF makes it possible to add new functionality to the system using additional RDF vocabularies describing new types of data sources or coordination primitives shared only by a subset of applications. Technologies used to implement the service serve a double purpose, they are based in platforms like Node.js and MongoDB well supported by cloud infrastructure providers and at the same time try to offer a simple interface to the RDF system for application developers using standard RESTful design and a JSON based data exchange format like JSON-LD.

Future lines of work include the integration of additional information into the RDF graph, specially additional data sources beside HDFS data, like relational databases as well as configuration and deployment information. This information can be retrieved from the source code repository information of the registered applications already stored in the RDF graph. The ultimate goal of this job would be to track data being processed in the cluster as higher level data entities being transformed despite their multiple possible representations, from text files stored in HDFS to tables in a relational database or documents in a MongoDB repository. Another area of research is the processing of the cluster log information being generated by the coordination service. This log consist of a stream of time-stamped RDF assertions describing the temporal evolution of the cluster configuration stored in the coordination service RDF graph in a modified Turtle RDF encoding. This source of temporal data can be processed to compute a number of valuable metrics on the historical performance of the system.

References

1. Sporny, K.: SON-LD Syntax 1.0, A Context-based JSON Serialization for Linking Data, online specification (2012)
2. Beckett, B.-L., Prud'hommeaux: Turtle Terse RDF Triple Language. W3C Working Draft (2012)

3. Kim, W.: Cloud architecture: a preliminary look. In: Proceedings of the 9th International Conference on Advances in Mobile Computing and Multimedia. ACM (2011)
4. Shvachko, K., Kuang, H., Radia, S., Chansler, R.: The Hadoop Distributed File System, Yahoo. In: IEEE 26th Symposium on Mass Storage Systems and Technologies (2010)
5. Dean, J., Ghemawat, S.: MapReduce: simplified data processing on large clusters. Google, Communications of the ACM (2008)
6. Prud'hommeaux, Seaborne: SPARQL Query Language for RDF. W3C Recommendation (2008)
7. Fensel, D.: Triple-Space Computing: Semantic Web Services Based on Persistent Publication of Information. In: Aagesen, F.A., Anutariya, C., Wuwongse, V. (eds.) INTELLCOMM 2004. LNCS, vol. 3283, pp. 43–53. Springer, Heidelberg (2004)
8. Fielding, R.T.: Architectural styles and the design of network-based software architectures. PhD. Thesis, University of California, Irvine (2000)

Analysis of Applying Enterprise Service Bus Architecture as a Cloud Interoperability and Resource Sharing Platform

Amirhossein Mohtasebi, Zuraini Ismail, and Bharanidharan Shanmugam

Advanced Informatics School, Universiti Teknologi Malaysia, Kuala Lumpur, Malaysia
mamirhossein2@live.utm.my, {zurainisma,bharani}@ic.utm.my

Abstract. Interoperability is one of the main elements affecting the adoption of a technology by businesses. Interoperability in the Cloud is crucial in the sense that it can guaranty inter-cloud communications between heterogeneous platforms. This paper identifies different aspect of interoperability in the Cloud. Moreover, it discusses Distributed Infrastructure architecture as a base for Enterprise Service Bus (ESB) model. The authors split ESB into different layers to increase the flexibility of the framework. This paper extends the concept of ESB to build a Cloud service model that facilitates secure interactions between different Cloud platforms. The proposed architecture is to use service repository and registry mechanisms to enhance flexibility and portability of the model.

Keywords: Cloud Computing, Interoperability, Mobility, Portability.

1 Introduction

The concept of Cloud computing can be simply defined as "Software deployed as a hosted service and accessed over the Internet" [1]. Cloud computing is somehow adopted from the concept of abstract computing. In other words, endpoints can connect to each other in a loosely coupled manner without knowing much about each other.

National Institute of Standards and Technology of United States (NIST) defines Cloud Computing as a pool of shared resources such as servers, storage, applications, and data that can be provisioned and released with minimum configuration and management efforts so that be accessible on-demand by consumers [2]. Firstly, Cloud consists of an abstract "unlimited" pool of resources. In other words, customers do not have to worry about provisioning of hardware, software, and bandwidth for future needs. Secondly, the pricing model is based on "per use" model instead of license model. Moreover, infrastructures are usually (or necessarily) distributed geographically. This helps customers to implement geo-redundant services as well as keeping network latency at the minimum level.

Configuration and provision of the systems are instant and automated based on pre-configured templates. The amount of effort to provision a new system or scale

L. Uden et al. (Eds.): 7th International Conference on KMO, AISC 172, pp. 581–588.
springerlink.com

an existing system is the matter of minutes or couple of hours instead of days or even months [3,4].

Finally, in the Cloud environment resources are virtualized completely or to a great extent. While infrastructures are kept intact and rarely changed, virtualized components such as disks, virtual machines, and virtual networks changed continuously.

One of the main hurdles ahead of migration to Cloud platforms is the lack of clear interoperability and inter-cloud communication standards. In other words, although current Cloud platforms expose some of their functionalities using standard protocols such as Representational State Transfer (REST) and Simple Object Access Protocol (SOAP), they are designed so proprietary and closed that makes inter-cloud migration and communication almost impossible or not justifiable [5]. The problem can be summarized as how to exchange data from one Cloud to another and how to move data back and forth between those Clouds.

Different solutions have been proposed by standard bodies (e.g. IEEE, NASA) and companies (e.g. IBM) in order to bring interoperability to the Cloud. While most of these blueprints are completely relevant to the idea of interoperability, they require vendors to either open source parts of their platforms, or make significant changes to them [6]. In this paper, the authors focus on adding new components to the Enterprise Service Bus (ESB) model in order to provide this middleware with flexibility and automation required for cloud interactions. In our research, we focus on using ESB concept as the Cloud orchestration model base because it has minimum impact on the current platforms. Moreover, we propose new components to be added to ESB, so that it can be extended from enterprise domain to the wider context of the Cloud.

In this paper, we first identify the attributes of interoperability in the Cloud. Secondly, we introduce the concept of ESB and then model it to work as a middleware platform that facilitates inter-cloud communications. Additional architectural layers are introduced in order to enhance service discovery and registry.

2 Interoperability in the Cloud

Interoperability covers a wide range of topics. It spans from hardware and software, to very conceptual notions such as data and its semantics. European Commission defines it as interlinking a system, information, or workflows across multiple domains such as enterprise sectors, geographic locations, administrations, etc. [7]. Generally speaking, enabling two or more different systems communicate to each other and exchange information makes those systems interoperable [8]. IEEE defines interoperability as "The capability of two or more networks, systems, devices, applications, or components to externally exchange and readily use information securely and effectively" [9].

Interoperability in Smart Grids is implemented in the following contexts [9]:

- Hardware, Software, Platforms that enable machine-to-machine communication. This context is more on physical layer and infrastructures.

- Data types and formats, that keeps the integrity of data and messages intact from system-to-system,
- Semantic context, that means both systems have the same understanding about sets of data.

Interoperability in the Cloud, which can be considered as mesh of smart grids, means that information should be able to go back and forth between platforms without losing its integrity, and data-type in a secure way.

3 Enterprise Service Bus Architecture

Distributed infrastructure is an architectural model or software implementation. It lets different services from different platforms connect to each other in a loosely coupled manner [10]. Based on Bean, enterprise service bus is more than a messaging system and it includes all the main parts of SOA such as network, transport, routing, content and message delivery and protocol conversion system [11].

Enterprise Service Bus (ESP) is a terminal for integrating different services and creating mesh of services through a messaging environment [12]. Service Bus sits between different service endpoints and helps them use each other's capabilities by inter-connecting them in a loosely coupled manner [13]. Basically the change ESB does to endpoints is that it makes them to have a one-to-one relationship to ESB instead of one-to-many relationship with all other endpoints [14]. Figure 1 illustrates the change ESB makes in the topology of a distributed system with multiple isolated endpoints.

Some researchers argue that the combination of SOAP (Simple Object Access Protocol) messages and WS-Addressing is equal service bus [15]. However, considering other methods of bounding such as Representational State Transfer (REST), defining Simple Object Access Protocol (SOAP) messages as the only transportation standard is somehow restrictive. Moreover, service bus should not

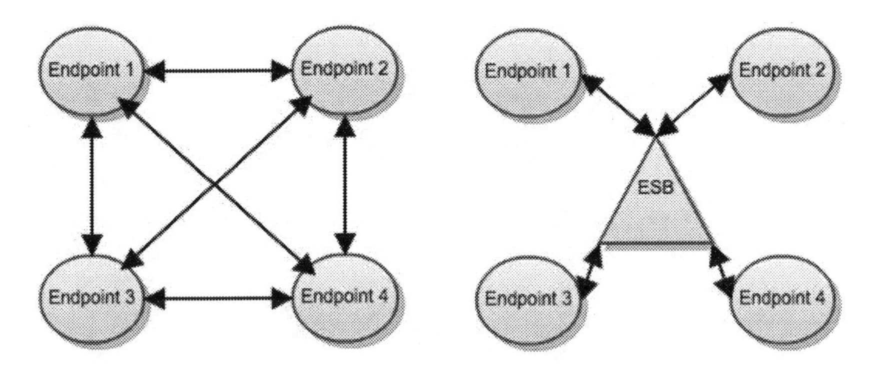

Fig. 1 A conventional topology (Left) An ESB topology (Right)

be limited to transferring and translating of messages between services but it should consider security and access control, Quality of Service (QoS), management, monitoring, orchestration, etc. It should be noted that service bus should work as an abstract layer to provide loosely coupled service interaction. In other words, its nature is bounded to the fact that services know nothing but an interface from each other, and it is platform, operation system, and language independent [10].

As the Figure 1 illustrates, the amount of customization required to connect conventional applications to each other is reduced. The amount of work is required to connect applications in conventional way is so high that companies sometimes better off righting the functionality from scratch instead of reusing the functionality available in other applications. To emphasize the significance of using ESB, the amount of customization required in the conventional model versus ESB model depicts as below:

$$\alpha = \frac{n(n-1)}{2}$$

$$\beta = n$$

Where α and β are the number of customizations required in conventional model and ESB model respectively and n is total number of endpoints. Using these formulas, it proves that ESB topology reduces the amount of customization by the factor of $\frac{n-3}{n-1}$.

Enterprise Service Bus (ESB) works as a middleware façade in order to create an interoperable environment. Wu and Tao discussed the main functionalities an ESB provides SOA environment with as location decoupling, transport protocol conversion, routing, transformation, security, and monitoring [14].

Location decoupling lets consumer and producer of messages to contact to each other without any need to bind them together tightly. In other words, instead of having two systems that are heavily dependent to each other, a lightweight binding layer manages the interaction between systems [16]. Figure 2 illustrates the differences between heavy dependency and lightweight binding architectures.

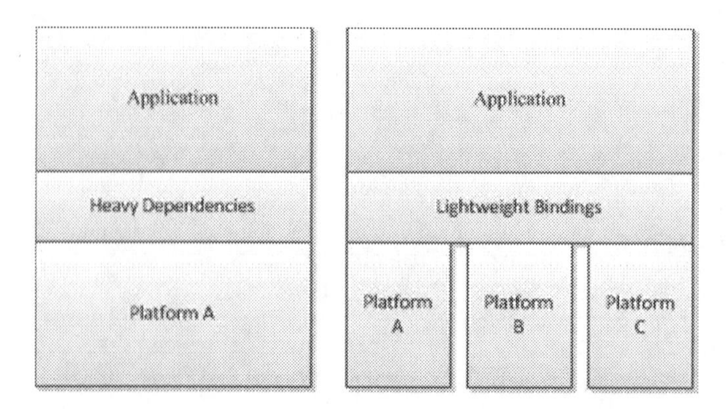

Fig. 2 Heavy Dependencies vs. Lightweight Bindings (Source: VMware, 2012)

Transport protocol conversion helps producers and consumers use their own transport protocol such as HTTP(s), FTP, TCP, and even proprietary formats such as variants of message queue protocols, and ESB take responsibility to translate protocols in the middle. Transport protocol aside; it helps translating different design principals' formats such as REST [17], and SOAP to each other.

Routing is using some set of business rules in ESB to route messages dynamically from producer to consumer and vice versa. It can resembles and work as a load balancer or workflow system to redirect messages to the correct endpoint base on different conditions.

Message transformation is the place where ESB validate and transform messages at the content level and make it understandable to the destination. One of the easiest and conventional ways to do that is to use Extensible Stylesheet Language Transformation (XSLT) to transform XML documents to other formats or field mappings.

Security is one the most important aspects of SOA. The security module can be configured to validate message both on transport and message level, communicating to identity servers, and transforming security tokens required to communicate to endpoints.

The last but not the least function of ESB is monitoring and management modules. Using these modules provides system admins and consumers with automatic notification, handling poison messages, Service Level agreement (SLA) violations alerts.

There may be other modules consolidated from the above modules such as load balancer, encryption/decryption services, etc.

4 Applying ESB to Cloud Architecture

The main advantage of enterprise service bus middleware in an SOA environment is to enable services to be reused from different applications. As discussed by Pouli et al. [18], the Composable Service Architecture (CSA) provides ESB with dynamic provisioning and integration. Based on the definition by Demchenko [19], CSA provides a layered architecture that consists of Network, Transport, Message, Virtualization, and application layer (Figure 3).

Network layer is responsible for handling networking mechanisms required for establishing connections such as Virtual Private Networks or handling virtual IP block translations. Network layer passes messages to transport layer where all the message transport functionalities such as Secure Socket Layer (SSL) or Transport Layer Security (TLS) are implemented.

Message layer is analogical to message transformation mechanism in ESB. Any data mapping and message transformation capability are implemented in this phase. In the virtualization layer, service composition and interaction are handled. Finally, Application layer refers to actual applications at both ends of connection.

Using composite service architecture allows ESB to span from specific number of services to unlimited number of services by implementing federated interfaces. Any application and third party that comply with this interface or at least build a

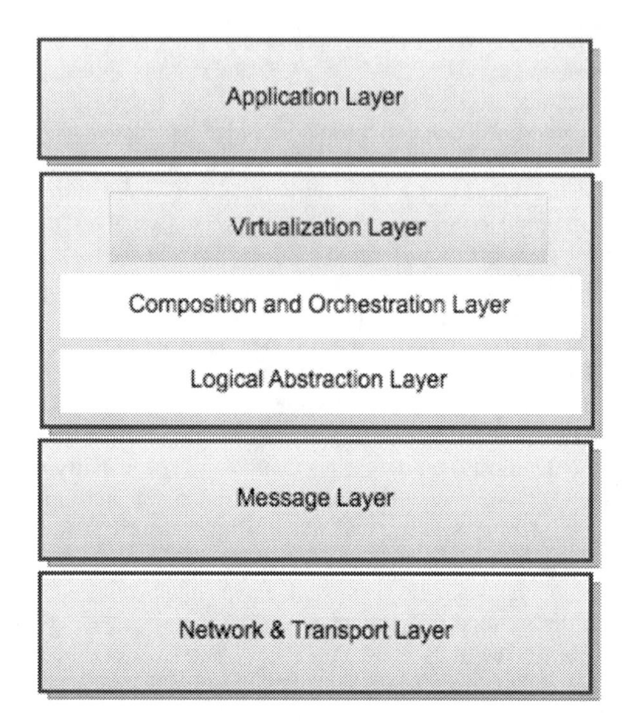

Fig. 3 CSA Layered Architecture (Source: Open Grid Forum, 2011)

plug-in component that can be applied to this abstract layer can communicate through ESB. Because of vast diversity of services in Cloud environment, applying the concept of CSA to ESB can extend them from a limited enterprise environment to unlimited Cloud environment.

As an example, Figure 3 illustrates a resource discovery and sharing mechanism between two cloud environments, namely Amazon and Azure, that both have compute and storage components.

Grammatikou et al. discussed about enterprise cloud bus and highlighted the two main elements of virtualization layer that are service registry and service repository [20]. The first step of communication is to connect to ESB registry and submit the metadata required for using resources hosted in the cloud platform. These can be but not limited to security requirements, and conventional service definition transferred using web service definition language (WSDL). The second step is that each cloud connects to repository and gets the latest service definition of the cloud that is intended to connect to using ESB. After configuration, the initiator, i.e. AWS, query for resources of target cloud, i.e. Azure, using ESB as a middleware. Communication between each Cloud platform and ESB is through a proxy called service component. Implementation of this proxy is the minimum requirement to connect to ESB. In other words, service component is an implementation of federation interface [20]. Figure 4 illustrates the architecture of using service repository and registry in ESB model to bring more flexibility to the ESB model.

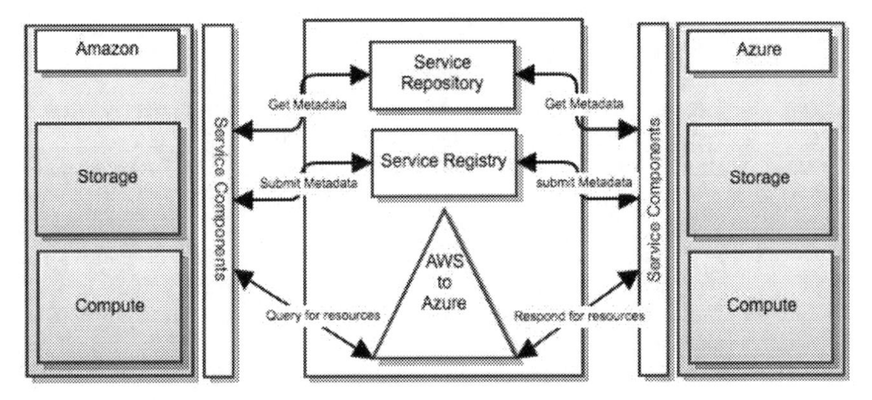

Fig. 4 Sample registration, discovery, and flow of information through ESB (Source: Grammatikou et al., 2011)

5 Conclusion and Future Works

The interoperability is not a new issue and it led to lots of problems and overworks before. There are lots of solutions proposed for bringing interoperability to different aspects of Cloud such as physical, data, and semantic context. In the context of data interoperability, using the concept of Enterprise Service Bus is one of the most viable solutions. ESBs can help well established but not interoperable Cloud platforms connect to each other and share resources. Moreover, using composite service architecture increases the ability to have platform agnostic, configurable service bus.

ESBs can be placed as a middleware between different platforms and act as a translator between them. They can implement transportation protocol conversion, message routing, and data mapping, as well as providing message security and monitoring mechanisms.

By making Cloud platforms connect to each other, data is transferred successfully and its integrity kept intact. However, the question is whether data has the same meaning for destination Cloud as it has in source Cloud. The next phase of research should be how to make a semantic ESB in order to give meaning to the data that has been transferred. Applying Web Ontology Language (OWL) or other similar techniques in future researches can enhance the semantic structure of data.

References

1. Carraro, G., Chong, F.: Architecture Strategies for Catching the Long Tail. Microsoft Developer Networks (2006)
2. Mell, P., Grace, T.: NIST Definition of Cloud Computing. National Institute of Standards and Technology (2009)
3. Corp. D Dell Unveils Industry's First OpenStack Infrastructure-as-a-Service Cloud Solution. Dell (2011)

4. Hirschfeld, R.: Unboxing OpenStack clouds with Crowbar and Chef [in just over 9,000 seconds!]. Agile in the Cloud (2011)
5. Armbrust, M., Fox, A., Joseph, A.D., Kats, R.H., Konwinski, A., Lee, G., Patterson, D.A., Rabkin, A., Stoica, I., Zaharia, M.: Above the Clouds: A Berkeley View of Cloud Computing. University of Berkeley, California (2009)
6. Rimal, B.P., Choi, E., Lumb, I.: A Taxonomy and Survey of Cloud Computing Systems. In: Fifth International Joint Conference on INC, IMS and IDC, NCM 2009, pp. 44–51 (2009)
7. Directorate-General E Linking up Europe: the importance of interoperability for e-government services. The Commission of The European Communities (2003)
8. Teixeira, T., Maló, P., Almeida, B., Mateus, M.: Towards an Interoperability Management System. Information Systems and Technologies (CISTI), pp. 1–4 (2011)
9. IEEE, IEEE Guide for Smart Grid Interoperability of Energy Technology and Information Technology Operation with the Electric Power System (EPS), End-Use Applications, and Loads. IEEE Std 2030-2011 (2011)
10. Robinson, R.: Understand Enterprise Service Bus scenarios and solutions in Service-Oriented Architecture, Part 1. IBM Developerworks (2004)
11. Bean, J.: Enterprise Service Bus. In: Service Interface Design, p. 10. Morgan Kaufmann (2009)
12. Lou, M., Goldshlager, B., Zhang, L.-J.: Designing and implementing Enterprise Service Bus (ESB) and SOA solutions. In: IEEE International Conference on Service Computing. IBM Global Services (2005)
13. Jizhe, L., YongJun, Y.: Research & Implementation of Light Weight ESB With Microsoft .NET. In: International Conference on Frontier of Computer Science and Technology (2009)
14. Wu, J., Tao, X.: Research of Enterprise Application Integration Based-on ESB. In: 2nd International Conference on Advanced Computer Control, ICACC (2010)
15. Webber, J.: Thought Works. Guerrilla SOA (2005)
16. VMWare, Multi-Language, Multi-Framework, what about Multi-Cloud? VMWare (2012)
17. Fielding RT Architectural Styles and the Design of Network-based Software Architectures. In: UC Irvine (2000)
18. Pouli, V., Demchenko, Y., Marinos, C., Lopez, D.R., Grammatikou, M.: Composable Services Architecture for Grids. In: Computer Communications and Networks, pp. 223–247 (2011)
19. Demchenko, Y.: Composable Services Architecture (CSA). OGF (2011)
20. Grammatikou, M., Marinos, C., Demchenko, Y., Lopez, D.R., Dombek, K., Jofre, J.: GEMBus as a Service Oriented Platform for Cloud-Based Composable Services. In: Third IEEE International Conference on Coud Computing Technology and Science (2011)

The Role of Knowledge in the Value Creation Process and Its Impact on Marketing Strategy

Anna Závodská[1], Veronika Šramová[1], and Anne-Maria Aho[2]

[1] University of Žilina, Slovak Republic
{anna.zavodska,veronika.sramova}@fri.uniza.sk
[2] Seinäjoki University of Applied Sciences, Finland
anne-maria.aho@seamk.fi

Abstract. The aim of this paper is to compare approaches for value creation and the role of knowledge in these processes. The framework for the value creation process shows the role of knowledge in different phases of this process. Knowledge is compared in each of the individual phases of the process and also between case companies. There is identified knowledge for marketing strategy. The methodology involves three single case studies from which data are derived and analyzed. The analysis shows that the framework for the value creation process can be used as an analytical tool for value overview in different phases and there is a need for different approaches to improve business. Based on the analyzed problems, proposed recommendations for improvement are made. These recommendations are based on the Principles of Blue Ocean Strategy. The Blue Ocean Strategy is considered a better approach to knowledge management in value creation and to the improvement of the companies in question.

Keywords: knowledge, marketing strategy, value creation process, Blue Ocean Strategy, value, processes, analysis, research, value innovation, customers.

1 Introduction

Many companies do not know anything about knowledge and they do not try to use it in organizational processes. They work exclusively with information pertaining to customers, competitors, suppliers, etc. Authors of this paper see knowledge as something what is very subjective and what is in the minds of employees based on their experiences, skills, beliefs and attitudes. We analyzed the role of knowledge in different phases of the value creation process and proposed improvements using the Principles of Blue Ocean Strategy.

Three different companies have been chosen for the analysis to show the importance of knowledge in marketing strategy focused on value creation. Two of them were typical consumer companies and one of them was a B2B company. We have identified the companies as A, B and C. Each case study is different as is the knowledge they use in value creation process. Companies cooperate with

L. Uden et al. (Eds.): 7th International Conference on KMO, AISC 172, pp. 589–600.
springerlink.com

customers in different phases of value creation process: Company A involves its customers in its product development process, Company B and C cooperate with their customers in operational and post-sale services, but they do not involve them in their product development process. These companies did not have anything in common before the study was carried out. The description of the cases and the data collection methods are presented in Table 1.

Company A is multinational machinery company. The empirical domain of the study is the company's software engineering process, which supplies embedded software in the machinery company. The software engineering process is geographically divided into the three different locations: Finland, Europe and Northern America.

Company B is the most popular and biggest seller of consumer electronics in Slovakia. It offers a broad portfolio of products, which include many famous and popular brands. The company was the first that started the wholesales and distribution of electronics of famous brands in Slovakia. The Company has 25 stores all over Slovakia. The present study focused only on one retail seller in Zilina in the Slovak Republic.

Company C is an international pharmaceutical company which produces and sells a variety of nutritional supplements. It was established shortly after the Velvet Revolution in the Czech Republic. Although it is not a large-sized pharmaceutical company, it has obtained a strong foothold in many European countries during its 20 years on the market. Companies in this industry do not sell products directly to their customers but via pharmacies, and therefore communication with customers is limited. The present study focused on one product for pregnant women.

Table 1 Description of the data collection techniques of case companies

	Case Company A	Case Company B	Case Company C
Data collection techniques	Modeling of the information flows Workshops Interviews	Observation Questionnaires Interview Benchmarking	Content analysis Interviews Expert opinions Benchmarking Observation

This paper begins with a brief overview of the value creation process and knowledge followed by a brief explanation of the Blue Ocean Strategy. This is followed by the analysis of the value creation process in the different company cases. The subsequent section proposes improvements on how to use knowledge management in value creation according to the Principles of Blue Ocean Strategy.

2 Value Creation Process and Knowledge

To understand value creation process it is important to define value. Value is generally considered to describe either the importance of something or its worth. Value creation can be defined in many ways because each company has its own value creation process – its own way of how to serve value. Value creation is grounded in the appropriate combination of human networks, social capital, intellectual capital, and technology assets, facilitated by a culture of change. By delineating the value creation strategy of an organization using the framework, marketers can clearly define product concepts, a new product key success factor. (Yusof, 2010) Companies need to think about two main issues if they want their own value creation process to work properly. Firstly, it is necessary to know which basic elements are important for this process. Secondly, who is the user of value (customer or consumer of services and products, inventor or fund provider, regulator, intermediary, distributor, employee, volunteer and supplier)? (Marr, 2009).

Each business has a unique set of processes for creating value for customers and producing financial results. Kaplan and Norton (1996) found a generic value-chain model, which provides a template that companies can customize in preparing their internal business process perspective. This model encompasses three principal business processes: innovation, operations and post-sale service. In the innovation process, the business unit surveys the emerging or latent needs of customers, and then creates the products or services that will meet these needs. The operations process is where existing products and services are produced and delivered to customers. Post-sale services include warranty and repair activities, treatment of defects and returns, and the processing of payments, such as credit card administration.

Knowledge is applied to create value more effectively. (Ruggles, 1999) A firm's competitive advantage depends more than anything on its knowledge: on what it knows, how it uses what it knows, and how fast it can find out new things. (Prusak, 1997) ***Knowledge is experience, everything else is just information.*** (Kucza, 2001; Zeleny, 2005) ***Knowledge is more than a simple collection of information.*** (Nonaka, Toyama and Hirata, 2008) "It cannot be handled as information, does not have the same uses. Having information is not the same as knowing. ***Not everybody, who read cookbooks, is necessarily a great chef.*** Knowledge is purposeful coordination of action, not description of action." (Zeleny, 2005) "Knowledge management must be brought onto the marketing scene to improve product development, communication and marketing-information processes. This allows marketers to apply knowledge more successfully, which will improve the efficiency of internal marketing research, deliver well thought-out products and services, and adopt a more effective customer approach." (Boersma, 2004)

Managers should first and foremost focus on knowledge *from* the customer – knowledge residing in customers, rather than focusing on knowledge *about* the customer, as is characteristic of customer relationship management. Customers are more knowledgeable than one might think. (Gibbert, Leibold and Probst, 2002)

3 Blue Ocean Strategy

The market universe is composed of two types of oceans: "red oceans" and "blue oceans". "Red oceans" represent all of the existing industries today. This is the known market space. "Blue oceans" denote all of the industries which are not in existence today. This is the unknown market space. (Kim and Mauborgne, 2005)

In the "red oceans", industry boundaries are defined and accepted, and the competitive rules of the game are known. Companies try to outperform their rivals grab a greater share of the existing demand. As the market space gets crowded, prospects for profits and growth decrease. (Kim and Mauborgne, 2005)

"Blue oceans", in contrast, are defined by untapped market space, demand creation, and the opportunity for highly profitable growth. Although some "blue oceans" are created well beyond existing industry boundaries. In "blue oceans", competition is irrelevant because the rules of the game are waiting to be set. (Kim and Mauborgne, 2005)

The companies caught in the "red ocean" follow a conventional approach, racing to beat their competition. The creators of the "blue oceans" do not use their competition as their benchmark. They follow a different strategic logic that is called value innovation. They focus on making competition irrelevant by creating a leap in value for buyers and their company, thereby opening up a new and uncontested market space. Value innovation places equal emphasis on value and innovation. Value without innovation tends to focus on value creation, something that improves value but is not sufficient to make a company stand out in the marketplace. Innovation without value tends to be technology-driven, market pioneering, futuristic, often shooting beyond what buyers are ready to accept and pay for. Value innovation is a new way of thinking about and executing strategy that results in a creation of a "blue ocean" and a break from the competition. There are paths by which companies can systematically create uncontested market space across diverse industry domains, hence attenuating search risk. Companies can make competition irrelevant by looking across the six conventional boundaries of competition to open up important "blue oceans". The six paths focus on looking across: alternative industries, strategic groups, buyer groups, complementary product and service offerings, functional-emotional orientation of an industry, and even across time. (Kim and Mauborgne, 2005)

4 Analysis and Proposed Improvements

(Kaplan and Norton, 1996) have developed the following framework (see Figure 1.) to depict the value creation process. Following this framework, the need of knowledge in different phases in the value creation process was analyzed. The analysis of the three different case companies indicated that this framework does not work for all companies without modification and the reason is because it does not provide any ideas to improve the companies. The companies are represented by the letters A, B and C.

Identifying customer needs. In Company A's software engineering process, the knowledge about customer needs is derived from sales, project leaders, service and start-up engineers. Although the importance of the customer's role is understood, the customer's role in relation to the software process is quite

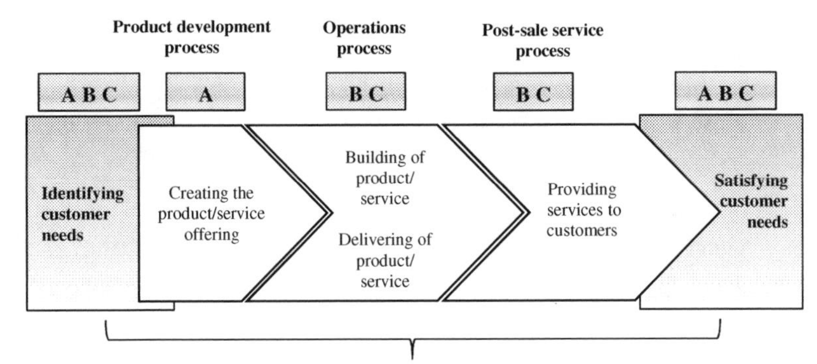

Fig. 1 Linkage between knowledge and the value creation process adopted by Kaplan & Norton (1996)

insignificant. The software engineering process as a part of multidisciplinary product development is not responsible for identifying customer needs. Based on the modeling of information flows, we can conclude that there is no reciprocal interaction between customer and software engineer, and contact is made only when problems occur. Sales have a key role in identifying customer needs in Company A.

Company B identifies customer needs according to market information and knowledge from customers. The parent company collected information about the market, i.e. customers, competitors, suppliers, etc. Based on this information, the company makes a decision about purchasing products for different brands.

Employees of retails were first in contact with customers and their task was to identify customer needs. There is a difference between previous information about market and identifying customer needs by employees: marketers try to estimate future customer needs but sellers need to identify current customer needs at the time when customers come to the retail and want to buy a product. Knowledge about customers was important in the process of identifying customer needs. This kind of knowledge was based on evaluated information obtained through a loyalty program, credit cards and questionnaires, experiences of employees, their skills and attitudes.

In Company C, the knowledge about customer needs was provided by the marketing and sales department. Brand managers endeavored to communicate with customers and gain information in various ways (questionnaires, interview, feedback from hotlines or loyalty programs, etc.), and they sent some of the ideas to R&D where new products were developed. Direct contact was limited and that was the reason why some products were not successful on the market. Marketers were not able to identify customer needs and because of that products produced were not according to customer needs. Because these products were of a special kind, it was difficult to find out what customers really needed. Usually signals for new products came from the market, from domestic or foreign competitors or findings from international researches. Customers have to buy what the company offers them, not what they really want or need because it might not be possible or not cost-effective to produce it.

It can be concluded that marketing and sales departments play a key role in identifying customer needs. Customer knowledge is very important because identifying customer needs and satisfying them is not possible without knowing your customers. Gaining knowledge of customers is difficult, and therefore companies should try to interact with their customers more. Even direct contact in Company A and C is not possible, and therefore they should try to obtain the customers' insights as much as they can. Indirect communication often causes many problems when the customers' requirements reach the company too late or in the wrong form so company cannot use them.

Creating the product/service offering. In Company A the use of knowledge was analyzed according to the phases of the software engineering process. Concerning the requirement phase, we can conclude that the main problem in the requirement phase was informality and lack of critical information. The project manager has a key role in the process; thus all important information in the software engineering process is passed through him/her. The specifications were quite unclear, which is the consequence of inappropriate information about software features required by customers.

Concerning the phases of design and re-design, most of the information about programming and interfaces was in an extremely informal format and was very subjective. There are no formal procedures or databases. The process involved a lot of tacit knowledge. Some weaknesses regarding knowledge management in the software engineering process are as follows: documentation was not sufficient, contacts to other functions of product development were based on personal characteristics, management of engineering was mechanically oriented, and therefore there was not enough knowledge about time management in the engineering process.

It can be concluded that, overall, communication was incomplete or inaccurate. Very often more information and details were requested. The difficulties were based largely on incomplete documentation and that the need to improve documentation was obvious.

Building of product/service and delivering of product/service. Company B and its retails were not responsible for product development but for building and delivering service, the sale of consumer electronics and installation. The Company offers products from different brands to customers. There are employees whose only task is to create services, and then these services are provided to customers in retails. Each company has a basic package of services which is common in the service industry (e. g. card payment). Companies want to be best in this area and try to develop something special for their customers. Their promotions are often not about products which they are selling but about services they can provide. Nowadays the Company is trying to involve customers in the process of developing services. The customers recognize the importance and appreciate the opportunity to be a part of the team.

Employees can use their knowledge from previous sales situation. The best example of using same knowledge is when an employee serves the same customer more than once. When a new customer comes, the seller needs to use knowledge (s)he already has and connect it to the new knowledge from customer (s)he serves. Using this knowledge is necessary when working with a problematic customer. Many employees find it difficult dealing with these types of customers. Retails have the possibility to cooperate together, which enables employees to share key knowledge about processes and help other retails to improve.

In Company C, communication among employees was poor; they did not cooperate. Also, communication with customers was limited because they do not have direct contact with them, only via telephone (customers who call to a hotline) and use information from loyalty programs, doctors, pharmacists or medical representatives. All information goes to brand managers who work with it. Every brand manager receives the information (s)he needs or refers to his/her brand portfolio. Brand managers examined each product category independently without sharing the results with each other.

Knowledge from customers is usually used for marketing purposes, especially in marketing strategy. Customers can help to make effective marketing strategies by giving the company insight into how to market this product. Involving customers can promote the success or cause the failure of a product on the market. Knowledge from customers and employees is important. Employees who hold knowledge about product building and delivering are brand managers, product developers, medical representatives, doctors and pharmacists.

Creating a product according to customer needs is difficult. Customers cannot be involved in the product development process because it may complicate this process (it can increase production costs, confuse product developers because every customer needs something different, brand managers can promise customers ingredients which cannot be in the product, customers may expect outcomes that cannot be promised, etc.). The only phase customers can be fully involved in is the delivering of the product.

Knowledge was also important in this phase. If they know how to gain customer knowledge and also knowledge from their employees, they are likely to succeed on the market. Companies should seek opportunities for development and ways to serve better value to their customers. If they create and deliver products according to customer requirements, customers will most likely buy the products they requested and helped to develop them further.

Providing post-sale services to customers. In Company B, post-sale services start when the customer pays for the products. An example of this kind of service would be the home delivery of a purchased product and the installation of it. Customers know that employees are available for them every time they need them. When a customer has problem with using some type of technology, (s)he can come to the retail and ask or call the helpdesk. Customers expect to be able to purchase additional or spare parts for their product(s) (e. g. cables for a computer, etc.) in the same retail. Employees should know what kind of services customers need and they must be helpful in every situation. Employees should inform customers about spare parts and advise them on how to use a product when they buy it.

Company C uses knowledge from customers in the phase of providing post-sale services also. They try to take care of customers by providing them with the possibility of becoming a member of their loyalty program. The loyalty program is a very good source of information about customers. Employees can see what kind of products the person has bought and then target their communication directly according to this information. Customers can also call the hotline if they have any problems with the nutritional supplements they took. Company C tried to interact with customers as much as possible but the problem was that this communication was not direct. The Company relies on advice from doctors and pharmacists given to customers. Pharmacists should advise customers on how to use purchased the product(s).

It can be concluded that in both case companies it is important to include customers in the post-sale service phase. In company B, it was easy to communicate with customers and provide the services they needed (installation of products, recycling of old products, membership to the loyalty program, etc.). In company C, it was more difficult to provide this type of service. Brand managers could encourage doctors and pharmacists to advise customers in the purchasing process, they could offer them membership to their loyalty program and encourage them to call hotline when needed. Both companies could provide additional services as a part of their marketing strategy. In these companies, services focus on satisfying customers to keep them.

Satisfying customer needs. According to workshops, the main reasons in Company A for elusive satisfaction of customer needs were due to insufficient information from sales. The reasons for the problem were the overlooking of information or insufficient requirements. In the designing phase, the reason for the elusiveness of complete customer benefit was the emphasis on the wrong issues in the design process. The cross-disciplinary debate in product development was important. In the sales phase, deficiencies in the manufacturing process were sometimes ignored, or unnecessary features were promised by the sales people. The activities carried out in the after-sales phase were essential, e.g. support and training of customers. Better follow-up was required. It is important to invest in production start-up and carefully listen to the customer.

Sometimes at the end of a customer project a new feature or changes had to be added and this created problems in time schedules. This led to a situation where time ran out and the needs of the customer were not fulfilled. The data in the beginning of the project may be incomplete and contradictory. Sometimes complete features may already exist, but due to the lack of a product launch process, the sales department was not aware of these. Also, incomplete testing environments were limited, which led to a situation where bugs were not discovered until customers used the software.

In Company B, it was important to have a variety of products to offer customers. When customers come to the retail, employees should be able to help them find a solution for their requirements. Company B provides approximately 22 services. Employees in retails were responsible for the quality of the services delivered. They were trained in the procedures for providing high-quality services. This company has the best training program for its employees. When a seller does not know any information about a product, (s)he can ask another colleague for help. For this reason, it was important to have a good team. It was necessary to provide high-quality services and after-sales services during the sales process. Customers must feel that employees and the company care about them. A good combination of a product portfolio and services promote success and competitive advantage.

In Company C, the main reasons for the customers' dissatisfaction were due to insufficient information from the marketing and sales department. In the product design phase, the problem could be overlooking information about customer needs, copying competitors without asking customers if they really want this type of product. Production costs play an important role too. There was a lack of cross-disciplinary or cross-department discussion in all of the processes. During the sales process, customers were promised impossible outcomes. There was a lack of information about customers, and doctors and pharmacists usually did not know anything about the products. Exclusive incentives were missing leading to a decrease in the motivation of experts and advisors. There was also a lack of interaction between the marketing department and customers during the product development process. In this area, it was difficult to satisfy all customers, because they requested something different contradictory to their health. The way to satisfy customers in this company was to offer additional value, in the form of services or price strategies, e.g. more tablets for the same price or another nutritional supplement for free.

We can conclude that possible problems of dissatisfaction in different companies can be due to:

- not identifying customer needs precisely,
- making products that are not requested by market, copying competitors,
- not involving customers in the product development process,
- indirect communication with customers,
- ignorance of customers' suggestions, and
- promising features or outcomes that are not possible.

Customer knowledge is important for value creation. In the phase of identifying customer needs, understanding customer knowledge about product and the services offered is important. In the product development process, customer knowledge needs to be involved to produce products according to customer requirements. Expertise among the software engineers is also important to be able to develop products to meet customer requirements. In the process of creating and delivering products, customer knowledge is important especially in Company B and C because this is the only way to implement the customers' insights into improving products and the services associated with these products. Post-sales services are the competitive advantage of Company B and C. If these companies do not use customer knowledge and do not interact with customers it will lead to customer dissatisfaction. These services provide customers with added value, i.e. they get what they asked for or, more importantly, they get more than what they expected. To satisfy customer needs, companies had to know customer needs and their knowledge about products. Mutual cooperation between the Company and customer and integration of customer knowledge into the process of value creation provides the companies with added value.

Table 2 Summary of the research findings

	Case Company A	Case Company B	Case Company C
Type of marketing strategy	Customization	Be better than competition	Differentiation, innovation
The use of CK* in marketing strategy	Explicit CK mostly used in product development	Explicit CK mostly used for improvement of customer services	Explicit CK mainly involved in marketing strategy
Value innovation possible in	Products and services	Services	Products and services
Department responsible for value innovation	R&D: Software engineering, marketing and sales	Marketing and sales	R&D, marketing and sales

* Customer knowledge.

Adopting a knowledge-based approach in marketing strategy can create many advantages to an organization. Some advantages of using knowledge in marketing strategy are as follows (Boersma, 2004): ability to identify the most important target groups, increased knowledge about products and services – building the brand and elevating brand perception by increasing brand awareness, ability to bring about product innovation according to customer needs, ability to create and provide targeted value proposition, increased efficiency of the internal processes by using relevant information and reducing information overload, better value to customers by selling products based on customer knowledge, explicit and tacit knowledge from customers optimizes organizational processes.

Based on the analysis of the value creation process, we propose to extend the analysis through the ideas expressed in the Blue Ocean Strategy (See Table 2). According to the ideas of Kim and Mauborgne (2005), case companies were in the "red ocean", i.e. there were many competitors on the market and the market was oversaturated. The authors of the Blue Ocean Strategy recommend focusing especially on new customer segments and seeking value innovation for customers. The new marketing strategy for companies should be to find an opportunity, the "blue ocean" - part of the market that has not been served already. Companies should choose their own path to the "blue ocean". Based on our analysis, the companies should implement different paths of the Blue Ocean Strategy. It is a tool to help companies find new opportunities on the market. A key element in Blue Ocean Strategy is value innovation. The Companies were not that successful because they were providing the same products or services as their competitors. Their strategy was about beating the competitors. Innovation chiefly involved technology or services with no accent on delivering higher value to customers. Analysis and research shows that companies have problems with identifying customer needs and if companies do not know what their customers' needs are, satisfying them and providing them with added value is also difficult. For any company, finding the "blue ocean" is difficult because it requires knowledgeable employees. The knowledge need was identified in every phase of the value creation process as were the obstacles for better satisfaction of customer needs. This was not enough to create value for customers and keep them. The framework proposed by Kaplan and Norton for analyzing the value chain works only when each company modifies it for its own purposes. As regards the case companies, it is better to use the Blue Ocean Strategy and the analysis of value creation just as a supporting tool. A possible solution and challenge for the companies is to create a leap in value for buyers.

5 Conclusion

This article focused on three companies representing different industries. There were several internal problems in these companies. An analysis of the value creation process in each company was used for comparison. The findings offer insight to what companies should do to improve their business. Companies should not try to beat their competitors but create new business opportunities via value innovation. Value innovation emphasizes both value and innovation at the same time. It means that companies should not focus on providing technological innovations but innovations which contribute to an increase in value for the customer. One useful tool that can be used to assist companies is the Blue Ocean Strategy. An analysis of the case studies indicated that the companies should use the Blue Ocean Strategy to improve their business rather than using the value creation process alone. The value creation process is a good tool for mapping value in organization. Companies can see which processes increase value and which do not. They can map their internal processes and see the potential for improvement. Not every company needs to analyze all of the phases of the value creation process. This process did not work for the case companies; each of the companies has different processes. Moreover, customer involvement was only possible in some parts of the processes. This framework should be modified by every company or used just as an example of the areas in which value can be analyzed. Having good knowledge in each phase of the value creation process is important but not enough for future success of these companies. When companies work with their customers and use their knowledge in all processes they have a chance to succeed. Companies should stop copying their competitors (being followers) and try to focus on value innovation.

Acknowledgments. This project has been supported by the Slovak VEGA n. 1/0992/11 and 1/0888/11.

References

1. Kaplan, R.S., Norton, D.P.: Balanced Scorecard. Harvard College, USA (1996)
2. Kim, W.C., Mauborgne, R.: Blue Ocean Strategy. Harvard Business School Press, USA (2005)
3. Kucza, T.: Knowledge Management Process Model. Technical research center of Finland (2001)
4. Marr, B.: Managing and delivering performance. Elevier Ltd., UK (2009)

5. Nonaka, I., Toyama, R., Hirata, T.: Managing flows - a process theory of the knowledge-based firms. Palgrave Macmilian, London (2008)
6. Prusak, L.: Knowledge in Organizations. Butterworth-Heinemann, UK (1997)
7. Ruggles, R., Holtshouse, D.: The Knowledge Advantage. Capstone, US (1999)
8. Zeleny, M.: HSM Integrating Knowledge, Management and Systems. World Scientific Publishing Co. Pte. Ltd., Singapore (2005)
9. Boersma, A.: The original knowledge management publication: Did someone say customer? Inside Knowledge magazine (2004), http://www.ikmagazine.com/xq/asp/sid.0/articleid.19B774B0-712C-4B28-ADD5-9E0EE916C781/eTitle.Embedding_KM_in_marketing_strategies/qx/display.htm (accessed February 29, 2004)
10. Gibbert, M., Leibold, M., Probst, G.: Five Styles of Customer Knowledge Management, and how Smart Companies Put them into Action [pdf] (2002), http://archive-ouverte.unige.ch/downloader/vital/pdf/tmp/9ecissm6hhc2ogkdavb6g2u221/out.pdf (accessed February 29, 2012)
11. Yusof, S.W.M.: Developing Strategy for Customer Value Creation Towards a Professional Services. [pdf] (2010), http://webs2002.uab.es/dep-economia-empresa/papers%20CPP%20angl%C3%A8s/3SitinorYusof_VIII-CPP-2010.pdf (accessed March 25, 2012)

Author Index